Series in BioEngineering

The Series in Bioengineering serves as an information source for a professional audience in science and technology as well as for advanced students. It covers all applications of the physical sciences and technology to medicine and the life sciences. Its scope ranges from bioengineering, biomedical and clinical engineering to biophysics, biomechanics, biomaterials, and bioinformatics.

Indexed by WTI Frankfurt eG, zbMATH.

Nirav J. Joshi · Gautam Patel
Editors

Nanoparticle Delivery in Cancer Care

Springer

Editors
Nirav J. Joshi
Department of Electronic and Biomedical Engineering
University of Barcelona
Barcelona, Spain

Gautam Patel
Department of Chemistry, Parul Institute of Applied Sciences
Parul University
Vadodara, Gujarat, India

ISSN 2196-8861 ISSN 2196-887X (electronic)
Series in BioEngineering
ISBN 978-981-95-0456-5 ISBN 978-981-95-0457-2 (eBook)
https://doi.org/10.1007/978-981-95-0457-2

This Springer imprint is published by the registered company Springer Nature Singapore Pte Ltd.
The registered company address is: 152 Beach Road, #21-01/04 Gateway East, Singapore 189721, Singapore

Preface

Cancer remains one of the most formidable challenges in modern medicine—a disease marked by its heterogeneity, dynamic progression, and ability to evade conventional therapeutic strategies. Despite significant advancements in surgical interventions, chemotherapeutic agents, and radiotherapy protocols, substantial limitations persist, including off-target toxicity, multidrug resistance, and suboptimal treatment response. In this context, nanotechnology-based drug delivery systems offer a way to overcome these barriers, enabling more precise, effective, and personalized oncological care.

Nanoparticle Delivery in Cancer Care provides an in-depth exploration of the principles, technologies, and translational aspects of nanoparticle-mediated therapeutic strategies. The opening chapters lay the groundwork by introducing the core concepts of nanotechnology and cancer biology. Chapter "Introduction to Nanoparticle Delivery in Cancer Therapy", offers an overview of the potential of nanocarriers in transforming cancer treatment. Chapter "Understanding Cancer Biology and Microenvironment", examines the pathophysiological mechanisms of cancer and the microenvironmental factors—such as hypoxia, pH gradients, and abnormal vasculature—that critically influence nanoparticle distribution and therapeutic efficacy.

Chapters "Functionalized Nanomaterials for Drug Delivery"–"Nanocarriers for Chemotherapy and Drug Delivery"—Functionalized Nanomaterials for Drug Delivery, Targeting Strategies in Nanoparticle Drug Delivery, and Nanocarriers for Chemotherapy and Drug Delivery—focus on the rational design and engineering of nanocarriers. Topics include surface modification strategies to enhance targeting specificity, passive and active targeting mechanisms, and pharmacokinetic considerations that can reduce systemic toxicity and improve therapeutic outcomes.

Chapters "Cancer Immunotherapy Advancements Using Nanoparticles"–"Nanoparticles in Cancer Imaging and Treatment Monitoring: Challenges and Perspectives"—Cancer Immunotherapy Advancements Using Nanoparticles,

Nanotechnology for RNA and DNA Delivery in Cancer Therapeutics, and Nanoparticles in Cancer Imaging and Treatment Monitoring: Challenges and Perspectives—explore clinical and experimental applications. These chapters cover nanoparticle-enabled immune modulation, nucleic acid delivery, and advances in multimodal imaging for real-time monitoring of tumor response. The remaining chapters address translational issues and emerging directions. Clinical Translation and Regulatory Landscape (Chapter "Clinical Translation and Regulatory Landscape") reviews the status of nanoparticle-based therapies in clinical trials and regulatory approval processes. Biocompatibility and Toxicity of Nanoparticles in Cancer Care (Chapter "Biocompatibility and Toxicity of Nanoparticles in Cancer Care") emphasizes safety evaluation. Personalized Cancer Treatment with Nanotechnology (Chapter "Personalized Cancer Treatment with Nanotechnology") discusses integrating omics technologies with nanomedicine. Emerging Nanotechnologies Transforming Cancer Therapeutics (Chapter "Emerging Nanotechnologies Transforming Cancer Therapeutic") presents innovative platforms such as stimuli-responsive systems, hybrid nanoparticles, and nanoscale biosensors. Finally, Economic and Ethical Considerations in Nanoparticle Therapies (Chapter "Economic and Ethical Considerations in Nanoparticle Therapies") examines cost-effectiveness, equitable access, and ethical challenges in adopting these technologies.

Barcelona, Spain — Dr. Nirav J. Joshi
Vadodara, India — Dr. Gautam Patel

Competing Interests The authors have no competing interests to declare that are relevant to the content of this manuscript.

Contents

Introduction to Nanoparticle Delivery in Cancer Therapy

Km Nikita, Srujal Sonera, and Sang Yong Nam

Abstract Extensive research and utilization of nanotechnology in cancer treatment have arisen because nanoparticles can serve as key agents for delivering drugs effectively. In contrast to traditional medications, utilizing nanoparticle-based drug delivery offers distinct benefits including enhanced stability and compatibility with biological systems, improved permeability and retention within the body, and particular directing capabilities. The advancement and utilization of hybrid nanoparticles, blending diverse nanoparticle characteristics, have propelled this form of drug carrier system to new heights. Cancer stands as a prominent contributor to both mortality and morbidity, characterized by a multifaceted pathophysiology. Conventional cancer treatments comprise chemotherapy, radiation therapy, targeted therapy, and immunotherapy. Nevertheless, challenges like non-specificity, cytotoxicity, and resistance to multiple drugs present significant obstacles for effective cancer therapy. Nanoparticles, ranging from 1 to 100 nm in size, offer unique advantages for cancer treatment, including biocompatibility, decreased toxicity, improved stability, enhanced permeability and retention, and accurate targeting. Furthermore, researchers have recently begun exploring the involvement of nanoparticles in immunotherapy, which holds increasingly significant promise in cancer treatment. This chapter examines various nanoparticle types, targeting strategies, and approved nanotherapeutics, focusing on their implications for cancer treatment. Additionally, we provide an overview of the present viewpoint, benefits, and obstacles in clinical application.

Keywords Nanoparticle · Cancer · Chemotherapy · Cellular targeting

K. Nikita (✉) · S. Y. Nam
Research Institute for Green Energy Convergence Technology, Gyeongsang National University, Jinju, Gyeongsangnam-do, Republic of Korea
e-mail: kmnikita1992@gmail.com

S. Y. Nam
Department of Materials Engineering and Convergence Technology, Gyeongsang National University, Jinju, Gyeongsangnam-do, Republic of Korea

S. Sonera
Parul Institute of Applied Sciences, Faculty of Applied Sciences, Parul University, Vadodara, Gujarat, India

N. J. Joshi and G. Patel (eds.), *Nanoparticle Delivery in Cancer Care*,
Series in BioEngineering, https://doi.org/10.1007/978-981-95-0457-2_1

1 Introduction

A group of illnesses collectively referred to as cancer is defined by invasiveness and unchecked, random cell division. Over several years, recognizing different cancer risk factors has been the subject of extensive work. Nonetheless, an unhealthy lifestyle that includes smoking, drinking alcohol, eating unbalanced food, experiencing stress, and not getting enough exercise has a significant influence on cancer risk assessment [1, 2]. Although these outside variables have been identified as important contributors to cancer, it has been difficult to determine how much of proto-oncogene mutations, tumour suppressor gene expression forms, and DNA restoration genes are involved. Hereditary factors are only associated with 5–10% of cancer cases [3]. Another important risk aspect for cancer and countless different types of cancer is getting older. There will likely be 1.9 million new cases, as per the American Cancer Society [4]. Operation, chemotherapy, radiation rehabilitation, targeted rehabilitation, immunotherapy, and hormone rehabilitation are some of the outmoded therapeutic modalities utilised in the healing of cancer [5, 6]. Though cytotoxicity and cytostasis can be caused by radiation and chemotherapy [7], these treatments are frequently associated by critical side effects and a great hazard of repetitions. The utmost often occurring adverse properties contain neuropathies, tiredness, gastrointestinal and skin diseases, destruction of bone marrow, and hair fall (Fig. 1). In addition, a few adverse effects are unique to a particular medicine, like bleomycin-induced cardiotoxicity, pulmonary toxicity and anthracyclines [8].

Precision therapy has expanded since the introduction of targeted therapy [9]. Therapeutic efficacy is however limited by several unavoidable side effects, such as multi-drug resistance [8]. Promising outcomes have been achieved with

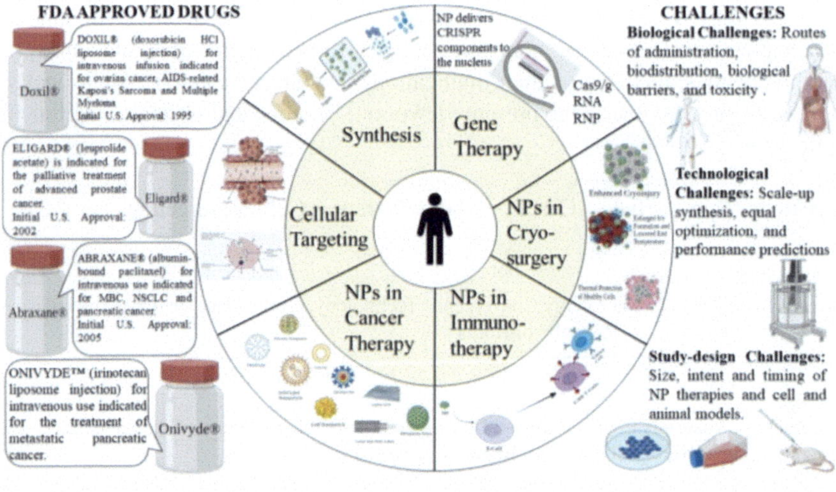

Fig. 1 Revolutionizing cancer treatment with nanoparticles [13]. Copyright 2021. Reproduced with permission from Springer Nature

immunotherapeutic drugs in treating primary cancer, avoiding distant metastases, and reducing the recurrence rate [10]. However, one of the main adverse effects of immunotherapy is autoimmune illness. Furthermore, according to studies and bits of data, immunotherapy appears to be less successful against solid tumours than lymphoma [11]. These tumours create a distinct connectivity tissue matrix, which makes it challenging for immune cells to pierce [12].

These newly developed targeted medications and immunotherapies produce dermatological adverse events (dAEs) by interfering with signalling pathways necessary for malignant behaviour and normal dermal and epidermal homeostasis [14].

To accurately treat cancer, there has been a growing demand in recent years for developing novel techniques. Recently, there have been attempts to overcome the drawbacks of current medical methods by using nanomaterials. Nanoparticle-based drug distribution classifications have demonstrated potential in the treatment and organization of cancer due to their favourable pharmacokinetics, accurate directing, decreased side effects, and decreased drug resistance [15, 16]. Many extensively marketed and commercialised nanotherapeutic drugs have been developed since the advent of nanotechnology, and since 2010, many more have advanced to the clinical stage. Drug distribution methods and anti-tumor multidrug resistance (MDR) have evolved thanks to nanotherapeutic drugs, which provide the option of drug combination therapy and the inhibition of drug resistance mechanisms [17]. ETH Zurich made a ground-breaking attempt to spread on nanotechnology in medication in the 1960s [18]. When it comes to developing a variety of diagnostic instruments and more potent therapies, this combination works better. The main topics of this review include the basic theories underlying the application of nanotherapeutics, as well as the current state of the field, potential future directions, and challenges.

The advancement of nanotechnology has made nanoparticles a feasible alternative for controlled drug delivery methods. Particles with a width between 10 and 1000 nm are commonly denoted as nanoparticles. As a drug delivery system (DDS), nanoparticles can upsurge the drug's half-life, make few hydrophobic medicines further soluble, and distribute the medication gradually or under controlled conditions. Stimulus-responsive nanoparticles can also help lower toxicity and control the distribution in the body. In the 1960s, liposomes the first nanoparticles to be discovered were employed as protein and medication carriers. Since then, an increasing number of materials have been created as DDSs by fabricating them into nanoparticles (Fig. 2). According to a 2016 assessment by Bobo et al., 51 nanoparticles have received FDA approval, while 77 items are undergoing clinical testing. Polymeric and liposomal materials make up a significant share of all the materials authorised for use in nanoparticles. Nonetheless, scientists think that more sophisticated materials, including protein-based, metallic, and micelle materials, can also be employed as DDSs for nanoparticles.

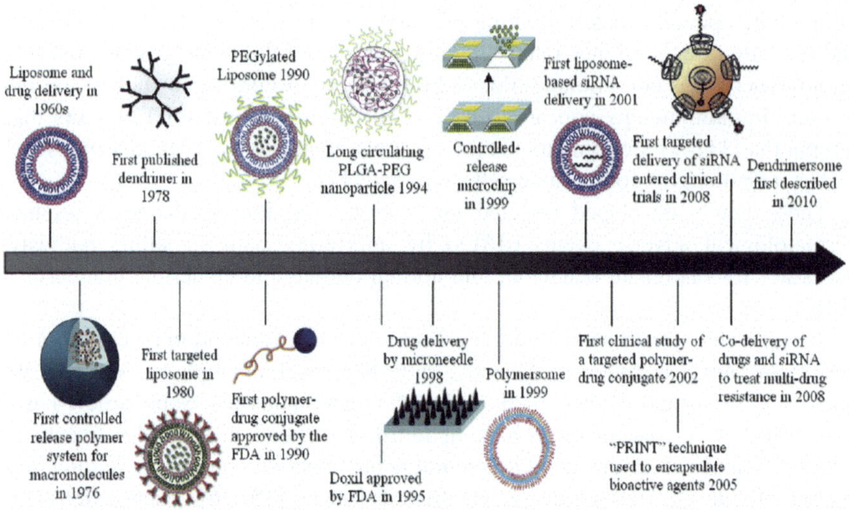

Fig. 2 History of the development of medication delivery systems using nanoparticles [19]. Copyright 2010. Reproduced with permission from American Chemical Society

2 Nanoparticles

Particles that have 1D and smaller than 100 nm and possess distinctive characteristics not observed in larger sections of the identical matter are known as nanoparticles (NPs) [20, 21]. The structure of nanoparticles is intricate, consisting of three main components: the exterior film, the shell film, and the core. The core, often referred to simply as the nanoparticle itself, forms the central part of the nanoparticle [22]. As of their remarkable potential such as a great exterior-to-volume proportion, diversity, sub-micrometre dimensions, and improved aiming capabilities, such substances have become highly significant across various interdisciplinary domains. Nanoparticles have been observed to possess the ability to penetrate deep tissues, thereby augmenting the enhanced permeability and retention (EPR) effect. Additionally, exterior properties influence the half-life and bioavailability by efficiently traversing surface cell fenestrations [23]. For instance, nanoparticles covered by polyethylene glycol (PEG), which is a hydrophilic material, reduce opsonic tagging and evade consent by the immune system [24].

Moreover, it's feasible to adjust the distribution scale of effective components via modifying the features of particle polymers. In summary, the unique properties of nanoparticles govern their healing influence in the management and therapy of cancer.

2.1 *Methods for the Synthesis of Nanoparticles (NPs)*

The nanoparticles come in several dimensions, shapes, and forms. These approaches can be broadly organized into 2 major classes: (I) top-down approach (II) bottom-up approach. Such methodologies can be subdivided into various subcategories depending upon reaction environment and operational parameters (Fig. 3).

2.1.1 Top-Down Approach

This technique, alternatively termed the destructive approach, involves reducing mass substances to produce nanoparticles. A substantial fragment undergoes breakdown or decomposition into tiny units, which are then transformed into nanoparticles [25]. It comprises methods e.g. chemical etching, thermal decomposition, nano-lithography, electro-explosion, laser ablation, sputtering and mechanical milling. Notably, the structural characteristics like dimension, charge, and shape of nanoparticles can be altered by adjusting the reaction environments and further synthesis factors [26]. Moreover, the outgrowth procedure likewise states the chemical properties of nanoparticles. Therefore, comprehending the outgrowth procedure is crucial for synthesizing the desired nanoparticles.

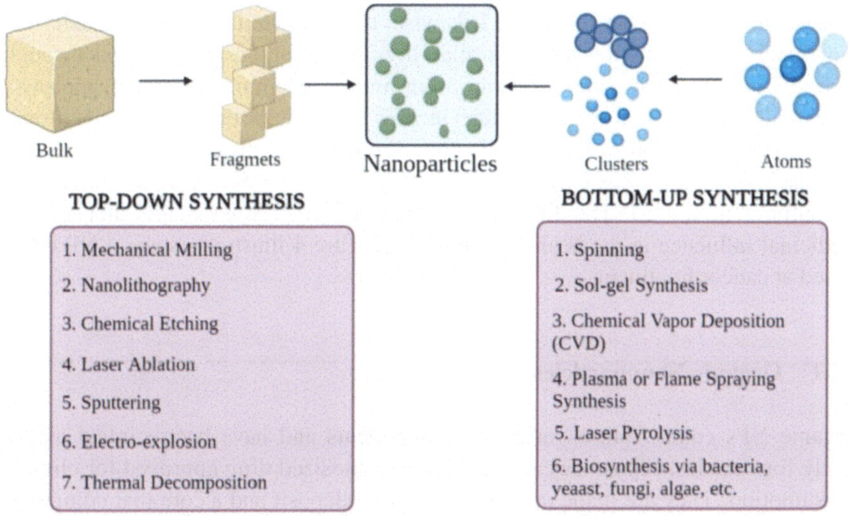

Fig. 3 Schematic representations for various kinds of top-down and bottom-up synthesis [13]. Copyright 2021. Reproduced with permission from Springer Nature

2.1.2 Bottom-Up Approach

This approach entails constructing materials from particles to aggregates to nanoparticles, essentially constructing from smaller components, therefore referred to as the productive approach [27]. Several frequently employed techniques include biosynthesis, sol-gel synthesis, chemical vapor deposition (CVD), laser pyrolysis, plasma scattering synthesis and spinning.

2.2 Nanoparticles in Cancer Therapy

As these 3 features have a substantial effect on the efficiency of the distribution of nanoparticle drugs and, accordingly, adjust medicinal effectiveness, the NPs employed in medicinal treatment often have certain dimensions, exterior, and formulae properties. As they may well carry medications and get EPR effect, NPs having a diameter scale of 10–100 nm are generally considered appropriate for cancer rehabilitation.

Atoms bigger than 100 nm are probable to be detached from movement by scavenger cell, but smaller atoms (<1–2 nm) can simply discharge from the ordinary vasculature to mutilation usual cells and can be certainly sieved via kidneys (fewer than 10 nm in length), whereas phagocytes are likely to remove particles larger than 100 nm from the veins. Likewise, the characteristics of NPs' exteriors can impact their half-life and bioavailability. For example, NPs treated through hydrophilic substances, lower opsonisation and evade immune system approval. To extend the time that medicines are in flow and amend their infringement and deposit in tumours, NPs are typically altered to happen to hydrophilic. The advantages and constraints associated with employing nanoparticles as drug administration are outlined in Table 1. The different properties of NPs taken together dictate their medicinal influence in the healing of tumour. Figure 4 illustrates several NP types aimed at cancer treatment.

2.2.1 Organic Nanoparticles

Organic NPs come from a variety of components and have been studied extensively for decades. Liposomes are the earliest nanosized drug approved for clinical consumption. They are made of an external lipid deposit and a core that comprises a hydrophilic or hydrophobic medicament. Via varying the lipid coating, liposomes can accomplish a multi tasks, such as mimicking the biophysical characteristics of living cells, which can support in the area of added competent healing drug transport. The creation of liposomes has ended beyond multiple inventions with spans of research. Regarding cancer treatment, liposomes offer an effective stage for the in vivo transport of several anti-tumour medicines, e.g. paclitaxel and doxorubicin

Table 1 The advantages and limitations of using nanoparticles as drugs transporters for cancer [28]. Copyright 2022. Reproduced with permission from MDPI

Nanoparticle	Advantage	Disadvantage
Dendrimers	Reduced index of polydispersity	Generalized toxicity; expensive synthesis expenditures
Small extracellular vesicles	Biodegradable, not hazardous, potential for cellular activation	Poor solubility in water
Polymer nanoparticles	Biodegradable, capacity to capture medicines that are both hydrophilic and hydrophobic, simple modifications; monitored release of medicines, prevent the medication from being broken down metabolically, easily integrated into additional drug delivery-related operations	Rupture outcome; little capacity to inject drugs; expensive; poor adhesion for cells; the hazardous nature of the byproducts; nonbiodegradable polymers frequently build up in tissue; encourage an allergic response; the process of digestion and excretion of in vivo remain uncertain
Polymer micelles	Extended periods of time spent in persistence; ease of synthesis; paired with ligands that target tumors to improve accessibility, minimize the adverse consequences; capacity to manage drug distribution over an extended length of time	Elevated toxicity throughout the system
Silica nanoparticles with mesoporosity	Outstanding capacity to load genes and drugs; variable pore size; substantial surface area; both biodegradable and biocompatible; managed porosity; adaptability; non-hazardous; heat and pH resistance, as well as simple endocytosis	High cost; inadequate knowledge regarding neurotoxicity; aggregates with low stability, anemia
Metallic and magnetic nanomaterials	Simple setup and operation; substantial surface area; application in several modes; various shapes (triangles, nanorods, and spheres); biocompatibility; adjustable dimensions; simple functionalization excellent in vivo biodegradability; no encapsulated medication leaking	Low levels of storage stability; inadequate knowledge of biocompatibility, minimal cytotoxicity, and absorption in vivo
Carbon nanotubes	Soluble in water; versatile; less harmful; biocompatibility; biodegradability; capable of capturing both hydrophobic and hydrophilic medications; substantial ability to load; a large variety of surface alteration alternatives; thermal conductivity, chemical resistance, large surface area, and needle-like shape	Expensive production costs; minimal deterioration; insufficient in vivo research

NPs used extensively in drug delivery systems include organic NPs, inorganic NPs, and hybrid NPs.

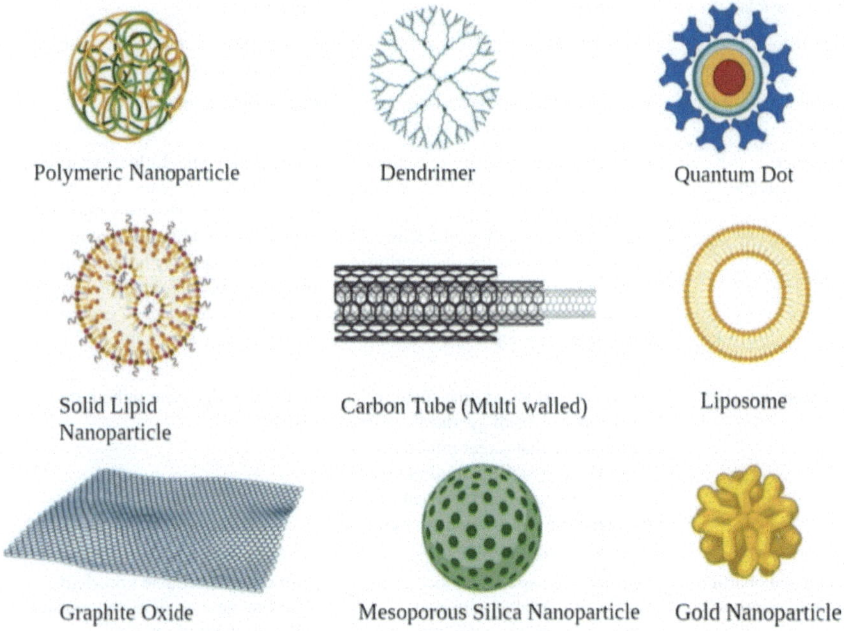

Fig. 4 Various types of nanoparticles utilized in cancer therapy [13]. Copyright 2021. Reproduced with permission from Springer Nature

[29, 30]. Liposome application has become further popular in the field of prostate and breast cancer. Table 2 comprises some drugs employed in cancer treatment.

It has been revealed that paclitaxel liposomes, in contrast to free paclitaxel, display greater bioavailability and superior anti-tumor efficiency [32]. This proved that liposomal doxorubicin lowers cardiotoxicity and is as active in healing breast cancer [33,

Table 2 Drugs comprised of nanoparticles and employed at different stages of cancer [31]. Copyright 2017. Reproduced with permission from Insight Medical Publishing

Product	Nanoparticles	Applications
NK105	Micellar nanoparticles	Breast cancer
AuroLase	Gold nanoparticles	Lung and neck cancer
CYT-6091	Gold nanoparticles	Breast and pancreatic cancer
Combidex	Iron oxide nanoparticle	Tumor imaging
C-dots	PEG-coated SiO_2 nanoparticles	Melanoma
Docetaxel PNP	Polymeric nanoparticles	Advanced solid malignancies

34]. Additionally, two or more drugs together have been made possible by liposome-based systems, which can advance healing impact and even overcome drug resistance [35, 36].

Another kind of NPs is polymer-based NPs, which is comprised of various monomers and have certain physical arrangements for drug distribution. Polylactic-co-glycolic acid (PLGA) is frequently active as a drug transfer transporter due to its improved biocompatibility, biodegradation, and EPR impact. Furthermore, an alternative polymer that has been employed in medicine is dendrimers. Their three-dimensional branch structure distinguishes them as biocompatible and adaptable macromolecules. Their surface possesses several functional groups that improve their ability to load and transport medicinal substances.

2.2.2 Inorganic Nanoparticles

One benefit of inorganic nanoparticles is their increased surface area-to-volume ratio. Although this typically derives at the cost of worse biodegradability and biocompatibility, they have a broad and simply changeable superficial coupling interaction and simple synthesis. Carbon nanotubes (CNTs), magnetic nanoparticles (MNPs), silica nanoparticles (SNPs), and quantum dots are among the inorganic NPs that have been examined. The maximum investigated inorganic NPs are AuNPs, and assorted mono-layer confined bunches through a gold core are supposed to be a decent option for the drug distribution arrangement. The surface modified AuNPs have been exposed to both increase drug accretion in tumours and overawed action confrontation; the gold foundation is non-reactive and non-poisonous. Furthermore, according to AuNPs may be employed in combination with photothermal treatment, gene therapy, and immunotherapy as part of multimodal cancer treatment. Since carbon nanotubes exhibit special biological, physical, and chemical characteristics, it has been demonstrated that they hold a great deal of promise for use in the drug delivery industry. They have therefore been employed to provide anticancer medications, such as doxorubicin and paclitaxel, for a series of cancer types [37]. In the meantime, CNTs that are treated with near-infrared radiation generate heat, which may be used in thermal excision as a tumour treatment [38].

One kind of SNPs that is appropriate for medication administration is mesoporous silica nanoparticle carriers. The supramolecular components function as a cap to allow drug capture and release, and the huge internal pore capacity lets them to compress the extreme quantity of antitumor medications.

SNPs are regarded as greatest drug delivery sources because of their superior pharmacokinetics, high stability, and treatment efficacy. Furthermore, because of their immunoadjuvant qualities, which comprise promoting polarising lymph cell, antigen cross-presentation and secreting permeable silicon NPs and interferon-g (IFN-g) have recognized incredible capacity in immunotherapy.

Drug delivery magnetic nanoparticles (MNPs) often include metal or metal oxide nanoparticles (NPs). Polymers and fatty acids are two examples of organic compounds that are frequently coated on MNPs to rise their steadiness and biocompatibility. They have been confirmed to have exceptional effectiveness in chemotherapy for the cure of cancer. Additionally, thermal excision of tumours can be accomplished via magnetic hyperthermy utilising MNPs, providing an alternate method of treating cancer.

2.2.3 Hybrid Nanoparticles

Although each type of nanoparticle has aids and drawbacks of its own, mixing inorganic and organic NPs into one hybrid drug transport arrangement offers the multifunctional transporter improved biological characteristics that can expand treatment efficacy and lesser drug resistance. It has been revealed that lipid-polymer combined nanoparticles (NPs), which have a lipid exterior and an internal polymeric core, are a capable drug distribution stage for the treatment of prostate tumour, pancreatic tumour, and breast tumour [39]. To produce a superior therapeutic impact, this type of hybrid NPs can encapsulate both hydrophilic and hydrophobic medicines as they join the large bio-tolerance of lipids through the physical reliability accessible by polymeric NPs. Concurrently, this system can be efficiently taken up through tumour and escape the reticuloendothelial structure's rapid clearance.

Hybrid nanoparticles made of both organic and inorganic materials are frequently used in NP design. A liposome silicon oxide hybrid (LSH) nanoparticle, for instance, has been synthesised and demonstrated to be effective in delivering medicines to heal breast and prostate tumours. It is comprised of a silicon oxide core and a surrounding lipid dual layer.

In a mouse model of pancreatic tumour, it has also been shown that the LSH nanoparticle provides a platform intended for the coactive transport of paclitaxel and gemcitabine. By assembling oversize liposomes and permeable silicon nanoparticles against a microfluidic chip, Kong et al. have demonstrated that co-delivering medicines and synthesised DNA nano molecules on this platform pointedly increase the bereavement of doxorubicin-resistant breast tumours. Moreover, methotrexate vectorization to lung tumours using chitosan and CNTs hybrid nanoparticles seems to enhance antitumor action though lowering medication noxiousness in regular cells.

Furthermore, it is possible to combine targeted drug delivery with hyperthermia in half-shells of metallic multilayers (e.g. gold and manganese) and PLGA hybrid nanoparticles, which can improve the killing of tumour cells. Another technique for designing NPs is the hybridizing normal biomaterial through either inorganic or organic NPs. For example, nanotechnology for covering cell membranes is novel and is receiving progressively more attention. According to Parodi et al. [40], covering nanoporous SiO_2 units through a leukocyte-purified cell membrane can stop the nano-transporter from being cleared using scavenger cell. The properties of this hybrid element also enable the drug to remain in circulation for longer, which increases tumour accumulation [40]. Similarly, some research has employed mesoporous silicon oxide NPs encased in cancer cell membranes to heal tumours, which enhances the permanency and aiming capacity of nano-carriers. Furthermore, the creation of dual-membrane covered NPs can improve NP performance even more.

For example, it has been demonstrated that membrane coated NPs with erythrocyte-cancer and erythrocyte-platelet coatings have greater constancy and longer movement lives.

Additionally, Wong et al. [41] suggested a multistep NP distribution organization that modifies the dimensions and properties of NPs at various steps to achieve deep penetration into tumours. In their investigation, 10 nm quantum dot NPs were released by protease-mediated degradation of the 100 nm gelatin NP cores inside the tumour microenvironment, resulting in a size shift of the NPs [41]. A notable metabolic descriptions of tumour cells is glycolysis, which works as the primary energy roots aimed at the development of tumour cells. The pH of the tumour microenvironment is lowered via glycolysis, creating an acidic atmosphere. The low pH level then causes some pH-sensitive NPs to become activated, allowing them to release medications near cancer cells. Nevertheless, there are certain disadvantages to passive targeting, including as Ambiguous drug delivery, a specific EPR effect, and variable blood vascular porousness among unlike kinds of tumours.

3 Mechanisms of Cellular Targeting

Nano-carriers perform a crucial part in drug transport via targeting cancer cells, enhancing healing effectiveness and defending regular cells from cytotoxicity. To investigate the directing design of NP-based medications, many research investigations have been performed. It is crucial to first comprehend tumour biology and the interface among tumour cells and nano-vesicles in order to more effectively tackle the issues of tumour directing and the strategy of the nano-vesicles device. In general, there are 2 kinds of directing processes: active targeting and passive targeting.

3.1 Active Targeting

Active targeting uses specific interactions among binders and sensors to aim at tumour cells in particular (Fig. 5). With the aim of differentiate among directed and regular cells, the binders on the exterior of NPs are specifically designed for targeting substances that are overemphasized on the exterior of cancer cells [42, 43]. Sensor moderated endocytosis is triggered via the interaction among binders on NPs and the sensors on the exterior of tumour cells. This process enables ingested NPs to efficiently deliver therapeutic medicines [44]. For this reason, active targeting is particularly fine suitable for the transport of macromolecular drugs such as siRNAs and proteins. Proteases, amino acids, vitamins, and carbohydrates are illustrations of directing fractions [45]. The targeted cells' receptors on which these binders particularly fix are the transferrin sensors, glycoproteins, the folate sensors, and the epidermal growth factor receptor (EGFR). These receptors have been the subject of extensive research.

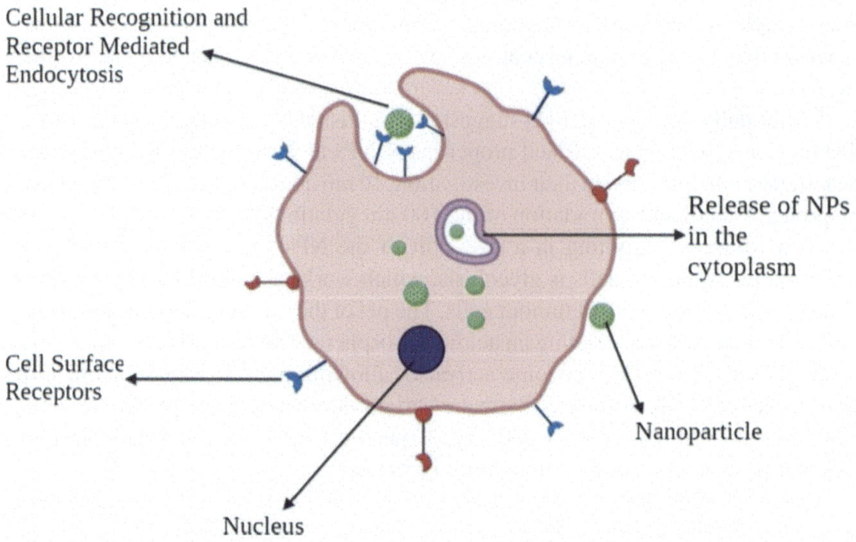

Fig. 5 Illustration of active targeting [13]. Copyright 2021. Reproduced with permission from Springer Nature

3.2 *Passive Targeting*

Passive targeting aims to make use of the alterations between normal and malignant tissue. When drugs are passively targeted, they are effectively delivered to the intended position to perform a therapeutic action (Fig. 6). The neovascularization caused by high cancer growth of cells and the increased perm selectivity of tumor vasculature relative to healthy vessels is caused by large vascular wall pores [46]. Accelerated and faulty angiogenesis allows macromolecules, such as NPs, to seep out of the blood vessels supporting the tumor to grow up inside the tissue of the tumor. In the meantime, the accumulation of NPs is increased by inadequate lymphatic drainage attributed to cancer, which permits the nanocarriers to transfer their contents into tumour cells. The EPR outcome, which is a factor behind passive targeting, is produced by such procedures [47]. The dimensions of NPs affect the EPR outcome because, as numerous studies showed, small NPs are more permeable yet do not escape into regular blood vessels [48, 49]. However, the immune system has a higher chance of eliminating bigger particles [50]. The microenvironment of tumours acts as a substantial part of the passive delivery of nanomedicines besides the EPR effect. Another metabolic feature of cancer cells is the process of glycolysis which assists as the main energy source for the growth of cancerous cells [51]. The pH of the tumor microenvironment gets reduced by glycolysis, generating a highly acidic atmosphere. Consequently, a low pH value stimulates some pH-sensitive NPs, enabling them to discharge drugs in close immediacy of tumour cells [52]. Passive

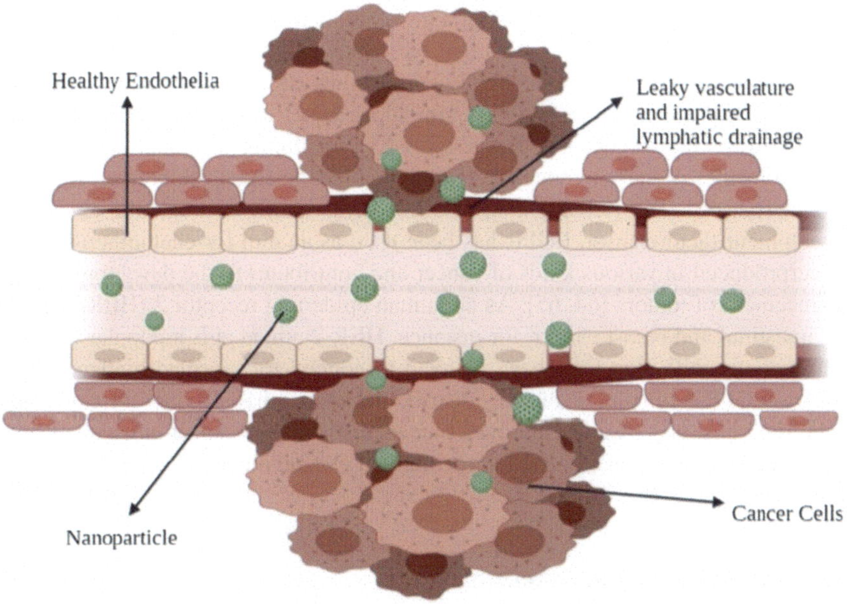

Fig. 6 Illustration of passive targeting [13]. Copyright 2021. Reproduced with permission from Springer Nature

targeting does have certain drawbacks, though, which include as ambiguous circulation of drugs, a particular EPR effect, and variable blood vascular absorbency throughout various kinds of tumors [53].

3.3 Targeting to Cancer Cells

A sort of serum glycoprotein called transferrin is responsible for bringing iron within cells. Typically, solid tumor cells enhance transferrin sensor, although healthy cells possess them at small levels. Consequently, drugs for cures of cancer are transported via transferrin-conjugated NPs as an effective directing method [54–56]. Transferrin decorated NPs have been established to have amended intracellular drug distribution and superior cellular absorption rate when associated to Pristine NPs [57]. Furthermore, results suggest that transferrin-linked polymeric nanoparticles are important in fighting back chemotherapy which is resilient to drugs [58]. Vitamins such as folic acid are necessary for the production of nucleotides. It is taken up by a folate sensor, which is found on an insignificant regular cell kind. Nonetheless, hematological malignancies produce FR-β on their exterior, however the alpha isoform of folate receptor (FR-α) is overproduced in around 40% of human tumours [59]. Folate-linked nanomaterials have so been broadly utilized in chemotherapy for cancer as a means of targeting the folate receptor [60, 61]. Furthermore, many glycoprotein forms, e.g. lectins-non-immunological proteins that recognize and attach selectively to some carbohydrates are also characteristically produced by cancer cells [62]. The straight lectin directing process comprises via lectins close to NPs to aim for carbohydrates on the outermost layer of cancer cells; the inverse lectin aiming route includes utilizing NPs' molecules of carbohydrates to bind lectins on cancerous cells in an reversed manner [62, 63]. The tyrosine kinase sensor ErbB family contains the epidermic development factor sensor. EGFR has been used as an aim for chemotherapy as it is overproduced in various kinds of cancer and contributes to the development and advancement of tumors [64, 65]. As an human epidermal receptor 2 (HER-2) positive intestinal and breast cancer, for instance, HER-2 aimed at is a popular therapy. Consequently, NPs that have been planned to contain transformed binders which attach to EGFR signify a capable drug distribution approach for EGFR-overexpressed cancerous cells [66]. Besides, uniting 2 binders distinctive to cancer into a on its own NP is an added technique of active targeting that can augment aim selectivity [67].

3.4 *Targeting to Endothelium*

Several NPs affect angiogenesis, that is a different way of treating cancer, somewhat than straight directing cancer cells. A crucial part of vascularization is the association among VEGF and VEGFRs, which is the biological endogenous growth factor [68]. Furthermore, it has been demonstrated that liposomes targeting both of the primary VEGF receptors, VEGFR-2 and VEGFR-3, simultaneously advance the effectiveness of therapy [69]. Tumor cell movement and spread are significantly influenced by these extracellular matrix receptors for proteins on the cell interface called integrins [70]. According to Nisato et al. [71], the $\alpha v\beta 3$ integrin plays a crucial role in the calcium dependent mechanism that triggers endothelial cell movement. It is more abundantly produced by cancer tumour neovascular endothelial cells than in inactive endothelium and usual cells [71]. Hood et al. have documented the beneficial therapeutic effectiveness of cationic nanoparticles in combination through an $\alpha v\beta 3$ integrin directing ligand intended for delivering gene therapy to mice harboring tumors [72]. Furthermore, $\alpha v\beta 3$ integrin is linked to VEGFR-2 signalling [73]. By preventing $\alpha v\beta 3$ integrin binding, VEGF signalling can be reduced, suggesting that integrin directing can advance the effectiveness of anti-VEGFR rehabilitation. Through its interactions with vascular endothelial cells, the glycoprotein also named as vascular cell adhesion molecule-1 (VCAM-1) which resembles an immune serum globulin is additionally produced on the outermost layer of tumour vascular endothelium and plays a role in neovascularization. VCAM-1 upregulation has been reported in numerous tumours [74], implying that it may play a part in actively directing NPs for drug delivery. In a cancerous breast model, a research by Pan et al. revealed the great efficacy of VCAM-1 directed NPs [75]. Additionally, the tumor microenvironment's metalloproteinase enzyme (MMP) is involved in the restructuring of the extracellular matrix and the neovascularization of the cancerous tissue [76]. There have been observations regarding MMP-sensitive NPs having a possible anticancer effect on melanoma, pancreatic carcinoma, and breast cancer [77–79].

4 Mechanism of Nanoparticles in Overcoming Drug Resistance

Drug resistance is a major problem in the diagnosis, treatment, and management of cancer. It is common in all cancer types and treatment modalities. When illnesses develop a resistance to prescribed therapy, it is referred to as "drug resistance". Drug resistance falls into two categories: innate and acquire [80]. Innate resistance is usually caused by changes in genes that are already engaged in cell death or proliferation. "Procure resistance" describes the kind of defiance that develops after receiving a certain anti-tumor medication. While the patient is undergoing treatment, the TME may alter or acquire additional mutations, which could result in resistance.

4.1 Targeting Efflux Transporters

Efflux transporters are classified under the "ATP-binding cassette (ABC) transporters" family. These are crucial to MDR. The primary job of these transporters is to reduce the concentration of drugs by pumping them out of cells. "P-glycoprotein (P-gp)" is one such efflux transporter that drug-resistant cancer cells overexpress [81, 82].

Treatment resistance has been associated with P-gp overexpression, particularly in cases of ovarian and breast cancer [83, 84]. NPs work well as an efflux pump therapy. NPs can avoid the efflux pumps because they enter the cell by "endocytosis" rather than "diffusion" and release the medication at the "perinuclear site," which is separated from the active efflux pumps [85]. Additionally, NPs can effectively circumvent efflux pumps by changing how drugs are released from the body, for instance by employing low pH and redox as triggers [86, 87].

Combination therapy is an additional approach to getting past MDR. A single drug carrier can hold many drugs placed onto NPs [88]. Rather than just avoiding them, another sensible strategy would be to limit the expression of the efflux transporter. This can be accomplished by designing NPs to carry efflux pump inhibitors and chemotherapeutic drugs [89]. A recent study found that doxorubicin and COX-2 inhibitor-delivering NPs effectively reversed MDR in breast cancer cells [90]. Similarly, it has been shown that silica NP containing miRNA-495 and doxorubicin can effectively overcome treatment resistance in lung cancer cells [91]. Another fascinating study found that using NPs to target KDR receptors in the neo-vasculature of tumors has a more powerful anti-tumour effect than using P-gp inhibitors in combo therapy.

Another way to overcome drug resistance is to deplete the ATP supply required for the ABC transporters to function. This can be achieved by limiting the amount of ATP produced, which is the target of the mitochondria.

4.2 *Targeting an Apoptotic Pathway*

When apoptotic machinery isn't working properly, cancer cells proliferate and grow more drug-resistant [92]. These anti-apoptotic proteins have been studied extensively and may be the key to reversing drug resistance. NPs can overcome MDR by the conventional means of "Bcl-2 siRNA and chemotherapeutics" codelivery [93]. NF-κB inhibitors have been used in conjunction with curcumin and "pyrrolidine dithiocarbamate (PDTC)" [94, 95]. One tactic to counter "apoptotic pathway-mediated drug resistance" is to increase pro-apoptotic factors while reducing anti-apoptotic ones. The concurrent use of ceramide and paclitaxel is one such instance [96]. Ceramide restores the expression of the p53 protein, a key tumour suppressor, via modifying alternative pre-mRNA splicing. Using NPs to deliver ceramide while correcting the p53 missense mutation is an excellent method [97]. Ceramide has potential; in cancer treatment resistance models, it has shown substantial therapeutic benefit when combined with paclitaxel. There have been documented cases of lung cancer where cationic silver nanoparticles transfected the p53 gene [98]. Analogously, strong activation of apoptosis and decreased tumour growth have been demonstrated in models of PLGA-transfected breast cancer cells [99].

Some DDS that are NP-based work by blocking efflux pumps and encouraging apoptosis [100]. The NP complex reduced the activation of apoptosis and the drug efflux mediated by P-gp.

Similarly, co-administration of "doxorubicin and resveratrol encapsulated in NPs" has been demonstrated to significantly damage doxorubicin-resistant breast cancer cells by downregulating NF-κB and Bcl-2 expression and inhibiting the expression of the efflux transporter, ultimately leading to cell death [101]. A related study used folic acid-conjugated planetary ball-milled nanoparticles encapsulating resveratrol and docetaxel to treat multidrug-resistant prostate cancer cells. This functioned by reducing ABC transporter indicators and suppressing the production of anti-apoptotic genes [102].

4.3 *Targeting Hypoxia*

Some tumour cells experience hypoxia regularly because of anomalous blood vessels adjacent to the tumour and the quickly expanding tumour's increased oxygen requirement. Chemotherapy medicines frequently escape from the hypoxic portion of the tumour. An oxygen ramp brought on by hypoxia increases tumour heterogeneity and enhances a more aggressive phenotype within the tumour. Furthermore, it has been shown that the oxygen-deprived environment promotes the overexpression of outflow proteins [103]. Therefore, one method of overcoming drug resistance is to target or silence the HIF-1α gene. Hypoxia-mediated medication resistance can be lessened by using NPs carrying HIF-1α siRNA [104]. By efficiently downregulating HIF-1α expression, inhibition of this mechanism increases MDR cells' susceptibility to cancer treatments [105]. Furthermore, HIF-1's transcriptional activity and HSP90's repression, which lowers HIF-1α production, depend on "heat shock protein 90 (HSP90)" [106]. The HSP90 inhibitor in "17AAG loaded NPs" has significantly enhanced MDR for the treatment of bladder cancer [107].

5 Advantages of Nanoparticles in Cancer Therapy

Cancer diagnosis, therapy, and care have entered a completely new age thanks to the application of nanotechnology. NPs increase the intracellular concentration of pharmaceuticals through active or passive targeting without posing a threat to healthy tissue. To manage and control the medication release, the targeted NPs could be made pH- or temperature-sensitive. Medication can be given inside the acidic TME using the pH-sensitive drug delivery system. When external sources such as magnetic fields and ultrasonic vibrations cause temperature changes in the target area, temperature-sensitive nanoparticles (NPs) release the drugs in that area. For the planned drug delivery strategy, NPs' "physicochemical characteristics," which include their molecular mass, surface chemistry, size, and shape, are also essential. Moreover, NPs can be employed to target a specific moiety and customized to the required function. Because of their cytotoxicity and uneven distribution, conventional chemotherapy and radiation therapy have several drawbacks in their effectiveness and adverse effects.

As a result, careful dosage is needed to eradicate cancer cells efficiently and without causing any negative side effects. The drug must travel through multiple defenses to reach its destination. Under physiological conditions, the drug must cross the BBB, renal infiltration, TME, RES, and RES. "Blood monocytes, macrophages, and other immune cells" make up the RES, or macrophage system [108]. "Macrophages or leukocytes" are triggered when drugs interact with MPS in the liver, spleen, or lungs; this leads to rapid drug elimination. Consequently, the drug has a short half-life [109]. NPs with "surface modification," like PEG, go around this process and lengthen the "drug half-life" to get around this. Renal infiltration is also a vital physiological function. Hence, sufficient renal infiltration lessens the toxicity that NPs cause. Since the primary function of the blood-brain barrier is to keep toxic substances out of the brain, the only effective chemotherapy therapies for brain cancer that are now available are intraventricular or intracerebral infusions [110]. Nowadays, NPs can be given by several techniques, such as targeted ultrasound, transcytosis, peptide-modified endocytosis, and the EPR effect. Rats that received glutathione PEGylated liposomes packed with methotrexate showed increased methotrexate uptake [111]. Au-NPs are widely used because of their capacity to assist in the transportation of medications that induce apoptosis [112]. Since NPs are transporters, they also stop the degradation of the contained cargo, which enhances medication stability. Moreover, many drugs can be encapsulated without generating a chemical reaction. Compared to dosage forms formed of nano-liquids, those made of dry solids are more stable [13, 113, 114]. Stabilizers are one tool for increasing stability. Porous NPs are another way to increase stability. Because tumour tissue has limited lymphatic clearance and a decreased venous return, NPs are efficiently maintained. This phenomenon is known as EPR. In a similar vein, concentrating on the surrounding tissues can help target tumours. There are several methods for administering NPs, such as oral, nasal, intraocular, and parenteral. NPs have a high surface-to-volume ratio and are extensively absorbed into cells. Studies show that NPs are a more effective drug transporter than microparticles.

5.1 Nanoparticles in Immunotherapy

The immune system is crucial to the development and proliferation of cancer cells. The advent of immunotherapy has revolutionized the treatment of cancer. Not only can NPs help with the targeted delivery of chemotherapy, but they can also be used in conjunction with immunotherapy. "Nanovaccines," "aAPCs (artificial antigen-presenting cells)," and "immunosuppressed TME targeting" are examples of NP-based immunotherapy.

Targeting vital TME cell types including "myeloid-derived suppressor cells (MDSCs)," "tumor-associated macrophages (TAMs)," and regulatory T cells accomplish this cell.

It has been shown that the combination of chemoimmunotherapy is a competent method for cancer treatment. As an illustration, research has demonstrated that co-loading the cytokine GM-CSF and chemotherapeutic drug Nutlin-3a into "spermine-modified acetylated dextran (AcDEX) NPs" enhanced the proliferation of cytotoxic CD8(+) T cells and triggered an immunological response.

Among the crucial immunological checkpoints are "programmed cell death ligand 1 (PD-L1)" and "programmed cell death protein 1 (PD-1)". Thus, immune checkpoint inhibitors are utilized in conjunction with NPs to target them. A study found that responses to conventional immune checkpoint inhibitors of PD-L1/PD-1 were not consistent. Immunological checkpoint inhibitors and immunological checkpoints were meant to be more likely to bind when multivalent poly (amidoamine) dendrimers were used. Utilizing these dendrimers improved PD-L1 blocking as well as medication accumulation at the tumour site.

An "arms race" where immunity is suppressed and rapid tumour cell repetition and transformation rates promote tumour development results from the immune structure being activated to slow down the growth of tumours. Crucially, immunity ultimately helps the suitable and slightest immunogenic tumour cells survive; this procedure is recognized as "tumour immunoediting." Immunity has the power to selectively change the cell composition of the tumour. The immune response can either destroy the tumour or solitary restrict its development, generating a subtle stalemate that can be broken by immunosuppressor states that occur over time and encourage tumour recurrence.

There are multiple ways in which tumours might avoid immune system identification (Fig. 7). Tumour cells may be able to survive longer if they downregulate the expression of MHC class I. This effectively inhibits the capacity of TAA specific CTLs to remove tumour cells by identifying peptides offered by MHC class I particles. However, it also makes it easier for natural killer (NK) cells to recognize and killing tumour cells. NK cells are a subdivision of lymph cell that activate once the body's total levels of MHC class I, which are stated by all nucleated cell categories, fall beneath a specific verge. Transformations in particles linked to TAA antigen processing and demonstration, such as proteasomes, or direct transformations in the MHC genes themselves can also result in changes in MHC class I expression on tumour cells.

Long peptides or whole proteins have been used in modified vaccination approaches that have been tested in recent years. These adjuvants, which include

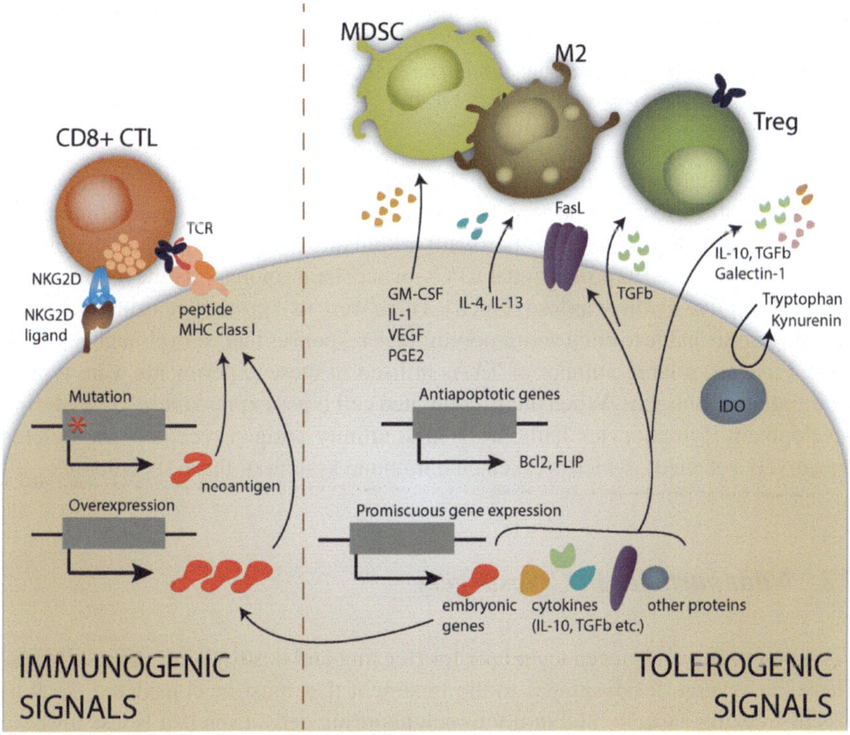

Fig. 7 Signals from tumours that either stimulate (immunogenic) or inhibit (tolerogenic) the immune system's ability to fight cancer [115]. Copyright 2015. Reproduced with permission from American Chemical Society

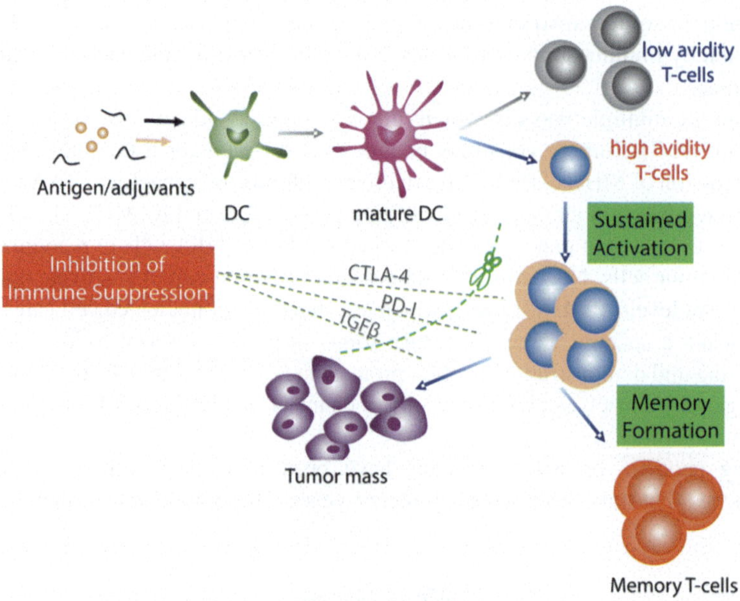

Fig. 8 Necessities for effective immune responses against tumours [115]. Copyright 2015. Reproduced with permission from American Chemical Society

cytokines similar to IL-2 or GM-CSF, like receptor (TLR) agonists (recognised by molecular outline recognition receptors), or recombinant lymphocyte co-stimulatory particles like CD80 (B7.1), are powerful stimulants that can stimulate distinctive immune cells, as well as professional APCs, or accelerate or magnify the immunogen-induced adaptive T-cell response (Fig. 8). There were two primary reasons why these methods were unable to elicit antitumor immune responses that were clinically significant. First off, a large number of TAAs utilised in these experiments were typical differentiation antigens. When undifferentiated cell types express these TAAs during development, lymphocytes harbouring high-affinity antigen receptors are deleted (negatively selected), which weakens the immune system of these specificities.

5.2 *Nanoparticles in Cryosurgery*

Cryosurgery is an advanced technique for freezing and destroying cancerous tissue. There are several disadvantages to the treatment that must be considered, such as loss of freezing capacity and injury to neighbouring cells, even if it is less invasive and does result in intraoperative bleeding and postoperative problems. The essential method of nano cryosurgery involves freezing cancer cells by injecting NPs with specific characteristics. It gets damaged during this process because ice forms inside the cells. This is a crucial procedure that can be completed successfully with NPs. In

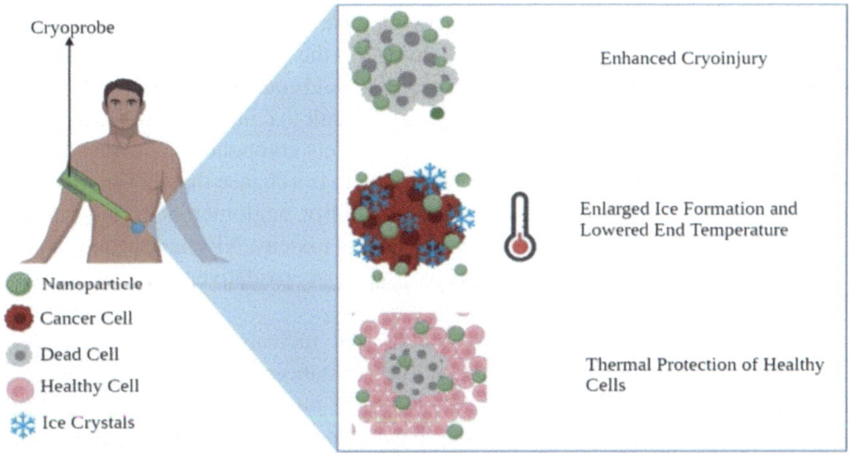

Fig. 9 Exemplary display of nanomaterials in cryogenic surgery [13]. Copyright 2021. Reproduced with permission from Springer Nature

addition, they cool down quickly, and the "growth direction" and "direction of the ice ball" may be controlled (Fig. 9).

There are possibilities that the freezing will harm healthy tissue if the tumour's position precludes cryosurgery or if other nearby organs are in danger. Phase change materials (PMs) composed of NPs have been utilized recently in cryosurgery to preserve the surrounding normal, healthy tissue. For example, liposome-based microencapsulated phase change nanoparticles have demonstrated remarkable outcomes in safeguarding adjacent healthy tissue.

6 Significant Challenges in the Clinical Application of Nanoparticles

The current state of nanoparticle research and expertise has increased sharply with the growth of nanotechnology. However, only a few of these proceed to clinical trials. The majority only get as far as the in vitro and in vivo phases. Although every distinct nanoformulation has its own set of challenges when it comes to clinical translation, most NPs face three main types of challenges that are generally connected to study design, biology, and technology. Some of the biological issues are toxicity, lack of delivery techniques, regulating bio-distribution, route of NPs over biological barriers, and their elimination. NPs are frequently injected intravenously into the blood, which removes them from the blood and makes it challenging for them to stay and interact with the target site. As a result, a high-concentration medicine that might not have the desired therapeutic effects is used. The application of 3D magnetic fields to control the movement of NPs against blood flow has been shown in many in vitro and in vivo studies. This can be done by using magnetic nanoparticles. However, more research is required to understand how magnetic fields affect the human body, how many NPs are present, and how crosstalk between magnetic fields occurs.

The biological aspects of nanoparticles are difficult to control and require careful attention. Even though NPs are composed of biosafety components and are adjusted to lengthen their half-life and retention period, there is a chance that they could harm the kidneys, liver, and lungs. Surface area, solubility, agglomeration, and particle size and shape are some of the variables that control toxicity. NPs have been found to deposit more heavily in the lung and to have cytotoxic, oxidative, and inflammatory effects.

Studies reveal that NPs typically create free radicals that damage healthy cells. Making NPs from more biocompatible substances, including chitosan, and substances that break down in the presence of near-infrared light are viable substitutes. Steering clear of the "mononuclear phagocytic system (MPS)" is another challenging undertaking. For NPs to be absorbed, PC must first break down MPS by adsorbing proteins from physiological fluids. To prevent this, NPs have been coated with materials that inhibit the synthesis of the corona protein. But they haven't produced any noteworthy outcomes. To get around this issue, NPs that are specifically designed to target "macrophages" could be used as novel drug delivery systems. At the moment, prevalent tactics include blocking "CD47-SIRPα pathways," decreasing and reprogramming TAMs, and inhibiting macrophage recruitment. Performance forecasts, equal optimisation, and scale-up synthesis are some of the technological problems facing NPs. These are essential for ensuring NPs' clinical success. The majority of NPs utilised in vitro and in vivo investigations are often manufactured in small batches, and scale-up for large amounts is not always possible due to equipment and other factors. There is no systematic design or optimisation process for the lead clinical candidates that perform best in animal models. To get around this, we can employ specific techniques that allow us to test multiple nanoformulations and choose one optimised formulation through a series of selective iterations. Such impacts shouldn't however be added straight into experiments with humans. It is difficult to predict the effectiveness and performance of nanoparticles, and it is extremely difficult to replicate the in vivo outcomes in human studies. It is possible to create theoretical or computational modelling that mimics physiological environments and tissue in conjunction with experimental data. Organs-on-chips, for example, are a subject of active research and have the potential to enhance NP predictions of efficacy and performance. Issues with study design, such as study size, purpose, and when to administer NP treatments have a big impact on clinical trials. The majority of the research uses "cell and animal models," which might not yield interpretable results in trials involving humans. As such, it is difficult to replicate natural reactions in the human body using a single model. Furthermore, given one of the important characteristics of cancer is metastasis, "models of cancer metastasis" ought to be thoroughly studied. Furthermore, if we concentrate on personalized treatment, $N = 1$ clinical investigation will be necessary. Numerous factors, including genetics, environment, and prior medical history, must be taken into account. NPs are never employed as first-line treatments, which presents another significant obstacle. Even though our approved nano-formulations work well, they are typically reserved for use in future treatments if a clinical trial reveals disease progression. The majority of patients have developed medication resistance or have advanced on several different treatment trajectories. These circumstances frequently distort the findings of clinical trials and reduce the chances that NPs treatment will help people who are probably still curable.

7 Conclusion and Future Perspective

With its ability to carry out tiny components for cancer identification, assessment, and treatment, nanotechnology has demonstrated an emerging new age in cancer treatment. In the therapeutic context of multiple kinds of cancer, treatments that utilize the distinctive characteristics of NPs are becoming widely employed. Especially compared to conventional medications, an NP-based drug transport system is coupled via advanced biological compatibility, pharmaceutical kinetics, tumour directing, and durability. Additionally, NPs offer an exceptional establishment for mixed therapy, that helps in the combating of multidrug resistance. Several NPs, containing metallic, hybrid, and polymeric NPs, have proven better drug deliverance efficacy through further investigation. Experts ought to pay particular attention to the characteristics of the therapeutic agents and the attributes of the nanoplatforms that have been selected. Nevertheless, there are several drawbacks, such as immune system damage, overtime toxicity, neurological damage, and the inadequacy of in vitro models to accurately recreate in vivo phase. The medical effectiveness of "nanovaccines" and "artificial APCs" is inadequate, regardless of their apparent augmented effectiveness over traditional immune therapy. It is necessary to examine these novel mechanisms' adaptability and toxicity. Furthermore, manufacturing "immune-modulating factor-loaded NPs" could improve vaccination efficacy for immune therapy. This field is still in its early years but as genomics studies on the "mechanism that causes cancer, multidrug resistance, recurrence" advance, further NP-based medicines should be competent to be developed. Of the countless studies conducted, very rare NP-based doses are really in consumption, just a limited are in clinical studies, and the majority are in the phase of exploration. Substantial research is required to be performed on "comprehending noxiousness, cellular and pharmacological variables that stimulus NP-based drug transport, EPR, and PC method" in human beings to develop nanotechnology effectively. The emergence of nanotechnology and cancer treatment is expected to eventually bring about an unprecedented shift in clinical implementation for NP-based treatment of cancer, as supported by the given findings.

Acknowledgement This research was supported by Basic Science Research Program through the National Research Foundation of Korea (NRF) funded by the Ministry of Education (2020R1A6A1A03038697).

Conflict of Interest There is no conflict of interest among the authors.

References

1. Wu, S., Zhu, W., Thompson, P., Hannun, Y.A.: Evaluating intrinsic and non-intrinsic cancer risk factors. Nat. Commun. **9**, 3490 (2018)
2. Quazi S (2021) Telomerase gene therapy: a remission towards cancer. Preprints **2021**, 2021100407
3. Anand, P., Kunnumakkara, A.B., Sundaram, C., Harikumar, K.B., Tharakan, S.T., Lai, O.S., Sung, B., Agrawal, B.B.: Cancer is a preventable disease that requires major lifestyle changes. Pharm. Res. **25**, 2097–2116 (2008)
4. Bauleth-Ramos, T., Shahbazi, M., Liu, D., Fontana, F., Correia, A., Figueiredo, P., Santos, H.A.: Nutlin-3a and cytokine co-loaded spermine-modified acetalated dextran nanoparticles for cancer chemo-immunotherapy. Adv. Funct. Mater. **27**, 1703303 (2017)
5. Park, W., Heo, Y.J., Han, D.K.: New opportunities for nanoparticles in cancer immunotherapy. Biomater. Res. **22**, 1–10 (2018)
6. Jovcevska, I., Muyldermans, S.: The therapeutic potential of nanobodies. BioDrugs Clin. Immunother. Biopharm. Gene Ther. **34**, 11–26 (2020)
7. Zitvogel, L., Apetoh, L., Ghiringhelli, F., Kroemer, G.: Immunological aspects of cancer chemotherapy. Nat. Rev. Immunol. **8**, 59–73 (2008)
8. Chan, H.-K., Ismail, S.: Side effects of chemotherapy among cancer patients in a Malaysian general hospital: experiences, perceptions and informational needs from clinical pharmacists. Asian Pac. J. Cancer Prev. **15**, 5305–5309 (2014)
9. Bu, J., Nair, A., Iida, M., Jeong, W.J., Poellmann, M.J., Mudd, K., Kubiatowicz, L.J., Liu, E.W., Wheeler, D.L., Hong, S.: An avidity-based PD-L1 antagonist using nanoparticle-antibody conjugates for enhanced immunotherapy. Nano Lett. **20**, 4901–4909 (2020)
10. Mahapatro, A., Singh, D.K.: Biodegradable nanoparticles are excellent vehicle for site directed in-vivo delivery of drugs and vaccines. J. Nanobiotechnol. **9**, 55 (2011)
11. Kroemer, G., Zitvogel, L.: The breakthrough of the microbiota. Nat. Rev. Immunol. **18**, 87–88 (2018)
12. Rosenberg, S.A., Restifo, N.P., Yang, J.C., Morgan, R.A., Dudley, M.E.: Adoptive cell transfer: a clinical path to effective cancer immunotherapy. Nat. Rev. Cancer **8**, 299–308 (2008)
13. Gavas, S., Quazi, S., Karpiński, T.M.: Nanoparticles for cancer therapy: current progress and challenges. Nanoscale Res. Lett. **16**, 173 (2021)
14. Melero, I., Rouzaut, A., Motz, G.T., Coukos, G.: T-cell and NK-cell infiltration into solid tumors: a key limiting factor for efficacious cancer immunotherapy. Cancer Discov. **4**, 522–526 (2014)
15. Lacouture, M., Sibaud, V.: Toxic side effects of targeted therapies and immunotherapies affecting the skin, oral mucosa, hair, and nails. Am. J. Clin. Dermatol. **19**, 31–39 (2018)
16. Nikita, Km., Karkare, P., Ray, D., Aswal, V.K., Singh, P.S., Murthy, C.N.: Understanding the morphology of MWCNT/PES mixed-matrix membranes using SANS: interpretation and rejection performance. Appl Water Sci **9**, 1–9 (2019)
17. Palazzolo, S., Bayda, S., Hadla, M., Caligiuri, I., Corona, G., Toffoli, G., Rizzolio, F.: The clinical translation of organic nanomaterials for cancer therapy: a focus on polymeric nanoparticles, micelles, liposomes and exosomes. Curr. Med. Chem. **25**, 4224–4268 (2018)
18. Li, W., Zhang, H., Assaraf, Y.G., Zhao, K., Xu, X., Xie, J., Yang, D.H., Chen, Z.S.: Overcoming ABC transporter-mediated multidrug resistance: molecular mechanisms and novel therapeutic drug strategies. Drug Resist. Updates **27**, 14–29 (2016)
19. Shi, J., Votruba, A.R., Farokhzad, O.C., Langer, R.: Nanotechnology in drug delivery and tissue engineering: from discovery to applications. Nano Lett. **10**, 3223–3230 (2010)
20. Boisseau, P., Loubaton, B.: Nanomedicine, nanotechnology in medicine. Comptes Rendus Phys. **12**, 620–636 (2011)
21. Laurent, S., Forge, D., Port, M., Roch, A., Robic, C., Vander Elst, L., Muller, R.: Magnetic iron oxide nanoparticles: synthesis, stabilization, vectorization, physicochemical characterizations, and biological applications. Chem. Rev. **10**, 2064–2110 (2008)

22. Tiwari, J.N., Tiwari, R.N., Kim, K.S.: Zero-dimensional, one-dimensional, two-dimensional and three-dimensional nanostructured materials for advanced electrochemical energy devices. Prog. Mater. Sci. **57**, 424–803 (2012)
23. Shin, W.K., Cho, J., Kannan, A., et al.: Cross-linked composite gel polymer electrolyte using mesoporous methacrylate-functionalized SiO_2 nanoparticles for lithium-ion polymer batteries. Sci. Rep. **6**, 26332 (2016)
24. Prokop, A., Davidson, J.M.: Nanovehicular intracellular delivery systems. J. Pharm. Sci. **97**, 3518–4359 (2008)
25. Shafey, A.: Green synthesis of metal and metal oxide nanoparticles from plant leaf extracts and their applications: a review. Green Process. Synth. **9**(1), 304–339 (2020)
26. Lassalle, V., Ferreira, M.: PLA nano- and microparticles for drug delivery: an overview of the methods of preparation. Macromol. Biosci. **7**, 767–783 (2007)
27. Yang, Q., Jones, S., Parker, C., Zamboni, W., Bear, J., Lai, S.: Evading immune cell uptake and clearance requires PEG grafting at densities substantially exceeding the minimum for brush conformation. Mol. Pharm. **11**, 1250–1258 (2014)
28. Krasteva, N., Georgieva, M.: Promising therapeutic strategies for colorectal cancer treatment based on nanomaterials. Pharmaceutics **14**, 1213 (2022)
29. Chen, Y., Zhu, X., Zhang, X., Liu, B., Huang, L.: Nanoparticles modified with tumor-targeting scFv deliver siRNA and miRNA for cancer therapy. Mol. Ther. **18**, 1650–1656 (2010)
30. Wang, Q., Alshaker, H., Bohler, T., Srivats, S., Chao, Y., Cooper, C., et al.: Core shell lipid-polymer hybrid nanoparticles with combined docetaxel and molecular targeted therapy for the treatment of metastatic prostate cancer. Sci. Rep. **7**, 5901 (2017)
31. Grigore, M.E.: Organic and inorganic nano-systems used in cancer treatment. J. Med. Res. Health Educ. **1**, 1 (2017)
32. Han, B., Yang, Y., Chen, J., Tang, H., Sun, Y., Zhang, Z., et al.: Preparation, characterization, and pharmacokinetic study of a novel long-acting targeted paclitaxel liposome with antitumor activity. Int. J. Nanomed. **15**, 553–571 (2020)
33. O'Brien, M.E., Wigler, N., Inbar, M., Rosso, R., Grischke, E., Santoro, A., et al.: Reduced cardiotoxicity and comparable efficacy in a phase III trial of PEGylated liposomal doxorubicin HCl (CAELYX/Doxil) versus conventional doxorubicin for first-line treatment of metastatic breast cancer. Ann. Oncol. **15**, 440–449 (2004)
34. Geisberg, C.A., Sawyer, D.B.: Mechanisms of anthracycline cardiotoxicity and strategies to decrease cardiac damage. Curr. Hypertens. Rep. **12**, 404–410 (2010)
35. Eloy, J.O., Petrilli, R., Topan, J.F., Antonio, H.M.R., Barcellos, J.P.A., Chesca, D.L., et al.: Co-loaded paclitaxel/rapamycin liposomes: development, characterization and in vitro and in vivo evaluation for breast cancer therapy. Colloids Surf. B Biointerfaces **141**, 74–82 (2016)
36. Nikita, Km., Ray, D., Aswal, V.K., Murthy, C.N.: Surface modification of functionalized multiwalled carbon nanotubes containing mixed matrix membrane using click chemistry. J. Memb. Sci. **596**, 117710 (2020)
37. Madani, S.Y., Naderi, N., Dissanayake, O., Tan, A., Seifalian, A.M.: A new era of cancer treatment: carbon nanotubes as drug delivery tools. Int. J. Nanomed. **6**, 2963–2979 (2011)
38. Luo, E., Song, G., Li, Y., Shi, P., Hu, J., Lin, Y.: The toxicity and pharmacokinetics of carbon nanotubes as an effective drug carrier. Curr. Drug Metab. **14**, 879–890 (2013)
39. Gao, F., Zhang, J., Fu, C., Xie, X., Peng, F., You, J., et al.: IRGD-modified lipid-polymer hybrid nanoparticles loaded with isoliquiritigenin to enhance anti-breast cancer effect and tumor-targeting ability. Int. J. Nanomed. **12**, 4147–4162 (2017)
40. Search Parodi, A., Quattrocchi, N., van de Ven, A.L., Chiappini, C., Evangelopoulos, M., Martinez, J.O., et al.: Synthetic nanoparticles functionalized with biomimetic leukocyte membranes possess cell-like functions. Nat. Nanotechnol. **8**, 61–68 (2013)
41. Wong, C., Stylianopoulos, T., Cui, J., Martin, J., Chauhan, V.P., Jiang, W., et al.: Multistage nanoparticle delivery system for deep penetration into tumor tissue. Proc. Natl. Acad. Sci. U.S.A. **108**, 2426–2431 (2011)
42. Shi, J., Xiao, Z., Kamaly, N., Farokhzad, O.C.: Self-assembled targeted nanoparticles: evolution of technologies and bench to bedside translation. ACC Chem. Res. **44**, 1123–1134 (2011)

43. Nikita, Km., Kumar, S., Aswal, V.K., Kanchan, D.K., Murthy, C.N.: Porous structure studies of the mixed-matrix polymeric membranes of polyether sulfone incorporated with functionalized multiwalled carbon nanotubes. Desal. Water Treat. **146**, 29–38 (2019)
44. Farokhzad, O.C., Langer, R.: Impact of nanotechnology on drug delivery. ACS Nano **3**, 16–20 (2009)
45. Danhier, F., Feron, O., Préat, V.: To exploit the tumor microenvironment: passive and active tumor targeting of nanocarriers for anti-cancer drug delivery. J. Control. Release **148**, 135–146 (2010)
46. Carmeliet, P., Jain, R.K.: Angiogenesis in cancer and other diseases. Nature **407**, 249–257 (2000)
47. Maeda, H.: The enhanced permeability and retention (EPR) effect in tumor vasculature: the key role of tumor-selective macromolecular drug targeting. Adv. Enzyme Regul. **41**, 189–207 (2001)
48. Torchilin, V.P.: Recent advances with liposomes as pharmaceutical carriers. Nat. Rev. Drug Discov. **4**, 145–160 (2005)
49. Carita, A.C., Eloy, J.O., Chorilli, M., Lee, R.J., Leonardi, G.R.: Recent advances and perspectives in liposomes for cutaneous drug delivery. Curr. Med. Chem. **25**, 606–635 (2018)
50. Sykes, E.A., Chen, J., Zheng, G., Chan, W.C.: Investigating the impact of nanoparticle size on active and passive tumor targeting efficiency. ACS Nano **8**, 5696–5706 (2014)
51. Pelicano, H., Martin, D.S., Xu, R.H., Huang, P.: Glycolysis inhibition for anticancer treatment. Oncogene **25**, 4633–4646 (2006)
52. Lim, E.K., Chung, B.H., Chung, S.J.: Recent advances in pH-sensitive polymeric nanoparticles for smart drug delivery in cancer therapy. Curr. Drug Targets **19**, 300–317 (2018)
53. Jain, R.K.: Barriers to drug delivery in solid tumors. Sci. Am. **271**, 58–65 (1994)
54. Amreddy, N., Muralidharan, R., Babu, A., Mehta, M., Johnson, E.V., Zhao, Y.D., et al.: Tumor-targeted and pH-controlled delivery of doxorubicin using gold nanorods for lung cancer therapy. Int. J. Nanomed. **10**, 6773–6788 (2015)
55. Nikita, Km., Chivukula, D.S., Murthy, C.N.: Uniquely modified polyether sulphone and f-CNTs mixed matrix membranes for enhanced water transport and reduced biofouling. Desal. Water Treat. **245**, 16–34 (2022)
56. Santi, M., Maccari, G., Mereghetti, P., Voliani, V., Rocchiccioli, S., Ucciferri, N., et al.: Rational design of a transferrin-binding peptide sequence tailored to targeted nanoparticle internalization. Bioconjug. Chem. **28**, 471–480 (2017)
57. Cui, Y.N., Xu, Q.X., Davoodi, P., Wang, D.P., Wang, C.H.: Enhanced intracellular delivery and controlled drug release of magnetic PLGA nanoparticles modified with transferrin. Acta Pharmacol. Sin. **38**, 943–953 (2017)
58. Soe, Z.C., Kwon, J.B., Thapa, R.K., Ou, W., Nguyen, H.T., Gautam, M., et al.: Transferrin-conjugated polymeric nanoparticle for receptor mediated delivery of doxorubicin in doxorubicin-resistant breast cancer cells. Pharmaceutics **11**, 63 (2019)
59. Low, P.S., Kularatne, S.A.: Folate-targeted therapeutic and imaging agents for cancer. Curr. Opin. Chem. Biol. **13**, 256–262 (2009)
60. Muralidharan, R., Babu, A., Amreddy, N., Basalingappa, K., Mehta, M., Chen, A., et al.: Folate receptor-targeted nanoparticle delivery of HuR-RNAi suppresses lung cancer cell proliferation and migration. J. Nanobiotechnol. **14**, 47 (2016)
61. Nikita, K., Patel, D., Patel, G.: Additive manufacturing of multifunctional polymer nanocomposites: from 3D to 4D. In: Nanotechnology-Based Additive Manufacturing: Product Design, Properties and Applications, vol. 1, pp. 277–313 (2023)
62. Minko, T.: Drug targeting to the colon with lectins and neoglycoconjugates. Adv. Drug Deliv. Rev. **56**, 491–509 (2004)
63. Obaid, G., Chambrier, I., Cook, M.J., Russell, D.A.: Cancer targeting with biomolecules: a comparative study of photodynamic therapy efficacy using antibody or lectin conjugated phthalocyanine-PEG gold nanoparticles. Photochem. Photobiol. Sci. **14**, 737–747 (2015)
64. Nicholson, R.I., Gee, J.M., Harper, M.E.: EGFR and cancer prognosis. Eur. J. Cancer **37**, S9–S15 (2001)

65. Sigismund, S., Avanzato, D., Lanzetti, L.: Emerging functions of the EGFR in cancer. Mol. Oncol. **12**, 3–20 (2018)
66. Alexis, F., Basto, P., Levy-Nissenbaum, E., Radovic-Moreno, A.F., Zhang, L., Pridgen, E., et al.: HER-2-targeted nanoparticle-affibody bioconjugates for cancer therapy. ChemMedChem **3**, 1839–1843 (2008)
67. Balasubramanian, S., Girija, A.R., Nagaoka, Y., Iwai, S., Suzuki, M., Kizhikkilot, V., et al.: Curcumin and 5-fluorouracil-loaded, folate- and transferrin decorated polymeric magnetic nanoformulation: a synergistic cancer therapeutic approach, accelerated by magnetic hyperthermia. Int. J. Nanomed. **9**, 437–459 (2014)
68. Apte, R.S., Chen, D.S., Ferrara, N.: VEGF in signaling and disease: beyond discovery and development. Cell **176**, 1248–1264 (2019)
69. Orleth, A., Mamot, C., Rochlitz, C., Ritschard, R., Alitalo, K., Christofori, G., et al.: Simultaneous targeting of VEGF-receptors 2 and 3 with immunoliposomes enhances therapeutic efficacy. J. Drug Target. **24**, 80–89 (2016)
70. Desgrosellier, J.S., Cheresh, D.A.: Integrins in cancer: biological implications and therapeutic opportunities. Nat. Rev. Cancer **10**, 9–22 (2010)
71. Nisato, R.E., Tille, J.C., Jonczyk, A., Goodman, S.L., Pepper, M.S.: Alphav beta 3 and alphav beta 5 integrin antagonists inhibit angiogenesis in vitro. Angiogenesis **6**, 105–119 (2003)
72. Hood, J.D., Bednarski, M., Frausto, R., Guccione, S., Reisfeld, R.A., Xiang, R., et al.: Tumor regression by targeted gene delivery to the neovasculature. Science **296**, 2404–2407 (2002)
73. Ruoslahti, E.: Specialization of tumour vasculature. Nat. Rev. Cancer **2**, 83–90 (2002)
74. Dienst, A., Grunow, A., Unruh, M., Rabausch, B., Nör, J.E., Fries, J.W., et al.: Specific occlusion of murine and human tumor vasculature by VCAM1-targeted recombinant fusion proteins. J. Natl. Cancer Inst. **97**, 733–747 (2005)
75. Pan, H., Myerson, J.W., Hu, L., Marsh, J.N., Hou, K., Scott, M.J., et al.: Programmable nanoparticle functionalization for in vivo targeting. FASEB J. **27**, 255–264 (2013)
76. Lia, N.G., Shib, Z.H., Tang, Y.P., Duan, J.A.: Selective matrix metalloproteinase inhibitors for cancer. Curr. Med. Chem. **16**, 3805–3827 (2009)
77. Mansour, A.M., Drevs, J., Esser, N., Hamada, F.M., Badary, O.A., Unger, C., et al.: A new approach for the treatment of malignant melanoma: enhanced antitumor efficacy of an albumin-binding doxorubicin prodrug that is cleaved by matrix metalloproteinase 2. Cancer Res. **63**, 4062–4066 (2003)
78. Xiao, Q., Zhu, X., Yuan, Y., Yin, L., He, W.: A drug-delivering-drug strategy for combined treatment of metastatic breast cancer. Nanomedicine **14**, 2678–2688 (2018)
79. Cun, X., Chen, J., Li, M., He, X., Tang, X., Guo, R., et al.: Tumor associated fibroblast-targeted regulation and deep tumor delivery of chemotherapeutic drugs with a multifunctional size-switchable nanoparticle. ACS Appl. Mater. Interfaces **11**, 39545–39559 (2019)
80. Gary-Bobo, M., Hocine, O., Brevet, D., Maynadier, M., Raehm, L., Richeter, S., Charasson, V., Loock, B., Morère, A., Maillard, P., Garcia, M., Durand, J.O.: Cancer therapy improvement with mesoporous silica nanoparticles combining targeting, drug delivery and PDT. Int. J. Pharm. **423**, 509–515 (2012)
81. Housman, G., Byler, S., Heerboth, S., Lapinska, K., Longacre, M., Snyder, N., Sarkar, S.: Drug resistance in cancer: an overview. Cancers **6**, 1769–1792 (2014)
82. Schneider, E., Hunke, S.: ATP-binding-cassette (ABC) transport systems: functional and structural aspects of the ATP-hydrolyzing subunits/domains. FEMS Microbiol. Rev. **22**, 1–20 (1998)
83. Allen, J.D., Brinkhuis, R.F., van Deemter, L., Wijnholds, J., Schinkel, A.H.: Extensive contribution of the multidrug transporters P-glycoprotein and Mrp1 to basal drug resistance. Can. Res. **60**, 5761–5766 (2000)
84. Singh, J.P., Mittal, M.K., Saxena, S., Bansal, A., Bhatia, A., Kulshreshtha, P.: Role of p-glycoprotein expression in predicting response to neoadjuvant chemotherapy in breast cancer—a prospective clinical study. World J. Surg. Oncol. **14**(3), 61 (2005)
85. Agarwal, R., Kaye, S.B.: Ovarian cancer: strategies for overcoming resistance to chemotherapy. Nat. Rev. Cancer **3**, 502–516 (2003)

86. Murakami, M., Cabral, H., Matsumoto, Y., Wu, S., Kano, M.R., Yamori, T., Nishiyama, N., Kataoka, K.: Improving drug potency and efficacy by nanocarrier-mediated subcellular targeting. Sci. Transl. Med. **3**, 64ra2 (2011)
87. Yu, B., Song, N., Hu, H., Chen, G., Shen, Y., Cong, H.: A degradable triple temperature-, pH-, and redox-responsive drug system for cancer chemotherapy. J. Biomed. Mater. Res. Part A **106**, 3203–3210 (2018)
88. Kundu, M., Sadhukhan, P., Ghosh, N., Chatterjee, S., Manna, P., Das, J., Sil, P.C.: PH-responsive and targeted delivery of curcumin via phenylboronic acid-functionalized ZnO nanoparticles for breast cancer therapy. J. Adv. Res. **18**, 161–172 (2019)
89. Cuvier, C., Roblot-Treupel, L., Millot, J.M., Lizard, G., Chevillard, S., Manfait, M., Couvreur, P., Poupon, M.F.: Doxorubicin-loaded nanospheres bypass tumor cell multidrug resistance. Biochem. Pharmacol. **44**, 509–517 (1992)
90. Soma, C.E., Dubernet, C., Bentolila, D., Benita, S., Couvreur, P.: Reversion of multidrug resistance by co-encapsulation of doxorubicin and cyclosporine A in polyalkylcyanoacrylate nanoparticles. Biomaterials **21**, 1–7 (2000)
91. Zhang, S., Guo, N., Wan, G., Zhang, T., Li, C., Wang, Y., Wang, Y., Liu, Y.: PH and redox dual-responsive nanoparticles based on disulfide-containing poly(β-amino ester) for combining chemotherapy and COX-2 inhibitor to overcome drug resistance in breast cancer. J. Nanobiotechnol. **17**, 109 (2019)
92. He, J., Gong, C., Qin, J., Li, M., Huang, S.: Cancer cell membrane decorated silica nanoparticle loaded with miR495 and doxorubicin to overcome drug resistance for effective lung cancer therapy. Nanoscale Res. Lett. **14**, 339 (2019)
93. Viktorsson, K., Lewensohn, R., Zhivotovsky, B.: Apoptotic pathways and therapy resistance in human malignancies. Adv. Cancer Res. **94**, 143–196 (2005)
94. Fan, L., Li, F., Zhang, H., Wang, Y., Cheng, C., Li, X., Gu, C.H., Yang, Q., Wu, H., Zhang, S.: Co-delivery of PDTC and doxorubicin by multifunctional micellar nanoparticles to achieve active targeted drug delivery and overcome multidrug resistance. Biomaterials **31**, 5634–5642 (2010)
95. Choi, K.Y., Correa, S., Min, J., Li, J., Roy, S., Laccetti, K.H., Dreaden, E., Kong, S., Heo, R., Roh, Y.H., Lawson, E.C., Palmer, P.A., Hammond, P.T.: Binary targeting of siRNA to hematologic cancer cells in vivo using layer-by-layer nanoparticles. Adv. Funct. Mater. **29**, 1900018 (2019)
96. Zhao, M.D., Li, J.Q., Chen, F.Y., Dong, W., Wen, L.J., Fei, W.D., Zhang, X., Yang, P.L., Zhang, X.M., Zheng, C.H.: Co-delivery of curcumin and paclitaxel by "core-shell" targeting amphiphilic copolymer to reverse resistance in the treatment of ovarian cancer. Int. J. Nanomed. **14**, 9453–9467 (2019)
97. Van Vlerken, L.E., Duan, Z., Little, S.R., Seiden, M.V., Amiji, M.M.: Augmentation of therapeutic efficacy in drug-resistant tumor models using ceramide coadministration in temporal-controlled polymer-blend nanoparticle delivery systems. AAPS J. **12**, 171–180 (2010)
98. Choi, S.H., Jin, S.E., Lee, M.K., Lim, S.J., Park, J.S., Kim, B.G., Ahn, W.S., Kim, C.K.: Novel cationic solid lipid nanoparticles enhanced p53 gene transfer to lung cancer cells. Eur. J. Pharm. Biopharm. **68**, 545–554 (2008)
99. Khiste, S.K., Liu, Z., Roy, K.R., Uddin, M.B., Hosain, S.B., Gu, X., Nazzal, S., Hill, R.A., Liu, Y.Y.: Ceramide-rubusoside nanomicelles, a potential therapeutic approach to target cancers carrying p53 missense mutations. Mol. Cancer Ther. **19**, 564–574 (2020)
100. Prabha, S., Labhasetwar, V.: Nanoparticle-mediated wild-type p53 gene delivery results in sustained antiproliferative activity in breast cancer cells. Mol. Pharm. **1**, 211–219 (2004)
101. Cheng, H., Wu, Z., Wu, C., et al.: Overcoming STC2 mediated drug resistance through drug and gene co-delivery by PHB-PDMAEMA cationic polyester in liver cancer cells. Mater. Sci. Eng. C Mater. Biol. Appl. **83**, 210–217 (2018)
102. Zhao, Y., Huan, M.L., Liu, M., Cheng, Y., Sun, Y., Cui, H., Liu, D.Z., Mei, Q.B., Zhou, S.Y.: Doxorubicin and resveratrol co-delivery nanoparticle to overcome doxorubicin resistance. Sci. Rep. **6**, 35267 (2016)

103. Jing, X., Yang, F., Shao, C., Wei, K., Xie, M., Shen, H., Shu, Y.: Role of hypoxia in cancer therapy by regulating the tumor microenvironment. Mol. Cancer **18**, 157 (2019)
104. Luo, B., Li, W., Wang, R., et al.: Effect of hypoxia on expression of multidrug resistance protein 2 and its regulation mechanism. J. Cent. South Univ. Med. Sci. **42**, 98–107 (2017)
105. Hajizadeh, F., Moghadaszadeh Ardebili, S., Baghi Moornani, M., Masjedi, A., Atyabi, F., Kiani, M., Namdar, A., Karpisheh, V., Izadi, S., Baradaran, B., Azizi, G., Ghalamfarsa, G., Sabz, G., Yousefi, M., Jadidi-Niaragh, F.: Silencing of HIF-1α/CD73 axis by siRNA-loaded TAT-chitosan-spion nanoparticles robustly blocks cancer cell progression. Eur. J. Pharmacol. **882**, 173235 (2020)
106. Zhang, J., Zhang, Q., Lou, Y., Fu, Q., Chen, Q., Wei, T., Yang, J., Tang, J., Wang, J., Chen, Y., Zhang, X., Zhang, J., Bai, X., Liang, T.: Hypoxia-inducible factor-1α/interleukin-1β signaling enhances hepatoma epithelial-mesenchymal transition through macrophages in a hypoxic-inflammatory microenvironment. Hepatology (Baltimore, Md.) **67**, 1872–1889 (2018)
107. Semenza, G.L.: Evaluation of HIF-1 inhibitors as anticancer agents. Drug Discov. Today **12**, 853–859 (2007)
108. Rezvantalab, S., Drude, N.I., Moraveji, M.K., Güvener, N., Koons, E.K., Shi, Y., Kiessling, F.: PLGA-based nanoparticles in cancer treatment. Front. Pharmacol. **9**, 1260 (2018)
109. Yona, S., Gordon, S.: From the reticuloendothelial to mononuclear phagocyte system—the unaccounted years. Front. Immunol. **6**, 328 (2015)
110. Liang, T., Zhang, R., Liu, X., Ding, Q., Wu, S., Li, C., Lin, Y., Ye, Y., Zhong, Z., Zhou, M.: Recent advances in macrophage-mediated drug delivery systems. Int. J. Nanomed. **16**, 2703–2714 (2021)
111. Tran, S., DeGiovanni, P.J., Piel, B., Rai, P.: Cancer nanomedicine: a review of recent success in drug delivery. Clin. Transl. Med. **6**, 44 (2017)
112. Hu, Y., Gaillard, P.J., de Lange, E., Hammarlund-Udenaes, M.: Targeted brain delivery of methotrexate by glutathione PEGylated liposomes: how can the formulation make a difference? Eur. J. Pharm. Biopharm. **139**, 197–204 (2019)
113. Feng, Q., Shen, Y., Fu, Y., Muroski, M.E., Zhang, P., Wang, Q., Xu, C., Lesniak, M.S., Li, G., Cheng, Y.: Self-assembly of gold nanoparticles shows microenvironment-mediated dynamic switching and enhanced brain tumor targeting. Theranostics **7**, 1875–1889 (2017)
114. Yao, Y., Zhou, Y., Liu, L., Xu, Y., Chen, Q., Wang, Y., Wu, S., Deng, Y., Zhang, J., Shao, A.: Nanoparticle-based drug delivery in cancer therapy and its role in overcoming drug resistance. Front. Mol. Biosci. **7**, 193 (2020)
115. Shao, K., Singha, S., Casares, X.C., Tsai, S., Yang, Y., Santamaria, P.: Nanoparticle-based immunotherapy for cancer. ACS Nano **9**, 16–30 (2015)

Understanding Cancer Biology and Microenvironment

Arnab Saha, Prakriti Mahlawat, Mohit Kumar, and Naval V. Koralkar

Abstract Cancer is characterized by uncontrolled cell growth and the spread of abnormal cells, leading to tumor formation. While genetic mutations are central to cancer initiation, the tumor microenvironment (TME), consisting of immune cells, fibroblasts, blood vessels, and the extracellular matrix, plays a crucial role in supporting tumor progression, metastasis, and therapy resistance. The dynamic interactions between cancer cells and the TME facilitate tumor growth, immune evasion, and resistance to treatment, posing significant challenges in cancer therapy. Emerging strategies targeting the TME, including immunotherapies and anti-angiogenic treatments, have shown promise in disrupting these pro-tumorigenic interactions. Advances in technologies such as CRISPR, organ-on-chip systems, and nanoparticle-based drug delivery provide new opportunities to enhance cancer treatment efficacy. A comprehensive understanding of the reciprocal interactions between cancer cells and the TME is essential for developing effective, personalized therapies that address both the tumor and its supportive ecosystem, thereby overcoming current treatment limitations.

Keywords Cancer · Metastasis · Tumor microenvironment · Cancer therapy

---

A. Saha
School of Biological Sciences, UM-DAE CEBS, Mumbai, India

P. Mahlawat
Department of Biochemistry, Kurukshetra University, Kurukshetra, Haryana, India

M. Kumar
Department of Pharmacology and Experimental Neuroscience, University of Nebraska Medical Center, Omaha, USA

N. V. Koralkar (✉)
Department of Chemical Engineering, ITM SLS Baroda University, Vadodara, Gujarat, India
e-mail: naval.koralkar@itmbu.ac.in

N. J. Joshi and G. Patel (eds.), *Nanoparticle Delivery in Cancer Care*,
Series in BioEngineering, https://doi.org/10.1007/978-981-95-0457-2_2

1 Introduction

Cancer is a group of diseases characterized by the uncontrolled growth and spread of abnormal cells within the body. It arises when normal cells undergo genetic mutations that cause them to evade the body's regulatory mechanisms governing cell division, growth, and death [1]. As a result, these cells proliferate uncontrollably, forming a mass of abnormal tissue called a tumor (Fig. 1). Some tumors remain localized and are benign, while malignant tumors have the ability to invade nearby tissues and metastasize to distant sites through the bloodstream or lymphatic system [2]. Metastasis, the spread of cancer cells from the primary tumor to other parts of the body, is a significant cause of cancer-related morbidity and mortality, contributing to the lethality of the disease [3]. The term "metastasis" comes from the Greek words "meta," meaning change or alteration, and "stasis," meaning a state of equilibrium, thus describing both the process of spreading and its outcome (Fig. 2). Cancer is triggered by a combination of genetic factors, environmental exposures (such as radiation and chemicals like tobacco, benzene etc.), and lifestyle choices like diet and physical activity [4]. Different types of cancer are classified based on the origin of the abnormal cells. For example, carcinomas originate in the epithelial tissues, which include the skin and the linings of internal organs, while sarcomas arise in connective tissues such as bone and muscle [5]. Leukemias develop in the blood and bone marrow, and lymphomas target the lymphatic system [6]. Each type of cancer exhibits unique behaviors, and thus, treatment strategies vary depending on the cancer's type, stage, and molecular characteristics. The biology of cancer is complex, involving not only the genetic mutations that initiate and drive tumor growth but also the cellular and molecular adaptations that cancer cells use to survive, proliferate, and metastasize [7, 8]. These genetic mutations often affect key regulatory genes, such as oncogenes and tumor suppressor genes, which normally control cell growth and repair mechanisms [9]. When these genes are mutated, cancer cells gain the ability to continuously signal for growth and evade apoptosis, leading to unchecked proliferation. Additionally, cancer cells can recruit the body's own systems, such as blood vessels (through angiogenesis), to support their growth by supplying oxygen and nutrients [10]. Equally important in the progression of cancer is the tumor microenvironment (TME), the surrounding cellular environment in which cancer cells reside [11]. The TME includes a variety of non-cancerous components, such as immune cells, fibroblasts, blood vessels, signaling molecules, and extracellular matrix components, all of which interact dynamically with the cancer cells [12] (Fig. 3). These elements of the TME are not merely passive bystanders but actively influence cancer progression. For example, immune cells within the TME can play a dual role—they may either attack cancer cells or, under certain conditions, be co-opted by the tumor to suppress the immune response and promote inflammation, thereby aiding cancer growth [13]. Understanding the interactions between cancer cells and the TME is essential for developing effective cancer treatments. The TME provides not only structural support but also biochemical signals that can promote tumor growth, metastasis, and resistance to therapy [14] (Fig. 3). In recent years, therapies targeting the TME, such as immunotherapies that aim to reinvigorate the body's immune system to fight cancer, have shown promising results [15]. For instance, immune checkpoint inhibitors have been developed to block inhibitory signals that prevent immune cells from attacking cancer, allowing the immune system to recognize and destroy cancer cells [16]. It highlights the importance of understanding not only the genetic mutations that drive cancer but also the broader context of how cancer cells interact with their surrounding environment. A comprehensive understanding of both cancer biology and the TME can lead to more effective, targeted treatments. This holistic approach, which considers both the cancer cells and their microenvironment, is crucial for developing personalized therapies that can improve outcomes and provide more durable responses for cancer patients. As research continues to progress, further insights into the interplay between cancer cells and their environment will likely lead to innovative strategies for preventing, diagnosing, and treating cancer. By emphasizing the role of both cancer cells and their microenvironment, this chapter aims to provide a detailed understanding of how tumors develop, progress, and respond to therapy, laying the foundation for new therapeutic approaches that go beyond targeting cancer cells alone and include strategies to modulate the TME for better treatment outcomes.

2 Basic Concepts of Cancer Biology

2.1 Carcinogenesis: A Multistage Process

Carcinogenesis is a multistage process involving genetic and cellular alterations that disrupt the balance between cell proliferation and apoptosis, essential processes for maintaining cellular integrity and metabolic regulation [1]. Two critical drivers of this process are genomic instability and chromosomal instability (CIN) [17, 18]. Genomic instability, which encompasses mutations in oncogenes, tumor suppressor genes, and DNA repair genes, leads to progressive accumulation of genetic errors [18]. These mutations can activate pathways that drive uncontrolled cell growth and survival, while inactivating those that suppress tumor development [19] (Fig. 1). Complementarily, CIN arises from persistent errors in chromosome segregation during mitosis, resulting in aneuploidy and structural abnormalities, such as deletions and translocations, which promote tumor heterogeneity and cancer progression [20, 21]. Complementarily, CIN arises from persistent errors in chromosome segregation during mitosis, resulting in aneuploidy and structural abnormalities, such as deletions and translocations, which promote tumor heterogeneity and cancer progression [20, 21].

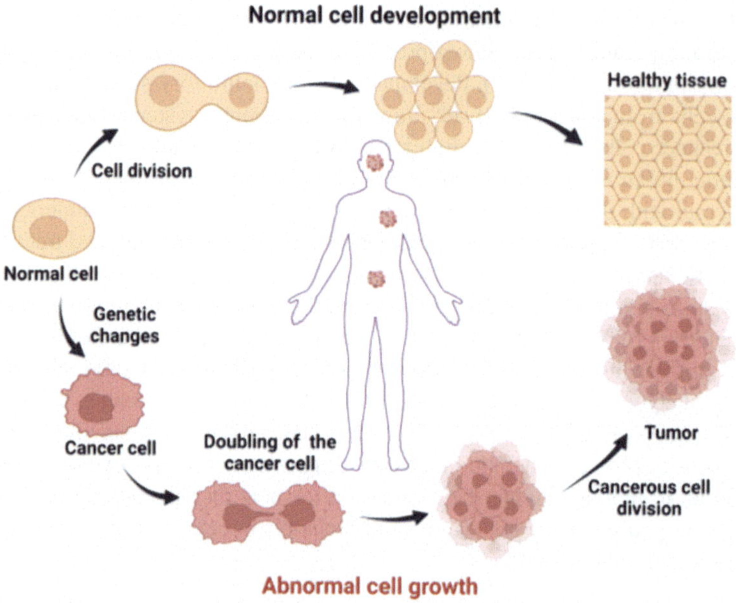

Fig. 1 Progression from normal cell development to abnormal cell growth and tumor formation. The schematic depicts the transition from normal cell development to abnormal cancerous cell growth. Under physiological conditions, a normal cell undergoes cell division, resulting in organized tissue formation (healthy tissue). However, genetic changes such as mutations in the DNA can lead to the transformation of normal cells into cancer cells. These cancer cells bypass regulatory mechanisms and undergo uncontrolled proliferation, characterized by doubling of the cancer cell. This abnormal growth leads to continuous cancerous cell division, forming a tumor. The tumor represents abnormal cell growth, which disrupts normal tissue architecture and function, ultimately resulting in pathological conditions (created by Biorender.com and modified from [22])

2.2 *Cell Cycle Regulation and Dysregulation in Cancer*

The dysregulation of the cell cycle is a critical aspect of carcinogenesis, influenced heavily by genomic instability [23]. Differentiated cells predominantly reside in the G0 phase, remaining metabolically active without undergoing DNA replication or division. Upon mitogenic stimulation, the cell cycle is initiated, progressing through four phases: G1, S, G2, and M. Continuous mitogenic signaling drives daughter cells through these cycles, whereas the absence of such signaling causes cells to revert to G0. Key transitions between these phases are tightly regulated by molecular mechanisms that relieve inhibitory checkpoints [24]. A central regulator of the G1-to-S phase transition is the retinoblastoma protein (RB), encoded by the RB1 gene [25]. In its hypophosphorylated state, RB inhibits progression by binding transcription factors such as E2F1 and DP1, which are essential for the transcription of genes required for DNA replication and cell cycle progression. Phosphorylation of RB by cyclin-dependent kinases (CDKs), specifically the CDK4/6-cyclin D complex, followed by

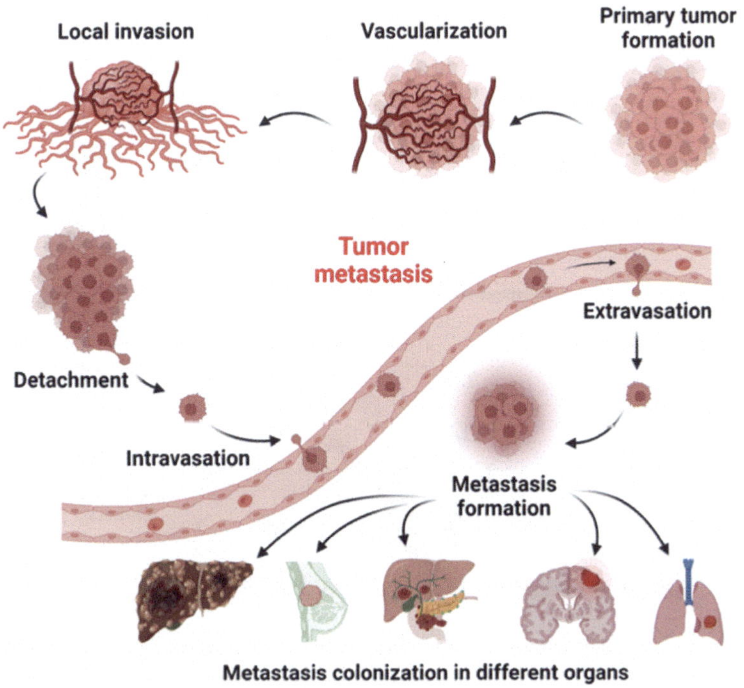

Fig. 2 The process of tumor metastasis and colonization in distant organs. This figure illustrates the sequential steps of tumor metastasis, the process through which cancer cells disseminate from a primary tumor to establish secondary tumors in distant organs. The process begins with primary tumor formation, followed by vascularization, where new blood vessels form to provide essential nutrients and oxygen for tumor growth. Tumor cells then undergo local invasion, infiltrating surrounding tissues, leading to detachment and entry into the bloodstream via intravasation. Circulating tumor cells migrate through the vasculature and exit into distant tissues in a process called extravasation. Once in new tissues, these cells proliferate, forming secondary tumors through metastasis formation. Finally, metastatic cells colonize different organs, such as the liver, breast, pancreas, brain, and lungs, as shown in the figure (created by Biorender.com and modified from [70, 71])

the CDK2-cyclin E complex, releases these transcription factors, enabling advancement into the S phase [26]. The G2-to-M transition is regulated by the Maturation Promoting Factor (MPF), a complex of CDK1 and cyclin B, which phosphorylates key substrates essential for mitosis [27]. Cyclins also regulate critical processes such as meiosis, differentiation, and apoptosis [28]. Cyclin-dependent kinase inhibitors (CKIs), including the Cip/Kip (p21, p27, p57) and Ink4 (p15, p16, p18, p19) families, modulate CDK activity by forming complexes with CDKs, inhibiting their kinase activity [29]. The tumor suppressor protein p53, encoded by the TP53 gene, plays a pivotal role in both cell cycle regulation and apoptosis [30]. The tumor suppressor protein p53, encoded by the TP53 gene, plays a pivotal role in both cell cycle regulation and apoptosis [30]. In response to stressors like DNA damage or hypoxia, p53

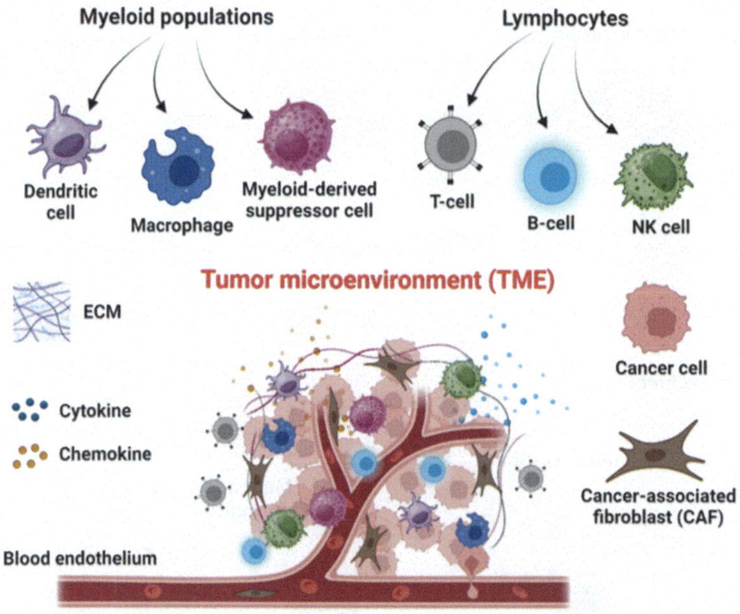

Fig. 3 The tumor microenvironment (TME): overview of cancer-associated changes. The figure highlights the complex composition of the TME, which plays a crucial role in tumor progression and immune modulation. The TME comprises various myeloid populations including dendritic cells, macrophages, and myeloid-derived suppressor cells (MDSCs), as well as lymphocytes such as T-cells, B-cells, and NK cells. Additionally, the TME contains cancer cells, cancer-associated fibroblasts (CAFs), the extracellular matrix (ECM), and the blood endothelium. Signaling molecules, such as cytokines and chemokines, are secreted within this microenvironment, facilitating immune cell recruitment, tumor growth, angiogenesis, and metastasis. The interplay between these components and the dynamic cellular environment contributes to tumor progression and immune evasion (created by Biorender.com and modified from [138])

induces cell cycle arrest by upregulating p21WAF-1, which inhibits CDK activity. Additionally, p53 promotes DNA repair and apoptosis if the damage is irreparable, preventing the propagation of defective cells. However, mutations in TP53, frequently observed in cancers, compromise these protective mechanisms and contribute to genomic instability [31].

2.3 Role of Mitogenic Signals in Cancer

Mitogenic signals, including growth factors such as Platelet-Derived Growth Factor (PDGF), Epidermal Growth Factor (EGF), and Fibroblast Growth Factor (FGF), along with hormones and cytokines, are essential in triggering cellular proliferation [32]. These signals often originate from the expression of proto-oncogenes, genes that normally regulate cell growth, division, and differentiation. These signals can also be relayed from the extracellular matrix (ECM) and through integrins, which act as cellular adhesion molecules and mechanosensors [33]. When proto-oncogenes such as RAS (small GTPase involved in signaling), MYC (a transcription factor regulating growth genes), or ERBB2 (HER2) (a receptor tyrosine kinase) undergo mutations or dysregulation, they can become oncogenes, driving uncontrolled cell growth and contributing to carcinogenesis [34–36]. Furthermore, growth factors bind specific transmembrane receptors, propagating signals via phosphorylation cascades involving various kinases [37]. For example, Tumor growth factor-α (TGF-α) and PDGF activate cytoplasmic proteins (SOS-Ras-Raf-MEK) and mitogen-activated protein kinases (MAPKs), which subsequently activate nuclear transcription factors such as fos, jun, and myc [38]. Moreover, integrins can initiate the Ras-MAPK cascade, adding complexity to the signaling network [38]. Inhibitory signals also play a critical role in regulating cell division [39]. Tumor growth factor-β (TGF-β), for instance, activates its receptor, leading to downstream signaling via Smad proteins and the transcription of cell cycle inhibitors such as p15 and p21, which prevent cyclin-CDK complex formation and RB phosphorylation [32]. Tumor growth factor-β (TGF-β), for instance, activates its receptor, leading to downstream signaling via Smad proteins and the transcription of cell cycle inhibitors such as p15 and p21, which prevent cyclin-CDK complex formation and RB phosphorylation [32].

2.4 Genetic and Molecular Insights into Breast Cancer

Several oncogenes and tumor suppressor genes play critical roles in the development and progression of breast cancer. Among the frequently deregulated oncogenes are ErbB2 (HER2), PI3KCA, MYC, and CCND1, which encodes cyclin D1 [34–36, 40]. These genes drive abnormal cell proliferation, survival, and metabolism. Initial evidence of tumor suppressor gene involvement in breast cancer, such as RB1 mutations, emerged in the late 1980s from studies on breast cancer cell lines and primary tumors [41]. However, a more comprehensive understanding of the genetic landscape of breast cancer evolved through investigations into hereditary breast cancer [42]. To date, ten tumor suppressor genes have been identified as significantly associated with hereditary breast cancer [43]. BRCA1 and BRCA2 mutations are particularly notable, as they dramatically increase the risk of both breast and ovarian cancers [44]. Rare germline mutations in other tumor suppressors, including TP53 and PTEN, also elevate the risk of breast and other cancers [45, 46]. Moreover, moderate-risk genes such as ATM, CHK2, NBS1, RAD50, PALB2, and BRIP1 contribute to breast cancer susceptibility [47]. Among these, ATM and CHK2 are integral to the DNA damage response, coordinating cell cycle arrest and repair mechanisms [48]. RAD50 is a critical component of the protein complex responsible for double-strand break end processing [49]. PALB2, which interacts with BRCA2, and BRIP1, a BRCA1-associated helicase, also play pivotal roles in maintaining genomic stability [50]. Many of these genes are implicated in other malignancies, such as pancreatic and prostate cancers, underscoring their broad significance in cancer biology. Mutations in these oncogenes and tumor suppressor genes result in the deregulation of several key signaling pathways in breast cancer [51, 52]. These include pathways involving lipid and protein kinases, such as the PI3K/AKT/mTOR pathway, which governs cellular growth and survival; cell cycle regulation mechanisms influenced by cyclins and CDKs; and the DNA damage response pathways, which ensure genomic integrity [53, 54]. Apoptosis pathways are also frequently altered, allowing cancer cells to evade programmed cell death [55].

3 Mechanisms and Dynamics of Metastasis: The Hallmark of Cancer Progression

Metastasis, the hallmark of advanced cancer, is a complex and multistep process in which cancer cells spread from the primary tumor to distant sites, leading to the formation of secondary tumors and colonization of other organs [3] (Fig. 2). This process begins with local invasion, where tumor cells breach the basement membrane and infiltrate the surrounding stroma [56]. Epithelial-mesenchymal transition (EMT) plays a critical role in this stage by enabling epithelial cancer cells to acquire mesenchymal traits such as motility and invasiveness [57]. EMT is regulated by growth factors like transforming growth factor-beta (TGF-β) and signaling pathways such as WNT, Notch, and Hedgehog, which are influenced by the tumor microenvironment, including hypoxia, ECM stiffness, and interactions with tumor-associated fibroblasts [58]. Once the cancer cells invade locally, they intravasate into the bloodstream or lymphatic vessels, aided by the expression of adhesion molecules such as L1CAM and integrins. Within the circulation, cancer cells face various challenges, including shear stress and immune surveillance. To survive, they often aggregate with platelets, forming emboli that shield them from immune attacks. These circulating tumor cells (CTCs) then adhere to the endothelium at distant sites, extravasate, and colonize new tissues [59, 60].

The successful colonization of secondary sites depends on the ability of metastatic cells to adapt to the foreign microenvironment [61]. This often involves the reverse process of EMT, known as mesenchymal-epithelial transition (MET), which is essential for the formation of metastatic colonies [62]. For instance, in bone metastases, WNT signaling and bone-derived factors like E-selectin induce MET, enabling cancer cells to proliferate and establish secondary tumors [63]. The organ-specific nature of metastasis, known as the "seed and soil" hypothesis, highlights the importance of compatibility between cancer cells ("seed") and the target organ microenvironment ("soil") [64]. Tumor-secreted factors and exosomes can precondition distant sites, forming a pre-metastatic niche that supports colonization. This niche is characterized by vascular remodeling, recruitment of immune cells, and ECM remodeling, all of which facilitate metastatic outgrowth [65, 66].

Genetic and epigenetic alterations further enhance the metastatic potential of cancer cells. Mutations in genes like TP53, CDKN2A, and PIK3CA, as well as CIN, drive the acquisition of traits necessary for metastasis including the generation of genomic diversity and promoting adaptation to different microenvironments [67]. Moreover, metastatic cells often exhibit resistance to therapy due to their genetic heterogeneity and activation of survival pathways such as PI3K/AKT/mTOR and MAPK [68]. Epigenetic changes, including DNA methylation and histone modifications, also play a role in the plasticity of cancer cells, enabling them to switch between epithelial and mesenchymal states as needed [69].

4 Mechanism of Premetastatic Niche Formation and Exosome-Mediated Metastasis

Metastasis remains the leading cause of cancer-related deaths due to its complexity and resistance to conventional therapies [3]. Emerging therapeutic strategies aim to target key steps in the metastatic cascade, including EMT, CTC survival, and premetastatic niche (PMN) formation [57, 59, 60, 72]. Before metastasis occurs, primary tumors often prime secondary sites, creating what is known as the PMN [73]. Tumor-secreted exosomes play a key role in this process by transferring invasion-promoting factors, such as microRNAs (miRNAs), to cells at distant sites. These exosomes prepare target organs for colonization by remodeling the ECM, recruiting immune cells, and promoting angiogenesis [74, 75]. These exosomes prepare target organs for colonization by remodeling the ECM, recruiting immune cells, and promoting angiogenesis [74, 75]. For instance, exosome integrin patterns help determine metastatic cells' organotropism, guiding their colonization in specific organs, such as the liver or lungs [76]. Exosomes also influence the formation of a permissive microenvironment by interacting with stromal cells, fibroblasts, and immune cells. In liver metastasis, for example, cancer cells secrete exosomes that promote IL-6-mediated STAT3 signaling in hepatocytes, enhancing the establishment of the premetastatic niche and supporting cancer cell survival and growth [77].

Furthermore, Intravasation, the process by which cancer cells enter the bloodstream, is critical for metastasis. It can occur actively or passively, depending on the TME and type of vasculature. Mechanical stress during intravasation, such as nuclear squeezing, can cause genomic rearrangements that enhance the metastatic potential of tumor cells [59]. Once in circulation, tumor cells must survive various challenges, including immune surveillance and physical stress [78]. CTCs travel either as single cells or in clusters, with clusters having a higher likelihood of forming metastases [79]. CTCs travel either as single cells or in clusters, with clusters having a higher likelihood of forming metastases [79]. These clusters often contain immune and stromal cells from the primary tumor site, which enhance their survival. Platelets and neutrophils play a protective role by shielding CTCs from immune detection and providing mechanical support [80]. In contrast, extravasation is the process by which circulating tumor cells exit the bloodstream and invade distant organs. This step is critical for establishing metastases and is facilitated by interactions between cancer cells, endothelial cells, and the surrounding microenvironment [81]. Once lodged in small capillaries, cancer cells extravasate through transendothelial migration or microvascular rupture. Organ-specific factors, such as calcium flux in bone metastasis or weaning-induced liver involution in postpartum breast cancer patients, help dictate where metastasis occurs [82, 83]. After extravasation, cancer cells must adapt to the microenvironment of the distant organ to establish metastatic colonies. Tumor-derived factors and exosomes help educate stromal cells and bone marrow-derived cells to support the growth of metastatic tumors [84]. Cancer cells interact with host cells, such as hepatocytes in the liver, which secrete factors like IL-6 that promote cancer cell survival and colonization. Additionally, vascular mimicry, where cancer cells form blood vessel-like structures and supports the development of a functional vasculature within the metastatic site, further facilitating tumor growth and progression [85]. Understanding these mechanisms provides valuable insights into therapeutic strategies aimed at preventing metastasis or targeting the metastatic cells at each stage of their journey from primary tumor to distant colonization.

5 The Tumor Microenvironment (TME)

The tumor microenvironment (TME) is a complex and dynamic network that surrounds cancer cells, playing a critical role in tumor initiation, progression, and metastasis [11]. The TME is not a passive background for tumor growth but an active participant in cancer biology, helping to shape the cancer's growth patterns, its response to therapies, and its ability to evade the immune system [86]. The TME is composed of several cellular and acellular components, including immune cells, cancer-associated fibroblasts (CAFs), endothelial cells, pericytes, and the extracellular matrix (ECM) [87] (Fig. 3). These components form a dynamic and reciprocal network of interactions with tumor cells. The TME provides cancer cells with biochemical signals and structural support, influencing tumor growth, survival, invasion, and resistance to therapies. Additionally, the TME evolves, responding to changes in the tumor's metabolic needs, genetic mutations, and treatments, making it a critical factor in both cancer progression and treatment resistance [88]. A key aspect of the TME is its heterogeneity, as different tumors and even different regions within a single tumor can have distinct microenvironments [89]. The TME can vary depending on the type of cancer, its stage, and its location in the body, and these variations can influence how a tumor behaves and responds to treatment [90]. Researchers are increasingly focused on the TME as a potential target for new therapies that can disrupt the supportive role it plays in cancer progression.

The ECM is a critical non-cellular component of the TME, providing structural support and biochemical cues that regulate cell behavior. The ECM is composed of proteins such as collagen, fibronectin, laminin, and proteoglycans, which create a scaffold for cells to adhere to and migrate through [91, 92]. In cancer, the ECM undergoes significant remodeling, often becoming stiffer and more fibrotic. This altered ECM can promote cancer cell proliferation, invasion, and metastasis. One of the key roles of the ECM in cancer is to facilitate cell migration and invasion, which are essential steps in metastasis [93–95]. Cancer cells often produce enzymes, such as matrix metalloproteinases (MMPs), that degrade the ECM and create pathways for them to invade surrounding tissues and spread to distant organs [96]. The stiffening of the ECM, known as desmoplasia, can also create mechanical forces that promote cancer cell proliferation and survival [97, 98]. Moreover, the ECM is involved in signaling pathways that regulate cell differentiation, proliferation, and survival [93]. It interacts with integrins and other cell surface receptors on cancer cells, transmitting signals that can promote tumor growth [76, 99]. The ECM can also sequester growth factors and cytokines, creating a reservoir of pro-tumorigenic signals that further enhance cancer cell survival and resistance to treatment [100–102]. Given its central role in cancer progression, the ECM has become a target for therapies aimed at disrupting its structure or inhibiting the enzymes that degrade it [103].

CAFs are one of the most abundant and important cellular components of the TME [104]. CAFs are activated fibroblasts that have been co-opted by cancer cells to support tumor growth and metastasis. Unlike normal fibroblasts, which help maintain tissue integrity and repair, CAFs promote cancer progression by secreting growth factors, cytokines, and ECM components that enhance tumor cell survival, proliferation, and invasion [105–107]. CAFs also contribute to ECM remodeling by producing enzymes like MMPs, which degrade the ECM and create a more favorable environment for cancer cell invasion. Additionally, CAFs play a role in angiogenesis, the formation of new blood vessels, by secreting pro-angiogenic factors such as vascular endothelial growth factor (VEGF). These new blood vessels supply the growing tumor with oxygen and nutrients, facilitating its continued growth and expansion [108]. Moreover, CAFs can influence the immune response within the TME. They secrete immunosuppressive cytokines that help tumors evade detection by the immune system. By creating a more immunosuppressive environment, CAFs contribute to the tumor's ability to escape immune surveillance, allowing cancer cells to grow unchecked [109, 110]. Targeting CAFs and their interactions with cancer cells and the immune system is a promising area of research for developing new cancer therapies. By creating a more immunosuppressive environment, CAFs contribute to the tumor's ability to escape immune surveillance, allowing cancer cells to grow unchecked [109, 110]. Targeting CAFs and their interactions with cancer cells and the immune system is a promising area of research for developing new cancer therapies.

Furthermore, other component of TME, the immune system plays a dual role in cancer, capable of both attacking and supporting tumor growth [111]. In the TME, immune cells such as macrophages, T cells, dendritic cells, and myeloid-derived suppressor cells (MDSCs) can either mount an anti-tumor response or be co-opted by cancer cells to promote tumor progression. This ability of cancer cells to manipulate the immune system is a hallmark of cancer, allowing tumors to evade immune detection and destruction [112–114]. Tumor-associated macrophages (TAMs) are one of the most prominent immune cell types in the TME. These macrophages can adopt either a pro-inflammatory, anti-tumor phenotype (M1) or an anti-inflammatory, pro-tumor phenotype (M2) [115–117]. In many cancers, TAMs predominantly exhibit the M2 phenotype, which supports tumor growth by promoting angiogenesis, suppressing anti-tumor immune responses, and enhancing tissue remodeling and metastasis. High levels of M2 macrophages are often associated with poor prognosis in cancer patients [118, 119]. T cells, particularly cytotoxic T lymphocytes (CTLs), are essential for recognizing and killing cancer cells. However, in the TME, cancer cells can suppress T cell activity by expressing immune checkpoint molecules, such as PD-L1, that inhibit T cell function [120]. Immune checkpoint inhibitors, such as anti-PD-1 and anti-PD-L1 antibodies, have revolutionized cancer therapy by reactivating T cells to attack tumors. However, not all tumors respond to these therapies, highlighting the complexity of the TME's influence on immune cells [121, 122]. Myeloid-derived suppressor cells (MDSCs) are another immune cell type that contributes to the immunosuppressive environment in tumors [123]. These cells inhibit the activity of T cells and natural killer (NK) cells, further dampening the immune response and allowing cancer cells to evade destruction [124–126].

Next, endothelial cells line the blood vessels and play a central role in angiogenesis, the process by which new blood vessels form from pre-existing ones. In cancer, angiogenesis is essential for supplying the growing tumor with oxygen and nutrients [127]. Tumor cells and stromal cells in the TME secrete pro-angiogenic factors, such as VEGF, that stimulate endothelial cells to form new blood vessels. These new vessels are often irregular and leaky, contributing to a chaotic and poorly organized tumor vasculature [128]. Angiogenesis not only supports tumor growth by providing nutrients but also facilitates metastasis by allowing cancer cells to enter the bloodstream and spread to distant organs. Inhibiting angiogenesis is a therapeutic strategy aimed at starving tumors of the nutrients they need to grow [129, 130]. Anti-angiogenic therapies, that target VEGF and have shown efficacy in several cancers, although resistance can develop over time [130].

Moreover, biochemical factors such as growth factors, cytokines, and chemokines are critical mediators of communication within the TME. These signaling molecules regulate the behavior of both cancer cells and stromal cells, shaping the tumor's growth, survival, and ability to evade the immune system [131, 132]. Growth factors, such as VEGF, FGF, and TGF-β, are often overexpressed in cancer. They promote cell proliferation, survival, and angiogenesis [133–135]. For instance, TGF-β plays a dual role in cancer, acting as a tumor suppressor in early-stage cancers but promoting tumor progression and metastasis in advanced cancers [134]. Cytokines and chemokines are key regulators of immune cell recruitment and activity within the TME. Pro-inflammatory cytokines, such as TNF-α and interleukin-6 (IL-6), can promote tumor growth by inducing chronic inflammation. Chemokines, such as CXCL12, regulate the recruitment of immune cells to the tumor site and can either enhance or suppress anti-tumor immune responses [136, 137].

6 Tumor-Microenvironment Interactions: Mechanisms Driving Metastasis

Tumor cells interact with their microenvironment through several mechanisms, including autocrine and paracrine signaling, direct cell-to-cell contact, and interactions with the ECM. The complexity of these interactions is key to tumor progression, allowing cancer cells to manipulate surrounding stromal and immune cells to create a pro-tumorigenic environment [85]. Autocrine signaling allows tumor cells to produce and respond to their own growth factors, fostering an environment that promotes survival and proliferation [139]. Paracrine signaling, in contrast, involves the secretion of signaling molecules by tumor cells to influence nearby stromal and immune cells [140]. These signals can activate fibroblasts and immune cells, transforming them into CAFs or TAMs, respectively, which support tumor growth [141]. The TME also plays a crucial role in therapeutic resistance [142]. CAFs secrete ECM components and growth factors that protect tumor cells from chemotherapy and radiation. The dense stroma created by CAFs can act as a physical barrier to drug delivery, while the secretion of survival factors like hepatocyte growth factor (HGF) and FGF helps tumor cells resist apoptosis [143]. Additionally, CAFS creates tracks and niches that facilitate tumor cell invasion. Through secretion of MMPs, CAFs degrade ECM barriers, promoting tumor cell motility and invasion.

Furthermore, hypoxia (low oxygen levels) within the TME can trigger adaptive responses that enhance tumor cell survival and drive angiogenesis, further supporting tumor growth [144]. One of the hallmarks of cancer is the ability to evade immune detection and destruction [145]. The TME plays a crucial role in facilitating this immune evasion by modulating immune cell behavior. Tumor cells can manipulate immune cells within the TME to create an immunosuppressive environment. For example, TAMs are often polarized towards an M2 phenotype, which supports tumor growth by promoting angiogenesis and suppressing anti-tumor immune responses. MDSCs and Tregs also contribute to immune suppression by inhibiting the activity of cytotoxic T cells and NK cells, which are critical for recognizing and eliminating tumor cells [11, 118, 119].

Tumor cells undergo significant metabolic reprogramming to meet the high energy demands of rapid growth and proliferation. This metabolic shift is heavily influenced by the TME, which provides nutrients and signaling cues that support altered metabolic pathways in tumor cells [146]. One of the most well-known metabolic adaptations in cancer is the Warburg effect, where tumor cells preferentially utilize glycolysis for energy production, even in the presence of oxygen [147]. This allows tumor cells to rapidly generate ATP and biomass for proliferation while producing lactate, which acidifies the TME and promotes immune evasion and tissue invasion [148]. The TME also influences other aspects of tumor cell metabolism, such as glutamine metabolism and lipid synthesis. Tumor cells often rely on glutamine as an alternative energy source and for the synthesis of macromolecules required for cell growth. CAFs and other stromal cells in the TME can supply glutamine to tumor cells, fueling their metabolic needs [149]. In addition, the altered ECM in the TME can promote lipid synthesis by providing structural components that are used for membrane production and energy storage [150]. Hypoxia within the TME is another key factor driving metabolic reprogramming. In response to low oxygen levels, tumor cells activate hypoxia-inducible factors (HIFs), which promote glycolysis and angiogenesis. Hypoxia also enhances the expression of enzymes involved in lipid metabolism and drives the accumulation of lactate. The buildup of lactate within the TME contributes to acidification, which not only promotes tumor invasion by degrading ECM components but also suppresses the anti-tumor activity of immune cells. Additionally, metastatic cancer cells exploit lactate as an energy source through monocarboxylate transporter 1 (MCT1). Lactate uptake via MCT1 enables these cells to fuel oxidative phosphorylation, alleviating oxidative stress and supporting the energy demands of metastasis. Preclinical studies have demonstrated that pharmacological inhibition of MCT1 disrupts lactate metabolism, impairing the metastatic potential of cancer cells. This approach holds promise as a therapeutic strategy to target metastatic progression by blocking a key metabolic adaptation of disseminating tumor cell [151]. In addition to lactate metabolism, metastatic progression involves alterations in cellular energy states, such as changes in ATP/ADP ratios and phosphocreatine production. In pancreatic cancer, these metabolic shifts are closely linked with ECM remodeling and modified adhesion properties, which drive tumor cell migration, invasion, and organ-specific colonization, such as liver metastases [83].

Recent studies have also highlighted the role of metabolic crosstalk between tumor cells and immune cells in the TME [152–154]. Tumor cells can secrete metabolites, such as adenosine and kynurenine, that suppress the activity of immune cells. Conversely, immune cells in the TME can influence tumor cell metabolism by producing cytokines and growth factors that modulate metabolic pathways. This bidirectional metabolic communication between tumor cells and immune cells is a critical aspect of tumor progression and immune evasion [155]. In conclusion, the tumor microenvironment is a highly complex and dynamic system that plays a critical role in cancer progression. Its components, including the ECM, CAFs, immune cells, and endothelial cells, interact with cancer cells to promote tumor growth, invasion, and resistance to therapy. Targeting the TME offers a promising strategy for developing new cancer therapies that disrupt the supportive environment tumors rely on, potentially leading to more effective and durable treatments for cancer patients.

7 Innovative Strategies for Overcoming Tumor Microenvironment Challenges in Cancer Therapy

Contrary to the traditional view of cancer as a disease driven solely by unchecked cell proliferation, cancer is now understood as a complex, dynamic ecosystem [156]. Central to this ecosystem is the TME, a varied and interactive milieu composed of different cells, all of which play crucial roles in tumor development, progression, and resistance to therapy [85]. The functional state and cellular composition of the TME vary depending on the tumor's location, characteristics of the cancer cells, tumor stage, and patient-specific factors. The intricate interplay between cancer cells and non-cancerous components within the TME is crucial for tumor growth and treatment resistance. Therefore, targeting the TME has become a key focus in cancer treatment, and novel strategies are emerging to modulate the TME and improve the efficacy of traditional therapies [157, 158]. Current therapeutic strategies aim to disrupt the supportive functions of the TME while enhancing the delivery and efficacy of conventional therapies such as chemotherapy, targeted therapy, and immunotherapy. This section explores several key therapeutic approaches that target the TME, including the depletion of immunosuppressive cells, modulation of angiogenesis, and emerging cell-based therapies [159].

8 Chemotherapy, Targeted Therapy, and Immunotherapy

Traditional cytotoxic agents, such as paclitaxel, cisplatin, and 5-fluorouracil, remain central to cancer therapy because of their ability to disrupt critical cellular processes like DNA replication and cell division [160]. However, the protective nature of the TME, particularly in solid tumors where the ECM and abnormal vasculature hinder drug delivery, limits the efficacy of these agents [161]. However, the protective nature of the TME, particularly in solid tumors where the ECM and abnormal vasculature hinder drug delivery, limits the efficacy of these agents [161]. To address these challenges, combination therapies integrating chemotherapy with targeted therapies and immunotherapies are being developed to improve overall treatment outcomes [162]. Targeting TAMs, a key component of the immunosuppressive TME that promotes tumor survival, angiogenesis, and metastasis, has emerged as a promising strategy. TAMs support tumor growth by secreting cytokines that inhibit immune responses and protect cancer cells from apoptosis [163]. Therapeutic approaches, such as blocking the CSF-1/CSF-1R signaling pathway to deplete TAMs or reprogramming them into an antitumor M1-like phenotype, have shown promise [164]. Preclinical studies have demonstrated that CSF-1R inhibition, when combined with chemotherapy, enhances tumor cell death and improves outcomes, as seen in breast cancer models where CSF-1R inhibitors combined with paclitaxel increased DNA damage and tumor regression [165]. Combining TAM depletion with immune checkpoint inhibitors like PD-1/PD-L1 blockade is also being explored to overcome immunotherapy resistance and boost antitumor immune responses [166]. Similarly, MDSCs, another population of immunosuppressive cells recruited to the TME, contribute to immune evasion by inhibiting cytotoxic T cells [123]. Targeting MDSCs through chemotherapeutic agents like gemcitabine or 5-fluorouracil or using targeted therapies that block MDSC recruitment or function, has emerged as a way to enhance the effectiveness of immunotherapies [167]. Inhibitors of STAT3, a key regulator of MDSC activity, have shown promise in reducing MDSC numbers and enhancing immune responses in preclinical models [168]. Inhibitors of STAT3, a key regulator of MDSC activity, have shown promise in reducing MDSC numbers and enhancing immune responses in preclinical models [168]. Furthermore, combining MDSC-targeting therapies with immune checkpoint inhibitors is being investigated in clinical trials to improve outcomes in advanced cancers [169]. Another key target in the TME is Tregs, which maintain immune tolerance and suppress antitumor immune responses [170]. Tregs inhibit cytotoxic T cells, allowing tumors to evade immune surveillance, making their depletion or inhibition a valuable strategy for enhancing immunotherapies like PD-1 and CTLA-4 checkpoint inhibitors. However, systemic Treg depletion can cause autoimmune toxicity, so selective targeting of Tregs within the TME is essential [171]. Approaches include antibodies against Treg-specific markers such as CCR4 or CD25 and small-molecule inhibitors that block Treg recruitment to the tumor. Combining Treg depletion with therapies that activate cytotoxic T cells is being explored to enhance immune responses and improve clinical outcomes [172, 173].

9 Targeting Tumor Angiogenesis

Angiogenesis is essential for tumor growth, as it provides the tumor with the oxygen and nutrients it needs to survive and proliferate [174]. The abnormal, leaky blood vessels formed in tumors also contribute to therapeutic resistance by reducing drug delivery and creating regions of hypoxia that promote tumor cell survival. Inhibiting angiogenesis by targeting the VEGF signaling pathway has been an effective strategy in several cancers, including colorectal, lung, and renal cancers [175]. Bevacizumab, a monoclonal antibody that targets VEGF, has been shown to improve survival when combined with chemotherapy in several cancers. However, resistance to anti-angiogenic therapies often develops, leading to tumor recurrence [175]. To address this, combination strategies that target both the VEGF pathway and other components of the TME are being explored. For example, combining VEGF inhibitors with immune checkpoint inhibitors or agents that normalize the abnormal vasculature may enhance treatment efficacy and overcome resistance [176, 177].

10 Targeting Hypoxia and Tumor Metabolism

Hypoxia, or low oxygen levels, is a characteristic feature of solid tumors and contributes to resistance to chemotherapy, radiation therapy, and immunotherapy. Tumor cells adapt to hypoxia by upregulating hypoxia-inducible factors (HIFs), which promote angiogenesis, metabolic reprogramming, and immune evasion [151]. Targeting hypoxia and its downstream effects is a promising therapeutic approach in cancer. Several strategies are being developed to target hypoxia, including the use of HIF inhibitors, which block the activity of HIF-1α and HIF-2α, and hypoxia-activated prodrugs, which are selectively activated in hypoxic tumor regions [178]. Additionally, therapies that target the altered metabolism of tumor cells, such as inhibitors of glycolysis or glutaminolysis, are being tested in combination with other treatments to exploit the metabolic vulnerabilities of cancer cells in hypoxic environments [179–182].

11 Emerging Cell-Based Therapies

Chimeric Antigen Receptor (CAR)-T cell therapy has revolutionized the treatment of blood cancers and is now gaining attention as a potential approach for treating solid tumors. CAR-T cell therapy involves the extraction of T cells from cancer patients, genetically engineering them to express CARs that can recognize tumor-specific antigens and reinfusing these enhanced T cells back into the patient [183]. These engineered T cells are designed to recognize and kill cancer cells more effectively than the patient's natural immune cells. While CAR-T therapy has shown significant success in hematologic malignancies, its efficacy in treating solid tumors remains limited due to the unique challenges posed by the TME [184–186]. One of the major obstacles in the success of CAR-T therapy for solid tumors is the immunosuppressive nature of the TME. Different TME components create physical and biochemical barriers that inhibit the infiltration and activity of CAR-T cells, resulting in diminished therapeutic efficacy. The TME not only restricts the entry of CAR-T cells but also induces T cell exhaustion, which further weakens the immune response. Therefore, new strategies that can overcome these obstacles and disrupt the TME are crucial for improving the effectiveness of CAR-T therapy in solid tumors [187–189]. One such strategy is the targeting of TAMs, which can comprise up to 50% of the tumor mass in certain cancers. TAMs play a dual role in the TME by acting as a physical barrier that blocks immune cell infiltration and by expressing high levels of immunosuppressive molecules such as PD-L1, which contributes to T cell exhaustion. Traditional CAR-T therapy primarily relies on conventional αβ T cells, which are effective in killing tumor cells but are limited in their ability to remodel the TME [190]. To address this limitation, researchers have explored the use of unconventional T cells, such as invariant natural killer (iNKT) cells, mucosal-associated invariant T (MAIT) cells, and gamma-delta T (γδT) cells. These innate-like T cells are not restricted by MHC molecules and possess T cell receptors (TCRs) that recognize non-peptide antigens presented by conserved antigen-presenting molecules, such as CD1d, MR1, and BTN3A1 [191]. These molecules are often expressed on TAMs, allowing these unconventional T cells to target and eliminate TAMs, thereby reducing immunosuppression and promoting a proinflammatory environment within the TME. By engineering these unconventional T cells with CARs, it becomes possible to simultaneously target the tumor cells and modulate the TME, thereby enhancing the overall efficacy of CAR-T therapy in solid tumors [192–194].

Besides, another major challenge to CAR-T therapy is the presence of MDSCs within the TME [195]. MDSCs are recruited to the TME and suppress the activity of cytotoxic T cells by secreting immunosuppressive cytokines and promoting the expansion of Tregs [196]. MDSCs play a crucial role in dampening the immune response and facilitating tumor growth. One promising approach to overcoming this challenge is the development of CAR-T cells engineered to target both MDSCs and tumor cells [196]. Agonists of TRAIL receptor 2 (TR2) have been shown to selectively induce apoptosis in MDSCs [197]. By engineering CAR-T cells to express TR2, researchers have demonstrated enhanced killing of MDSCs and increased cytotoxicity against tumor cells in preclinical models. This dual-targeting approach has the potential to improve the efficacy of CAR-T therapy by simultaneously depleting MDSCs and attacking tumor cells, thereby overcoming one of the key immunosuppressive elements of the TME [193, 198]. In addition to cellular components, the ECM presents a significant physical barrier to CAR-T cell infiltration. The ECM is a complex network of proteins and glycoproteins that surrounds the tumor and regulates many aspects of tumor biology, including cell migration, proliferation, and invasion. It also impedes the infiltration of immune cells, including CAR-T cells. One innovative solution to this problem is the overexpression of heparanase (HPSE) in CAR-T cells [199]. Heparanase is an enzyme that degrades heparan sulfate proteoglycans, a key component of the ECM, thereby disrupting the ECM and promoting immune cell infiltration [200]. In preclinical studies, CAR-T cells engineered to overexpress HPSE demonstrated enhanced infiltration into tumor tissues and greater antitumor efficacy in models of neuroblastoma. This approach highlights the importance of targeting both the tumor cells and the surrounding microenvironment to maximize the therapeutic potential of CAR-T therapy [199].

The immunosuppressive molecules present in the TME, such as PD-L1, also pose a significant challenge to CAR-T therapy. PD-L1 is expressed by various cells within the TME, including TAMs, MDSCs, fibroblasts, and tumor cells, and it interacts with PD-1 on immune cells to suppress their activity. Blocking this interaction using immune checkpoint inhibitors (ICBs) has shown great promise in boosting immune responses against tumors [201, 202]. Combining CAR-T therapy with checkpoint blockade offers a synergistic approach, as it prevents T cell exhaustion and enhances the cytotoxic activity of CAR-T cells. In preclinical models of Her2 + sarcoma and breast cancer, combining anti-PD-1 antibodies with Her2-targeting CAR-T cells resulted in enhanced CAR-T cell function and improved tumor killing compared to CAR-T therapy alone. This combination strategy holds great potential for overcoming the immunosuppressive effects of the TME and improving the efficacy of CAR-T therapy in solid tumors [203–205].

In conclusion, while CAR-T therapy has achieved remarkable success in the treatment of blood cancers, its application in solid tumors is hindered by the immunosuppressive nature of the TME. Innovative strategies that target key components of the TME, such as TAMs, MDSCs, the ECM, and immunosuppressive molecules, are essential for improving the efficacy of CAR-T therapy in solid tumors. By simultaneously targeting both the tumor cells and the TME, these approaches have the potential to overcome the barriers that limit the effectiveness of current CAR-T therapies, opening new avenues for treating solid tumors with this groundbreaking technology.

12 Oncolytic Viruses and Immunotherapy

Oncolytic viruses (OVs) represent an emerging class of cancer therapeutics that selectively infect and kill tumor cells while stimulating a robust immune response. OVs can be engineered to express immune-stimulatory molecules or checkpoint inhibitors, enhancing their ability to transform the immunosuppressive TME into an immune-active environment [206]. The key mechanisms of oncolytic viruses are the selective replication within tumor cells, leading to cell lysis and the release of tumor-associated antigens (TAAs) [207]. The lytic destruction of cancer cells can convert a "cold" tumor, one that is not recognized by the immune system, into a "hot" tumor, capable of being targeted by immune cells. For example, talimogene laherparepvec (T-VEC), a genetically modified herpes simplex virus, has been approved for the treatment of melanoma. T-VEC not only causes direct oncolysis but also expresses the immune-stimulating protein GM-CSF, enhancing the anti-tumor immune response [208]. The combination of OVs with immune checkpoint inhibitors or other immunotherapies is being explored as a way to enhance antitumor immunity and improve treatment outcomes in patients with solid tumors [209–211]. The combination of OVs with immune checkpoint inhibitors or other immunotherapies is being explored as a way to enhance antitumor immunity and improve treatment outcomes in patients with solid tumors [209–211]. Despite promising outcomes, challenges remain, including the potential for antiviral immune responses to limit the therapeutic efficacy of OVs and the difficulty in achieving effective delivery of viruses to solid tumors. Nonetheless, the integration of oncolytic viruses with immunotherapy holds significant potential in advancing cancer treatment, offering new strategies for overcoming immune evasion by tumors.

13 Nanoparticle-Based Therapy and Associated Challenges

Nanoparticle-based cancer care represents an exciting frontier in oncology, offering novel therapeutic strategies designed to improve treatment efficacy while minimizing adverse effects [212]. Nanoparticles (NPs) are minute particles, typically ranging from 1 to 100 nm, and have unique physical, chemical, and biological properties that make them ideal for medical applications. Their small size and ability to be functionalized with various biomolecules allow for targeted drug delivery, improved pharmacokinetics, and the potential for both diagnostic and therapeutic roles, commonly referred to as theranostics. These nanoparticles can be engineered to carry chemotherapeutic agents, RNA/DNA therapeutics, or immune modulators directly to the tumor site, enhancing drug accumulation while minimizing systemic side effects. One of the most promising applications of nanoparticles in cancer therapy is targeted drug delivery [213]. Traditional chemotherapy often affects both cancerous and healthy cells, leading to significant side effects. Nanoparticles can be engineered to specifically target tumor cells, sparing healthy tissues. For example, liposomal doxorubicin (Doxil) is a formulation in which the chemotherapeutic drug doxorubicin is encapsulated within liposomes—spherical vesicles made from lipid bilayers. This nanoparticle-based drug reduces cardiotoxicity and enhances the accumulation of the drug in tumor tissues due to the enhanced permeability and retention (EPR) effect, where nanoparticles accumulate in tumor tissues more than normal tissues due to leaky vasculature and impaired lymphatic drainage in tumors [1]. Another example is albumin-bound paclitaxel (Abraxane), which is used for the treatment of breast, lung, and pancreatic cancers. By binding the chemotherapeutic agent paclitaxel to albumin, a naturally occurring protein in the human body, Abraxane improves solubility, allows for better drug delivery to tumors, and reduces some of the hypersensitivity reactions commonly associated with paclitaxel in its non-nanoparticle form [214]. Additionally, gold nanoparticles (AuNPs) have been explored for both imaging and therapeutic applications due to their ability to absorb light and convert it into heat, which can be used in photothermal therapy to destroy cancer cells. In combination with radiation therapy, gold nanoparticles can also enhance the radiative damage to cancerous tissues while minimizing the dose to surrounding healthy tissues [1].

Furthermore, despite these promising applications, NP-based cancer therapies face several significant challenges. One of the foremost obstacles is achieving efficient and specific tumor targeting. While the EPR effect allows for enhanced accumulation of nanoparticles in tumors, this effect varies between patients and cancer types, limiting the universal application of NP-based therapies [215]. Additionally, the development of multidrug resistance (MDR) in cancer cells remains a critical issue. Cancer cells may upregulate drug efflux pumps or develop mutations that render them less susceptible to the therapeutic agents delivered via nanoparticles [216, 217]. Additionally, Toxicity and long-term safety also present significant concerns. Some nanoparticles, particularly those made from non-biodegradable materials, may accumulate in organs like the liver, spleen, or kidneys, leading to potential toxicity. Biocompatibility and the immune system's response to nanoparticles remain areas requiring further study, as NPs can be recognized and cleared by the immune system before they reach their target. Moreover, the production of nanoparticles at a clinical scale with precise control over size, shape, and functionalization remains technically challenging and costly [218, 219]. In conclusion, nanoparticle-based cancer therapeutic strategies offer numerous advantages over conventional therapies, including improved drug delivery, reduced side effects, and the potential for theranostic applications. However, issues such as targeting specificity, multidrug resistance, long-term toxicity, and scalable production need to be addressed to fully realize the potential of nanoparticles in cancer care.

Future therapeutic strategies will likely involve combination approaches that target multiple components of the TME, such as immunosuppressive cells, angiogenesis, and metabolic adaptations, while simultaneously enhancing the antitumor immune response. Advances in cell-based therapies, oncolytic viruses, and nanoparticle-based therapy hold great promise for overcoming the challenges posed by the TME and improving the overall success of cancer treatment. As our understanding of the TME and its interactions with cancer cells continues to evolve, the development of more effective and personalized therapies will follow, offering new hope for patients with advanced or resistant cancers.

14 Different Models Used in Tumor Microenvironment Research

The study of cancer biology and the TME involves various research methods that enable scientists to explore the molecular and cellular mechanisms underlying cancer initiation, progression, and response to therapies. These methods range from traditional biological techniques to cutting-edge technologies that enhance our ability to model the complex interactions between tumor cells and their surrounding environment.

14.1 In Vitro Models

Cell culture models play a crucial role in studying the TME, each offering different insights into cancer biology and its interaction with surrounding tissues [220]. Traditional two-dimensional (2D) cultures, where cancer cells grow as a monolayer on flat surfaces, provide simplicity and high throughput, making them suitable for drug screening. However, they lack the spatial architecture and complexity of the TME, limiting their ability to mimic in vivo conditions [221]. Three-dimensional (3D) culture models, such as spheroids and organoids, offer a more physiologically relevant environment [222]. Spheroid cultures, where cancer cells self-assemble into spherical aggregates, replicate conditions like oxygen and nutrient gradients similar to avascular tumor regions, making them useful for studying tumor growth and drug response. For example, spheroid cultures are used to model the hypoxic core found in tumors, which is relevant in drug resistance research [223]. Organoids, derived from stem cells or patient tumor biopsies, mimic the tumor architecture and cellular heterogeneity more closely. Patient-derived organoids (PDOs) have been instrumental in personalized medicine, as demonstrated by their application in modeling colorectal cancer for individualized drug testing [224]. Another advanced 3D model is hydrogel-based cultures, where cancer cells are embedded in hydrogels that simulate the ECM, allowing researchers to study cell-ECM interactions and tumor invasion dynamics [225]. Co-culture systems are another significant advancement, where tumor cells are grown alongside stromal components such as fibroblasts, immune cells, or endothelial cells to explore interactions between different cell types within the TME. For instance, tumor-stroma co-cultures are employed to study the role of CAFs in promoting tumor growth and invasion [226, 227]. Immune-tumor co-cultures allow the investigation of immune evasion mechanisms, such as the interaction between tumor cells and TAMs [228]. More sophisticated models, like microfluidic-based "tumor-on-a-chip" systems, enable precise control over the tumor microenvironment by replicating dynamic conditions such as blood flow, oxygen gradients, and mechanical forces. These models are particularly useful in real-time studies of cancer cell behavior and drug response [229]. For example, breast cancer cells interacting with endothelial cells on a microfluidic chip have been used to study metastasis and angiogenesis [230]. Together, these various cell culture models offer researchers a toolkit for studying cancer's complex interactions with the microenvironment, aiding in the development of more effective cancer therapies.

14.2 In Vivo Models

Animal models play an indispensable role in cancer research, particularly in studying the complex interactions between tumors and their surrounding TME [231]. Among animal models, mouse models are most commonly used due to their genetic similarity to humans and the ease with which they can be genetically engineered [232]. Nowadays, different types of mouse model has been used in the TME research.

Genetically Engineered Mouse Models (GEMMs): GEMMs are widely used to study the role of specific genes in cancer progression and to understand how mutations in oncogenes or tumor suppressor genes drive tumor development [233]. In GEMMs, genes can be mutated, deleted, or overexpressed to mimic the genetic alterations found in human cancers. These models are valuable for studying the interactions between cancer cells and their microenvironment because they allow researchers to explore how genetic mutations in cancer cells affect the behavior of stromal cells, immune cells, and the ECM [233, 234]. For example, in KRAS-driven lung cancer models, the mutation of the KRAS oncogene in mice leads to the development of lung tumors. These models have been instrumental in studying how the mutated KRAS gene interacts with immune cells in the TME, such as TAMs and MDSCs, which can promote tumor growth by suppressing the anti-tumor immune response [235]. Additionally, GEMMs for BRCA1- and BRCA2-mutated breast cancer have provided valuable insights into how these mutations affect the DNA damage response in tumor cells and how these cells interact with the immune microenvironment to evade detection by T cells [44].

Xenograft Models: Xenograft models involve the transplantation of human cancer cells or tumor tissues into immunocompromised mice, which lack a functional immune system [236]. These models are useful for studying human cancer in a living organism while maintaining the specific genetic and phenotypic characteristics of the human tumor. By implanting human tumors into mice, researchers can observe how the tumor interacts with its environment, including how it induces angiogenesis (the formation of new blood vessels) and invades surrounding tissues [237, 238]. There are two main types of xenograft models used in cancer research.

Cell line-derived xenografts (CDX): Human cancer cell lines are implanted into immunocompromised mice to study tumor growth and metastasis. These models are commonly used to test new cancer therapies, as they allow researchers to evaluate drug efficacy in a human-like tumor system [236]. For example, CDX models of colorectal cancer have been used to study the role of tumor angiogenesis and its modulation by anti-angiogenic drugs like bevacizumab [239].

Patient-derived xenografts (PDX): Tumor tissues from cancer patients are directly implanted into immunocompromised mice, providing a more accurate representation of the patient's tumor and its microenvironment. PDX models maintain the genetic diversity and heterogeneity of human tumors, making them valuable tools for personalized medicine [240]. For instance, PDX models of breast cancer have been used to explore how different molecular subtypes of breast cancer respond to chemotherapy and targeted therapies [241]. In PDX models, the human tumor microenvironment, including stromal cells and extracellular matrix components, is partially retained, allowing for a more physiologically relevant study of tumor-stroma interactions [242]. These models have been particularly useful in studying drug resistance mechanisms in cancers like pancreatic adenocarcinoma and glioblastoma, where the dense stromal environment contributes to therapeutic resistance [242].

Syngeneic Models: Unlike xenograft models, where human cancer cells are transplanted into immunocompromised mice, syngeneic models involve the transplantation of mouse cancer cells into immunocompetent mice of the same genetic background. These models are particularly useful for studying the immune response to cancer, as they allow for the interaction between cancer cells and a fully functional immune system [243]. Researchers can use syngeneic models to study how immune cells, such as T cells, macrophages, and NK cells, respond to tumor growth and how tumors develop strategies to evade immune surveillance [244]. An example of a syngeneic model is the B16 melanoma model, in which mouse melanoma cells are injected into syngeneic C57BL/6 mice. This model has been extensively used to study tumor-immune interactions and has provided critical insights into how tumors suppress immune responses through immune checkpoint molecules like PD-L1. Syngeneic models have also been used to evaluate the efficacy of immunotherapies, such as anti-PD-1 and anti-CTLA-4 antibodies, which work by reactivating the immune system to attack the tumor [245, 246].

Humanized Mouse Models: Humanized mouse models are a more advanced type of xenograft model in which immunocompromised mice are engrafted with human immune cells or tissues [247]. These models are valuable for studying the interaction between human tumors and human immune cells within a living organism. By reconstituting the mouse's immune system with human hematopoietic stem cells or immune cells, researchers can study how human tumors interact with human T cells, B cells, and myeloid cells in the context of immunotherapy [248, 249]. For instance, humanized mouse models have been used to evaluate the efficacy of CAR T-cell therapies in preclinical settings. In these models, researchers can investigate how engineered T cells target and eliminate tumors in the presence of a functional human immune system [250, 251].

15 Emerging Technologies in Cancer Research

Emerging technologies are revolutionizing cancer research by offering new ways to study tumor biology, gene editing, and therapeutic responses. **Single-cell RNA sequencing (scRNA-seq)** is a transformative technology that enables researchers to analyze gene expression at the individual cell level, providing insights into the heterogeneous composition of tumors, including cancer cells, immune cells, and stromal cells. This technology has been instrumental in understanding the roles of cancer stem cells, immune evasion mechanisms, and metabolic heterogeneity within tumors [252]. **CRISPR-Cas9 gene editing** has opened new avenues for cancer research by allowing precise manipulation of genes. It is widely used to introduce mutations into animal models and organoid systems to create more accurate representations of human cancers. CRISPR is also being explored as a potential therapeutic tool to directly target and disable cancer-driving genes within tumors [253]. **Organs-on-chips**, micro-engineered platforms that replicate the structure and function of human organs, offer a dynamic and physiologically relevant model for studying tumor interactions with blood vessels, immune cells, and other components of the TME. These systems have the potential to improve drug testing by providing more accurate predictions of how therapies will perform in humans compared to traditional models [254]. Lastly, **artificial intelligence (AI) and machine learning** are becoming integral to cancer research, helping analyze large genomic, transcriptomic, and imaging datasets. Machine learning algorithms are being used to predict patient responses to immunotherapies, while AI-driven models are enhancing the accuracy of cancer diagnosis and prognosis by automating the analysis of histopathological images. Together, these technologies are pushing the boundaries of cancer research and offering new avenues for personalized medicine and therapeutic development [255].

16 Summary and Future Directions

Cancer is characterized by uncontrolled cell growth driven by genetic mutations that allow evasion of the body's regulatory mechanisms. The tumor microenvironment, comprising immune cells, fibroblasts, blood vessels, and other stromal components, plays a critical role in cancer progression, metastasis, and treatment resistance. The TME is not passive but actively influences cancer cell behavior, including growth, survival, and metastasis. Tumor cells interact with various components of the TME through signaling mechanisms, which create an environment favorable for tumor progression and therapeutic resistance. Recent research has focused on developing therapies that target the TME, such as immunotherapies and anti-angiogenic treatments, to improve therapeutic outcomes. This includes immune checkpoint inhibitors and agents that modulate angiogenesis and immune cell functions. Different models, including in vitro cultures, animal models, and emerging technologies like organoids, are used to study cancer biology and TME interactions, providing insights into treatment strategies. Various case studies were discussed to illustrate how the TME influences cancer progression and treatment, leading to a better understanding of resistance mechanisms and new avenues for therapy.

The complexity and variability of the TME across different cancer types and even within individual tumors remain significant challenges. Understanding and addressing the heterogeneity of the TME is crucial for developing effective, personalized therapies. The protective nature of the TME contributes to resistance against current therapies. Developing strategies to modulate or disrupt the TME's protective functions is key to overcoming therapeutic resistance. Current in vitro and in vivo models have limitations in fully recapitulating the complexity of human tumors and their environments. Advances in modeling the TME more accurately are needed for better translational outcomes.

Integrating TME-targeting therapies with traditional cancer treatments and immunotherapies is a promising strategy to enhance efficacy and overcome resistance. Further research into combination approaches that simultaneously target multiple components of the TME may lead to improved outcomes. Advances in understanding the genetic and molecular landscape of tumors and their microenvironments will enable the development of personalized therapies that are tailored to the specific characteristics of a patient's tumor and TME. Identifying new molecular targets within the TME, such as novel signaling pathways, immune cells, and stromal components, will drive the next generation of therapeutic strategies. Targeting metabolic reprogramming and the interactions between tumor cells and stromal cells is an emerging area of research. Leveraging emerging technologies like CRISPR for gene editing, single-cell sequencing, organ-on-a-chip systems, and nanoparticle-based system will provide deeper insights into the TME and accelerate drug discovery. Combining these technologies with artificial intelligence may facilitate more precise predictions of therapeutic outcomes. Overall, continued exploration of the interactions between cancer cells and the TME, alongside technological advancements, holds significant potential for improving cancer treatment and patient outcomes.

References

1. Krieghoff-Henning, E., Folkerts, J., Penzkofer, A., Weg-Remers, S.: Cancer—an overview. Med. Monatsschr. Pharm. **40**(2), 48–54 (2017)
2. Chemotherapy, S.-R.: Preventing competitive release. Nat. Rev. Cancer **16**(4), 199 (2016)
3. Welch, D.R., Hurst, D.R.: Defining the hallmarks of metastasis. Cancer Res. **79**(12), 3011–3027 (2019)

4. Khoury, J.D.: The evolving potential of companion diagnostics. Scand. J. Clin. Lab. Invest. Suppl. **245**, S22–S25 (2016)
5. Carbone, A.: Cancer classification at the crossroads. Cancers (Basel) **12**(4) (2020)
6. Zoine, J.T., Moore, S.E., Velasquez, M.P.: Leukemia's next top model? Syngeneic models to advance adoptive cellular therapy. Front. Immunol. **13**, 867103 (2022)
7. Huntsman, D.G., Ladanyi, M.: The molecular pathology of cancer: from pan-genomics to post-genomics. J. Pathol. **244**(5), 509–511 (2018)
8. Mundy, G.R.: Metastasis to bone: causes, consequences and therapeutic opportunities. Nat. Rev. Cancer **2**(8), 584–593 (2002)
9. Cooper, L.A., Demicco, E.G., Saltz, J.H., Powell, R.T., Rao, A., Lazar, A.J.: PanCancer insights from The Cancer Genome Atlas: the pathologist's perspective. J. Pathol. **244**(5), 512–524 (2018)
10. Roda, N., Blandano, G., Pelicci, P.G.: Blood vessels and peripheral nerves as key players in cancer progression and therapy resistance. Cancers (Basel) **13**(17) (2021)
11. Neophytou, C.M., Panagi, M., Stylianopoulos, T., Papageorgis, P.: The role of tumor microenvironment in cancer metastasis: molecular mechanisms and therapeutic opportunities. Cancers (Basel) **13**(9) (2021)
12. Whipple, C.A.: Tumor talk: understanding the conversation between the tumor and its microenvironment. Cancer Cell Microenviron. **2**(2), e773 (2015)
13. McAllister, S.S., Weinberg, R.A.: The tumour-induced systemic environment as a critical regulator of cancer progression and metastasis. Nat. Cell Biol. **16**(8), 717–727 (2014)
14. Ma, X.J., Dahiya, S., Richardson, E., Erlander, M., Sgroi, D.C.: Gene expression profiling of the tumor microenvironment during breast cancer progression. Breast Cancer Res. **11**(1), R7 (2009)
15. Bejarano, L., Jordao, M.J.C., Joyce, J.A.: Therapeutic targeting of the tumor microenvironment. Cancer Discov. **11**(4), 933–959 (2021)
16. Meng, L., Wu, H., Wu, J., Ding, P., He, J., Sang, M., et al.: Mechanisms of immune checkpoint inhibitors: insights into the regulation of circular RNAS involved in cancer hallmarks. Cell Death Dis. **15**(1), 3 (2024)
17. Hosea, R., Hillary, S., Naqvi, S., Wu, S., Kasim, V.: The two sides of chromosomal instability: drivers and brakes in cancer. Signal Transduct. Target. Ther. **9**(1), 75 (2024)
18. Moon, J.J., Lu, A., Moon, C.: Role of genomic instability in human carcinogenesis. Exp. Biol. Med. (Maywood) **244**(3), 227–240 (2019)
19. Aguilera, A., Gomez-Gonzalez, B.: Genome instability: a mechanistic view of its causes and consequences. Nat. Rev. Genet. **9**(3), 204–217 (2008)
20. Balkwill, F.R., Capasso, M., Hagemann, T.: The tumor microenvironment at a glance. J. Cell Sci. **125**(Pt 23), 5591–5596 (2012)
21. Wang, R.A., Li, Z.S., Zhang, H.Z., Zheng, P.J., Li, Q.L., Shi, J.G., et al.: Invasive cancers are not necessarily from preformed in situ tumours—an alternative way of carcinogenesis from misplaced stem cells. J. Cell. Mol. Med. **17**(7), 921–926 (2013)
22. Chmurska, A., Matczak, K., Marczak, A.: Two faces of autophagy in the struggle against cancer. Int. J. Mol. Sci. **22**(6) (2021)
23. Eymin, B., Gazzeri, S.: Role of cell cycle regulators in lung carcinogenesis. Cell Adh. Migr. **4**(1), 114–123 (2010)
24. Matthews, H.K., Bertoli, C., de Bruin, R.A.M.: Cell cycle control in cancer. Nat. Rev. Mol. Cell Biol. **23**(1), 74–88 (2022)
25. Bertoli, C., Skotheim, J.M., de Bruin, R.A.: Control of cell cycle transcription during G1 and S phases. Nat. Rev. Mol. Cell Biol. **14**(8), 518–528 (2013)
26. Dick, F.A., Rubin, S.M.: Molecular mechanisms underlying RB protein function. Nat. Rev. Mol. Cell Biol. **14**(5), 297–306 (2013)
27. Lemonnier, T., Dupre, A., Jessus, C.: The G2-to-M transition from a phosphatase perspective: a new vision of the meiotic division. Cell Div. **15**, 9 (2020)
28. Wolgemuth, D.J.: Function of cyclins in regulating the mitotic and meiotic cell cycles in male germ cells. Cell Cycle **7**(22), 3509–3513 (2008)

29. Hume, S., Dianov, G.L., Ramadan, K.: A unified model for the G1/S cell cycle transition. Nucleic Acids Res. **48**(22), 12483–12501 (2020)
30. Wang, H., Guo, M., Wei, H., Chen, Y.: Targeting p53 pathways: mechanisms, structures, and advances in therapy. Signal Transduct. Target. Ther. **8**(1), 92 (2023)
31. Hernandez Borrero, L.J., El-Deiry, W.S.: Tumor suppressor p53: biology, signaling pathways, and therapeutic targeting. Biochim. Biophys. Acta Rev. Cancer **1876**(1), 188556 (2021)
32. Robson, C.N., Gnanapragasam, V., Byrne, R.L., Collins, A.T., Neal, D.E.: Transforming growth factor-beta1 up-regulates p15, p21 and p27 and blocks cell cycling in G1 in human prostate epithelium. J. Endocrinol. **160**(2), 257–266 (1999)
33. Madamanchi, N.R., Moon, S.K., Hakim, Z.S., Clark, S., Mehrizi, A., Patterson, C., et al.: Differential activation of mitogenic signaling pathways in aortic smooth muscle cells deficient in superoxide dismutase isoforms. Arterioscler. Thromb. Vasc. Biol. **25**(5), 950–956 (2005)
34. Sears, R., Nuckolls, F., Haura, E., Taya, Y., Tamai, K., Nevins, J.R.: Multiple Ras-dependent phosphorylation pathways regulate Myc protein stability. Genes Dev. **14**(19), 2501–2514 (2000)
35. Casacuberta-Serra, S., Gonzalez-Larreategui, I., Capitan-Leo, D., Soucek, L.: MYC and KRAS cooperation: from historical challenges to therapeutic opportunities in cancer. Signal Transduct. Target. Ther. **9**(1), 205 (2024)
36. Raghav, K.P.S., Moasser, M.M.: Molecular pathways and mechanisms of HER2 in cancer therapy. Clin. Cancer Res. **29**(13), 2351–2361 (2023)
37. Nussinov, R., Zhang, W., Liu, Y., Jang, H.: Mitogen signaling strength and duration can control cell cycle decisions. Sci. Adv. **10**(27), eadm9211 (2024)
38. Bahar, M.E., Kim, H.J., Kim, D.R.: Targeting the RAS/RAF/MAPK pathway for cancer therapy: from mechanism to clinical studies. Signal Transduct. Target. Ther. **8**(1), 455 (2023)
39. Besson, A., Dowdy, S.F., Roberts, J.M.: CDK inhibitors: cell cycle regulators and beyond. Dev. Cell **14**(2), 159–169 (2008)
40. Osborne, C., Wilson, P., Tripathy, D.: Oncogenes and tumor suppressor genes in breast cancer: potential diagnostic and therapeutic applications. Oncologist **9**(4), 361–377 (2004)
41. Zhang, S., Xiao, X., Yi, Y., Wang, X., Zhu, L., Shen, Y., et al.: Tumor initiation and early tumorigenesis: molecular mechanisms and interventional targets. Signal Transduct. Target. Ther. **9**(1), 149 (2024)
42. Lee, E.Y., To, H., Shew, J.Y., Bookstein, R., Scully, P., Lee, W.H.: Inactivation of the retinoblastoma susceptibility gene in human breast cancers. Science **241**(4862), 218–221 (1988)
43. Foulkes, W.D.: The ten genes for breast (and ovarian) cancer susceptibility. Nat. Rev. Clin. Oncol. **18**(5), 259–260 (2021)
44. Godet, I., Gilkes, D.M.: BRCA1 and BRCA2 mutations and treatment strategies for breast cancer. Integr. Cancer Sci. Ther. **4**(1) (2017)
45. Hollander, M.C., Blumenthal, G.M., Dennis, P.A.: PTEN loss in the continuum of common cancers, rare syndromes and mouse models. Nat. Rev. Cancer **11**(4), 289–301 (2011)
46. Olivier, M., Hollstein, M., Hainaut, P.: TP53 mutations in human cancers: origins, consequences, and clinical use. Cold Spring Harb. Perspect. Biol. **2**(1), a001008 (2010)
47. Breast Cancer Association, C., Dorling, L., Carvalho, S., Allen, J., Gonzalez-Neira, A., Luccarini, C., et al.: Breast cancer risk genes—association analysis in more than 113,000 women. N. Engl. J. Med. **384**(5), 428–439 (2021)
48. Smith, J., Tho, L.M., Xu, N., Gillespie, D.A.: The ATM-Chk2 and ATR-Chk1 pathways in DNA damage signaling and cancer. Adv. Cancer Res. **108**, 73–112 (2010)
49. Rojowska, A., Lammens, K., Seifert, F.U., Direnberger, C., Feldmann, H., Hopfner, K.P.: Structure of the Rad50 DNA double-strand break repair protein in complex with DNA. EMBO J. **33**(23), 2847–2859 (2014)
50. Zhang, F., Ma, J., Wu, J., Ye, L., Cai, H., Xia, B., et al.: PALB2 links BRCA1 and BRCA2 in the DNA-damage response. Curr. Biol. **19**(6), 524–529 (2009)
51. Saal, L.H., Gruvberger-Saal, S.K., Persson, C., Lovgren, K., Jumppanen, M., Staaf, J., et al.: Recurrent gross mutations of the PTEN tumor suppressor gene in breast cancers with deficient DSB repair. Nat. Genet. **40**(1), 102–107 (2008)

52. Holstege, H., Joosse, S.A., van Oostrom, C.T., Nederlof, P.M., de Vries, A., Jonkers, J.: High incidence of protein-truncating TP53 mutations in BRCA1-related breast cancer. Cancer Res. **69**(8), 3625–3633 (2009)
53. Hassan, B., Akcakanat, A., Holder, A.M., Meric-Bernstam, F.: Targeting the PI3-kinase/Akt/mTOR signaling pathway. Surg. Oncol. Clin. N. Am. **22**(4), 641–664 (2013)
54. Huang, R.X., Zhou, P.K.: DNA damage response signaling pathways and targets for radiotherapy sensitization in cancer. Signal Transduct. Target. Ther. **5**(1), 60 (2020)
55. Jan, R., Chaudhry, G.E.: Understanding apoptosis and apoptotic pathways targeted cancer therapeutics. Adv. Pharm. Bull. **9**(2), 205–218 (2019)
56. Langley, R.R., Fidler, I.J.: The seed and soil hypothesis revisited—the role of tumor-stroma interactions in metastasis to different organs. Int. J. Cancer **128**(11), 2527–2535 (2011)
57. Thiery, J.P.: Epithelial-mesenchymal transitions in tumour progression. Nat. Rev. Cancer **2**(6), 442–454 (2002)
58. Kim, B.N., Ahn, D.H., Kang, N., Yeo, C.D., Kim, Y.K., Lee, K.Y., et al.: TGF-beta induced EMT and stemness characteristics are associated with epigenetic regulation in lung cancer. Sci. Rep. **10**(1), 10597 (2020)
59. Chiang, S.P., Cabrera, R.M., Segall, J.E.: Tumor cell intravasation. Am. J. Physiol. Cell Physiol. **311**(1), C1–C14 (2016)
60. Shenoy, A.K., Lu, J.: Cancer cells remodel themselves and vasculature to overcome the endothelial barrier. Cancer Lett. **380**(2), 534–544 (2016)
61. Liu, M., Yang, J., Xu, B., Zhang, X.: Tumor metastasis: mechanistic insights and therapeutic interventions. MedComm (2020) **2**(4), 587–617 (2021)
62. Pei, D., Shu, X., Gassama-Diagne, A., Thiery, J.P.: Mesenchymal-epithelial transition in development and reprogramming. Nat. Cell Biol. **21**(1), 44–53 (2019)
63. Ben-Ghedalia-Peled, N., Vago, R.: Wnt signaling in the development of bone metastasis. Cells **11**(23) (2022)
64. Akhtar, M., Haider, A., Rashid, S., Al-Nabet, A.: Paget's, "Seed and Soil" theory of cancer metastasis: an idea whose time has come. Adv. Anat. Pathol. **26**(1), 69–74 (2019)
65. Guo, Y., Ji, X., Liu, J., Fan, D., Zhou, Q., Chen, C., et al.: Effects of exosomes on pre-metastatic niche formation in tumors. Mol. Cancer **18**(1), 39 (2019)
66. Peinado, H., Zhang, H., Matei, I.R., Costa-Silva, B., Hoshino, A., Rodrigues, G., et al.: Pre-metastatic niches: organ-specific homes for metastases. Nat. Rev. Cancer **17**(5), 302–317 (2017)
67. Chen, J.F., Yan, Q.: The roles of epigenetics in cancer progression and metastasis. Biochem. J. **478**(17), 3373–3393 (2021)
68. Eslami, M., Memarsadeghi, O., Davarpanah, A., Arti, A., Nayernia, K., Behnam, B.: Overcoming chemotherapy resistance in metastatic cancer: a comprehensive review. Biomedicines **12**(1) (2024)
69. Miller, J.L., Grant, P.A.: The role of DNA methylation and histone modifications in transcriptional regulation in humans. Subcell. Biochem. **61**, 289–317 (2013)
70. Valastyan, S., Weinberg, R.A.: Tumor metastasis: molecular insights and evolving paradigms. Cell **147**(2), 275–292 (2011)
71. Fu, A., Yao, B., Dong, T., Cai, S.: Emerging roles of intratumor microbiota in cancer metastasis. Trends Cell Biol. **33**(7), 583–593 (2023)
72. Wang, H., Pan, J., Barsky, L., Jacob, J.C., Zheng, Y., Gao, C., et al.: Characteristics of pre-metastatic niche: the landscape of molecular and cellular pathways. Mol. Biomed. **2**(1), 3 (2021)
73. Tyagi, A.: Editorial: cellular and molecular characteristics of the pre-metastatic niche. Front. Oncol. **13**, 1298958 (2023)
74. Rankin, E.B., Giaccia, A.J.: Hypoxic control of metastasis. Science **352**(6282), 175–180 (2016)
75. Dai, J., Su, Y., Zhong, S., Cong, L., Liu, B., Yang, J., et al.: Exosomes: key players in cancer and potential therapeutic strategy. Signal Transduct. Target. Ther. **5**(1), 145 (2020)

76. Hoshino, A., Costa-Silva, B., Shen, T.L., Rodrigues, G., Hashimoto, A., Tesic Mark, M., et al.: Tumour exosome integrins determine organotropic metastasis. Nature **527**(7578), 329–335 (2015)
77. Grigoryeva, E.S., Savelieva, O.E., Popova, N.O., Cherdyntseva, N.V., Perelmuter, V.M.: Do tumor exosome integrins alone determine organotropic metastasis? Mol. Biol. Rep. **47**(10), 8145–8157 (2020)
78. Pantel, K., Alix-Panabieres, C.: Crucial roles of circulating tumor cells in the metastatic cascade and tumor immune escape: biology and clinical translation. J. Immunother. Cancer **10**(12) (2022)
79. Schuster, E., Taftaf, R., Reduzzi, C., Albert, M.K., Romero-Calvo, I., Liu, H.: Better together: circulating tumor cell clustering in metastatic cancer. Trends Cancer **7**(11), 1020–1032 (2021)
80. Gu, X., Wei, S., Lv, X.: Circulating tumor cells: from new biological insights to clinical practice. Signal Transduct. Target. Ther. **9**(1), 226 (2024)
81. Lin, D., Shen, L., Luo, M., Zhang, K., Li, J., Yang, Q., et al.: Circulating tumor cells: biology and clinical significance. Signal Transduct. Target. Ther. **6**(1), 404 (2021)
82. Breuksch, I., Weinert, M., Brenner, W.: The role of extracellular calcium in bone metastasis. J. Bone Oncol. **5**(3), 143–145 (2016)
83. Fares, J., Fares, M.Y., Khachfe, H.H., Salhab, H.A., Fares, Y.: Molecular principles of metastasis: a hallmark of cancer revisited. Signal Transduct. Target. Ther. **5**(1), 28 (2020)
84. Lambert, A.W., Zhang, Y., Weinberg, R.A.: Cell-intrinsic and microenvironmental determinants of metastatic colonization. Nat. Cell Biol. **26**(5), 687–697 (2024)
85. de Visser, K.E., Joyce, J.A.: The evolving tumor microenvironment: from cancer initiation to metastatic outgrowth. Cancer Cell **41**(3), 374–403 (2023)
86. Tiwari, A., Trivedi, R., Lin, S.Y.: Tumor microenvironment: barrier or opportunity towards effective cancer therapy. J. Biomed. Sci. **29**(1), 83 (2022)
87. Binnewies, M., Roberts, E.W., Kersten, K., Chan, V., Fearon, D.F., Merad, M., et al.: Understanding the tumor immune microenvironment (TIME) for effective therapy. Nat. Med. **24**(5), 541–550 (2018)
88. Desai, S.A., Patel, V.P., Bhosle, K.P., Nagare, S.D., Thombare, K.C.: The tumor microenvironment: shaping cancer progression and treatment response. J. Chemother. 1–30 (2024)
89. Walsh, L.A., Quail, D.F.: Decoding the tumor microenvironment with spatial technologies. Nat. Immunol. **24**(12), 1982–1993 (2023)
90. Shen, M., Kang, Y.: Complex interplay between tumor microenvironment and cancer therapy. Front. Med. **12**(4), 426–439 (2018)
91. Winkler, J., Abisoye-Ogunniyan, A., Metcalf, K.J., Werb, Z.: Concepts of extracellular matrix remodelling in tumour progression and metastasis. Nat. Commun. **11**(1), 5120 (2020)
92. Naba, A., Clauser, K.R., Ding, H., Whittaker, C.A., Carr, S.A., Hynes, R.O.: The extracellular matrix: tools and insights for the "omics" era. Matrix Biol. **49**, 10–24 (2016)
93. Hastings, J.F., Skhinas, J.N., Fey, D., Croucher, D.R., Cox, T.R.: The extracellular matrix as a key regulator of intracellular signalling networks. Br. J. Pharmacol. **176**(1), 82–92 (2019)
94. Egeblad, M., Rasch, M.G., Weaver, V.M.: Dynamic interplay between the collagen scaffold and tumor evolution. Curr. Opin. Cell Biol. **22**(5), 697–706 (2010)
95. Yurchenco, P.D.: Basement membranes: cell scaffoldings and signaling platforms. Cold Spring Harb. Perspect. Biol. **3**(2) (2011)
96. Kessenbrock, K., Plaks, V., Werb, Z.: Matrix metalloproteinases: regulators of the tumor microenvironment. Cell **141**(1), 52–67 (2010)
97. Whatcott, C.J., Diep, C.H., Jiang, P., Watanabe, A., LoBello, J., Sima, C., et al.: Desmoplasia in primary tumors and metastatic lesions of pancreatic cancer. Clin. Cancer Res. **21**(15), 3561–3568 (2015)
98. Iacobuzio-Donahue, C.A., Argani, P., Hempen, P.M., Jones, J., Kern, S.E.: The desmoplastic response to infiltrating breast carcinoma: gene expression at the site of primary invasion and implications for comparisons between tumor types. Cancer Res. **62**(18), 5351–5357 (2002)
99. Kai, F., Drain, A.P., Weaver, V.M.: The extracellular matrix modulates the metastatic journey. Dev. Cell **49**(3), 332–346 (2019)

100. Lai, J., Chien, J., Staub, J., Avula, R., Greene, E.L., Matthews, T.A., et al.: Loss of HSulf-1 up-regulates heparin-binding growth factor signaling in cancer. J. Biol. Chem. **278**(25), 23107–23117 (2003)
101. Biffi, G., Oni, T.E., Spielman, B., Hao, Y., Elyada, E., Park, Y., et al.: IL1-Induced JAK/stat signaling is antagonized by TGFbeta to shape CAF heterogeneity in pancreatic ductal adenocarcinoma. Cancer Discov. **9**(2), 282–301 (2019)
102. Ohlund, D., Handly-Santana, A., Biffi, G., Elyada, E., Almeida, A.S., Ponz-Sarvise, M., et al.: Distinct populations of inflammatory fibroblasts and myofibroblasts in pancreatic cancer. J. Exp. Med. **214**(3), 579–596 (2017)
103. He, X., Lee, B., Jiang, Y.: Extracellular matrix in cancer progression and therapy. Med. Rev. (2021) **2**(2), 125–139 (2022)
104. Genomic map of the cellular composition of the tumor microenvironment. Nat. Immunol. **25**(10), 1789–1790 (2024)
105. Sahai, E., Astsaturov, I., Cukierman, E., DeNardo, D.G., Egeblad, M., Evans, R.M., et al.: A framework for advancing our understanding of cancer-associated fibroblasts. Nat. Rev. Cancer **20**(3), 174–186 (2020)
106. Calvo, F., Ege, N., Grande-Garcia, A., Hooper, S., Jenkins, R.P., Chaudhry, S.I., et al.: Mechanotransduction and YAP-dependent matrix remodelling is required for the generation and maintenance of cancer-associated fibroblasts. Nat. Cell Biol. **15**(6), 637–646 (2013)
107. Ishihara, S., Haga, H.: Matrix stiffness contributes to cancer progression by regulating transcription factors. Cancers (Basel) **14**(4) (2022)
108. Kalluri, R.: The biology and function of fibroblasts in cancer. Nat. Rev. Cancer **16**(9), 582–598 (2016)
109. Arpinati, L., Carradori, G., Scherz-Shouval, R.: CAF-induced physical constraints controlling T cell state and localization in solid tumours. Nat. Rev. Cancer **24**(10), 676–693 (2024)
110. Bottcher, J.P., Bonavita, E., Chakravarty, P., Blees, H., Cabeza-Cabrerizo, M., Sammicheli, S., et al.: NK cells stimulate recruitment of cDC1 into the tumor microenvironment promoting cancer immune control. Cell **172**(5), 1022–1037, e14 (2018)
111. Bindea, G., Mlecnik, B., Galon, J.: Immune sunrise: from the immunome to the cancer immune landscape. Oncoimmunology **11**(1), 2019896 (2022)
112. Jayasingam, S.D., Citartan, M., Thang, T.H., Mat Zin, A.A., Ang, K.C., Ch'ng, E.S.: Evaluating the polarization of tumor-associated macrophages into M1 and M2 phenotypes in human cancer tissue: technicalities and challenges in routine clinical practice. Front. Oncol. **9**, 1512 (2019)
113. Welte, T., Kim, I.S., Tian, L., Gao, X., Wang, H., Li, J., et al.: Oncogenic mTOR signalling recruits myeloid-derived suppressor cells to promote tumour initiation. Nat. Cell Biol. **18**(6), 632–644 (2016)
114. Peng, G.L., Li, L., Guo, Y.W., Yu, P., Yin, X.J., Wang, S., et al.: CD8(+) cytotoxic and FoxP3(+) regulatory T lymphocytes serve as prognostic factors in breast cancer. Am. J. Transl. Res. **11**(8), 5039–5053 (2019)
115. Basak, U., Sarkar, T., Mukherjee, S., Chakraborty, S., Dutta, A., Dutta, S., et al.: Tumor-associated macrophages: an effective player of the tumor microenvironment. Front. Immunol. **14**, 1295257 (2023)
116. Genard, G., Lucas, S., Michiels, C.: Reprogramming of tumor-associated macrophages with anticancer therapies: radiotherapy versus chemo- and immunotherapies. Front. Immunol. **8**, 828 (2017)
117. Hao, N.B., Lu, M.H., Fan, Y.H., Cao, Y.L., Zhang, Z.R., Yang, S.M.: Macrophages in tumor microenvironments and the progression of tumors. Clin. Dev. Immunol. **2012**, 948098 (2012)
118. Lee, C., Jeong, H., Bae, Y., Shin, K., Kang, S., Kim, H., et al.: Targeting of M2-like tumor-associated macrophages with a melittin-based pro-apoptotic peptide. J. Immunother. Cancer **7**(1), 147 (2019)
119. Gratchev, A., Guillot, P., Hakiy, N., Politz, O., Orfanos, C.E., Schledzewski, K., et al.: Alternatively activated macrophages differentially express fibronectin and its splice variants and the extracellular matrix protein betaIG-H3. Scand. J. Immunol. **53**(4), 386–392 (2001)

120. Lv, Y., Zhao, Y., Wang, X., Chen, N., Mao, F., Teng, Y., et al.: Increased intratumoral mast cells foster immune suppression and gastric cancer progression through TNF-alpha-PD-L1 pathway. J. Immunother. Cancer **7**(1), 54 (2019)
121. Cordova-Bahena, L., Velasco-Velazquez, M.A.: Anti-PD-1 and anti-PD-L1 antibodies as immunotherapy against cancer: a structural perspective. Rev. Invest. Clin. **73**(1), 008–016 (2020)
122. Wang, D.R., Wu, X.L., Sun, Y.L.: Therapeutic targets and biomarkers of tumor immunotherapy: response versus non-response. Signal Transduct. Target. Ther. **7**(1), 331 (2022)
123. Li, K., Shi, H., Zhang, B., Ou, X., Ma, Q., Chen, Y., et al.: Myeloid-derived suppressor cells as immunosuppressive regulators and therapeutic targets in cancer. Signal Transduct. Target. Ther. **6**(1), 362 (2021)
124. Pak, A.S., Wright, M.A., Matthews, J.P., Collins, S.L., Petruzzelli, G.J., Young, M.R.: Mechanisms of immune suppression in patients with head and neck cancer: presence of CD34(+) cells which suppress immune functions within cancers that secrete granulocyte-macrophage colony-stimulating factor. Clin. Cancer Res. **1**(1), 95–103 (1995)
125. Mielcarek, M., Martin, P.J., Torok-Storb, B.: Suppression of alloantigen-induced T-cell proliferation by CD14+ cells derived from granulocyte colony-stimulating factor-mobilized peripheral blood mononuclear cells. Blood **89**(5), 1629–1634 (1997)
126. Li, H., Han, Y., Guo, Q., Zhang, M., Cao, X.: Cancer-expanded myeloid-derived suppressor cells induce energy of NK cells through membrane-bound TGF-beta 1. J. Immunol. **182**(1), 240–249 (2009)
127. Carmeliet, P., Jain, R.K.: Molecular mechanisms and clinical applications of angiogenesis. Nature **473**(7347), 298–307 (2011)
128. Ghalehbandi, S., Yuzugulen, J., Pranjol, M.Z.I., Pourgholami, M.H.: The role of VEGF in cancer-induced angiogenesis and research progress of drugs targeting VEGF. Eur. J. Pharmacol. **949**, 175586 (2023)
129. Liang, W., Ferrara, N.: The complex role of neutrophils in tumor angiogenesis and metastasis. Cancer Immunol. Res. **4**(2), 83–91 (2016)
130. Shojaei, F., Singh, M., Thompson, J.D., Ferrara, N.: Role of Bv8 in neutrophil-dependent angiogenesis in a transgenic model of cancer progression. Proc. Natl. Acad. Sci. U. S. A. **105**(7), 2640–2645 (2008)
131. Majorini, M.T., Colombo, M.P., Lecis, D.: Few, but efficient: the role of mast cells in breast cancer and other solid tumors. Cancer Res. **82**(8), 1439–1447 (2022)
132. Canli, O., Nicolas, A.M., Gupta, J., Finkelmeier, F., Goncharova, O., Pesic, M., et al.: Myeloid cell-derived reactive oxygen species induce epithelial mutagenesis. Cancer Cell **32**(6):869–883, e5 (2017)
133. Wang, F., Peng, L., Wang, Y., Liu, X.: A meta-analysis of vascular endothelial growth factor for nasopharyngeal cancer prognosis. Front. Oncol. **8**, 486 (2018)
134. Batlle, E., Massague, J.: Transforming growth factor-beta signaling in immunity and cancer. Immunity **50**(4), 924–940 (2019)
135. Turner, N., Grose, R.: Fibroblast growth factor signalling: from development to cancer. Nat. Rev. Cancer **10**(2), 116–129 (2010)
136. Yi, M., Li, T., Niu, M., Zhang, H., Wu, Y., Wu, K., et al.: Targeting cytokine and chemokine signaling pathways for cancer therapy. Signal Transduct. Target. Ther. **9**(1), 176 (2024)
137. Lippitz, B.E.: Cytokine patterns in patients with cancer: a systematic review. Lancet Oncol. **14**(6), e218–e228 (2013)
138. Zhang, J., Veeramachaneni, N.: Targeting interleukin-1beta and inflammation in lung cancer. Biomark. Res. **10**(1), 5 (2022)
139. Ungefroren, H.: Autocrine TGF-beta in cancer: review of the literature and caveats in experimental analysis. Int. J. Mol. Sci. **22**(2) (2021)
140. Scheel, C., Eaton, E.N., Li, S.H., Chaffer, C.L., Reinhardt, F., Kah, K.J., et al.: Paracrine and autocrine signals induce and maintain mesenchymal and stem cell states in the breast. Cell **145**(6), 926–940 (2011)

141. Mao, X., Xu, J., Wang, W., Liang, C., Hua, J., Liu, J., et al.: Crosstalk between cancer-associated fibroblasts and immune cells in the tumor microenvironment: new findings and future perspectives. Mol. Cancer **20**(1), 131 (2021)
142. Junttila, M.R., de Sauvage, F.J.: Influence of tumour micro-environment heterogeneity on therapeutic response. Nature **501**(7467), 346–354 (2013)
143. Song, T., Dou, C., Jia, Y., Tu, K., Zheng, X.: TIMP-1 activated carcinoma-associated fibroblasts inhibit tumor apoptosis by activating SDF1/CXCR4 signaling in hepatocellular carcinoma. Oncotarget **6**(14), 12061–12079 (2015)
144. Lee, J.G., Wu, R.: Erlotinib-cisplatin combination inhibits growth and angiogenesis through c-MYC and HIF-1alpha in EGFR-mutated lung cancer in vitro and in vivo. Neoplasia **17**(2), 190–200 (2015)
145. Ghorani, E., Swanton, C., Quezada, S.A.: Cancer cell-intrinsic mechanisms driving acquired immune tolerance. Immunity **56**(10), 2270–2295 (2023)
146. Tufail, M., Jiang, C.H., Li, N.: Altered metabolism in cancer: insights into energy pathways and therapeutic targets. Mol. Cancer **23**(1), 203 (2024)
147. Vander Heiden, M.G., Cantley, L.C., Thompson, C.B.: Understanding the Warburg effect: the metabolic requirements of cell proliferation. Science **324**(5930), 1029–1033 (2009)
148. Wu, L., Jin, Y., Zhao, X., Tang, K., Zhao, Y., Tong, L., et al.: Tumor aerobic glycolysis confers immune evasion through modulating sensitivity to T cell-mediated bystander killing via TNF-alpha. Cell Metab. **35**(9), 1580–1596, e9 (2023)
149. Jin, J., Byun, J.K., Choi, Y.K., Park, K.G.: Targeting glutamine metabolism as a therapeutic strategy for cancer. Exp. Mol. Med. **55**(4), 706–715 (2023)
150. Jin, H.R., Wang, J., Wang, Z.J., Xi, M.J., Xia, B.H., Deng, K., et al.: Lipid metabolic reprogramming in tumor microenvironment: from mechanisms to therapeutics. J. Hematol. Oncol. **16**(1), 103 (2023)
151. Shi, R., Liao, C., Zhang, Q.: Hypoxia-driven effects in cancer: characterization, mechanisms, and therapeutic implications. Cells **10**(3) (2021)
152. Wei, F., Wang, D., Wei, J., Tang, N., Tang, L., Xiong, F., et al.: Metabolic crosstalk in the tumor microenvironment regulates antitumor immunosuppression and immunotherapy resistance. Cell. Mol. Life Sci. **78**(1), 173–193 (2021)
153. Yue, M., Jiang, J., Gao, P., Liu, H., Qing, G.: Oncogenic MYC activates a feedforward regulatory loop promoting essential amino acid metabolism and tumorigenesis. Cell Rep. **21**(13), 3819–3832 (2017)
154. Jeon, Y.J., Khelifa, S., Ratnikov, B., Scott, D.A., Feng, Y., Parisi, F., et al.: Regulation of glutamine carrier proteins by RNF5 determines breast cancer response to ER stress-inducing chemotherapies. Cancer Cell **27**(3), 354–369 (2015)
155. Izzo, L.T., Affronti, H.C., Wellen, K.E.: The bidirectional relationship between cancer epigenetics and metabolism. Annu. Rev. Cancer Biol. **5**(1), 235–257 (2021)
156. Chen, X., Song, E.: The theory of tumor ecosystem. Cancer Commun. (London) **42**(7), 587–608 (2022)
157. Li, Y.R., Wilson, M., Yang, L.: Target tumor microenvironment by innate T cells. Front. Immunol. **13**, 999549 (2022)
158. Li, Y.R., Yu, Y., Kramer, A., Hon, R., Wilson, M., Brown, J., et al.: An ex vivo 3D tumor microenvironment-mimicry culture to study TAM modulation of cancer immunotherapy. Cells **11**(9) (2022)
159. El-Tanani, M., Rabbani, S.A., Babiker, R., Rangraze, I., Kapre, S., Palakurthi, S.S., et al.: Unraveling the tumor microenvironment: insights into cancer metastasis and therapeutic strategies. Cancer Lett. **591**, 216894 (2024)
160. Bailly, C., Thuru, X., Quesnel, B.: Combined cytotoxic chemotherapy and immunotherapy of cancer: modern times. NAR Cancer **2**(1), zcaa002 (2020)
161. Giraldo, N.A., Sanchez-Salas, R., Peske, J.D., Vano, Y., Becht, E., Petitprez, F., et al.: The clinical role of the TME in solid cancer. Br. J. Cancer **120**(1), 45–53 (2019)
162. Jin, H., Wang, L., Bernards, R.: Rational combinations of targeted cancer therapies: background, advances and challenges. Nat. Rev. Drug Discov. **22**(3), 213–234 (2023)

163. Olson, O.C., Kim, H., Quail, D.F., Foley, E.A., Joyce, J.A.: Tumor-associated macrophages suppress the cytotoxic activity of antimitotic agents. Cell Rep. **19**(1), 101–113 (2017)
164. Holmgaard, R.B., Brachfeld, A., Gasmi, B., Jones, D.R., Mattar, M., Doman, T., et al.: Timing of CSF-1/CSF-1R signaling blockade is critical to improving responses to CTLA-4 based immunotherapy. Oncoimmunology **5**(7), e1151595 (2016)
165. Ries, C.H., Cannarile, M.A., Hoves, S., Benz, J., Wartha, K., Runza, V., et al.: Targeting tumor-associated macrophages with anti-CSF-1R antibody reveals a strategy for cancer therapy. Cancer Cell **25**(6), 846–859 (2014)
166. Yi, M., Zheng, X., Niu, M., Zhu, S., Ge, H., Wu, K.: Combination strategies with PD-1/PD-L1 blockade: current advances and future directions. Mol. Cancer **21**(1), 28 (2022)
167. Schirrmacher, V.: From chemotherapy to biological therapy: a review of novel concepts to reduce the side effects of systemic cancer treatment (review). Int. J. Oncol. **54**(2), 407–419 (2019)
168. Hellsten, R., Lilljebjorn, L., Johansson, M., Leandersson, K., Bjartell, A.: The STAT3 inhibitor galiellalactone inhibits the generation of MDSC-like monocytes by prostate cancer cells and decreases immunosuppressive and tumorigenic factors. Prostate **79**(14), 1611–1621 (2019)
169. Wu, Y., Yi, M., Niu, M., Mei, Q., Wu, K.: Myeloid-derived suppressor cells: an emerging target for anticancer immunotherapy. Mol. Cancer **21**(1), 184 (2022)
170. Yang, J., Bae, H.: Drug conjugates for targeting regulatory T cells in the tumor microenvironment: guided missiles for cancer treatment. Exp. Mol. Med. **55**(9), 1996–2004 (2023)
171. Gonzalez-Navajas, J.M., Fan, D.D., Yang, S., Yang, F.M., Lozano-Ruiz, B., Shen, L., et al.: The impact of Tregs on the anticancer immunity and the efficacy of immune checkpoint inhibitor therapies. Front. Immunol. **12**, 625783 (2021)
172. Trinh, J.Q., Acosta, C., Easwar, A., Galamaga, R., Tan, A.: Durable and dramatic response to checkpoint inhibition combined with COX-2 inhibitor celecoxib in a patient with p16+ metastatic sinonasal undifferentiated carcinoma: a case study. Cancer Rep. (Hoboken) **7**(1), e1915 (2024)
173. Mahapatra, A.K., Tandon, P.N.: Post-traumatic intradiploic pseudomeningocele in children. Acta Neurochir. (Wien) **100**(3–4), 120–126 (1989)
174. Lugano, R., Ramachandran, M., Dimberg, A.: Tumor angiogenesis: causes, consequences, challenges and opportunities. Cell. Mol. Life Sci. **77**(9), 1745–1770 (2020)
175. Liu, Z.L., Chen, H.H., Zheng, L.L., Sun, L.P., Shi, L.: Angiogenic signaling pathways and anti-angiogenic therapy for cancer. Signal Transduct. Target. Ther. **8**(1), 198 (2023)
176. Nishida, N., Yano, H., Nishida, T., Kamura, T., Kojiro, M.: Angiogenesis in cancer. Vasc. Health Risk Manag. **2**(3), 213–219 (2006)
177. Boere, I.A., Hamberg, P., Sleijfer, S.: It takes two to tango: combinations of conventional cytotoxics with compounds targeting the vascular endothelial growth factor-vascular endothelial growth factor receptor pathway in patients with solid malignancies. Cancer Sci. **101**(1), 7–15 (2010)
178. Jing, X., Yang, F., Shao, C., Wei, K., Xie, M., Shen, H., et al.: Role of hypoxia in cancer therapy by regulating the tumor microenvironment. Mol. Cancer **18**(1), 157 (2019)
179. Wang, B., Zhao, Q., Zhang, Y., Liu, Z., Zheng, Z., Liu, S., et al.: Targeting hypoxia in the tumor microenvironment: a potential strategy to improve cancer immunotherapy. J. Exp. Clin. Cancer Res. **40**(1), 24 (2021)
180. Bader, J.E., Voss, K., Rathmell, J.C.: Targeting metabolism to improve the tumor microenvironment for cancer immunotherapy. Mol. Cell **78**(6), 1019–1033 (2020)
181. Kim, S.W., Kim, I.K., Lee, S.H.: Role of hyperoxic treatment in cancer. Exp. Biol. Med. (Maywood) **245**(10), 851–860 (2020)
182. Chafe, S.C., McDonald, P.C., Saberi, S., Nemirovsky, O., Venkateswaran, G., Burugu, S., et al.: Targeting hypoxia-induced carbonic anhydrase IX enhances immune-checkpoint blockade locally and systemically. Cancer Immunol. Res. **7**(7), 1064–1078 (2019)
183. Hernandez-Gamarra, M., Salgado-Roo, A., Dominguez, E., Goiricelaya Seco, E.M., Veiga-Rua, S., Pedrera-Garbayo, L.F., et al.: CARTAR: a comprehensive web tool for identifying

potential targets in chimeric antigen receptor therapies using TCGA and GTEx data. Brief. Bioinform. **25**(4) (2024)
184. Li, Y.R., Dunn, Z.S., Yu, Y., Li, M., Wang, P., Yang, L.: Advancing cell-based cancer immunotherapy through stem cell engineering. Cell Stem Cell **30**(5), 592–610 (2023)
185. Pickup, M.W., Mouw, J.K., Weaver, V.M.: The extracellular matrix modulates the hallmarks of cancer. EMBO Rep. **15**(12), 1243–1253 (2014)
186. Schaaf, M.B., Garg, A.D., Agostinis, P.: Defining the role of the tumor vasculature in antitumor immunity and immunotherapy. Cell Death Dis. **9**(2), 115 (2018)
187. Yan, T., Zhu, L., Chen, J.: Current advances and challenges in CAR T-cell therapy for solid tumors: tumor-associated antigens and the tumor microenvironment. Exp. Hematol. Oncol. **12**(1), 14 (2023)
188. Chen, Q., Lu, L., Ma, W.: Efficacy, safety, and challenges of CAR T-cells in the treatment of solid tumors. Cancers (Basel) **14**(23) (2022)
189. Xu, J., Wang, Y., Shi, J., Liu, J., Li, Q., Chen, L.: Combination therapy: a feasibility strategy for CAR-T cell therapy in the treatment of solid tumors. Oncol. Lett. **16**(2), 2063–2070 (2018)
190. Sterner, R.C., Sterner, R.M.: CAR-T cell therapy: current limitations and potential strategies. Blood Cancer J. **11**(4), 69 (2021)
191. Kurioka, A., Klenerman, P.: Aging unconventionally: gammadelta T cells, iNKT cells, and MAIT cells in aging. Semin. Immunol. **69**, 101816 (2023)
192. Sanchez-Paulete, A.R., Mateus-Tique, J., Mollaoglu, G., Nielsen, S.R., Marks, A., Lakshmi, A., et al.: Targeting macrophages with CAR T cells delays solid tumor progression and enhances antitumor immunity. Cancer Immunol. Res. **10**(11), 1354–1369 (2022)
193. Luo, W., Napoleon, J.V., Zhang, F., Lee, Y.G., Wang, B., Putt, K.S., et al.: Repolarization of tumor-infiltrating myeloid cells for augmentation of CAR T cell therapies. Front. Immunol. **13**, 816761 (2022)
194. Gao, G., Liao, W., Shu, P., Ma, Q., He, X., Zhang, B., et al.: Targeting sphingosine 1-phosphate receptor 3 inhibits T-cell exhaustion and regulates recruitment of proinflammatory macrophages to improve antitumor efficacy of CAR-T cells against solid tumor. J. Immunother. Cancer **11**(8) (2023)
195. Kankeu Fonkoua, L.A., Sirpilla, O., Sakemura, R., Siegler, E.L., Kenderian, S.S.: CAR T cell therapy and the tumor microenvironment: current challenges and opportunities. Mol. Ther. Oncolytics **25**, 69–77 (2022)
196. Lindau, D., Gielen, P., Kroesen, M., Wesseling, P., Adema, G.J.: The immunosuppressive tumour network: myeloid-derived suppressor cells, regulatory T cells and natural killer T cells. Immunology **138**(2), 105–115 (2013)
197. Dai, X., Zhang, J., Arfuso, F., Chinnathambi, A., Zayed, M.E., Alharbi, S.A., et al.: Targeting TNF-related apoptosis-inducing ligand (TRAIL) receptor by natural products as a potential therapeutic approach for cancer therapy. Exp. Biol. Med. (Maywood) **240**(6), 760–773 (2015)
198. Nalawade, S.A., Shafer, P., Bajgain, P., McKenna, M.K., Ali, A., Kelly, L., et al.: Selectively targeting myeloid-derived suppressor cells through TRAIL receptor 2 to enhance the efficacy of CAR T cell therapy for treatment of breast cancer. J. Immunother. Cancer **9**(11) (2021)
199. Caruana, I., Savoldo, B., Hoyos, V., Weber, G., Liu, H., Kim, E.S., et al.: Heparanase promotes tumor infiltration and antitumor activity of CAR-redirected T lymphocytes. Nat. Med. **21**(5), 524–529 (2015)
200. Jayatilleke, K.M., Hulett, M.D.: Heparanase and the hallmarks of cancer. J. Transl. Med. **18**(1), 453 (2020)
201. Kumagai, S., Togashi, Y., Kamada, T., Sugiyama, E., Nishinakamura, H., Takeuchi, Y., et al.: The PD-1 expression balance between effector and regulatory T cells predicts the clinical efficacy of PD-1 blockade therapies. Nat. Immunol. **21**(11), 1346–1358 (2020)
202. Chen, S., Crabill, G.A., Pritchard, T.S., McMiller, T.L., Wei, P., Pardoll, D.M., et al.: Mechanisms regulating PD-L1 expression on tumor and immune cells. J. Immunother. Cancer **7**(1), 305 (2019)
203. Li, H., Yuan, W., Bin, S., Wu, G., Li, P., Liu, M., et al.: Overcome trastuzumab resistance of breast cancer using anti-HER2 chimeric antigen receptor T cells and PD1 blockade. Am. J. Cancer Res. **10**(2), 688–703 (2020)

204. Li, P., Yang, L., Li, T., Bin, S., Sun, B., Huang, Y., et al.: The third generation anti-HER2 chimeric antigen receptor mouse T cells alone or together with anti-PD1 antibody inhibits the growth of mouse breast tumor cells expressing HER2 in vitro and in immune competent mice. Front. Oncol. **10**, 1143 (2020)
205. Szoor, A., Toth, G., Zsebik, B., Szabo, V., Eshhar, Z., Abken, H., et al.: Trastuzumab derived HER2-specific CARs for the treatment of trastuzumab-resistant breast cancer: CAR T cells penetrate and eradicate tumors that are not accessible to antibodies. Cancer Lett. **484**, 1–8 (2020)
206. Chiocca, E.A., Rabkin, S.D.: Oncolytic viruses and their application to cancer immunotherapy. Cancer Immunol. Res. **2**(4), 295–300 (2014)
207. Shalhout, S.Z., Miller, D.M., Emerick, K.S., Kaufman, H.L.: Therapy with oncolytic viruses: progress and challenges. Nat. Rev. Clin. Oncol. **20**(3), 160–177 (2023)
208. Johnson, D.B., Puzanov, I., Kelley, M.C.: Talimogene laherparepvec (T-VEC) for the treatment of advanced melanoma. Immunotherapy **7**(6), 611–619 (2015)
209. Nakao, S., Arai, Y., Tasaki, M., Yamashita, M., Murakami, R., Kawase, T., et al.: Intratumoral expression of IL-7 and IL-12 using an oncolytic virus increases systemic sensitivity to immune checkpoint blockade. Sci. Transl. Med. **12**(526) (2020)
210. Oronsky, B., Gastman, B., Conley, A.P., Reid, C., Caroen, S., Reid, T.: Oncolytic adenoviruses: the cold war against cancer finally turns hot. Cancers (Basel) **14**(19) (2022)
211. Moehler, M., Goepfert, K., Heinrich, B., Breitbach, C.J., Delic, M., Galle, P.R., et al.: Oncolytic virotherapy as emerging immunotherapeutic modality: potential of parvovirus h-1. Front. Oncol. **4**, 92 (2014)
212. Niculescu, A.G., Grumezescu, A.M.: Novel tumor-targeting nanoparticles for cancer treatment—a review. Int. J. Mol. Sci. **23**(9) (2022)
213. Hristova-Panusheva, K., Xenodochidis, C., Georgieva, M., Krasteva, N.: Nanoparticle-mediated drug delivery systems for precision targeting in oncology. Pharmaceuticals (Basel) **17**(6) (2024)
214. Yuan, H., Guo, H., Luan, X., He, M., Li, F., Burnett, J., et al.: Albumin nanoparticle of paclitaxel (Abraxane) decreases while Taxol increases breast cancer stem cells in treatment of triple negative breast cancer. Mol. Pharm. **17**(7), 2275–2286 (2020)
215. Thambi, T., Park, J.H.: Recent advances in shell-sheddable nanoparticles for cancer therapy. J. Biomed. Nanotechnol. **10**(9), 1841–1862 (2014)
216. Sharifi, M., Cho, W.C., Ansariesfahani, A., Tarharoudi, R., Malekisarvar, H., Sari, S., et al.: An updated review on EPR-based solid tumor targeting nanocarriers for cancer treatment. Cancers (Basel) **14**(12) (2022)
217. Subhan, M.A., Parveen, F., Filipczak, N., Yalamarty, S.S.K., Torchilin, V.P.: Approaches to improve EPR-based drug delivery for cancer therapy and diagnosis. J. Pers. Med. **13**(3) (2023)
218. Feng, L., Wang, Y., Luo, Z., Huang, Z., Zhang, Y., Guo, K., et al.: Dual stimuli-responsive nanoparticles for controlled release of anticancer and anti-inflammatory drugs combination. Chemistry **23**(39), 9397–9406 (2017)
219. Wu, B., Li, M., Wang, L., Iqbal, Z., Zhu, K., Yang, Y., et al.: Size-transformable nanohybrids with pH/redox/enzymatic sensitivity for anticancer therapy. J. Mater. Chem. B **9**(21), 4319–4328 (2021)
220. Abuwatfa, W.H., Pitt, W.G., Husseini, G.A.: Scaffold-based 3D cell culture models in cancer research. J. Biomed. Sci. **31**(1), 7 (2024)
221. Lovitt, C.J., Shelper, T.B., Avery, V.M.: Advanced cell culture techniques for cancer drug discovery. Biology (Basel) **3**(2), 345–367 (2014)
222. Chaicharoenaudomrung, N., Kunhorm, P., Noisa, P.: Three-dimensional cell culture systems as an in vitro platform for cancer and stem cell modeling. World J. Stem Cells **11**(12), 1065–1083 (2019)
223. Breslin, S., O'Driscoll, L.: The relevance of using 3D cell cultures, in addition to 2D monolayer cultures, when evaluating breast cancer drug sensitivity and resistance. Oncotarget **7**(29), 45745–45756 (2016)

224. Zhou, Z., Cong, L., Cong, X.: Patient-derived organoids in precision medicine: drug screening, organoid-on-a-chip and living organoid biobank. Front. Oncol. **11**, 762184 (2021)
225. Fernando, K., Kwang, L.G., Lim, J.T.C., Fong, E.L.S.: Hydrogels to engineer tumor microenvironments in vitro. Biomater Sci. **9**(7), 2362–2383 (2021)
226. Dhem, A., Lengele, B.: Morphologic aspects of the central nervous system. J. Pharm. Belg. **41**(4), 239–255 (1986)
227. Gunti, S., Hoke, A.T.K., Vu, K.P., London Jr., N.R.: Organoid and spheroid tumor models: techniques and applications. Cancers (Basel) **13**(4) (2021)
228. Jeong, S.R., Kang, M.: Exploring tumor-immune interactions in co-culture models of T cells and tumor organoids derived from patients. Int. J. Mol. Sci. **24**(19) (2023)
229. Wan, L., Neumann, C.A., LeDuc, P.R.: Tumor-on-a-chip for integrating a 3D tumor microenvironment: chemical and mechanical factors. Lab Chip **20**(5), 873–888 (2020)
230. Tan, M.L., Jenkins-Johnston, N., Huang, S., Schutrum, B., Vadhin, S., Adhikari, A., et al.: Endothelial cells metabolically regulate breast cancer invasion toward a microvessel. APL Bioeng. **7**(4), 046116 (2023)
231. Li, Z., Zheng, W., Wang, H., Cheng, Y., Fang, Y., Wu, F., et al.: Application of animal models in cancer research: recent progress and future prospects. Cancer Manag. Res. **13**, 2455–2475 (2021)
232. Bryda, E.C.: The Mighty Mouse: the impact of rodents on advances in biomedical research. Mo. Med. **110**(3), 207–211 (2013)
233. Kersten, K., de Visser, K.E., van Miltenburg, M.H., Jonkers, J.: Genetically engineered mouse models in oncology research and cancer medicine. EMBO Mol. Med. **9**(2), 137–153 (2017)
234. Lampreht Tratar, U., Horvat, S., Cemazar, M.: Transgenic mouse models in cancer research. Front. Oncol. **8**, 268 (2018)
235. Kostyrko, K., Roman, M., Lee, A.G., Simpson, D.R., Dinh, P.T., Leung, S.G., et al.: UHRF1 is a mediator of KRAS driven oncogenesis in lung adenocarcinoma. Nat. Commun. **14**(1), 3966 (2023)
236. Jin, J., Yoshimura, K., Sewastjanow-Silva, M., Song, S., Ajani, J.A.: Challenges and prospects of patient-derived xenografts for cancer research. Cancers (Basel) **15**(17) (2023)
237. Georges, L.M.C., De Wever, O., Galvan, J.A., Dawson, H., Lugli, A., Demetter, P., et al.: Cell line derived xenograft mouse models are a suitable in vivo model for studying tumor budding in colorectal cancer. Front. Med. (Lausanne) **6**, 139 (2019)
238. Brown, K.M., Xue, A., Mittal, A., Samra, J.S., Smith, R., Hugh, T.J.: Patient-derived xenograft models of colorectal cancer in pre-clinical research: a systematic review. Oncotarget **7**(40), 66212–66225 (2016)
239. Burtin, F., Mullins, C.S., Linnebacher, M.: Mouse models of colorectal cancer: past, present and future perspectives. World J. Gastroenterol. **26**(13), 1394–1426 (2020)
240. Abdolahi, S., Ghazvinian, Z., Muhammadnejad, S., Saleh, M., Asadzadeh Aghdaei, H., Baghaei, K.: Patient-derived xenograft (PDX) models, applications and challenges in cancer research. J. Transl. Med. **20**(1), 206 (2022)
241. Marangoni, E., Vincent-Salomon, A., Auger, N., Degeorges, A., Assayag, F., de Cremoux, P., et al.: A new model of patient tumor-derived breast cancer xenografts for preclinical assays. Clin. Cancer Res. **13**(13), 3989–3998 (2007)
242. Garrido-Laguna, I., Uson, M., Rajeshkumar, N.V., Tan, A.C., de Oliveira, E., Karikari, C., et al.: Tumor engraftment in nude mice and enrichment in stroma-related gene pathways predict poor survival and resistance to gemcitabine in patients with pancreatic cancer. Clin. Cancer Res. **17**(17), 5793–5800 (2011)
243. Lulla, P.D., Mamonkin, M., Brenner, M.K.: Adoptive cell therapy for acute myeloid leukemia and T-cell acute lymphoblastic leukemia. Cancer J. **25**(3), 199–207 (2019)
244. Orlando, E.J., Han, X., Tribouley, C., Wood, P.A., Leary, R.J., Riester, M., et al.: Genetic mechanisms of target antigen loss in CAR19 therapy of acute lymphoblastic leukemia. Nat. Med. **24**(10), 1504–1506 (2018)
245. Pettan-Brewer, C., Morton, J., Coil, R., Hopkins, H., Fatemie, S., Ladiges, W.: B16 melanoma tumor growth is delayed in mice in an age-dependent manner. Pathobiol. Aging Age Relat. Dis. **2** (2012)

246. Taylor, M.A., Hughes, A.M., Walton, J., Coenen-Stass, A.M.L., Magiera, L., Mooney, L., et al.: Longitudinal immune characterization of syngeneic tumor models to enable model selection for immune oncology drug discovery. J. Immunother. Cancer **7**(1), 328 (2019)
247. Shan, L., Flavell, R.A., Herndler-Brandstetter, D.: Development of humanized mouse models for studying human NK cells in health and disease. Methods Mol. Biol. **2463**, 53–66 (2022)
248. Yin, L., Wang, X.J., Chen, D.X., Liu, X.N., Wang, X.J.: Humanized mouse model: a review on preclinical applications for cancer immunotherapy. Am. J. Cancer Res. **10**(12), 4568–4584 (2020)
249. Wege, A.K., Ernst, W., Eckl, J., Frankenberger, B., Vollmann-Zwerenz, A., Mannel, D.N., et al.: Humanized tumor mice—a new model to study and manipulate the immune response in advanced cancer therapy. Int. J. Cancer **129**(9), 2194–2206 (2011)
250. Mhaidly, R., Verhoeyen, E.: Humanized mice are precious tools for preclinical evaluation of CAR T and CAR NK cell therapies. Cancers (Basel) **12**(7) (2020)
251. Dotti, G., Gottschalk, S., Savoldo, B., Brenner, M.K.: Design and development of therapies using chimeric antigen receptor-expressing T cells. Immunol. Rev. **257**(1), 107–126 (2014)
252. Sun, G., Li, Z., Rong, D., Zhang, H., Shi, X., Yang, W., et al.: Single-cell RNA sequencing in cancer: applications, advances, and emerging challenges. Mol. Ther. Oncolytics **21**, 183–206 (2021)
253. Chira, S., Nutu, A., Isacescu, E., Bica, C., Pop, L., Ciocan, C., et al.: Genome editing approaches with CRISPR/Cas9 for cancer treatment: critical appraisal of preclinical and clinical utility, challenges, and future research. Cells **11**(18) (2022)
254. Yu, Y., Zhou, T., Cao, L.: Use and application of organ-on-a-chip platforms in cancer research. J. Cell Commun. Signal. **17**(4), 1163–1179 (2023)
255. Lin, G., Wang, X., Ye, H., Cao, W.: Radiomic models predict tumor microenvironment using artificial intelligence—the novel biomarkers in breast cancer immune microenvironment. Technol. Cancer Res. Treat. **22**, 15330338231218228 (2023)

Functionalized Nanomaterials for Drug Delivery

Noble George, Komal Janiyani, Matthew Ogoe, Anjali Gupta, Kwasi Kantanka Safo, Harjeet Singh, and Shivani R. Pandya

Abstract This chapter explores the role of nanotechnology in improving drug delivery systems through functionalized nanomaterials. These materials address limitations of traditional methods, such as poor solubility and low bioavailability. The chapter reviews various types of functionalized nanomaterials, including liposomes, polymeric nanoparticles, dendrimers, and inorganic nanoparticles, focusing on their synthesis, surface modifications, and mechanisms of action. It discusses both passive and active targeted drug delivery principles, as well as controlled release strategies utilizing stimuli-responsive systems. Additionally, the chapter examines the translational potential of these nanocarriers for treating complex diseases like cancer, cardiovascular disorders, and neurological conditions, while addressing safety, biocompatibility, and regulatory challenges. Recent case studies and emerging research trends are presented to highlight the future directions of functionalized nanomaterials in drug delivery, emphasizing their potential to transform therapeutic strategies and enhance patient outcomes.

Keywords Nanotechnology · Targeted delivery · Controlled release · Biocompatibility · Therapeutic efficacy

N. George · K. Janiyani · M. Ogoe · A. Gupta · K. K. Safo
Research and Development Cell, Parul University, Vadodara, Gujarat, India

H. Singh · S. R. Pandya (✉)
Narnarayan Shastri Institute of Technology-Institute of Forensic Science and Cyber Security, National Forensic Science University, Jetalpur, Ahmedabad, Gujarat, India
e-mail: shivpan02@gmail.com

N. J. Joshi and G. Patel (eds.), *Nanoparticle Delivery in Cancer Care*,
Series in BioEngineering, https://doi.org/10.1007/978-981-95-0457-2_3

1 Mechanisms of Drug Delivery

1.1 Overview of Drug Delivery Systems

An active pharmaceutical ingredient (API) is defined by the FDA as a substance included in official pharmacopoeias, intended for the diagnosis, treatment, mitigation, or prevention of disease. Drug delivery systems (DDS) encompass various techniques designed to selectively enhance drug concentration in targeted areas of the body while minimizing exposure in non-target regions. The primary objective of DDS is to confine and direct therapeutic agents to diseased tissues, ensuring a safe and effective interaction. Dosage forms comprise APIs, which provide therapeutic effects, and excipients, which are non-drug components that facilitate drug delivery and enhance efficacy [1].

Drug delivery systems (DDS) enhance the administration and effectiveness of pharmaceutical compounds, including antibodies, peptides, and vaccines. While oral pills offer convenience, they struggle with larger molecules, and injections effectively deliver macromolecules but are invasive. Research is exploring alternative routes like pulmonary, transdermal, ocular, and nasal delivery, leading to innovative DDS in clinical trials. Successful translation of these systems requires safety, efficacy, ease of use, and manufacturability. Key DDS categories—microsphere-depots, tumor-targeting nanoparticles, transdermal patches, advanced oral pills, inhalers, implants, and antibody–drug conjugates—demonstrate significant clinical activity and inform future advancements in drug delivery.

1.2 The Role of Nanotechnology in Drug Delivery Enhancement

Nanotechnology has become a vital technology in recent decades, significantly impacting human health. Operating at the nanoscale, it influences essential biological processes such as cell signaling, metabolism, energy production, and nutrient transport. Its potential to innovate therapeutic solutions at the molecular level is increasingly recognized in biology and medicine.

Nanomedicine, a key application of nanotechnology, focuses on developing nanoscale medicinal tools to enhance healthcare systems. This field has advanced significantly, leading to various nanoscale materials with notable clinical benefits. Current research aims to optimize the synthesis of nanomaterials such as polymers, micelles, dendrimers, liposomes, emulsions, nanocapsules, and metal nanoparticles [2]. Applications of nanomedicine include imaging, disease diagnosis, tissue engineering, and precise drug delivery systems that effectively target drugs, minimizing off-target effects and toxicities [3]. This approach improves treatment outcomes for severe diseases like cancer and cardiovascular disorders. As of now, 51 nanomedicines have been approved by the US Food and Drug Administration

(FDA), with 77 additional products in pre-clinical stages, of which around 40% are in clinical trials. The majority of approved formulations are based on polymeric liposomal and nanocrystal technologies, while ongoing research is exploring more efficient materials, including micelles and protein-based nanoparticles [4].

2 Functionalization of Nanomaterials

2.1 Surface Modification Techniques

Nanoparticles designed for biomedical applications must possess several critical attributes, including biocompatibility, non-toxicity, and the ability to evade detection by the immune system while minimizing side effects [5]. To achieve these characteristics, it is recommended that the particle size is maintained at less than 100 nm, with surface modifications using hydrophilic polymers or other functional groups. Functionalization of NPs altogether influences the interaction with surrounding environments, the dispersion, the colloidal stability and controlled assembly [6]. Surface functionalization of biomolecules can offer great cellular internalisation capacity, effective intracellular delivery, silent characteristics and general biocompatibility for biomedical applications. Surface modification offers selective identification in biosensing and targeted distribution in biomedicine [7].

Nanoparticles can bypass the reticuloendothelial system for prolonged bloodstream circulation and enhanced targeting of specific tissues. Effective site-specific delivery requires surface modification with functional ligands that bind to disease sites, improving targeting precision. This functionalization utilizes diverse ligands and strategies, including surfactants, small molecules, polymers, dendrimers, and biomolecules. Many studies have investigated these techniques to refine synthetic procedures for nanoparticles [8]. These tactics are numerous and include the following:

A. Physisorption (grinding and mixing) and chemisorption (such as adding thiol to the surface of NPs).
B. Positively charged NPs attract anionic molecules by electrostatic interactions, or vice versa.
C. The functional groups of NPs are attached to other molecules through conjugation chemistry in covalent contacts.
D. Weak interactions between receptors and ligands are the foundation of non-covalent or supramolecular affinity.
E. Integrating heteroatoms and defect engineering are two aspects of intrinsic surface engineering.

2.1.1 Covalent Bonding

Covalent bonding is a robust technique that involves the formation of stable covalent bonds between the nanoparticle surface and functional groups or ligands. This method is particularly advantageous because it provides a strong and permanent attachment, ensuring that the functionalized molecules remain bound to the nanoparticle during circulation in the body. Biomolecules, polymers, inorganic materials, and small organic molecules bind to the surface of nanomaterials to provide a variety of useful characteristics, such as improved dispersion and increased colloidal stability [7]. Common nanomaterials such as metal oxide nanoparticles, two-dimensional nanomaterials, and nano clays are frequently modified by covalent functionalization in order to suit their intended uses best.

Hetero bifunctional cross-linker molecules have been used to development of numerous functionalization procedures. Covalent functionalization is most often accomplished with organofunctional alkoxysilanes, such as (3-mercaptopropyl)-trimethoxysilane (MPTMS), (3-aminopropyl)-Triethoxysilane (APTES), (3-glycidoxypropyl)-dimethyl-ethoxy silane (GPMES), *N*-hydroxy succinimide (NHS), glutaraldehyde (GA) and 1-ethyl-3-(3 dimethylaminopropyl)carbodiimide (EDC) chemistry [9]. These techniques can be used to functionalize oligonucleotides, peptides, proteins, and many nanoparticles, on the activated surfaces through covalent interactions. Pristine nanoparticles have unique characteristics and uses, while functionalized nanomaterials offer new opportunities for changing their structures and functionalities.

2.1.2 Noncovalent Functionalization

Molecular attraction and repulsion forces lead to noncovalent interactions and functionalization. Noncovalent interactions are prevalent in both synthetic and natural systems, allowing for effortless discovery and detection [10]. Noncovalent interactions include hydrogen bonding, p-p interactions, polymer wrapping, weak contacts (van der Waals), and electron donor/acceptor ligand complexes. Noncovalent functionalization is reversible and kinetically labile, creating a self-correcting system. These interactions enable systems to respond well to external stimuli, such as pH and temperature [11].

Solvents including phenylethyl alcohol, *N*-methyl pyrrolidone (NMP), and *N,N*-dimethylformamide may effectively disperse 2D nanomaterials such as CNTs, graphene, and h-BN [12]. Noncovalent interactions do not disrupt sp^2 carbon networks, resulting in nanomaterials with novel properties while keeping their intrinsic ones. Noncovalent functionalization enhances nanomaterials' reactivity, biocompatibility, dispersibility, catalytic activity, binding capacity, and sensing properties [13, 14].

2.2 *Impact of Functionalization on Drug Loading and Release Profiles*

Functionalization, the process of changing the surface properties of drug carriers by adding certain functional groups, has a substantial impact on drug loading and release profiles in drug delivery systems. It improves medication compatibility, loading capacity, and therapeutic agent release control by altering carrier surface chemistry [15]. Functionalization is essential for increasing the carrier's drug-loading capacity. The level of drug loading is determined by a variety of parameters, including carrier surface characteristics, drug solubility, and drug-carrier interaction [16]. Functional groups such as carboxyl, amino, and hydroxyl groups can be added to the carrier surface, creating particular binding sites for medicinal molecules. For instance, the addition of hydrophilic or hydrophobic groups can enhance the carrier's compatibility with hydrophilic or hydrophobic medicines, respectively [17]. This can result in porous structures or enhance the surface area of carriers, such as mesoporous silica nanoparticles. This increased surface area creates more binding sites, which improves medication loading. Functional groups can provide the carrier surface with a specific charge, allowing for electrostatic interactions with oppositely charged drug molecules [18]. This is especially useful for ionic medications, as electrostatic forces can improve the binding strength and stability of the drug-carrier complex, resulting in increased loading efficiency. Covalent bonding offers a strong method of drug attachment, which might be beneficial for medications that are prone to premature release or degradation [19].

The critical for delivering regulated, prolonged, and targeted drug release. It will affect drug release kinetics, release mechanism, and site-specific delivery by changing the surface chemistry. Functionalized carriers can be designed to release medications at a specific rate. It can stabilise drug-carrier interactions and prevent premature release [20]. Hydrophobic functionalization, for example, can improve the interaction between hydrophobic medicines and carriers, lowering the risk of early release in non-target locations. Furthermore, covalent functionalization ensures that the medication remains attached to the carrier until it reaches the target location, where it can be released through enzyme cleavage or environmental changes [21].

3 Types of Functionalized Nanomaterials

3.1 *Liposomes: Structure, Properties, and Applications*

A liposome is a microscopic vesicle that encapsulates a fluid core, characterized by a phospholipid bilayer [22]. The thickness of this lipid bilayer typically ranges from 3 to 6 nm, while the diameter of the resulting liposomes can vary between 50 nm and 50 μm [23]. Liposomes enhance the delivery of hydrophilic and lipophilic drugs due to the amphipathic properties of their constituent lipids [24]. Their characteristics, including low intrinsic toxicity, biodegradability, and lack of immunogenicity, render liposomes highly suitable as carriers in contemporary drug delivery systems. These small sac-like structures can encapsulate drugs, enabling targeted delivery throughout the body. Consequently, one of the primary applications of liposomes is in drug delivery [25].

Different applications require liposomes with specific properties. To achieve liposomes with distinct and desirable characteristics, superficial modifications can be employed [26]. The range of applications for liposomes as carriers in advanced drug delivery systems has significantly expanded through surface modification and the introduction of new features. Figure 1 illustrates the classification of liposomes and the techniques for surface modification used in each category. The attachment of polymers or macromolecules to the liposome surface can occur through chemical bonds (covalent bonding) or physical interactions (such as van der Waals forces or hydrophobic interactions), depending on the nature of the polymers or macromolecules, as well as their surface or terminal functional groups [27].

Liposomes are utilized in drug delivery systems to transport vaccines, enzymes, and medications like insulin and anticancer agents. They protect healthy cells from drug toxicity and minimize exposure in sensitive tissues, such as the liver and kidneys [30]. This method can reduce chemotherapy side effects like nausea and hair loss. However, a significant drawback is the slow release of encapsulated drugs, which may result in insufficient dosages reaching tumor cells and contribute to drug resistance in cancer treatment [31].

Drug stabilization is affected by multidrug resistance (MDR) transporters, which pump chemotherapy drugs out of cancer cells, lowering their effective concentration. If drug release occurs too late, only drug-sensitive cells without MDR will be impacted [32]. Thus, the rate of anticancer drug release is critical for overcoming

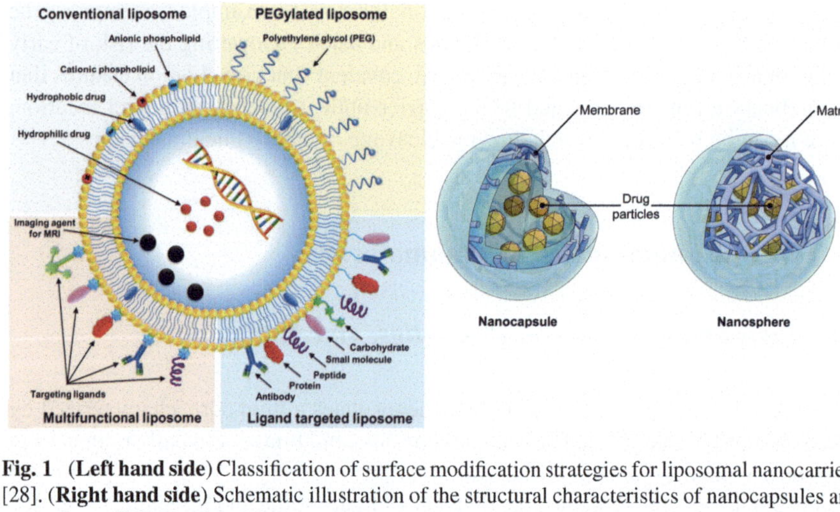

Fig. 1 (**Left hand side**) Classification of surface modification strategies for liposomal nanocarriers [28]. (**Right hand side**) Schematic illustration of the structural characteristics of nanocapsules and nanospheres [29]. Reprinted from open sources

drug stability in cancer populations. While faster release can enhance treatment efficacy, it may compromise safety. Controlling the temporal and spatial release rates of encapsulated drugs remains a significant challenge in liposomal and other nanocarrier systems [33]. Additionally, liposomes can effectively treat diseases that impair the immune system's phagocytic functions by accumulating in phagocytes recognized as foreign invaders.

Certain liposomes are used to deliver anti-HIV medications by targeting human T cells. These liposomes are recognized as foreign by the blood endothelial system, leading to their phagocytosis by immune cells, which then release the encapsulated drugs internally. In gene therapy, liposomes can transport intact genes into cells for repairing defective genes, facilitating DNA delivery to specific tissues [34]. Liposomes serve as drug carriers for various applications, including enzymes, anticancer drugs, antifungals, antiparasitics, antibacterials, antivirals, and nucleic acids. They are also used in diagnostic imaging, vaccine adjuvants, cosmetic formulations, and ophthalmology. Currently, 11 liposomal drug formulations have received regulatory approval, with six more in clinical trials [35]. Notably, the Iranian drug Sina Doxosome, developed for treating ovarian cancer, breast cancer, multiple myeloma, and Kaposi's sarcoma, is in the final stages of clinical trials. This innovative liposomal delivery system has demonstrated increased drug efficacy across various treatments [36].

3.2 *Polymeric Nanoparticles: Synthesis and Functionalization Strategies*

Polymeric nanoparticles have gained significant attention due to their unique properties and small size. As drug carriers, they offer several advantages, including controlled release, protection of bioactive compounds from environmental degradation, and enhanced bioavailability and therapeutic index [37].

Nanocapsules feature an oily core containing the drug, surrounded by a polymeric shell that regulates the release of the drug from the core. In contrast, nanospheres consist of a continuous polymeric matrix, where the drug can be either encapsulated within or adsorbed onto the surface [38]. These two categories of polymeric NPs are classified as reservoir systems (nanocapsules) and matrix systems (nanospheres) [39], as illustrated in Fig. 1. Table 1 lists examples of drugs and bioactive substances incorporated into polymeric nanoparticles.

Polymeric nanoparticles are characterized by biocompatibility, biodegradability, and non-toxicity. They are considered safe for human use and can enhance the bioavailability of encapsulated drugs [48]. These nanocarriers are notable for their excellent stability and capacity to encapsulate significant amounts of therapeutic agents. Recent studies indicate that both biodegradable and non-biodegradable polymeric nanocarriers are suitable for oral drug delivery. For instance, curcumin-loaded polymeric nanoparticles demonstrated a 5.6-fold increase in oral bioavailability

Table 1 Examples of therapeutic agents and bioactive compounds encapsulated in polymeric nanoparticles

Drug/bioactive	Type of polymers	Type of polymeric nanoparticles	Applications	Refs.
Paclitaxel (PTX)	PLGA-PEG	Nanocapsules; PTX-loaded PLGA-PEG NCs	Breast, pancreatic, ovarian, and brain cancers	[40]
Paclitaxel (PTX)	PCL-PEG-PCL	Nanocapsules; PTX-loaded PCL-PEG-PCL NCs	Lung cancers in combination with chrono-modulated chemotherapy	[40]
Pegademase bovine	PEG	Colloidal nanocapsules; pegademase bovine-loaded PEG NCs	Severe combined immunodeficiency disease	[41]
Curcumin (Cur)	F108: PEG-PPG-PEG	Colloidal nanocapsules; Cur-loaded PEG-PPG-PEG NCs	Anticancer	[42]
Curcumin (Cur)	PLGA	Nanocapsules; Cur-loaded PLGA NCs	Antibacterial activity, pancreatic cancer	[43]
Ciprofloxacin	PLGA, PCL	Nanocapsules; ciprofloxacin-loaded PLGA NCs	In situ tissue regeneration and accelerated healing, anti-inflammatory activity	[44]
Fenofibrate (FF)	Anionic copolymers based on methacrylic acid and methyl methacrylate	Nanocapsules; FF-loaded Eudragit L 100 NCs, obtained by nanoprecipitation method	Undefined oral delivery	[45]
Coumarin-6 (C-6)	PCL, PLA, PLGA	Nanospheres; C-6-loaded polymeric core–shell NPs prepared by spontaneous emulsification solvent evaporation method	Drug delivery, theranostics, or bioimaging	[46]
Rapamycin	PLGA	Nanospheres; rapamycin-loaded polysorbate 80-coated PLGA NPs	Anti-glioma activity	[47]

compared to pure curcumin, while silymarin delivered via polymeric nanoemulsion showed a four-fold enhancement in efficacy compared to conventional silymarin suspension [49].

Natural polymers such as chitosan, dextran, heparin, and hyaluronan are widely used in drug delivery due to their biodegradable, biocompatible, and mucoadhesive properties. The concept of biomimetic design has emerged to enhance nanocarrier development by incorporating suitable ligands onto carbon nanotubes or fabricating chitosan nanoparticles [50].

Polymeric nanogels are notable for their minimal toxicity, serum stability, and stimuli responsiveness, making them suitable for applications in biosensors, drug delivery systems, tissue engineering, and biomimetic materials [48]. Key properties such as size, surface charge, and hydrophilicity significantly influence the applications of polymeric nanoparticles. Functionalization of polymers is facilitated by the presence of alcohols, amines, and thiols, although the use of isocyanates is limited due to their toxicity and instability in mixtures.

3.3 Inorganic Nanoparticles: Types, Functionalization, and Therapeutic Uses

Inorganic nanoparticles offer superior stability and hydrophilicity compared to organic nanomaterials. They possess remarkable intrinsic physicochemical properties, including magnetic, thermal, optical, and catalytic functionalities, which allow for the integration of multiple dopants for multifunctionality. Inorganic nanoparticles also exhibit enhanced drug loading capacity, outstanding stability, and adjustable degradation rates [51]. These nanoparticles are broadly classified into metal-based and metal oxide-based categories. Metal-based nanoparticles are derived from metals like Fe, Cu, and Ag, while metal oxide nanoparticles, such as ZnO and iron oxides, are synthesized to enhance the properties of their metal counterparts. This review will explore recent advancements in the application of various inorganic nanoparticles, including gold, silver, graphene-based materials, iron oxide, zinc oxide, hydroxyapatite, and cerium oxide, for efficient drug delivery (Table 2).

4 Targeted Drug Delivery Mechanisms

Delivering therapeutic drugs to certain cells or tissues to achieve optimal therapeutic efficacy while simultaneously lessening their unwanted side effects on normal cells or tissues is the motive of targeted drug delivery [64]. Targeted drug delivery aims to improve treatment outcomes by overcoming treatment constraints and achieving precision medicine, efficacy enhancement, elimination of adverse effects, enhanced retention and circulation time of therapeutic agents [65] discuss in detail about the

Table 2 Inventory of diverse inorganic nanoparticles encapsulating various pharmacological agents

Nanoparticle type	Additional materials	Loaded drug(s)	Application	Refs.
Carbon-based nanoparticles	Polydopamine, amino-containing copolymers (poly(PEGMA-co-NAPAM))	Doxorubicin (DOX)	Intracellular drug delivery	[52]
	Galactosylated chitosan	Doxorubicin (DOX)	Drug delivery for tumor therapy	[53]
	Polycaprolactone (PCL), gelatin	Bortezomib (BTZ) and temozolomide (TMZ)	Drug delivery for glioblastoma	[54]
	Porphyrin	–	Delivery across the blood–brain barrier (BBB)	[55]
	Cyclophosphamide	–	Anticancer drug delivery	[56]
Cerium oxide nanoparticles	Polyethylenimine (PEI)	pDNA	Gene and drug delivery to cancer cells	[57]
Mesoporous silica nanoparticles	Glutathione	Doxorubicin (DOX)	Controlled drug delivery for breast cancer cells	[58]
	Triphenylphosphine (TPP)/ hyaluronic acid (HA)	Doxorubicin (DOX)	Targeted, enzyme-responsive drug delivery focusing on mitochondria and tumor cells	[59]
	Poly(amidoamine) (PAMAM)	Doxorubicin (DOX)	Mucoadhesive drug delivery system for bladder cancer therapy	[60]
Calcium carbonate ($CaCO_3$) nanoparticles	Poly(acrylic acid) (PAA)	Doxorubicin (DOX)	pH-responsive drug delivery	[61]
	PEG	Various protein-drug combinations	Targeted co-delivery of proteins and drugs	[62]
Gold nanoparticles	PEG	Varlitinib	Targeted drug delivery for pancreatic cancer cells	[63]

two delivery mechanisms for drugs; passive and active mechanisms of targeted drug delivery in this section.

4.1 Passive Drug Delivery Mechanisms

Drugs can be delivered to particular locations throughout the body passively—that is, without the need of targeting ligands. This methodology is predicated on the physical and chemical attributes of the drug carrier, including its dimensions and surface features, as well as the distinct features of diseased tissues and the tissue. The concept of passive targeting relies on nanoparticles formulation and modifications that can take advantage of physiological variations between healthy and sick tissues—like tumors—without the need for ligand interactions [66]. The mechanisms exploited in passive targeting drug delivery relies of the concept of Enhanced Permeability and Retention (EPR) Effect [67]. The hyper-permeable convoluted vasculature of cancer tissue, signified by compromised lymphatic drainage causes nanoparticles and/or drug carriers to easily extravasate into the tumor microenvironment and prevents them from being cleared as a result of the EPR effect [68]. Covalently attaching PEG to the nanoparticles has additionally been employed to accomplish passive targeting an approach that has been shown to circumvent desorption and aggregation of the attached ligand [68]. Another element that supports passive targeting is the milieu around targeted tissues [69]. Nano-formulations have targeted the acidic microenvironment of tumors for optimal drug efficiency in cancer treatment. Designed to remain at a pH of 7.4, liposomes break down in the acidic atmosphere of tumor cells, releasing the active medication [70]. Dynamic Covalent Chemistry (DCC) approach allows for reversible interactions between drug carriers and their payload, facilitating selective drug release in target tissues [71]. Doxorubicin delivery to target site based on DCC using DOX-loaded drug-sponge showed improved detoxification with cell lines and tumor inhibition in xenograft mice [72]. Figure 2 demonstrates various mechanisms of passive and active targeting in drug delivery of functionalized nanomaterials.

4.2 Active Targeting Mechanisms

In lieu of passive targeting, active targeting maximizes selectivity by means of receptor-mediated endocytosis, giving rise to greater accuracy in drug delivery to cells with disease, such as cancer cells [66]. The active targeting technique is predicated on the surface decorating of nanoparticles with molecules specific to target cells including as aptamers, antibodies, and receptors that are overexpressed in cancerous cells [67]. This interplay guarantees a more targeted delivery of the medication to the afflicted cells, enhancing absorption and raising the therapeutic index of the drug.

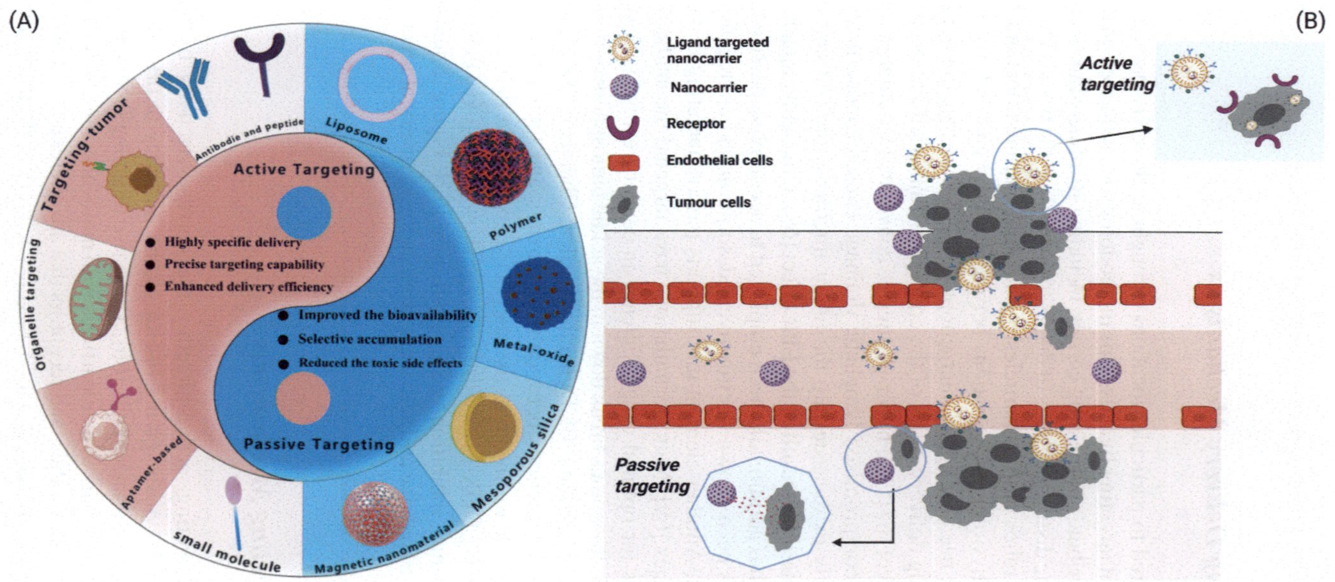

Fig. 2 **A** Diagrammatic representation of active and passive drug delivery systems with NP. Reproduced from open-source [73]. **B** Illustrative representation of active and passive delivery of drugs using nanocarriers

Because of recent advancements in molecular biology and genetic engineering, the usage of antibodies as primary targeting entities has become commonplace [74].

A. *Antibody-Functionalized Nanoparticles*

Wide variety of antibodies with excellent binding affinities and high specificity are being exploited as a mechanism of active drug delivery for targeted therapy thanks to recent advancements in antibody research. In order to reduce immunological reactions from full antibodies, antibody fragments are often used in NP functionalization for active drug delivery [75]. The best antigen must be identified and deliberately overexpressed on target cells rather than normal cells in order to optimize the usage of antibodies for targeted therapy. A number of antibody–drug couples (ADC) have been registered and approved; cetuximab for advanced therapy in colorectal cancer, breast cancer therapy with trastuzumab [76].

B. *Peptide-Functionalized Nanoparticles*

Drugs are associated with peptides in the form of short peptide sequences, which consist of a chain of three to fifteen amino acids connected by peptide bonds [77]. Due to peptides' numerous benefits, including their tiny size, low immunogenicity, affordability, and ease of synthesis, they have emerged as very desirable drug targeting molecules [78]. The first cell-targeting polypeptide discovered was the RGD, (arginine, glycine, and aspartic acid) that exhibited great affinity for $\alpha_v\beta_3$ integrins excessively expressed by both angiogenic endothelial as well as tumor cells [77]. The trans-activator of transcription (TAT) peptide has been coupled to silver nanoparticles to enhance the promotion of apoptosis within breast cancer cells [79].

C. *Carbohydrate-Conjugated Nanoparticles*

In recent times, targeted administration has made use of carbohydrate-linked polymer-drug conjugates due to their selective and reversible interaction with endogenous lectins found on the surface of tumor cells [80]. Carbohydrate (glycolipids, glycoproteins, and polysaccharides)—lectin specific binding is followed by a lectin mediated internalization into the target specific cells/tissues minimizing non-specific binding and unwarranted side effect of drug [75]. Doxorubicin and fucoidan have been conjugated covalently into a P-selectin-targeted drug delivery vesicle for breast cancer treatment [79].

D. *Aptamer-Based Targeted Nanoparticles*

Aptamers, or single-stranded DNA or RNA, have several advantages over antibodies, including more effective screening, ease of modification, and greater chemical and thermal stability lower molecular weight and enhanced tissue permeability [81]. Nucleic acid aptamer NP functionalization have been involved in targeting diverse proteins, sugars, and phospholipids as a delivery approach [79]. RNA aptamers specific for CD133 have been utilized to invade CD133 expressing tumors [82].

In targeted medication delivery systems, macromolecules like proteins, polysaccharides, and nucleic acids are essential, functioning as parts of the nanocarriers themselves as well as the active targeting ligands. Macromolecules therefore improve the functional characteristics and targeting specificity of drug delivery systems, resulting in increased therapeutic efficacy and fewer side effects.

5 Controlled Release Mechanisms

5.1 *Drug Release Kinetics: Mathematics Models and Approaches*

In order to optimize the design of a therapeutic agents and obtain data on the effectiveness of different release models, mathematical modeling has been used to assess the kinematics of drug. These models provide a comprehensive framework for understanding and predicting drug release kinetics, aiding in the design and optimization of pharmaceutical formulations [83]. We briefly describe the various mathematical models that describe how drugs are released and underscores the importance of these models.

5.1.1 Zero Order Kinetics

This model describes medications dissolving from dose forms with slow and non-disaggregating release [83]. Drug release is constant over time, independent of concentration.

$$Q_t = Q_0 + K_{0t} \tag{1}$$

where, Q_t is the amount of drug dissolved in time t, Q_0 is the initial amount of drug in the solution (most times, $Q_0 = 0$) and K_0 is the zero order release constant expressed in units of concentration/time.

5.1.2 First-Order Kinetics

Often used for drugs in porous matrices and describes absorption and elimination processes. This drug release, which adhered to first order kinetics, can be represented by the following equation:

$$\log C = \log C_0 n K t / 2.303 \tag{2}$$

where C_0 is the initial concentration of drug, K is the first order rate constant, and t is the time.

5.1.3 Higuchi Model

This model assumes drug diffusion is the primary mechanism of release and is applicable to planar systems. It is particularly useful for predicting release from matrix systems where the drug solubility is much lower than its concentration in the matrix [84].

$$\mathrm{ft} = Q = \sqrt{}\mathrm{D}\delta/\tau(2\mathrm{C} - \delta\mathrm{Cs})t \qquad (3)$$

where Q is the quantity of drug released in time t per unit area, δ is the porosity and τ is the tortuisity of the matrix, and D is the drug molecule's diffusion coefficient in the solvent. The drug's starting concentration is denoted by A, its solubility in the matrix media by Cs.

5.1.4 Hixson-Crowell Model

This model accounts for changes in surface area and diameter of particles as they dissolve useful for describing the release from systems where particle size decreases over time [84].

$$\mathrm{W}_0^{1/3} - W_{\mathrm{t}}^{1/3} = \kappa t \qquad (4)$$

where κ (kappa) is a constant including the surface-volume relation, W_0 is the starting amount of medication in the pharmaceutical dosage form, and W_t is the amount of drug left in the pharmaceutical dosage form at time t.

5.1.5 Korsmeyer-Peppas Model

This model is used for systems where drug release is governed by both diffusion and erosion processes.

$$\mathrm{M_t/M_\infty = Kt^n} \qquad (5)$$

where n is the release exponent, k is the release rate constant, and M_t/M_∞ is the fraction of medication released at time t.

Apart from the above popular models, drug-release models include: *Baker-Lonsdale Model* which focuses on the release of drugs from swelling and eroding matrices, *Weibull Model* a flexible model that can describe various release kinetics by adjusting parameters useful for characterizing complex release patterns, *Gompertz Model*; A sigmoidal model that can describe growth processes and drug release kinetics Commonly applied in biological and pharmaceutical systems, *Hopfenberg Model*, and *Regression Model* [83].

5.2 Stimuli-Responsive Drug Delivery Systems

Stimuli-responsive drug delivery formulations have improved treatment efficacy and mitigated side effects because they change physically or chemically in reaction to environmental stimuli or particular triggers, releasing their payload at the intended location only in the presence of the desired stimuli [85]. Nanocarriers have been engineered to release their cargoes in response to alterations in pH, temperature, ultrasound, magnetic, light and electrical stimuli among others (Table 3) [86].

The pH of the diseased tissues for instance inflammation, infection and cancer is markedly contrasting to the pH found in healthy tissues [85]. Topotecan (TPT) and Quercetin (QT) combination therapy was assessed using nanocarrier decorated with pH dependent polymers poly(acrylic acid) (PAA)-Chitosan (CS) (PAA-SC) [87]. The release profile of both drugs, assessed under physiological, and acidic pH of the tumor microenvironment and lysosomal pH (pH: 7.4, 6.0 and 5.0 respectively) showed pH dependent release of drug (48 h) at room temp: (pH 7.4: TPT 8.7%, QT 12.03%), (pH 6.0: TPT 46.98%, QT 54.77%) and (pH 5.0: TPT 76.24%, QT 80.72%). The controlled release was attributed to the pH dependent polymer (PAA-CS) coating which undergoes hydrolysis at low pH (pH 6.0 and 5.0) causing dissociation of polymer and subsequent release of drugs (Fig. 3) [87]. Doxorubicin loaded PLGA core, BSA shell NP, functionalized with (ATRAM) acidity-triggered rational membrane peptide demonstrated pH dependent internalization into cancer cells due to protonation of glutamic acid residues of ATRAM (Fig. 4) [88].

The HA-g-DEAP0.40 coupled docetaxel (DTX) loaded liposome has been designed to facilitate the effective release of DTX at a particular pH of 6.5 due to the acidic pH-induced protonation of the DEAP attached to the vesicular bilayers of lipids [98]. Temperature responsive drug release exploits the temperature variations typically found in diseased environments (e.g. tumour or inflammation), where local temperatures can reach around 40–42 °C compared to body temperature (~37 °C) [99]. Strong anticancer efficacy against brain tumor cells has been found with co-loaded multipurpose responsive thermal and magnetic liposomes containing cetuximab (CET) and camptosar (CPT-11) [99].

Among the various stimuli-responsive delivery methods, photodynamic therapy has aided in the development of a successful treatment plan that combines the use of a photosensitizing agent with a light source [86]. Naproxen sodium salt (NAP) was loaded into mesoporous particles (MCM-14) enclosed with photoactive molecule methyl-(2*E*)-3-(4-(triethoxysilyl)-propoxyphenyl)-2-propenoate (CA) capable of reversible dimerization by cycloaddition $[2\pi + 2\pi]$ upon irradiation with UV light of particular wavelength NAP release showed a 68 wt% release upon UV irradiation from 13 wt% (in the absence of UV) attributed to the opening of the pores of MCM-14 (Fig. 5) [100].

Folic acid lipid nanoparticles mediated by photothermal/photo-sonodynamic processes for combined treatment of ovarian cancer has been described [101]. Recent nano-engineering studies have designed carriers that responds to various other stimuli as ultrasound [102], enzymes [103], and Magnetic/electric field [104].

Table 3 Drug release mechanism of various nanocarriers

Nanocarriers	Functionalizing agent	Drug	Drug release mechanism	Refs.
HPG@Fe_3O_4 nanoparticles	Hyper branched glycerol (HPG) and folic acid (FA)	Curcumin	Temperature = 37 °C and at pH = 7.0, HPG@Fe_3O_4 shows 75% release within 8 days but, FA HPG@Fe_3O_4 shows only 45–60% drug release	[89]
Chitosan functionalized MNPs	Chitosan	Telmisartan	pH controlled drug release mechanism at pH = 5.5, 82% within 50 h, but only 25% at pH = 7.4	[90]
Graphene quantum dots based MNPs (Fe_3O_4@PEG@GQD)	PEG-poly ethylene glycol and graphene quantum dots-GQD	Doxorubicin	pH responsive drug release system at pH = 5.0, 85% within 90 h but at pH = 7.4, less than 21% release of drug is noted	[91]
MTX conjugated multifunctional MNPs (MTX-CSC@MNPs)	Coated with chitosan copolymer and conjugated with methotrexate (MTX)	Erlotinib	pH and temperature dependent drug release At 40 °C, pH = 5, and at 37 °C, pH = 7.4, the drug release was higher and fast at pH = 5 than pH = 7 due to swelling behaviour and protonation of NH_2 groups of chitosan	[92]
Superparamagnetic iron oxide nanoparticles (CS-SPIONS)	Chitosan	Epirubicin	Diffusion controlled drug release process at pH = 4.6, 94.06% release within 24 h but at pH = 5.6, 65.03% pH = 6.6 = 30.12% pH = 7.4, only 10.2% drug has been released due to EPI base CS–SPIONS interaction i.e. breakage of amide bond	[93]
		Telozolomide	At pH = 7.4, 48.89% drug release is noted after 12 h but at pH = 6.6, 63.51% pH = 5.6, 57.89% pH = 4.6, 87.68% is observed due to solubility of drug in acidic conditions	
Amino-functionalized Fe_3O_4 magnetic nanospheres modified by using hyperbranched phenylboronic acid	Hyperbranched phenylboronic acid	Doxorubicin	Stimuli triggered drug release mechanism room temperature, at pH = 4.0, 72.55% drug release pH = 5.0, 66.52% pH = 6.0, 50.58% and at pH = 7.0, 35.36% drug release was observed during 6.6 h	[94]
MNP@PEGlyted nanoparticles	PEG	Curcumin	pH controlled drug release mechanism temperature = 37 °C at pH = 5.4, 53.5% drug release is noted within 144 h but at pH = 7.4, only 43.7% is observed	[95]
Magnetic graphene oxide nanocarrier	MUCI aptamer	Paclitaxel (PAC)	pH responsive drug release mechanism Temperature = 37 °C at pH = 2.0, 67.59% drug release is observed but at pH = 5.5, 38.38% and at pH = 7.4, 17.89% is noted due to strong hydrogen bonding interaction among PAC and MGO at alkaline and neutral conditions	[96]
Bovine serum albumin coated iron oxide MNPs	BSA (bovine serum albumin)	Curcumin	At pH = 5.5, 55% drug release during 160 h but at pH = 7.4, only 35% is released	[97]

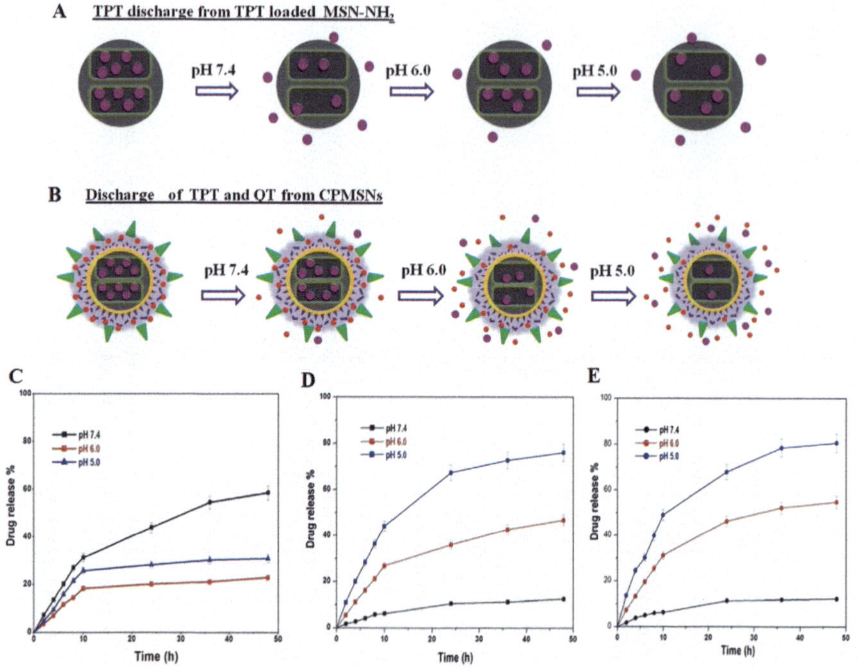

Fig. 3 Illustration of pH-dependent drug release behavior of TPT-loaded nanocarrier (TPT-loaded $MSN-NH_2$ and CPMSNs) investigated at the endo/lysosomal pH (5.0), physiological pH (7.4), and tumor ambient pH (6.0) at 37 °C (**A**, **B**). The release of drugs curves for TPT from $MSNs-NH_2$, TPT released from CPMSNs, and QT released from CPMSNs are shown in (**C**–**E**). Reproduced from open-source [87]

6 Applications in Disease Treatment

Nanomaterials like nanoparticles, nanorods, and nanoshells enhance disease treatment by improving therapeutic efficacy and specificity. In oncology, they serve as drug delivery vehicles, imaging agents, and therapeutic enhancers, enabling targeted delivery and reducing side effects, ultimately improving patient outcomes.

6.1 Nanomaterials in Oncology: Targeted Therapy

Cancer therapy employs various strategies, such as irradiation and chemical inhibition, to target specific cancer types, optimizing outcomes while minimizing harm to healthy tissues. The diversity of treatment modalities highlights the need for personalized regimens based on individual cancer characteristics and patient health. The integration of novel technologies, particularly nanomaterials, is advancing cancer therapy by enhancing the precision and efficacy of treatments across different malignancies.

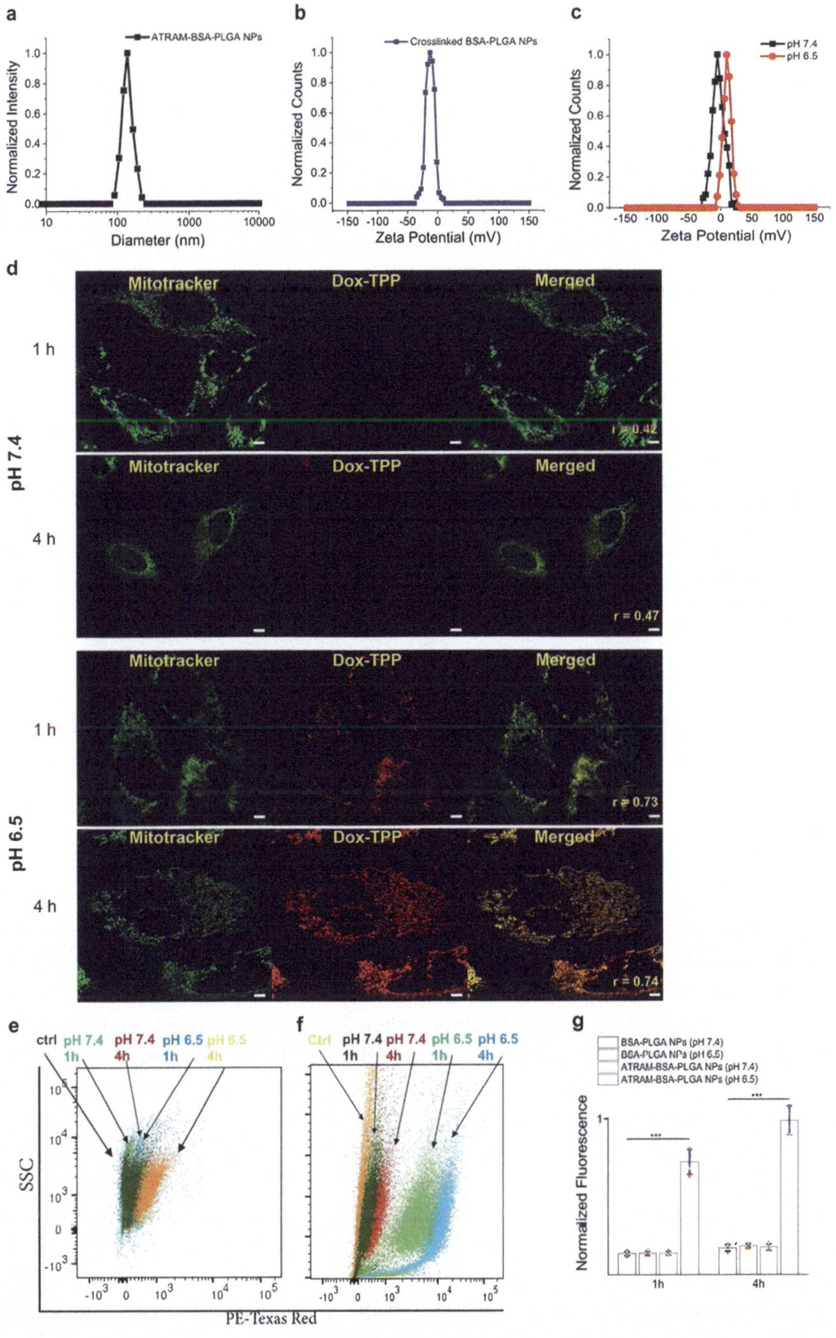
a
ATRAM-BSA-PLGA NPs
Normalized Intensity
Diameter (nm)
b
Crosslinked BSA-PLGA NPs
Normalized Counts
Zeta Potential (mV)
c
pH 7.4
pH 6.5
Normalized Counts
Zeta Potential (mV)
d
Mitotracker
Dox-TPP
Merged
1 h
4 h
pH 7.4
pH 6.5
r = 0.42
r = 0.47
r = 0.73
r = 0.74
e
ctrl pH 7.4 1h pH 7.4 4h pH 6.5 1h pH 6.5 4h
SSC
f
Ctrl pH 7.4 1h pH 7.4 4h pH 6.5 1h pH 6.5 4h
PE-Texas Red
g
BSA-PLGA NPs (pH 7.4)
BSA-PLGA NPs (pH 6.5)
ATRAM-BSA-PLGA NPs (pH 7.4)
ATRAM-BSA-PLGA NPs (pH 6.5)
Normalized Fluorescence
1h
4h

◂**Fig. 4** **a** Analysis of the size of ATRAM-BSA-PLGA nanoparticles (NPs) in aqueous solution. Zeta potential assessments for BSA-PLGA (**b**) and ATRAM-BSA-PLGA (**c**) NPs in solution. **d** Confocal laser scanning microscopy images showing MCF-7 cells incubated with Dox-TPP loaded ATRAM-BSA-PLGA NPs for durations of 1 or 4 h at either physiological or acidic pH levels. The colocalization of the Dox-TPP cargo with mitochondria was quantified using the Pearson's correlation coefficient. The scale bar represents 10 μm. **e–g** Flow cytometric analysis of cellular uptake of Dox-TPP loaded BSA-PLGA and ATRAM-BSA-PLGA NPs in MCF-7 cells: a scatter plot of side scatter (SSC) against fluorescence signal for MCF-7 cells that were either left untreated (ctrl) or treated with Dox-TPP loaded BSA-PLGA (**e**) or ATRAM-BSA-PLGA (**f**) NPs for 1 or 4 h at pH levels of 7.4 or 6.5; **g** quantification of the cellular uptake of BSA-PLGA and ATRAM-BSA-PLGA NPs at various incubation times and pH conditions based on flow cytometry results (n = 4). ***P < 0.001 compared to NPs at pH 7.4. Reproduced from open-source [88]

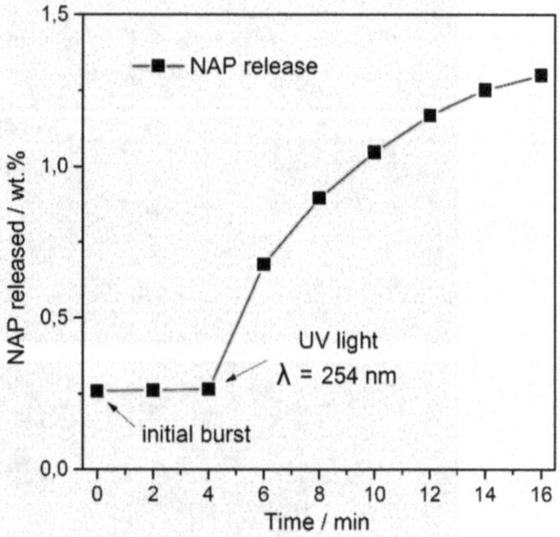

Fig. 5 NAP released from MCM-41-CA/NAP (in vitro) both before (0–4 min) and after (4 from 4 min) being exposed to UV light at λ = 254 nm. Reproduced from open-source [100]

6.1.1 Chemotherapy

Chemotherapeutic drugs are vital in cancer treatment, with doxorubicin being one of the most widely used. Paclitaxel is particularly effective for breast cancer, alongside other agents like cisplatin, tamoxifen, trastuzumab, and docetaxel. Targeted drug delivery methods can significantly enhance the effectiveness of these therapies. Nanoparticle-based carriers improve drug delivery precision, extending circulation half-life and maximizing drug accumulation in tumor tissues [105]. Common nanocarriers include dendrimers, liposomes, and polymeric micelles, with many formulations exploring active targeting and triggered release mechanisms. Approved nanoparticle formulations include pegylated liposomal doxorubicin and nab-paclitaxel, while proteins and peptides are often conjugated to nanoparticle surfaces to enhance selectivity, utilizing serum glycoproteins as targeting ligands [106].

Gold nanoparticles (AuNPs) offer significant advantages for drug delivery in cancer therapy due to their nontoxic, nonimmunogenic nature, high permeability,

SPR, tunable sizes (1–150 nm), and retention capabilities, facilitating drug accumulation at tumor sites [107]. Their negatively charged surfaces facilitate the functionalization with biomolecules such as drugs and targeting ligands. Chen et al. to enhance the efficacy of MTX, developed a formulation of MTX conjugated to AuNPs to improve drug retention in tumor cells. The MTX-AuNP conjugate was successfully formed and demonstrated faster and higher accumulation of MTX in tumor cells compared to free MTX, resulting in significantly increased cytotoxicity across various tumor cell lines. In vivo studies showed that MTX-AuNP effectively suppressed tumor growth in a Lewis lung carcinoma mouse model, while free MTX had no antitumor efficacy [108]. Similarly, doxorubicin (DOX) linked to Au NPs via an acid-labile linker shows improved toxicity against multidrug-resistant breast cancer cells through enhanced uptake and targeted release [109]. Furthermore, conjugating peptide-drug conjugates (PDCs) to Au NPs significantly increases their circulation half-life while retaining cytotoxic effects. Additionally, phytochemicals like kaempferol exhibit increased apoptotic and anti-angiogenic effects when conjugated to Au NPs, addressing challenges such as low specificity and rapid clearance associated with these compounds [110]. Rozalen et al. synthesized MTX silver nanoparticles (AgNPs-MTX) using borohydride and citrate, resulting in nanoparticles with a mean size of 13 nm. The drug loading capacities ranged from 28 to 40%, with in vitro release showing 77–85% of MTX released. AgNPs-MTX exhibited a rapid initial release followed by a plateau, consistent with a Higuchi diffusion model. Flow cytometry demonstrated significant anticancer effects in the colon cancer cell line HTC-116, reducing viable cells to 40%, while effects on lung cancer cells (A-549) were less pronounced [111].

Organic nanoparticles are increasingly employed for chemotherapeutic drug delivery, enhancing drug circulation half-life and tumor accumulation [112]. They play a vital role in combination chemotherapy, facilitating the targeted delivery of various therapeutic agents. For example, polymer-lipid hybrid NPs coencapsulating doxorubicin (DOX) and mitomycin C have shown efficacy against multidrug-resistant breast cancer cells, while micellar formulations delivering paclitaxel, 17-AAG (Triolimus), and rapamycin have been effective in MDA-MB-231 tumor-bearing mice [113].

To address tumor hypoxia and associated chemoresistance, Manli et al. presents MnO_2 NPs for alleviating tumor hypoxia by targeting TAMs and generating oxygen from H_2O_2. Hyaluronic acid (HA) modification reprograms M2 TAMs into pro-inflammatory M1 macrophages, enhancing MnO_2 NP efficacy. Treatment with HA-coated, mannan-conjugated MnO_2 (Man-HA-MnO_2) significantly improved tumor oxygenation, decreased hypoxia-inducible factor-1α (HIF-1α) and vascular endothelial growth factor (VEGF), and, when combined with doxorubicin, more effectively inhibited tumor growth than chemotherapy alone. Additionally, Man-HA-MnO_2 NPs enhanced T1- and T2-MRI imaging for tumor detection [114]. Targeting carriers, such as nanomicelles for curcumin and baicalin, utilize mannose to increase uptake by tumor-associated macrophages, while oligomeric hyaluronic acid can improve cellular uptake. Nanodandelions, due to their small size, effectively penetrate tumor tissue, demonstrating significant antitumor activity with reduced side effects in A549 tumor-bearing mice. The sustained-release characteristics of NPs further enhance chemotherapy effectiveness by maintaining drug concentrations at the tumor site, making them particularly suitable for lung cancer treatment [115].

6.1.2 PDT

Photodynamic therapy (PDT) is a non-invasive treatment that employs molecular oxygen, visible light, and photosensitizers (PS) to selectively destroy cancer cells and pathogenic bacteria through mechanisms such as apoptosis, necrosis, and autophagy [105]. The excited triplet state of the photosensitizer transfers energy to molecular oxygen, producing singlet oxygen, which is highly oxidizing. This process leads to oxidative damage and destruction of cellular structures, while also stimulating the immune system to combat cancer. Furthermore, photodynamic therapy (PDT) enhances its antitumor effects by occluding blood vessels around the tumor (Fig. 6A) [116]. Blassan et al. developed the combination of AuNP conjugated PDT with cannabidiol (CBD), a non-psychoactive compound from *Cannabis sativa.* AuNP-PDT offers enhanced specificity and fewer side effects than conventional methods [117]. As illustrated in Fig. 6B, co-loading photosensitizers (PSs) with antitumor chemotherapeutics in a single delivery system can enhance the efficacy of monotherapy. Co-encapsulation of small-molecule inhibitors with PSs significantly improves PDT outcomes by effectively destroying primary tumors while reducing recurrence and metastasis rates. Additionally, blocking angiogenic molecules or their receptors during PDT may upregulate the expression of vascular endothelial growth factor (VEGF) and cyclooxygenase-2 (COX-2) [118].

Woźniak et al. studied curcumin encapsulated in hydrogenated soy phosphatidylcholine (HSPC) liposomes for PDT. The stable liposomes limited curcumin diffusion and enhanced phototoxicity in melanoma (MugMel2) and squamous cell carcinoma (SCC-25) cells compared to normal keratinocytes (HaCaT). Liposomal curcumin PDT improved the apoptotic-to-necrotic cell ratio and reduced malignant cell motility. These results indicate that liposomal encapsulation enhances curcumin's photosensitizing effects in skin cancers while reducing toxicity to normal cells [119]. Zhang et al. developed phycocyanin-functionalized black phosphorus quantum dots (PC@BPQDs), which exhibit strong phototherapeutic properties due to their optical and photothermal effects. The PC layer enhances stability, reduces plasma protein adsorption, and boosts reactive oxygen species (ROS) release, promoting apoptosis. PC@BPQDs selectively accumulate in tumor tissues, efficiently enter tumor cells via endocytosis, and escape to the cytosol. In vivo, they effectively eliminate tumors with minimal toxicity, highlighting their promising therapeutic potential [120].

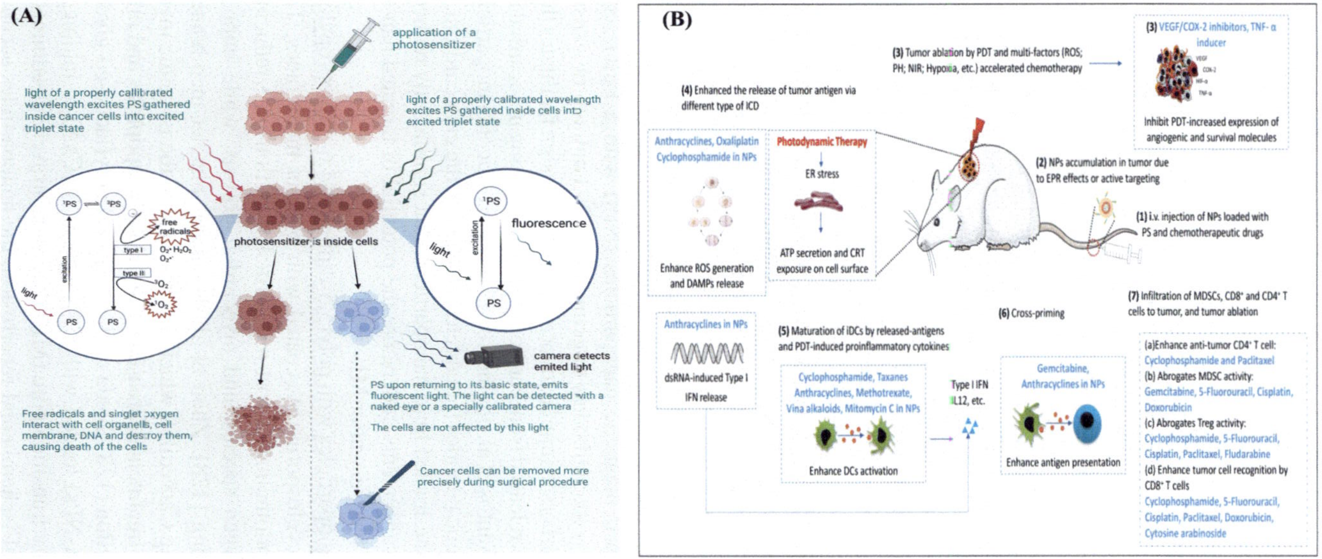

Fig. 6 **A** The mechanism of action of PDT and PDD [116]. **B** The mechanisms of NP-mediated chemo-photodynamic therapy enhance monotherapy indices and synergistically trigger robust antitumor immune responses for anti-primary and metastatic tumors. Reprinted from open sources [118]

6.2 Nanomaterials in Cardiovascular Applications

Current conventional treatments for cardiovascular diseases (CVDs) often require high drug concentrations, leading to systemic toxicity. Targeted drug delivery systems (DDSs), utilizing nanomedicine, offer a promising alternative by selectively accumulating drugs in the heart. These DDSs, composed of biocompatible polymers, lipids, or inorganic compounds, enhance drug stability, solubility, and biodistribution while minimizing degradation and excretion [121]. Unlike cancer treatments, where nanoparticles aim to induce tumor cell death, DDSs for CVDs focus on tissue regeneration and cardiomyocyte proliferation. Nanoparticles need to remain stable in the bloodstream and avoid triggering immune responses, as they typically reach the myocardium within 20–25 min post-administration, leveraging the heart's blood-pumping capacity. This targeted approach holds great potential for improving the safety and efficacy of CVD therapies [122].

Targeting the angiotensin II type 1 receptor (AT1R), which is upregulated in ischemic cardiomyocytes, has been investigated for cardiac-specific therapies. Xue et al. developed AT1 peptide-conjugated nanoparticles (AT1-AMO-1-NPs) to deliver anti-miRNA-1 (AMO-1), effectively reducing apoptosis and infarct size post-MI [123]. To enhance targeting, McGuire et al. used phage display to isolate a tenascin-X-derived peptide that improved localization in cardiomyocytes [124]. Wei et al. created macrophage membrane-coated nanoparticles (PDA@M-NPs) that showed increased affinity for hypoxic cardiomyocytes and provided cardioprotective effects [125].

Liu et al. developed magnetic nanoparticles (NPs) with a Fe_3O_4 core and a silica shell, modified with antibodies targeting the CD63 antigen on exosomes and myosin light chain markers in injured myocardium. This design facilitates the specific targeting of exosomes released from cardiomyocytes following myocardial infarction [126]. Additionally, Santos et al. engineered PEGylated silicon nanoparticles conjugated to atrial natriuretic peptides for targeted delivery to the endocardium, while Yu et al. created adenosine-loaded lipid nanoparticles that significantly reduced infarct size [127]. These advancements underscore the potential of targeted nanomedicine in improving cardiac recovery after MI.

Li et al. created nanoparticles with poly(5,5-dimethyl-4,6-dithiopropylene glycol azelate) (PTK) and cyclosporine A, coated with platelet membranes. These NPs targeted the injured endothelium in post-myocardial infarction mice, reducing inflammation, reactive oxygen species, and myocardial fibrosis, while improving left ventricular remodelling [128]. Zhou et al. developed platelet membrane vesicles (PMVs) encapsulating Carvedilol, which accumulated in the ischemic myocardium of post-MI rats. This treatment decreased cardiomyocyte apoptosis and infarct size, preserving heart function [128]. Ikeda et al. developed poly(lactic-co-glycolic acid) nanoparticles (PLGA-NPs) loaded with Cyclosporin A (CsA) to inhibit the mitochondrial permeability transition pore (MPTP). After intravenous administration in mice with myocardial ischemia and reperfusion, the nanoparticles accumulated in cardiomyocyte mitochondria. CsA protected the heart from ischemia–reperfusion injuries by preserving mitochondrial structure and function, and mitigating ventricular remodelling [129]. Zhang et al. developed lipid-polymeric nanoparticles (NPs) for targeted delivery of Tanshinone IIA (TN). To enhance targeting specificity and therapeutic efficacy, the NPs were conjugated with triphenylphosphonium (TPP), facilitating mitochondrial accumulation due to its high negative membrane potential. Following intravenous injection in post-myocardial infarction (MI) rats, TN-TPP-NPs demonstrated increased uptake and bioavailability in the heart compared to free TN and unmodified TN-NPs, significantly reducing infarct size [130].

6.3 Nanomaterials in Neurological Applications

There is a pressing need for therapeutic strategies capable of overcoming the blood–brain barrier (BBB) and enhancing efficacy. Researchers are actively developing delivery methods to address this challenge. Various strategies have emerged, with nano-based approaches at the forefront of advancements in therapeutic delivery systems.

Alzheimer's disease (AD) is a leading neurodegenerative disorder, constituting over 80% of dementia cases worldwide and resulting in cognitive and functional decline. Lipoic acid, an antioxidant and anti-inflammatory compound found in mitochondria, exemplifies the potential of nanoparticles (NPs) to enhance AD treatment. These NPs are designed to be safe, biodegradable, and target-specific, effectively delivering medications and neuroprotective agents to the brain. The intranasal route allows for direct delivery by bypassing the BBB, with receptor-mediated endocytosis being the preferred transport mechanism [131]. Figure 7 depicts the mechanisms of action for various nanoparticles in neurons impacted by AD, classified into three categories [132]:

A. *Symptomatic Treatment*: NPs loaded with acetylcholinesterase (AchE) inhibitors and NMDA antagonists target specific sites, releasing their cargo to inhibit cholinergic transmission and block NMDA channels, thereby alleviating AD symptoms.
B. *Disease-Modifying Therapies*: NPs containing drugs targeting AD hallmarks, such as amyloid-beta (Aβ) and tau proteins, act as therapeutics that may modify disease progression. Anti-tau drugs prevent tau aggregation, while NP-conjugated anti-Aβ antibodies facilitate the clearance of Aβ fibrils from circulation, reducing plaque accumulation.
C. *Diagnostic Tools*: Modified NPs with conjugated capture antibodies and functionalized fluorescent markers enable visualization of Aβ fibrils and tau proteins, significantly improving diagnostic accuracy for AD.

Nanotechnology in diagnostics focuses on detecting amyloid-beta (Aβ) peptides, essential for developing Alzheimer's disease (AD) biomarkers. Polymeric nanoparticles encapsulating radio-labeled ^{125}I-clioquinol (CQ) enhance drug transport to

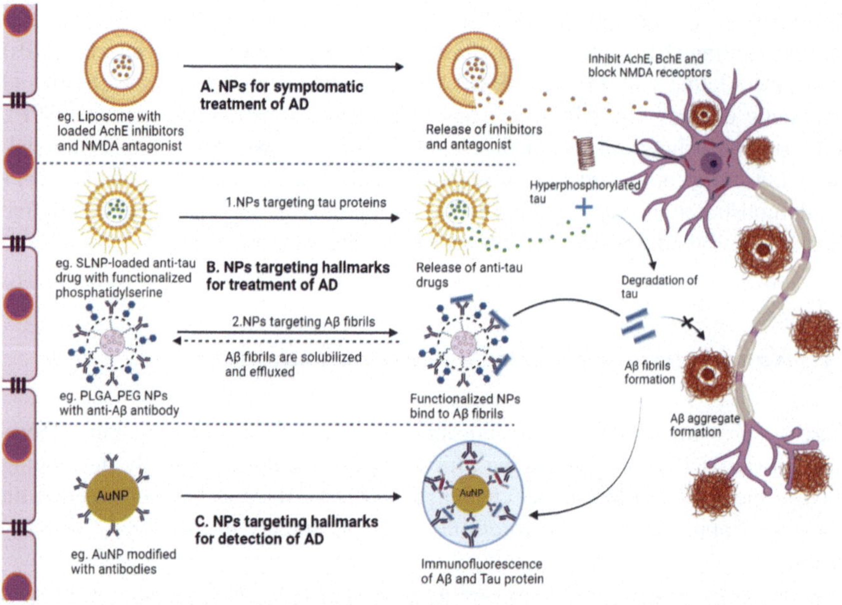

Fig. 7 Mechanisms of action of nanoparticles (NPs) in neurons affected by Alzheimer's disease (AD) following traversal of the blood–brain barrier (BBB). **A** A liposome containing acetylcholinesterase (AchE) inhibitors addresses cholinergic system dysfunction. **B** (1) Solid lipid nanoparticles (SLNPs) loaded with anti-tau therapeutics, functionalized with phosphatidylserine, selectively target hyper-phosphorylated tau proteins. **B** (2) Poly D,L-lactic-co-glycolic acid (PLGA) nanoparticles conjugated with polyethylene glycol (PEG) and anti-amyloid beta (Aβ) antibodies facilitate the targeting, solubilization, and clearance of Aβ fibrils. **C** Modified gold nanoparticles (AuNPs) with capture antibodies specifically target both Aβ and tau proteins, forming a sandwich complex with a secondary antibody for the diagnosis of AD-related biomarkers. Reproduced from an open source [132]

the brain and improve amyloid plaque retention. These radio-iodinated CQ NPs are promising as delivery vehicles for in vivo SPECT and PET imaging of amyloid deposits in AD [133]. Zhang et al. developed a dual-functional nanoparticle (NP) drug delivery system using a PEGylated poly(lactic acid) polymer with two targeting peptides, TGN and QSH. TGN targets ligands at the blood–brain barrier (BBB), while QSH binds specifically to Aβ1–42, a key component of amyloid plaques. The optimal maleimide/peptide molar ratio was determined to be 3 for both peptides, facilitating enhanced and precise delivery to amyloid plaques in Alzheimer's disease model mice [134, 135]. Brambilla et al. developed nanoparticles made of poly[(hexadecyl cyanoacrylate)-co-methoxypoly(ethylene glycol) cyanoacrylate] and investigated their effects on Aβ1–42 peptide aggregation through kinetic studies using capillary electrophoresis. The findings demonstrated that these NPs could bind to Aβ1–42 in both monomeric and soluble oligomeric forms, and they significantly influenced peptide aggregation, as confirmed by thioflavin-T assays [136]. Folic acid niosomes were formulated using various non-ionic surfactants and cholesterol through the lipid hydration technique, followed by ex vivo perfusion studies in a rat model. The results indicated that the drug was absorbed through the nasal cavity within six hours. Additionally, folic acid has been linked to enhanced responses to cholinesterase inhibitors in individuals with Alzheimer's disease [135, 137].

Parkinson's disease (PD), the second most common neurological disorder, suffers from inadequate drug delivery and diagnostic methods. Levodopa, the primary treatment, has poor brain transfer and low bioavailability. Drug delivery nanoparticles enhance therapeutic efficacy by enabling various administration routes to bypass the blood–brain barrier (BBB). Additionally, early PD diagnosis is hindered by traditional methods. Barcia et al. created a biodegradable drug delivery system for ropinirole (RP) utilizing poly(D,L-lactide-co-glycolide) (PLGA) nanoparticles formulated with 8 mg RP and 50 mg PLGA 502. This formulation achieved a mean encapsulation efficiency of 74.8%, a particle size of less than 155 nm, a zeta potential of −14.25 mV, and a zero-order in vitro release rate of 14.13 μg/h/10 mg NPs over 5 days. In a rat model of Parkinson's disease induced by rotenone, RP encapsulated in PLGA NPs (administered at 1 mg/kg/day every 3 days for 35 days) significantly improved neurodegeneration and alleviated PD-like symptoms, as demonstrated by histological and behavioral evaluations [138]. Narender et al. explored lipid nanoparticles and hydrogel formulations as delivery systems for ropinirole (RP) in Parkinson's disease. They developed stable solid lipid nanoparticles (RP-SLN) and nanostructured lipid carriers (RP-NLC) that demonstrated sustained release and enhanced permeability. Pharmacokinetic studies revealed 2.1- and 2.7-fold increases in bioavailability for oral administration and 3.0- and 3.3-fold increases for topical use, respectively. Additionally, RP-SLN-C and RP-NLC-C showed improved topical bioavailability compared to oral forms. These findings suggest that lipid nanoparticles and hydrogel formulations are effective alternatives for RP delivery in parkinsonism [139].

7 Safety and Biocompatibility of Nanomaterials

Nanomaterials' safety and biocompatibility are critical concerns in biomedical applications, particularly in drug delivery. Their unique physicochemical properties, such as a high surface area-to-volume ratio and small size, enable distinct biological interactions compared to bulk materials. Assessments focus on how these nanoscale materials interact with biological processes, including cellular absorption, distribution, metabolism, and excretion. It is essential to consider potential adverse effects, such as cytotoxicity, inflammation, or immunological reactions, to determine their suitability for biomedical use.

7.1 *Toxicological Assessment of Nanomaterials*

Research conducted over the years confirms the idea that nanoparticles can be an evil or a benefit. While nanoparticles have numerous benefits, there is a lack on their toxicity, non-specific protein interactions, and translocation to secondary target organs. As the nanoparticles employed in these investigations are similar to those used in biomedicine, it's important to evaluate their toxicological properties. Superparamagnetic iron oxide (SPIONS) is an FDA-approved nanoparticle used for diagnostics and therapy, including hyperthermia, magnetic medication delivery, and treatment of anaemia. Recently, there has been an increase in the use of SPIONs as medication delivery systems for cancer and other disorders [140]. But, the concerns related to toxicity of SPIONS have somehow limited their use until their property is enhanced. SPIONs can affect the complement system, causing adverse effects associated to innate immunity. Some recent investigations have found that the toxicity of SPIONs can be lowered by using coating agents. Polymers such as PEG, PV, dextran, and chitosan can increase the half-life and stability of SPIONs, preventing aggregation within cells. The study found that coating magnetic iron nanoparticles with zwitterionic ligands resulted in improved stability, biodistribution, and renal clearance. Before using SPIONs for drug administration, it's important to evaluate their toxicity [141, 142].

Liposomes are small, spherical objects with an enclosed compartment made of phospholipid layers. More than 40 liposome-based medications have been granted or are in clinical studies [143, 144]. This raises interest in exploring how liposomes interact with living systems and their potential harmful effects. Liposomes are PEGylated to inhibit opsonization and evade phagocytic cells. Gabizon et al. found that encapsulated doxorubicin affects the clearance of particles and the phagocytic function of liver cells [145].

7.2 *Biocompatibility Testing and Regulatory Considerations*

According to a consensus conference of the European Society for Biomaterials in 1986, biocompatibility is defined as the ability of a substance to provide a suitable host response in a specific application [146]. It is essential to remember that the interactions between a material and a host are influenced by various aspects, including the host parameters, the qualities of the material, and the site and length of the exposure. However, there are no harmonised standards for evaluating the toxicity and biocompatibility of nanomaterials in biological systems, and the guidelines are still being studied [147].

7.3 Regenerative Medicine and Tissue Engineering

Nanotechnology, in addition with tissue engineering, provides the fundamental scientific underpinning for the development of regenerative medicine. Nanomaterial applications in biomedical sciences include molecular delivery (drugs, growth factors, and DNA), imaging and monitoring iPSC, surface modification of implantable materials or nanodevices (biosensor), and nano-fibers for tissue scaffolds [148]. Nanoparticle research in regenerative medicine has been focused mostly on the creation of entrapment and delivery systems. Delivery methods can improve the success of therapeutic substances in nanoparticles for continuous release in a controlled manner, boosting the success rate of regeneration [149].

8 Future Directions and Research Challenges

8.1 Innovations in Functionalization Techniques and Materials

The application of nanomaterials in drug delivery has been greatly expanded by advances in functionalization methods, which allow for more accurate targeting, better bioavailability, and controlled release of therapeutic agents [150]. When a chemical group, ligand, or biomolecule is coated onto the surface of a nanomaterial to achieve desired interactions with biological systems, especially for precise drug delivery, this process is referred to as functionalization The ability of nanomaterials to selectively bind to overexpressed receptors on cancer cells by surface conjugation with targeting ligands, such as antibodies, peptides, or aptamers, is a significant breakthrough that maximizes nanoparticle accumulation in tumor tissue while reducing off-target effects [151]. To improve treatment specificity, folate or transferrin-ligand-functionalized nanoparticles [152] can be used to specifically target cancer cells that overexpress their respective receptors. A further discovery that enhances nanoparticle stability, extends circulation duration, and lowers immune recognition is PEGylation, or the addition of polyethylene glycol (PEG) chains to nanoparticles [153]. By hindering opsonization, PEGylation protects nanoparticles from the mononuclear phagocyte system's quick clearance [154]. Stimuli-responsive functionalization approaches provide regulated medication release at the site of illness by engineering nanoparticles to react to certain stimuli in the tumor microenvironment, such as pH, temperature, or enzyme activity [155, 156]. Systemic toxicity is decreased and therapeutic effectiveness is increased by this dynamic functionalization. Treatment approaches for cancer and other diseases will be revolutionized by these advancements in nanoparticle functionalization, which make it possible to create more accurate and efficient drug delivery systems.

8.2 Integrating Nanotechnology with Personalized Medicine

By offering individualized, patient-specific therapy solutions, the combination of personalized medicine with nanotechnology for medication delivery is transforming healthcare [157]. By addressing the distinct genetic, biochemical, and physiological aspects of every patient's illness, medication delivery systems can be designed with nanotechnology, to improve therapeutic efficacy and accuracy while reducing side effects. Functionalized surfaces of nanoparticles can be created to target certain biomarkers or receptors that are overexpressed in a patient's tumor or sick tissue [158]. For instance, medications can be delivered to HER2-positive cells only using nanoparticles functionalized with antibodies or ligands targeting HER2 receptors in breast cancer. This improves therapy results while protecting healthy tissue. Furthermore, gene modification according to a person's genetic profile is made possible by nanotechnology through the delivery of RNA- [159] and DNA-based [160] therapies, which are essential to personalized medicine. Major obstacles to the use of these treatments can be overcome by using nanoparticles to preserve nucleic acids from deterioration in the bloodstream and to guarantee effective cellular absorption and intracellular distribution. In addition, in dealing with cancer treatments, because each patient's tumor microenvironment is different, nanocarriers may be tailored to release drugs in a regulated manner in response to certain cues like pH or enzyme activity [161]. This dynamic regulation of medication release lowers systemic toxicity while improving therapeutic effectiveness.

8.3 Addressing Regulatory and Ethical Challenges in Nanomedicine

It is imperative to tackle the ethical and regulatory obstacles in nanomedicine for drug delivery to guarantee the secure, efficient, and conscientious advancement of these cutting-edge treatments. Compared to traditional medications, nanomedicines face different regulatory obstacles because of their special physicochemical characteristics [162], which include nanoscale size, increased permeability, and multifunctionality. Establishing standardized procedures for evaluating the safety, effectiveness, and quality of nanomedicines presents difficulties for regulatory organizations such as the FDA and EMA [163]. Conventional pharmacokinetic and toxicity models might not adequately take into consideration the intricate relationships that exist between biological systems and nanoparticles, including immune responses, biodistribution, and long-term toxicity. Consequently, in order to handle the unique properties of nanoparticles, such as their surface charge, shape, and functionalization, as well as any possible impacts on the environment and public health, regulatory frameworks must be adjusted. In order to guarantee consistent treatment effects, there must be guidelines to take into account the repeatability and scalability of nanoparticle manufacturing. Transparency in the creation and application of nanomedicine in clinical settings is essential to ethical issues [164]. Careful thought must be given to matters like patient permission, long-term safety data, and fair access to treatments using nanomedicine. Also, there are ethical questions regarding sustainability and the possible environmental effects of the production and disposal of nanoparticles. Establishing strong regulatory standards and addressing ethical issues are critical tasks for regulatory agencies, researchers, and industry stakeholders in order to enable the safe and fair application of nanomedicine in drug delivery [165, 166].

9 Conclusions

This chapter synthesizes the significant contributions of nanotechnology to modern drug delivery systems. Functionalized nanomaterials, including liposomes, polymeric nanoparticles, dendrimers, and inorganic nanoparticles, have revolutionized drug delivery by enhancing solubility, bioavailability, and targeted delivery, thereby improving therapeutic efficacy for complex diseases like cancer and cardiovascular disorders. These materials leverage surface modifications and functionalization for precise targeting and controlled drug release, minimizing side effects and improving patient compliance. While emphasizing the transformative potential of these technologies, the chapter also highlights the necessity for thorough safety assessments and navigating regulatory challenges to ensure successful clinical translation. Future research directions focus on advancing functionalization techniques, integrating with personalized medicine, and addressing regulatory and ethical considerations. In summary, functionalized nanomaterials are poised to redefine therapeutic strategies across various diseases, marking a significant advancement toward precision medicine and enhanced patient outcomes.

References

1. Anselmo, A.C., Mitragotri, S.: An overview of clinical and commercial impact of drug delivery systems. J. Control. Release **190**, 15–28 (2014). https://doi.org/10.1016/j.jconrel.2014.03.053
2. Undre B., Pandya, S.R., Kumar, V., Singh, M.: Dendrimers as smart materials for developing the various applications in the field of biomedical sciences. Adv. Mat. Lett. **7**, 502–516 (2016). https://doi.org/10.5185/amlett.2016.6153
3. Hauser-Kawaguchi, A.M.N., Luyt, L.G.: Nanomedicine—nanoparticles in cancer imaging and therapy. Cancer Metastasis - Biol. Treat. **20**, 205–244 (2015). https://doi.org/10.1007/978-3-319-12136-9_10
4. Sahu, T., Ratre, Y.K., Chauhan, S., Bhaskar, L.V.K.S., Nair, M.P., Verma, H.K.: Nanotechnology based drug delivery system: current strategies and emerging therapeutic potential for medical science. J. Drug Deliv. Sci. Technol. **63**, 102487 (2021). https://doi.org/10.1016/j.jddst.2021.102487
5. Luo, R., Neu, B., Venkatraman, S.S.: Surface functionalization of nanoparticles to control cell interactions and drug release. Small **8**, 2585–2594 (2012). https://doi.org/10.1002/smll.201200398

6. Chetty, R., Pandya, S.R., Singh, M.: Physicochemical interaction of cerium oxide nanoparticles with simulated biofluids, hemoglobin, insulin, and ds-DNA at 310.15 K. New J. Chem. **44**(5), 1825–1845 (2020). https://doi.org/10.1039/C9NJ04155A
7. Sperling, R.A., Parak, W.J.: Surface modification, functionalization and bioconjugation of colloidal Inorganic nanoparticles. Philos. Trans. R. Soc. A Math. Phys. Eng. Sci. **368**, 1333–1383 (2010). https://doi.org/10.1098/rsta.2009.0273
8. Díez-Pascual, A.M.: Surface engineering of nanomaterials with polymers, biomolecules, and small ligands for nanomedicine. Materials (Basel) **15** (2022). https://doi.org/10.3390/ma15093251
9. Treccani, L., Yvonne Klein, T., Meder, F., Pardun, K., Rezwan, K.: Functionalized ceramics for biomedical, biotechnological and environmental applications. Acta Biomater. **9**, 7115–7150 (2013). https://doi.org/10.1016/j.actbio.2013.03.036
10. Lehn, J.M.: Toward complex matter: supramolecular chemistry and self-organization. Proc. Natl. Acad. Sci. U. S. A. **99**, 4763–4768 (2002). https://doi.org/10.1073/pnas.072065599
11. Georgakilas, V., Tiwari, J.N., Kemp, K.C., Perman, J.A., Bourlinos, A.B., Kim, K.S., Zboril, R.: Noncovalent functionalization of graphene and graphene oxide for energy materials, biosensing, catalytic, and biomedical applications. Chem. Rev. **116**, 5464–5519 (2016). https://doi.org/10.1021/acs.chemrev.5b00620
12. Di Crescenzo, A., Ettorre, V., Fontana, A.: Non-covalent and reversible functionalization of carbon nanotubes. Beilstein J. Nanotechnol. **5**, 1675–1690 (2014). https://doi.org/10.3762/BJNANO.5.178
13. Kumar, N., Sinha Ray, S., Ngila, J.C.: Ionic liquid-assisted synthesis of Ag/Ag_2Te nanocrystals: via a hydrothermal route for enhanced photocatalytic performance. New J. Chem. **41**, 14618–14626 (2017). https://doi.org/10.1039/C7NJ03295A
14. Gusain, R., Singhal, N., Singh, R., Kumar, U., Khatri, O.P.: Ionic-liquid-functionalized copper oxide nanorods for photocatalytic splitting of water. ChemPlusChem **81**, 489–495 (2016). https://doi.org/10.1002/CPLU.201600047
15. Abd Ellah, N.H., Abouelmagd, S.A.: Surface functionalization of polymeric nanoparticles for tumor drug delivery: approaches and challenges. Expert Opin. Drug Deliv. **14**, 201–214 (2017). https://doi.org/10.1080/17425247.2016.1213238
16. Zeng, L., An, L., Wu, X.: Modeling drug-carrier interaction in the drug release from nanocarriers. J. Drug Deliv. **2011**, 1–15 (2011). https://doi.org/10.1155/2011/370308
17. Wang, B., He, C., Tang, C., Yin, C.: Effects of hydrophobic and hydrophilic modifications on gene delivery of amphiphilic chitosan based nanocarriers. Biomaterials **32**, 4630–4638 (2011). https://doi.org/10.1016/j.biomaterials.2011.03.003
18. An, F.F., Cao, W., Liang, X.J.: Nanostructural systems developed with positive charge generation to drug delivery. Adv. Healthc. Mater. **3**, 1162–1181 (2014). https://doi.org/10.1002/adhm.201300600
19. Ding, C., Li, Z.: A review of drug release mechanisms from nanocarrier systems. Mater. Sci. Eng. C **76**, 1440–1453 (2017). https://doi.org/10.1016/j.msec.2017.03.130
20. Such, G.K., Yan, Y., Johnston, A.P.R., Gunawan, S.T., Caruso, F.: Interfacing materials science and biology for drug carrier design. Adv. Mater. **27**, 2278–2297 (2015). https://doi.org/10.1002/adma.201405084
21. Ulbrich, K., Holá, K., Šubr, V., Bakandritsos, A., Tuček, J., Zbořil, R.: Targeted drug delivery with polymers and magnetic nanoparticles: covalent and noncovalent approaches, release control, and clinical studies. Chem. Rev. **116**, 5338–5431 (2016). https://doi.org/10.1021/ACS.CHEMREV.5B00589
22. Christensen, D., Korsholm, K.S., Andersen, P., Agger, E.M.: Cationic liposomes as vaccine adjuvants. Expert Rev. Vaccines **10**, 513–521 (2011). https://doi.org/10.1586/erv.11.17
23. Ebrahim, S., Peyman, G.A., Lee, P.J.: Applications of liposomes in ophthalmology. Surv. Ophthalmol. **50**, 167–182 (2005). https://doi.org/10.1016/j.survophthal.2004.12.006
24. Zhang, L., Gu, F.X., Chan, J.M., Wang, A.Z., Langer, R.S., Farokhzad, O.C.: Nanoparticles in medicine: therapeutic applications and developments. Clin. Pharmacol. Ther. **83**, 761–769 (2008). https://doi.org/10.1038/sj.clpt.6100400

25. Usković, V., Drofenik, M.: Reverse micelles: inert nano-reactors or physico-chemically active guides of the capped reactions. Adv. Colloid Interface Sci. **133**, 23–34 (2007). https://doi.org/10.1016/j.cis.2007.02.002
26. Jiang, Y., Smaisim, G.F., Mahmoud, M.Z., Li, Z., Aybar, H., Abed, A.M.: Simultaneous numerical investigation of the passive use of phase-change materials and the active use of a nanofluid inside a rectangular duct in the thermal management of lithium-ion batteries. J. Power Sources **541** (2022). https://doi.org/10.1016/j.jpowsour.2022.231610
27. Brontowiyono, W., AbdulHussein, W.A., Smaisim, G.F., Mahmoud, M.Z., Singh, S., Lafta, H.A., Hussein, S.A., Kadhim, M.M., Mustafa, Y.F., Aravindhan, S.: Annealing temperature effect on structural, magnetic properties and methyl green degradation of Fe_2O_3 nanostructures. Arab. J. Sci. Eng. **48**, 375–382 (2023). https://doi.org/10.1007/s13369-022-07118-4
28. Milan, A., Mioc, A., Prodea, A., Mioc, M., Buzatu, R., Ghiulai, R., Racoviceanu, R., Caruntu, F., Şoica, C.: The optimized delivery of triterpenes by liposomal nanoformulations: overcoming the challenges. Int. J. Mol. Sci. **23**(3), 1140 (2022). https://doi.org/10.3390/ijms23031140
29. Suffredini, G., East, J.E., Levy, L.M.: New applications of nanotechnology for neuroimaging. Am. J. Neuroradiol. **35**, 1246–1253 (2014). https://doi.org/10.3174/ajnr.A3543
30. Biswas, S., Dodwadkar, N.S., Deshpande, P.P., Torchilin, V.P.: Liposomes loaded with paclitaxel and modified with novel triphenylphosphonium-PEG-PE conjugate possess low toxicity, target mitochondria and demonstrate enhanced antitumor effects in vitro and in vivo. J. Control. Release **159**, 393–402 (2012). https://doi.org/10.1016/j.jconrel.2012.01.009
31. Bahadoran, A., Khoshnoudi Jabarabadi, M., Hameed Mahmood, Z., Bokov, D., Jushi Janani, B., Fakhri, A.: Quick and sensitive colorimetric detection of amino acid with functionalized-silver/copper nanoparticles in the presence of cross linker, and bacteria detection by using DNA-template nanoparticles as peroxidase activity. Spectrochim. Acta - Part A Mol. Biomol. Spectrosc. **268** (2022). https://doi.org/10.1016/j.saa.2021.120636
32. Rheima, A.M., Sabri Abbas, Z., Kadhim, M.M., Hashim Mohammed, S., Yahaia Alhameedi, D., Rasen, F.A., Dhari Jawad Al-Bayati, A., Ramadan, M.F., Talib Abed, Z., Salam Jaber, A., Hachim, S.K., Ali, F.K., Mahmoud, Z.H., Kianfar, E.: Aluminum oxide nano porous: synthesis, properties, and applications. Case Stud. Chem. Environ. Eng. **8** (2023). https://doi.org/10.1016/j.cscee.2023.100428
33. Mishra, G.P., Bagui, M., Tamboli, V., Mitra, A.K.: Recent applications of liposomes in ophthalmic drug delivery. J. Drug Deliv. **2011**, 1–14 (2011). https://doi.org/10.1155/2011/863734
34. Anderson, L.J.E., Hansen, E., Lukianova-Hleb, E.Y., Hafner, J.H., Lapotko, D.O.: Optically guided controlled release from liposomes with tunable plasmonic nanobubbles. J. Control. Release **144**, 151–158 (2010). https://doi.org/10.1016/j.jconrel.2010.02.012
35. Minko, T., Pakunlu, R.I., Wang, Y., Khandare, J.J., Saad, M.: New Generation of Liposomal Drugs for Cancer. Anticancer Agents Med. Chem. **6**, 537–552 (2008). https://doi.org/10.2174/187152006778699095
36. Drulis-Kawa, Z., Dorotkiewicz-Jach, A.: Liposomes as delivery systems for antibiotics. Int. J. Pharm. **387**, 187–198 (2010). https://doi.org/10.1016/j.ijpharm.2009.11.033
37. Anwar, M., Muhammad, F., Akhtar, B.: Biodegradable nanoparticles as drug delivery devices. J. Drug Deliv. Sci. Technol. **64**, 1–20 (2021). https://doi.org/10.1016/j.jddst.2021.102638
38. Crucho, C.I.C., Barros, M.T.: Polymeric nanoparticles: a study on the preparation variables and characterization methods. Mater. Sci. Eng. C **80**, 771–784 (2017). https://doi.org/10.1016/j.msec.2017.06.004
39. Christoforidis, J.B., Chang, S., Jiang, A., Wang, J., Cebulla, C.M.: Intravitreal devices for the treatment of vitreous inflammation. Mediat. Inflamm. **2012** (2012). https://doi.org/10.1155/2012/126463
40. Avramović, N., Mandić, B., Savić-Radojević, A., Simić, T.: Polymeric nanocarriers of drug delivery systems in cancer therapy. Pharmaceutics **12** (2020). https://doi.org/10.3390/pharmaceutics12040298

41. Dourado, D.: Pharmaceutical nanotechnology: a therapeutic revolution. Int. J. Pharm. Sci. Dev. Res. **6**, 009–011 (2020). https://doi.org/10.17352/ijpsdr.000027
42. Bechnak, L., Khalil, C., El Kurdi, R., Khnayzer, R.S., Patra, D.: Curcumin encapsulated colloidal amphiphilic block co-polymeric nanocapsules: colloidal nanocapsules enhance photodynamic and anticancer activities of curcumin. Photochem. Photobiol. Sci. **19**, 1088–1098 (2020). https://doi.org/10.1039/D0PP00032A
43. Gao, M., Long, X., Du, J., Teng, M., Zhang, W., Wang, Y., Wang, X., Wang, Z., Zhang, P., Li, J.: Enhanced curcumin solubility and antibacterial activity by encapsulation in PLGA oily core nanocapsules. Food Funct. 448–455 (2020). https://doi.org/10.1039/c9fo00901a
44. Günday, C., Anand, S., Gencer, H.B., Munafò, S., Moroni, L., Fusco, A., Donnarumma, G., Ricci, C., Hatir, P.C., Türeli, N.G., Türeli, A.E., Mota, C., Danti, S.: Ciprofloxacin-loaded polymeric nanoparticles incorporated electrospun fibers for drug delivery in tissue engineering applications. Drug Deliv. Transl. Res. **10**, 706–720 (2020). https://doi.org/10.1007/s13346-020-00736-1
45. Torres-Flores, G., Nazende, G.T., Emre, T.A.: Preparation of fenofibrate loaded Eudragit L100 nanoparticles by nanoprecipitation method. Mater. Today Proc. 428–435 (2019). https://doi.org/10.1016/j.matpr.2019.03.176
46. Szczęch, M., Szczepanowicz, K.: Polymeric core-shell nanoparticles prepared by spontaneous emulsification solvent evaporation and functionalized by the layer-by-layer method. Nanomaterials **10** (2020). https://doi.org/10.3390/nano10030496
47. Escalona-Rayo, O., Fuentes-Vázquez, P., Jardon-Xicotencatl, S., García-Tovar, C.G., Mendoza-Elvira, S., Quintanar-Guerrero, D.: Rapamycin-loaded polysorbate 80-coated PLGA nanoparticles: optimization of formulation variables and in vitro anti-glioma assessment. J. Drug Deliv. Sci. Technol. **52**, 488–499 (2019). https://doi.org/10.1016/j.jddst.2019.05.026
48. Pinelli, F., Perale, G., Rossi, F.: Coating and functionalization strategies for nanogels and nanoparticles for selective drug delivery. Gels **6** (2020). https://doi.org/10.3390/gels6010006
49. Maurya, A., Singh, A.K., Mishra, G., Kumari, K., Rai, A., Sharma, B., Kulkarni, G.T., Awasthi, R.: Strategic use of nanotechnology in drug targeting and its consequences on human health: a focused review. Interv. Med. Appl. Sci. **11**, 38–54 (2019). https://doi.org/10.1556/1646.11.2019.04
50. Lombardo, D., Kiselev, M.A., Caccamo, M.T.: Smart nanoparticles for drug delivery application: development of versatile nanocarrier platforms in biotechnology and nanomedicine. J. Nanomater. **2019** (2019). https://doi.org/10.1155/2019/3702518
51. Paul, W., Sharma, C.P.: Inorganic nanoparticles for targeted drug delivery. Biointegration Med. Implant Mater. 333–373 (2019). https://doi.org/10.1016/B978-0-08-102680-9.00013-5
52. Xu, H., Liu, M., Lan, M., Yuan, H., Yu, W., Tian, J., Wan, Q., Zhang, X., Wei, Y.: Mussel-inspired PEGylated carbon nanotubes: biocompatibility evaluation and drug delivery applications. Toxicol. Res. (Camb) **5**, 1371–1379 (2016). https://doi.org/10.1039/c6tx00094k
53. Wang, C., Zhang, Z., Chen, B., Gu, L., Li, Y., Yu, S.: Design and evaluation of galactosylated chitosan/graphene oxide nanoparticles as a drug delivery system. J. Colloid Interface Sci. **516**, 332–341 (2018). https://doi.org/10.1016/j.jcis.2018.01.073
54. Unal, S., Arslan, S., Gokce, T., Atasoy, B.M., Karademir, B., Oktar, F.N., Gunduz, O.: Design and characterization of polycaprolactone-gelatin-graphene oxide scaffolds for drug influence on glioblastoma cells. Eur. Polym. J. **115**, 157–165 (2019). https://doi.org/10.1016/j.eurpolymj.2019.03.027
55. Su, S., Wang, J., Qiu, J., Martinez-Zaguilan, R., Sennoune, S.R., Wang, S.: In vitro study of transportation of porphyrin immobilized graphene oxide through blood brain barrier. Mater. Sci. Eng. C **107** (2020). https://doi.org/10.1016/j.msec.2019.110313
56. Kakaei, A., Mirzaei, M.: Cyclophosphamide@CNT: in silico exploration of nano drug delivery system. Lab-in-Silico **2**, 9–14 (2021). https://sciengpub.ir/lab-in-silico/article/view/labinsilico21021009. Accessed 9 Oct 2024
57. Hasanzadeh, L., Darroudi, M., Ramezanian, N., Zamani, P., Aghaee-Bakhtiari, S.H., Nourmohammadi, E., Kazemi Oskuee, R.: Polyethylenimine-associated cerium oxide nanoparticles:

a novel promising gene delivery vector. Life Sci. **232** (2019). https://doi.org/10.1016/j.lfs.2019.116661
58. Hadipour Moghaddam, S.P., Yazdimamaghani, M., Ghandehari, H.: Glutathione-sensitive hollow mesoporous silica nanoparticles for controlled drug delivery. J. Control. Release **282**, 62–75 (2018). https://doi.org/10.1016/j.jconrel.2018.04.032
59. Naz, S., Wang, M., Han, Y., Hu, B., Teng, L., Zhou, J., Zhang, H., Chen, J.: Enzyme-responsive mesoporous silica nanoparticles for tumor cells and mitochondria multistage-targeted drug delivery. Int. J. Nanomedicine **14**, 2533–2542 (2019). https://doi.org/10.2147/IJN.S202210
60. Wang, B., Zhang, K., Wang, J., Zhao, R., Zhang, Q., Kong, X.: Poly(amidoamine)-modified mesoporous silica nanoparticles as a mucoadhesive drug delivery system for potential bladder cancer therapy. Colloids Surf. B Biointerfaces **189** (2020). https://doi.org/10.1016/j.colsurfb.2020.110832
61. Unnikrishnan, G., Joy, A., Megha, M., Kolanthai, E., Senthilkumar, M.: Exploration of inorganic nanoparticles for revolutionary drug delivery applications: a critical review. Discov. Nano **18** (2023). https://doi.org/10.1186/s11671-023-03943-0
62. Peng, J.Q., Fumoto, S., Suga, T., Miyamoto, H., Kuroda, N., Kawakami, S., Nishida, K.: Targeted co-delivery of protein and drug to a tumor in vivo by sophisticated RGD-modified lipid-calcium carbonate nanoparticles. J. Control. Release **302**, 42–53 (2019). https://doi.org/10.1016/j.jconrel.2019.03.021
63. Coelho, S.C., Reis, D.P., Pereira, M.C., Coelho, M.A.N.: Gold nanoparticles for targeting varlitinib to human pancreatic cancer cells. Pharmaceutics **10** (2018). https://doi.org/10.3390/pharmaceutics10030091
64. Di Stefano, A.: Nanotechnology in targeted drug delivery. Int. J. Mol. Sci. **24**(8194) (2023). https://doi.org/10.3390/IJMS24098194
65. Khizar, S., Alrushaid, N., Alam Khan, F., Zine, N., Jaffrezic-Renault, N., Errachid, A., Elaissari, A.: Nanocarriers based novel and effective drug delivery system. Int. J. Pharm. **632**, 122570 (2023). https://doi.org/10.1016/j.ijpharm.2022.122570
66. Tiwari, H., Rai, N., Singh, S., Gupta, P., Verma, A., Singh, A.K., Kajal, Salvi, P., Singh, S.K., Gautam, V.: Recent advances in nanomaterials-based targeted drug delivery for preclinical cancer diagnosis and therapeutics. Bioengineering **10**, 760 (2023). https://doi.org/10.3390/bioengineering10070760
67. Kang, H., Rho, S., Stiles, W.R., Hu, S., Baek, Y., Hwang, D.W., Kashiwagi, S., Kim, M.S., Choi, H.S.: Size-dependent EPR effect of polymeric nanoparticles on tumor targeting. Adv. Healthc. Mater. **9**, 1901223 (2020). https://doi.org/10.1002/ADHM.201901223
68. Ramamurthi, P., Zhao, Z., Burke, E., Steinmetz, N.F., Müllner, M.: Tuning the hydrophilic–hydrophobic balance of molecular polymer bottlebrushes enhances their tumor homing properties. Adv. Healthc. Mater. **11** (2022). https://doi.org/10.1002/adhm.202200163
69. Yadav, K., Singh, D., Singh, M.R., Pradhan, M.: Nano-constructs targeting the primary cellular energy source of cancer cells for modulating tumor progression. OpenNano **8**, 100107 (2022). https://doi.org/10.1016/j.onano.2022.100107
70. Bhattacharya, S., Prajapati, B.G., Singh, S.: A critical review on the dissemination of PH and stimuli-responsive polymeric nanoparticular systems to improve drug delivery in cancer therapy. Crit. Rev. Oncol. Hematol. **185**, 103961 (2023). https://doi.org/10.1016/j.critrevonc.2023.103961
71. Wang, Z., Ling, L., Xia, Q., Li, X.: Disulfide-crosslinked reduction-responsive prodrug micelles for on-demand paclitaxel release. J. Drug Deliv. Sci. Technol. **53**, 101168 (2019). https://doi.org/10.1016/j.jddst.2019.101168
72. Liu, F., Niko, Y., Bouchaala, R., Mercier, L., Lefebvre, O., Andreiuk, B., Vandamme, T., Goetz, J.G., Anton, N., Klymchenko, A.: Drug-sponge lipid nanocarrier for in situ cargo loading and release using dynamic covalent chemistry. Angew. Chemie Int. Ed. **60**, 6573–6580 (2021). https://doi.org/10.1002/anie.202014259
73. Li, J., Wang, Q., Xia, G., Adilijiang, N., Li, Y., Hou, Z., Fan, Z., Li, J.: Recent advances in targeted drug delivery strategy for enhancing oncotherapy. Pharmaceutics **15** (2023). https://doi.org/10.3390/pharmaceutics15092233

74. Seynhaeve, A.L.B., Amin, M., Haemmerich, D., van Rhoon, G.C., ten Hagen, T.L.M.: Hyperthermia and smart drug delivery systems for solid tumor therapy. Adv. Drug Deliv. Rev. **163–164**, 125–144 (2020). https://doi.org/10.1016/j.addr.2020.02.004
75. Biffi, S., Voltan, R., Bortot, B., Zauli, G., Secchiero, P.: Actively targeted nanocarriers for drug delivery to cancer cells. Expert Opin. Drug Deliv. **16**, 481–496 (2019). https://doi.org/10.1080/17425247.2019.1604679
76. Dixon, S.J., Lemberg, K.M., Lamprecht, M.R., Skouta, R., Zaitsev, E.M., Gleason, C.E., Patel, D.N., Bauer, A.J., Cantley, A.M., Yang, W.S., Morrison, B., Stockwell, B.R.: Ferroptosis: an iron-dependent form of nonapoptotic cell death. Cell **149**, 1060–1072 (2012). https://doi.org/10.1016/j.cell.2012.03.042
77. Gong, Z., Liu, X., Zhou, B., Wang, G., Guan, X., Xu, Y., Zhang, J., Hong, Z., Cao, J., Sun, X., Gao, Z., Lu, H., Pan, X., Bai, J.: Tumor acidic microenvironment-induced drug release of RGD peptide nanoparticles for cellular uptake and cancer therapy. Colloids Surf. B Biointerfaces **202**, 111673 (2021). https://doi.org/10.1016/j.colsurfb.2021.111673
78. Dmitrieva, M.D., Voitova, A.A., Dymova, M.A., Richter, V.A., Kuligina, E.V.: Tumor-targeting peptides search strategy for the delivery of therapeutic and diagnostic molecules to tumor cells. Int. J. Mol. Sci. **22**, 314 (2020). https://doi.org/10.3390/ijms22010314
79. Majeed, S., Saravanan, M., Danish, M., Zakariya, N.A., Ibrahim, M.N.M., Rizvi, E.H., un NisaAndrabi, S., Barabadi, H., Mohanta, Y.K., Mostafavi, E.: Bioengineering of green-synthesized TAT peptide-functionalized silver nanoparticles for apoptotic cell-death mediated therapy of breast adenocarcinoma. Talanta **253**, 124026 (2023). https://doi.org/10.1016/j.talanta.2022.124026
80. Anwer, M.K., Mohammad, M., Iqbal, M., Ansari, M.N., Ezzeldin, E., Fatima, F., Alshahrani, S.M., Aldawsari, M.F., Alalaiwe, A., Alzahrani, A.A., Aldayel, A.M.: Sustained release and enhanced oral bioavailability of rivaroxaban by PLGA nanoparticles with no food effect. J. Thromb. Thrombolysis **49**, 404–412 (2020). https://doi.org/10.1007/s11239-019-02022-5
81. Nezlin, R.: Aptamers in immunological research. Immunol. Lett. **162**, 252–255 (2014). https://doi.org/10.1016/j.imlet.2014.10.001
82. Jafari, M., Sriram, V., Xu, Z., Harris, G.M., Lee, J.Y.: Fucoidan-doxorubicin nanoparticles targeting P-selectin for effective breast cancer therapy. Carbohydr. Polym. **249** (2020). https://doi.org/10.1016/j.carbpol.2020.116837
83. Askarizadeh, M., Esfandiari, N., Honarvar, B., Sajadian, S.A., Azdarpour, A.: Kinetic modeling to explain the release of medicine from drug delivery systems. ChemBioEng Rev. **10**, 1006–1049 (2023). https://doi.org/10.1002/cben.202300027
84. Yakut, M.Z., Esen, S.: Impact of energy efficient design parameters on energy consumption in hot-humid climate zones. Res. Eng. Struct. Mater. (2020). https://doi.org/10.17515/resm2019.156ic2809
85. Tang, Q., Yu, B., Gao, L., Cong, H., Song, N., Lu, C.: Stimuli responsive nanoparticles for controlled anti-cancer drug release. Curr. Med. Chem. **25**, 1837–1866 (2018). https://doi.org/10.2174/0929867325666180111095913
86. Hossen, S., Hossain, M.K., Basher, M.K., Mia, M.N.H., Rahman, M.T., Uddin, M.J.: Smart nanocarrier-based drug delivery systems for cancer therapy and toxicity studies: a review. J. Adv. Res. **15**, 1–18 (2019). https://doi.org/10.1016/j.jare.2018.06.005
87. Murugan, C., Rayappan, K., Thangam, R., Bhanumathi, R., Shanthi, K., Vivek, R., Thirumurugan, R., Bhattacharyya, A., Sivasubramanian, S., Gunasekaran, P., Kannan, S.: Combinatorial nanocarrier based drug delivery approach for amalgamation of anti-tumor agents in breast cancer cells: an improved nanomedicine strategy. Sci. Rep. **6**, 34053 (2016). https://doi.org/10.1038/srep34053
88. Palanikumar, L., Al-Hosani, S., Kalmouni, M., Nguyen, V.P., Ali, L., Pasricha, R., Barrera, F.N., Magzoub, M.: PH-responsive high stability polymeric nanoparticles for targeted delivery of anticancer therapeutics. Commun. Biol. **3**, 95 (2020). https://doi.org/10.1038/s42003-020-0817-4
89. Ramezani Farani, M., Azarian, M., Heydari Sheikh Hossein, H., Abdolvahabi, Z., Mohammadi Abgarmi, Z., Moradi, A., Mousavi, S.M., Ashrafizadeh, M., Makvandi, P., Saeb, M.R., Rabiee,

N.: Folic acid-adorned curcumin-loaded iron oxide nanoparticles for cervical cancer. ACS Appl. Bio Mater. **5**, 1305–1318 (2022). https://doi.org/10.1021/acsabm.1c01311

90. Dhavale, R.P., Dhavale, R.P., Sahoo, S.C., Kollu, P., Jadhav, S.U., Patil, P.S., Dongale, T.D., Chougale, A.D., Patil, P.B.: Chitosan coated magnetic nanoparticles as carriers of anticancer drug telmisartan: pH-responsive controlled drug release and cytotoxicity studies. J. Phys. Chem. Solids **148**, 109749 (2021). https://doi.org/10.1016/J.JPCS.2020.109749
91. Javadian, S., Najafi, K., Sadrpoor, S.M., Ektefa, F., Dalir, N., Nikkhah, M.: Graphene quantum dots based magnetic nanoparticles as a promising delivery system for controlled doxorubicin release. J. Mol. Liq. **331**, 115746 (2021). https://doi.org/10.1016/j.molliq.2021.115746
92. Fathi, M., Barar, J., Erfan-Niya, H., Omidi, Y.: Methotrexate-conjugated chitosan-grafted pH- and thermo-responsive magnetic nanoparticles for targeted therapy of ovarian cancer. Int. J. Biol. Macromol. **154**, 1175–1184 (2020). https://doi.org/10.1016/J.IJBIOMAC.2019.10.272
93. Nalluri, L.P., Popuri, S.R., Lee, C.H., Terbish, N.: Synthesis of biopolymer coated functionalized superparamagnetic iron oxide nanoparticles for the pH-sensitive delivery of anti-cancer drugs epirubicin and temozolomide. Int. J. Polym. Mater. Polym. Biomater. 1–14 (2020). https://doi.org/10.1080/00914037.2020.1785449
94. Song, C., Sun, W., Xiao, Y., Shi, X.: Ultrasmall iron oxide nanoparticles: synthesis, surface modification, assembly, and biomedical applications. Drug Discov. Today **24**, 835–844 (2019). https://doi.org/10.1016/j.drudis.2019.01.001
95. Ayubi, M., Karimi, M., Abdpour, S., Rostamizadeh, K., Parsa, M., Zamani, M., Saedi, A.: Magnetic nanoparticles decorated with PEGylated curcumin as dual targeted drug delivery: synthesis, toxicity and biocompatibility study. Mater. Sci. Eng. C **104**, 109810 (2019). https://doi.org/10.1016/j.msec.2019.109810
96. Hussien, N.A., Işıklan, N., Türk, M.: Aptamer-functionalized magnetic graphene oxide nanocarrier for targeted drug delivery of paclitaxel. Mater. Chem. Phys. **211**, 479–488 (2018). https://doi.org/10.1016/j.matchemphys.2018.03.015
97. Nosrati, H., Sefidi, N., Sharafi, A., Danafar, H., Kheiri Manjili, H.: Bovine serum albumin (BSA) coated iron oxide magnetic nanoparticles as biocompatible carriers for curcumin-anticancer drug. Bioorg. Chem. **76**, 501–509 (2018). https://doi.org/10.1016/j.bioorg.2017.12.033
98. Lee, J.M., Park, H., Oh, K.T., Lee, E.S.: PH-Responsive hyaluronated liposomes for docetaxel delivery. Int. J. Pharm. **547**, 377–384 (2018). https://doi.org/10.1016/j.ijpharm.2018.06.028
99. Lu, Y.-J., Chuang, E.-Y., Cheng, Y.-H., Anilkumar, T.S., Chen, H.-A., Chen, J.-P.: Thermosensitive magnetic liposomes for alternating magnetic field-inducible drug delivery in dual targeted brain tumor chemotherapy. Chem. Eng. J. **373**, 720–733 (2019). https://doi.org/10.1016/j.cej.2019.05.055
100. Almáši, M., Matiašová, A.A., Šuleková, M., Beňová, E., Ševc, J., Váhovská, L., Lisnichuk, M., Girman, V., Zeleňáková, A., Hudák, A., Zeleňák, V.: In vivo study of light-driven naproxen release from gated mesoporous silica drug delivery system. Sci. Rep. **11**, 20191 (2021). https://doi.org/10.1038/s41598-021-99678-y
101. Rahim, M.A., Jan, N., Khan, S., Shah, H., Madni, A., Khan, A., Jabar, A., Khan, S., Elhissi, A., Hussain, Z., Aziz, H.C., Sohail, M., Khan, M., Thu, H.E.: Recent advancements in stimuli responsive drug delivery platforms for active and passive cancer targeting. Cancers (Basel) **13** (2021). https://doi.org/10.3390/cancers13040670
102. Sun, L., Xu, Y., Zhang, X., Gao, Y., Chen, J., Zhou, A., Lu, Q., Wang, Z., Shao, K., Wu, H., Ning, X.: Mesenchymal stem cells functionalized sonodynamic treatment for improving therapeutic efficacy and compliance of orthotopic oral cancer. Adv. Mater. **32** (2020). https://doi.org/10.1002/adma.202005295
103. Pourhassan, H., Clergeaud, G., Hansen, A.E., Østrem, R.G., Fliedner, F.P., Melander, F., Nielsen, O.L., O'Sullivan, C.K., Kjær, A., Andresen, T.L.: Revisiting the use of sPLA 2-sensitive liposomes in cancer therapy. J. Control. Release **261**, 163–173 (2017). https://doi.org/10.1016/j.jconrel.2017.06.024

104. Hernández-Gil, J., Cobaleda-Siles, M., Zabaleta, A., Salassa, L., Calvo, J., Mareque-Rivas, J.C.: An iron oxide nanocarrier loaded with a Pt(IV) prodrug and immunostimulatory dsRNA for combining complementary cancer killing effects. Adv. Healthc. Mater. **4**, 1034–1042 (2015). https://doi.org/10.1002/adhm.201500080
105. Siddique, S., Chow, J.C.L.: Application of nanomaterials in biomedical imaging and cancer therapy. Nanomaterials **10**, 1–41 (2020). https://doi.org/10.3390/nano10091700
106. Lee, M.S., Dees, E.C., Wang, A.Z.: Nanoparticle-delivered chemotherapy: old drugs in new packages, Oncology (Williston Park) **31**, 198–208 (2017). https://europepmc.org/article/med/28299757. Accessed 7 Oct 2024
107. Singh, P., Pandit, S., Mokkapati, V.R.S.S., Garg, A., Ravikumar, V., Mijakovic, I.: Gold nanoparticles in diagnostics and therapeutics for human cancer. Int. J. Mol. Sci. **19**, 1979 (2018). https://doi.org/10.3390/ijms19071979
108. Chen, Y.H., Tsai, C.Y., Huang, P.Y., Chang, M.Y., Cheng, P.C., Chou, C.H., Chen, D.H., Wang, C.R., Shiau, A.L., Wu, C.L.: Methotrexate conjugated to gold nanoparticles inhibits tumor growth in a syngeneic lung tumor model. Mol. Pharm. **4**, 713–722 (2007). https://doi.org/10.1021/mp060132k
109. Wang, F., Wang, Y.C., Dou, S., Xiong, M.H., Sun, T.M., Wang, J.: Doxorubicin-tethered responsive gold nanoparticles facilitate intracellular drug delivery for overcoming multidrug resistance in cancer cells. ACS Nano **5**, 3679–3692 (2011). https://doi.org/10.1021/nn200007z
110. Kalimuthu, K., Lubin, B.C., Bazylevich, A., Gellerman, G., Shpilberg, O., Luboshits, G., Firer, M.A.: Gold nanoparticles stabilize peptide-drug-conjugates for sustained targeted drug delivery to cancer cells. J. Nanobiotechnology **16**, 1–13 (2018). https://doi.org/10.1186/s12951-018-0362-1
111. Rozalen, M., Sánchez-Polo, M., Fernández-Perales, M., Widmann, T.J., Rivera-Utrilla, J.: Synthesis of controlled-size silver nanoparticles for the administration of methotrexate drug and its activity in colon and lung cancer cells. RSC Adv. **10**, 10646–10660 (2020). https://doi.org/10.1039/c9ra08657a
112. Meng, F., Han, N., Yeo, Y.: Organic nanoparticle systems for spatiotemporal control of multimodal chemotherapy. Expert Opin. Drug Deliv. **14**, 427–446 (2017). https://doi.org/10.1080/17425247.2016.1218464
113. Fisusi, F.A., Akala, E.O.: Drug combinations in breast cancer therapy. Pharm. Nanotechnol. **7**, 3–23 (2019). https://doi.org/10.2174/2211738507666190122111224
114. Song, M., Liu, T., Shi, C., Zhang, X., Chen, X.: Bioconjugated manganese dioxide nanoparticles enhance chemotherapy response by priming tumor-associated macrophages toward m1-like phenotype and attenuating tumor hypoxia. ACS Nano **10**, 633–647 (2016). https://doi.org/10.1021/acsnano.5b06779
115. Wang, B., Zhang, W., Zhou, X., Liu, M., Hou, X., Cheng, Z., Chen, D.: Development of dual-targeted nano-dandelion based on an oligomeric hyaluronic acid polymer targeting tumor-associated macrophages for combination therapy of non-small cell lung cancer. Drug Deliv. **26**, 1265–1279 (2019). https://doi.org/10.1080/10717544.2019.1693707
116. Olszowy, M., Nowak-Perlak, M., Woźniak, M.: Current strategies in photodynamic therapy (PDT) and photodynamic diagnostics (PDD) and the future potential of nanotechnology in cancer treatment. Pharmaceutics **15**, 1712 (2023). https://doi.org/10.3390/pharmaceutics15061712
117. Mokoena, D.R., George, B.P., Abrahamse, H.: Enhancing breast cancer treatment using a combination of cannabidiol and gold nanoparticles for photodynamic therapy. Int. J. Mol. Sci. **20**, 4771 (2019). https://doi.org/10.3390/ijms20194771
118. Hao, Y., Chung, C.K., Yu, Z., Huis In 'T Veld, R.V., Ossendorp, F.A., Ten Dijke, P., Cruz, L.J.: Combinatorial therapeutic approaches with nanomaterial-based photodynamic cancer therapy. Pharmaceutics **14**, 120 (2022). https://doi.org/10.3390/pharmaceutics14010120
119. Woźniak, M., Nowak, M., Lazebna, A., Więcek, K., Jabłońska, I., Szpadel, K., Grzeszczak, A., Gubernator, J., Ziółkowski, P.: The comparison of in vitro photosensitizing efficacy of curcumin-loaded liposomes following photodynamic therapy on melanoma mug-mel2, squamous cell carcinoma scc-25, and normal keratinocyte hacat cells. Pharmaceuticals **14**, 374 (2021). https://doi.org/10.3390/ph14040374

120. Zhang, W., Yu, L., Jiang, Y., Guo, C.: Phycocyanin-functionalized black phosphorus quantum dots enhance PDT/PTT therapy by inducing ROS and irreparable DNA damage. Biomater. Sci. **9**, 5302–5318 (2021). https://doi.org/10.1039/d1bm00106j
121. Skourtis, D., Stavroulaki, D., Athanasiou, V., Fragouli, P.G., Iatrou, H.: Nanostructured polymeric, liposomal and other materials to control the drug delivery for cardiovascular diseases. Pharmaceutics **12**, 1160 (2020). https://doi.org/10.3390/PHARMACEUTICS12121160
122. Gupta, P., Garcia, E., Sarkar, A., Kapoor, S., Rafiq, K., Chand, H.S., Jayant, R.D.: Nanoparticle based treatment for cardiovascular diseases. Cardiovasc. Hematol. Disord. Targets **19**, 33–44 (2018). https://doi.org/10.2174/1871529x18666180508113253
123. Xue, X., Shi, X., Dong, H., You, S., Cao, H., Wang, K., Wen, Y., Shi, D., He, B., Li, Y.: Delivery of microRNA-1 inhibitor by dendrimer-based nanovector: an early targeting therapy for myocardial infarction in mice. Nanomedicine Nanotechnol. Biol. Med. **14**, 619–631 (2018). https://doi.org/10.1016/j.nano.2017.12.004
124. McGuire, M.J., Samli, K.N., Johnston, S.A., Brown, K.C.: In vitro selection of a peptide with high selectivity for cardiomyocytes in vivo. J. Mol. Biol. **342**, 171–182 (2004). https://doi.org/10.1016/j.jmb.2004.06.029
125. Wei, Y., Zhu, M., Li, S., Hong, T., Guo, X., Li, Y., Liu, Y., Hou, X., He, B.: Engineered biomimetic nanoplatform protects the myocardium against ischemia/reperfusion injury by inhibiting pyroptosis. ACS Appl. Mater. Interfaces **13**, 33756–33766 (2021). https://doi.org/10.1021/acsami.1c03421
126. Liu, S., Chen, X., Bao, L., Liu, T., Yuan, P., Yang, X., Qiu, X., Gooding, J.J., Bai, Y., Xiao, J., Pu, F., Jin, Y.: Treatment of infarcted heart tissue via the capture and local delivery of circulating exosomes through antibody-conjugated magnetic nanoparticles. Nat. Biomed. Eng. **4**, 1063–1075 (2020). https://doi.org/10.1038/s41551-020-00637-1
127. Yu, J., Li, W., Yu, D.: Atrial natriuretic peptide modified oleate adenosine prodrug lipid nanocarriers for the treatment of myocardial infarction: in vitro and in vivo evaluation. Drug Des. Devel. Ther. **12**, 1697–1706 (2018). https://doi.org/10.2147/DDDT.S166749
128. Li, F., Liu, D., Liu, M., Ji, Q., Zhang, B., Mei, Q., Cheng, Y., Zhou, S.: Tregs biomimetic nanoparticle to reprogram inflammatory and redox microenvironment in infarct tissue to treat myocardial ischemia reperfusion injury in mice. J. Nanobiotechnology **20**, 1–21 (2022). https://doi.org/10.1186/s12951-022-01445-2
129. Ikeda, G., Matoba, T., Nakano, Y., Nagaoka, K., Ishikita, A., Nakano, K., Funamoto, D., Sunagawa, K., Egashira, K.: Nanoparticle-mediated targeting of cyclosporine a enhances cardioprotection against ischemia-reperfusion injury through inhibition of mitochondrial permeability transition pore opening. Sci. Rep. **61**(6), 1–13 (2016). https://doi.org/10.1038/srep20467
130. Wang, J., Zhang, S., Di, L.: Acute myocardial infarction therapy: in vitro and in vivo evaluation of atrial natriuretic peptide and triphenylphosphonium dual ligands modified, baicalin-loaded nanoparticulate system. Drug Deliv. **28**, 2198–2204 (2021). https://doi.org/10.1080/10717544.2021.1989086
131. Agarwal, M., Alam, M.R., Haider, M.K., Malik, M.Z., Kim, D.K.: Alzheimer's disease: an overview of major hypotheses and therapeutic options in nanotechnology. Nanomaterials **11**, 1–18 (2021). https://doi.org/10.3390/nano11010059
132. Poudel, P., Park, S.: Recent advances in the treatment of Alzheimer's disease using nanoparticle-based drug delivery systems. Pharmaceutics **14**, 835 (2022). https://doi.org/10.3390/pharmaceutics14040835
133. Kulkarni, P.V., Roney, C.A., Antich, P.P., Bonte, F.J., Raghu, A.V., Aminabhavi, T.M.: Quinoline-n-butylcyanoacrylate-based nanoparticles for brain targeting for the diagnosis of Alzheimer's disease. Wiley Interdiscip. Rev. Nanomedicine NanoBiotechnology **2**, 35–47 (2010). https://doi.org/10.1002/wnan.59
134. Zhang, C., Wan, X., Zheng, X., Shao, X., Liu, Q., Zhang, Q., Qian, Y.: Dual-functional nanoparticles targeting amyloid plaques in the brains of Alzheimer's disease mice. Biomaterials **35**, 456–465 (2014). https://doi.org/10.1016/j.biomaterials.2013.09.063

135. Fonseca-Santos, B., Gremião, M.P.D., Chorilli, M.: Nanotechnology-based drug delivery systems for the treatment of Alzheimer's disease. Int. J. Nanomedicine **10**, 4981–5003 (2015). https://doi.org/10.2147/IJN.S87148
136. Brambilla, D., Verpillot, R., Taverna, M., De Kimpe, L., Le Droumaguet, B., Nicolas, J., Canovi, M., Gobbi, M., Mantegazza, F., Salmona, M., Nicolas, V., Scheper, W., Couvreur, P., Andrieux, K.: New method based on capillary electrophoresis with laser-induced fluorescence detection (CE-LIF) to monitor interaction between nanoparticles and the amyloid-β peptide. Anal. Chem. **82**, 10083–10089 (2010). https://doi.org/10.1021/ac102045x
137. Malouf, R., Evans, J.G.: Folic acid with or without vitamin B12 for the prevention and treatment of healthy elderly and demented people. Cochrane Database Syst. Rev. (2008). https://doi.org/10.1002/14651858.CD004514.pub2
138. Barcia, E., Boeva, L., García-García, L., Slowing, K., Fernández-Carballido, A., Casanova, Y., Negro, S.: Nanotechnology-based drug delivery of ropinirole for Parkinson's disease. Drug Deliv. **24**, 1112–1123 (2017). https://doi.org/10.1080/10717544.2017.1359862
139. Dudhipala, N., Gorre, T.: Neuroprotective effect of ropinirole lipid nanoparticles enriched hydrogel for Parkinson's disease: in vitro, ex vivo, pharmacokinetic and pharmacodynamic evaluation. Pharmaceutics **12**, 448 (2020). https://doi.org/10.3390/pharmaceutics12050448
140. Bakhtiary, Z., Saei, A.A., Hajipour, M.J., Raoufi, M., Vermesh, O., Mahmoudi, M.: Targeted superparamagnetic iron oxide nanoparticles for early detection of cancer: possibilities and challenges. Nanomedicine Nanotechnol. Biol. Med. **12**, 287–307 (2016). https://doi.org/10.1016/j.nano.2015.10.019
141. Dulińska-Litewka, J., Łazarczyk, A., Hałubiec, P., Szafrański, O., Karnas, K., Karewicz, A.: Superparamagnetic iron oxide nanoparticles-current and prospective medical applications. Materials (Basel) **12** (2019). https://doi.org/10.3390/ma12040617
142. Unterweger, H., Dézsi, L., Matuszak, J., Janko, C., Poettler, M., Jordan, J., Bäuerle, T., Szebeni, J., Fey, T., Boccaccini, A.R., Alexiou, C., Cicha, I.: Dextran-coated superparamagnetic iron oxide nanoparticles for magnetic resonance imaging: evaluation of size-dependent imaging properties, storage stability and safety. Int. J. Nanomedicine **13**, 1899–1915 (2018). https://doi.org/10.2147/IJN.S156528
143. Baden, L.R., El Sahly, H.M., Essink, B., Kotloff, K., Frey, S., Novak, R., Diemert, D., Spector, S.A., Rouphael, N., Creech, C.B., McGettigan, J., Khetan, S., Segall, N., Solis, J., Brosz, A., Fierro, C., Schwartz, H., Neuzil, K., Corey, L., Gilbert, P., Janes, H., Follmann, D., Marovich, M., Mascola, J., Polakowski, L., Ledgerwood, J., Graham, B.S., Bennett, H., Pajon, R., Knightly, C., Leav, B., Deng, W., Zhou, H., Han, S., Ivarsson, M., Miller, J., Zaks, T.: Efficacy and safety of the mRNA-1273 SARS-CoV-2 vaccine. N. Engl. J. Med. **384**, 403–416 (2021). https://doi.org/10.1056/nejmoa2035389
144. Polack, F.P., Thomas, S.J., Kitchin, N., Absalon, J., Gurtman, A., Lockhart, S., Perez, J.L., Pérez Marc, G., Moreira, E.D., Zerbini, C., Bailey, R., Swanson, K.A., Roychoudhury, S., Koury, K., Li, P., Kalina, W.V., Cooper, D., Frenck, R.W., Hammitt, L.L., Türeci, Ö., Nell, H., Schaefer, A., Ünal, S., Tresnan, D.B., Mather, S., Dormitzer, P.R., Şahin, U., Jansen, K.U., Gruber, W.C.: Safety and efficacy of the BNT162b2 mRNA Covid-19 vaccine. N. Engl. J. Med. **383**, 2603–2615 (2020). https://doi.org/10.1056/nejmoa2034577
145. Gabizon, A.A.: Pegylated liposomal doxorubicin: metamorphosis of an old drug into a new form of chemotherapy. Cancer Invest. **19**, 424–436 (2001). https://doi.org/10.1081/CNV-100103136
146. Duncan, R., Izzo, L.: Dendrimer biocompatibility and toxicity. Adv. Drug Deliv. Rev. **57**, 2215–2237 (2005). https://doi.org/10.1016/j.addr.2005.09.019
147. Dobrovolskaia, M.A., McNeil, S.E.: Immunological properties of engineered nanomaterials: an introduction. In: Handbook of Immunological Properties of Engineered Nanomaterials, 2nd ed., vol. 1, pp. 1–24 (2016). https://doi.org/10.1142/9677

148. Engel, E., Michiardi, A., Navarro, M., Lacroix, D., Planell, J.A.: Nanotechnology in regenerative medicine: the materials side. Trends Biotechnol. **26**, 39–47 (2008). https://doi.org/10.1016/j.tibtech.2007.10.005
149. Reddy, S.T., Rehor, A., Schmoekel, H.G., Hubbell, J.A., Swartz, M.A.: In vivo targeting of dendritic cells in lymph nodes with poly(propylene sulfide) nanoparticles. J. Control. Release **112**, 26–34 (2006). https://doi.org/10.1016/j.jconrel.2006.01.006
150. Real, D.A., Bolaños, K., Priotti, J., Yutronic, N., Kogan, M.J., Sierpe, R., Donoso-González, O.: Cyclodextrin-modified nanomaterials for drug delivery: classification and advances in controlled release and bioavailability. Pharmaceutics **13**, 2131 (2021). https://doi.org/10.3390/pharmaceutics13122131
151. Chehelgerdi, M., Chehelgerdi, M., Allela, O.Q.B., Pecho, R.D.C., Jayasankar, N., Rao, D.P., Thamaraikani, T., Vasanthan, M., Viktor, P., Lakshmaiya, N., Saadh, M.J., Amajd, A., Abo-Zaid, M.A., Castillo-Acobo, R.Y., Ismail, A.H., Amin, A.H., Akhavan-Sigari, R.: Progressing nanotechnology to improve targeted cancer treatment: overcoming hurdles in its clinical implementation. Mol. Cancer **22**, 1–103 (2023). https://doi.org/10.1186/S12943-023-01865-0/TABLES/15
152. Sadri, E., Khoee, S., Moayeri, S., Haji Ali, B., Pirhajati Mahabadi, V., Shirvalilou, S., Khoei, S.: Enhanced anti-tumor activity of transferrin/folate dual-targeting magnetic nanoparticles using chemo-thermo therapy on retinoblastoma cancer cells Y79. Sci. Rep. **13**, 22358 (2023). https://doi.org/10.1038/S41598-023-49171-5
153. Shi, L., Zhang, J., Zhao, M., Tang, S., Cheng, X., Zhang, W., Li, W., Liu, X., Peng, H., Wang, Q.: Effects of polyethylene glycol on the surface of nanoparticles for targeted drug delivery. Nanoscale **13**, 10748–10764 (2021). https://doi.org/10.1039/D1NR02065J
154. Mills, J.A., Liu, F., Jarrett, T.R., Fletcher, N.L., Thurecht, K.J.: Nanoparticle based medicines: approaches for evading and manipulating the mononuclear phagocyte system and potential for clinical translation. Biomater. Sci. **10**, 3029–3053 (2022). https://doi.org/10.1039/D2BM00181K
155. Mi, P.: Stimuli-responsive nanocarriers for drug delivery, tumor imaging, therapy and theranostics. Theranostics **10**, 4557 (2020). https://doi.org/10.7150/THNO.38069
156. Li, L., Yang, W.W., Xu, D.G.: Stimuli-responsive nanoscale drug delivery systems for cancer therapy. J. Drug Target. **27**, 423–433 (2019). https://doi.org/10.1080/1061186X.2018.1519029
157. Alghamdi, M.A., Fallica, A.N., Virzì, N., Kesharwani, P., Pittalà, V., Greish, K.: The promise of nanotechnology in personalized medicine. J. Pers. Med. **12**, 673 (2022). https://doi.org/10.3390/JPM12050673
158. Rahman, M.M., Islam, M.R., Akash, S., Harun-Or-Rashid, M., Ray, T.K., Rahaman, M.S., Islam, M., Anika, F., Hosain, M.K., Aovi, F.I., Hemeg, H.A., Rauf, A., Wilairatana, P.: Recent advancements of nanoparticles application in cancer and neurodegenerative disorders: at a glance. Biomed. Pharmacother. **153**, 113305 (2022). https://doi.org/10.1016/J.BIOPHA.2022.113305
159. Pozharov, V.P., Minko, T.: Nanotechnology-based RNA vaccines: fundamentals, advantages and challenges. Pharmaceutics **15**, 194 (2023). https://doi.org/10.3390/PHARMACEUTICS15010194
160. Mohammadi, A., Gill, P., Ebrahimnejad, P., Abediankenari, S., Kashi, Z.: DNA-based nanostructures as novelty in biomedicine. Nov. Biomed. **10**, 43–75 (2022). https://doi.org/10.22037/NBM.V10I1.35408
161. Das, S.S., Bharadwaj, P., Bilal, M., Barani, M., Rahdar, A., Taboada, P., Bungau, S., Kyzas, G.Z.: Stimuli-responsive polymeric nanocarriers for drug delivery, imaging, and theragnosis. Polymers **12**, 1397 (2020). https://doi.org/10.3390/POLYM12061397
162. Wu, L.P., Wang, D., Li, Z.: Grand challenges in nanomedicine. Mater. Sci. Eng. C **106**, 110302 (2020). https://doi.org/10.1016/J.MSEC.2019.110302
163. Musazzi, U.M., Franzè, S., Condorelli, F., Minghetti, P., Caliceti, P., Musazzi, U.M., Franzè, S., Minghetti, P., Caliceti, P.: Feeding next-generation nanomedicines to Europe: regulatory and quality challenges. Adv. Healthc. Mater. **12**, 2301956 (2023). https://doi.org/10.1002/ADHM.202301956

164. Satalkar, P., Elger, B.S., Shaw, D.M.: Defining nano, nanotechnology and nanomedicine: why should it matter? Sci. Eng. Ethics **22**, 1255–1276 (2016). https://doi.org/10.1007/S11948-015-9705-6/METRICS
165. Souto, E.B., Blanco-Llamero, C., Krambeck, K., Kiran, N.S., Yashaswini, C., Postwala, H., Severino, P., Priefer, R., Prajapati, B.G., Maheshwari, R.: Regulatory insights into nanomedicine and gene vaccine innovation: safety assessment, challenges, and regulatory perspectives. Acta Biomater. **180**, 1–17 (2024). https://doi.org/10.1016/J.ACTBIO.2024.04.010
166. Mehta, N., Shetty, S., Prajapati, B.G., Shetty, S.: Regulatory and ethical concerns in the use of nanomaterials. Alzheimer's Dis. Adv. Drug Deliv. Strateg. 197–212 (2024). https://doi.org/10.1016/B978-0-443-13205-6.00002-9

Targeting Strategies in Nanoparticle Drug Delivery

Mazaher Ahmadi

Abstract Nanoparticle-based drug delivery systems have significantly transformed the field of medicine by providing more targeted, efficient, and safer delivery of therapeutic agents. These systems address the limitations of traditional drug delivery methods, such as off-target effects, low solubility, and rapid clearance. At the heart of nanoparticle drug delivery systems are targeting strategies that direct nanoparticles to specific tissues, cells, or organs. This chapter explores the three primary targeting strategies: passive targeting, active targeting, and stimuli-responsive targeting. This chapter discuss the mechanisms, types of nanoparticles, design considerations, and clinical applications for each strategy, focusing on their use in cancer, neurological disorders, infectious diseases, and chronic conditions. The chapter also highlights recent advances, challenges in clinical translation, and the potential for personalized medicine using these systems.

Keywords Nanomedicine · Active targeting · Passive targeting · Stimuli-responsive · Drug delivery systems

1 Introduction

The goal of targeted drug delivery is to deliver therapeutic agents directly to disease sites, thereby maximizing therapeutic efficacy while minimizing the systemic side effects associated with conventional drug delivery methods. Unlike traditional approaches, where drugs are delivered to the entire body, targeted drug delivery focuses on improving the bioavailability and localization of drugs in tissues, organs, or cells that are most in need of treatment. Targeted therapies, especially in cancer and neurological disorders, have the potential to significantly improve treatment outcomes and reduce adverse effects, such as toxicity and drug resistance [1].

M. Ahmadi (✉)
Department of Analytical Chemistry, Faculty of Chemistry and Petroleum Sciences, Bu-Ali Sina University, Hamedan, Iran
e-mail: m.ahmadi@basu.ac.ir; ahmadi.mazaher@yahoo.com

N. J. Joshi and G. Patel (eds.), *Nanoparticle Delivery in Cancer Care*,
Series in BioEngineering, https://doi.org/10.1007/978-981-95-0457-2_4

Nanoparticles (NPs) are microscopic particles that range from 1 to 200 nm in size and can be engineered to carry and release therapeutic agents in a controlled manner. Due to their high surface area-to-volume ratio, NPs are highly effective at carrying drugs, genes, or other biological molecules. NPs can be made from various materials, including lipids, polymers, proteins, and inorganic compounds, each providing distinct advantages depending on the application. For example, liposomal NPs can encapsulate both hydrophilic and hydrophobic drugs, while polymeric NPs offer flexibility in drug release rates. The versatility of NPs enables them to address several challenges associated with conventional drug delivery, such as poor solubility, rapid metabolism, and short circulation times [2, 3].

Traditional drug delivery methods often face challenges such as poor targeting, non-specific distribution of drugs, and limited ability to cross biological barriers, such as the blood–brain barrier (BBB) or the tumor vasculature. Additionally, many drugs have a narrow therapeutic window, meaning the concentration of the drug must be carefully controlled to avoid toxicity. Conventional drug delivery systems are not well-suited to delivering drugs in a controlled and localized manner. NPs, however, can be designed to overcome these limitations by enabling targeted, controlled release of therapeutic agents directly at the disease site [4–7].

2 Types of Targeting Strategies

Targeting strategies in nanoparticle drug delivery are classified into three main categories: passive targeting, active targeting, and stimuli-responsive targeting [2]. Each of these strategies offers unique advantages and challenges in terms of drug delivery efficiency, specificity, and controlled release. A comparison of targeting strategies has been summarized in Table 1.

Table 1 Comparison of passive, active, and stimuli-responsive targeting

Feature	Passive targeting	Active targeting	Stimuli-responsive targeting
Mechanism	EPR effect	Receptor-ligand binding	Environmental or external triggers
Specificity	Low	High	Moderate to high
Complexity	Low	High	High
Applications	Tumors, inflammation	Cancer, infections	Cancer, neurological disorders
Challenges	Heterogeneity of EPR effect	Receptor downregulation, cost	Design complexity, patient variability

2.1 Passive Drug Targeting Using Nanoparticles

In recent decades, the field of drug delivery has undergone a significant transformation with the advent of nanotechnology. NPs, owing to their small size and unique physicochemical properties, offer a powerful solution to the limitations of conventional drug delivery systems. Traditional drug delivery methods often face challenges such as poor bioavailability, low solubility, and non-specific distribution of therapeutic agents. The ability of NPs to accumulate at specific sites within the body, due to their size, surface characteristics, and drug release profiles, has made them a highly effective tool in overcoming these challenges.

One of the key strategies in nanoparticle-based drug delivery is passive targeting, which relies on the Enhanced Permeability and Retention (EPR) effect to direct NPs to the site of disease [8]. Passive targeting does not require the use of ligands or receptors for specific recognition of target cells or tissues. Instead, it exploits the natural characteristics of certain pathological tissues, such as tumors or inflammatory sites, which have leaky vasculature and impaired lymphatic drainage.

2.1.1 Mechanism of Passive Targeting

The EPR effect is the cornerstone of passive targeting, particularly in the context of cancer therapy. The EPR effect refers to the phenomenon by which NPs tend to accumulate in tumor tissues due to the unique properties of the tumor vasculature and the tumor microenvironment. The primary factors contributing to the EPR effect include:

(I) Leaky Tumor Vasculature: Tumor blood vessels are often abnormal, with large endothelial gaps (ranging from 100 to 600 nm) due to rapid and disorganized angiogenesis. These gaps allow NPs in the nanometer size range (typically 10–200 nm) to extravasate from the bloodstream into the tumor interstitial space.

(II) Impaired Lymphatic Drainage: Tumors also have impaired lymphatic drainage, which prevents the efficient removal of NPs from the tumor site. As a result, NPs tend to accumulate and persist in the tumor microenvironment, further enhancing the drug's effectiveness at the targeted site.

(III) Increased Vascular Permeability: Tumors often exhibit higher vascular permeability compared to normal tissues. This is due to the absence of tight junctions between endothelial cells, which normally regulate the flow of molecules in and out of blood vessels. As a result, NPs can penetrate these permeable vessels and accumulate in the tumor.

(IV) Immunosuppressive Tumor Microenvironment: The tumor microenvironment often exhibits an immunosuppressive nature, with poor immune cell infiltration. This can help maintain nanoparticle accumulation at the tumor site, preventing their removal by the immune system.

These properties make the EPR effect a highly effective mechanism for passively targeting drugs to tumors and other pathological sites, such as inflammation and infected tissues. NPs, including liposomes, polymeric NPs, and micelles, can exploit the EPR effect to deliver drugs in a controlled manner, improving therapeutic efficacy and minimizing systemic toxicity.

The size and surface characteristics of NPs are crucial for optimizing the EPR effect and enhancing drug delivery. The size of NPs influences their ability to extravasate from blood vessels and accumulate in tumor tissues. NPs in the 10–200 nm size range are particularly effective for EPR-based targeting. Particles smaller than 10 nm may be cleared rapidly by the kidneys, while larger particles (above 200 nm) are more likely to be taken up by the reticuloendothelial system (RES), particularly the liver and spleen. The surface charge of NPs affects their interaction with the biological environment. Neutral or slightly negative NPs tend to evade rapid clearance by the immune system, leading to longer circulation times in the bloodstream. On the other hand, positively charged NPs can interact with cell membranes but may be subject to rapid clearance and toxicity due to their interaction with serum proteins and immune cells. Surface modification with biocompatible polymers, such as polyethylene glycol (PEG), can significantly enhance nanoparticle stability and prolong circulation time. PEGylation reduces protein adsorption and shields NPs from immune recognition, making them more likely to accumulate at the target site via the EPR effect [9].

The main advantage of passive targeting through the EPR effect is that it does not require target-specific ligands or complex receptor-based targeting mechanisms. This makes it a simpler and more versatile approach for drug delivery. Some additional advantages include:

(I) Reduced Toxicity: By targeting the diseased tissue, NPs can reduce systemic toxicity and improve the therapeutic index of drugs.
(II) Improved Drug Bioavailability: NPs can enhance the solubility and bioavailability of poorly soluble drugs, making them more effective.
(III) Longer Circulation Times: NPs can be engineered to have long circulation times, allowing for sustained drug delivery and reduced dosing frequency.
(IV) Reduced Drug Resistance: NPs are less likely to induce the development of drug resistance compared to traditional antibiotics or chemotherapy agents.

Despite its numerous advantages, passive targeting using NPs also has certain challenges and limitations. The EPR effect is not uniform across all tumor types or individual patients, leading to variable drug delivery and effectiveness. Tumors with poor blood supply or hypoxic regions may not benefit from passive targeting as effectively as tumors with well-vascularized tissue. NPs may be recognized by the immune system and cleared before they can exert their therapeutic effect, particularly if they are not adequately shielded or modified. Manufacturing NPs on a large scale while maintaining their quality, size, and functionality remains a significant challenge for clinical applications [8].

Passive targeting using NPs is a promising strategy for the targeted delivery of therapeutic agents to sites of disease, particularly in cancer, cardiovascular, and inflammatory diseases. By exploiting the EPR effect, NPs can accumulate in diseased tissues, such as tumors, atherosclerotic plaques, and inflamed tissues, leading to improved drug efficacy and reduced toxicity. While there are challenges related to tumor heterogeneity, immune clearance, and scalability, the potential benefits of nanoparticle-based passive targeting make it a powerful tool in modern drug delivery systems. Ongoing advancements in nanoparticle design, surface modifications, and manufacturing processes will likely overcome these challenges and further improve the efficacy of passive targeting strategies in the treatment of chronic diseases.

2.2 *Active Drug Targeting Using Nanoparticles*

Nanoparticle-based drug delivery systems have revolutionized modern medicine by enabling more targeted, efficient, and controlled delivery of therapeutic agents. While passive targeting leverages the inherent properties of tumors or pathological tissues, active targeting involves functionalizing NPs with specific ligands to direct them to desired targets, such as tumor cells, infected tissues, or inflammatory sites [10]. This targeted approach is designed to maximize therapeutic efficacy while minimizing systemic toxicity, making it particularly useful for treating complex diseases like cancer, infectious diseases, and autoimmune disorders. Active drug targeting using NPs exploits the specific interactions between ligands and receptors overexpressed on disease-specific tissues.

2.2.1 Mechanism of Active Drug Targeting

Active drug targeting involves the attachment of specific biomolecules to the surface of NPs, enabling them to bind selectively to cell surface receptors or molecules that are overexpressed in diseased tissues. This receptor-ligand interaction facilitates the internalization of NPs via endocytosis or phagocytosis, allowing for the targeted release of therapeutic agents at the site of action.

Receptor-mediated endocytosis is one of the most common mechanisms by which NPs enter target cells. When NPs are functionalized with ligands that bind to specific receptors on the surface of target cells, the NPs are internalized through receptor-mediated endocytosis. Once inside the cell, the NPs can release their therapeutic payload in the cytoplasm or deliver it directly to the nucleus, depending on the type of drug or therapy. Clathrin-Mediated Endocytosis is the most common form of endocytosis involved in receptor-ligand interactions. Clathrin proteins coat the invaginated portion of the plasma membrane, which then buds off and forms vesicles containing the NPs. Caveolae-Mediated Endocytosis involves caveolin proteins, which help form caveolae (small invaginations in the plasma membrane). Caveolae-mediated endocytosis plays a role in lipid raft-mediated nanoparticle internalization, particularly for liposomes and lipid-based NPs. Macropinocytosis is an energy-dependent form of endocytosis in which NPs are engulfed by the cell in large vesicles. It is particularly common in immune cells such as macrophages [11].

One of the main strategies for active targeting is to functionalize NPs with specific ligands that can bind to disease-associated receptors on target cells or tissues. These ligands can be proteins, antibodies, peptides, aptamers, or small molecules. The ligand-receptor interaction enables selective binding and internalization of the NPs into the desired cells, minimizing exposure to healthy tissues and reducing systemic side effects. Monoclonal antibodies (mAbs) are commonly used in active targeting due to their high specificity and affinity for their target receptors. Antibodies can target overexpressed antigens, such as HER2 in breast cancer, EGFR in glioblastomas, or CD19 in certain leukemias [12]. Short peptides such as RGD peptides or TAT peptides are widely used for targeting specific receptors like integrins on tumor vasculature or tumor cells [13]. These peptides can bind to receptors involved in tumor cell migration, angiogenesis, and tumor growth. Aptamers are short oligonucleotides (DNA or RNA) that can bind to specific proteins with high affinity and specificity. They can be used to target cell surface markers or tumor-associated antigens, providing a non-immunogenic alternative to antibodies. Small molecules can be used to target specific receptors on disease sites. For example, folic acid or folate can be conjugated to NPs to target folate receptors overexpressed in some cancer cells, such as ovarian cancer.

Active targeting using NPs overs some advantages. Active targeting ensures that NPs are delivered specifically to the disease site, improving therapeutic efficacy and reducing off-target toxicity. By concentrating the therapeutic agent at the target site, active targeting improves the effectiveness of drugs, particularly those with low bioavailability or rapid clearance. Active targeting minimizes exposure to healthy tissues, reducing the risk of systemic side effects commonly associated with traditional drug delivery methods. NPs can be loaded with multiple therapeutic agents or functionalized with various ligands, enabling combination therapies that target multiple pathways simultaneously [14]. However, active targeting using NPs faces some challenges and limitations. Tumors are often genetically and phenotypically heterogeneous, and the receptor expression that is targeted in one part of the tumor may not be present throughout the entire tumor, leading to variable targeting efficiency. The immune system may recognize the NPs as foreign entities, potentially leading to their rapid clearance or immune-mediated toxicity. The production of targeted NPs on a large scale with consistent size, surface modification, and drug encapsulation efficiency remains a significant challenge for clinical translation. The approval process for nanoparticle-based therapies is complex, requiring extensive preclinical and clinical studies to ensure safety and efficacy.

Active drug targeting using NPs represents one of the most promising strategies in modern medicine, offering enhanced selectivity, efficacy, and safety in the treatment of various diseases. By functionalizing NPs with specific ligands, researchers can achieve precise targeting of tumor cells, infected tissues, immune cells, and other disease sites, thus improving treatment outcomes and minimizing side effects.

Although challenges such as tumor heterogeneity, immunogenicity, and scalability remain, ongoing advances in nanoparticle design, surface functionalization, and personalized medicine are likely to overcome these barriers, paving the way for more effective and targeted therapies in the future.

2.3 Stimuli-Responsive Drug Delivery Using Nanoparticles

The field of drug delivery has witnessed significant advancements with the introduction of NPs, which offer the ability to enhance drug efficacy, improve bioavailability, and reduce off-target effects. However, traditional drug delivery methods often fall short in providing precise control over the release of therapeutic agents, leading to issues such as non-specific distribution and systemic toxicity. Stimuli-responsive drug delivery systems using NPs have emerged as an innovative approach to address these limitations. These systems leverage the unique properties of the disease site and the body's internal environment to trigger the release of therapeutic agents in response to specific external or internal stimuli. Stimuli-responsive NPs are designed to release their payloads upon exposure to specific environmental cues such as pH changes, temperature fluctuations, magnetic fields, ultrasound, enzymatic activity, or redox changes. This ensures spatial and temporal control over drug delivery, making it possible to release drugs specifically at the target site while minimizing side effects on healthy tissues [15].

2.3.1 Mechanisms of Stimuli-Responsive Drug Delivery

Stimuli-responsive drug delivery systems are engineered to release their drug payloads in response to specific physiological or external stimuli. These systems can be designed to respond to both internal factors (e.g., pH, temperature, redox) and external factors (e.g., ultrasound, magnetic fields, light) that are unique to the disease site or therapeutic requirements [2].

Internal stimuli involve factors present in the disease microenvironment or the body's physiology. These factors can trigger a conformational change in the nanoparticle, leading to drug release at the target site. The following internal stimuli are commonly used in stimuli-responsive drug delivery systems:

(I) pH-Responsive Drug Delivery: Tumor tissues, inflammatory sites, and infected tissues often exhibit an acidic microenvironment compared to normal tissues. This difference in pH is commonly exploited in the design of pH-sensitive NPs. The drug release from these NPs is triggered when the pH of the surrounding environment reaches a certain threshold, causing a structural change in the nanoparticle's surface or core.

(II) Redox-Responsive Drug Delivery: The concentration of glutathione (GSH), a reducing agent, is significantly higher in the cytoplasm of cancer cells and tumors compared to normal tissues. Redox-sensitive NPs are designed with disulfide bonds or other redox-sensitive linkers that undergo cleavage in the presence of high GSH levels, facilitating the release of the encapsulated drug [16].

(III) Enzyme-Responsive Drug Delivery: Enzymes play a critical role in various biological processes, including tumor metastasis, infection, and inflammation. Enzyme-responsive NPs are designed to release their drug payload upon enzymatic cleavage of specific linkers or functional groups that are incorporated into the nanoparticle structure. These enzymes, such as matrix metalloproteinases (MMPs), cathepsins, and lipases, are overexpressed at the site of disease, particularly in tumors and inflamed tissues [17].

(IV) Temperature-Responsive Drug Delivery: Certain tissues, especially tumors, can exhibit increased temperature due to rapid metabolic activity or external heating, such as that induced by hyperthermia therapy. NPs can be engineered to undergo phase transition or structural rearrangement at a specific temperature, triggering the release of encapsulated drugs.

External stimuli provide controlled, non-invasive means to trigger drug release in a specific location. These stimuli can be applied to precisely modulate drug delivery and release at the target site. Key external stimuli include:

(I) Magnetic Field-Responsive Drug Delivery: Magnetic NPs can be guided to the site of disease using an external magnetic field. This enables targeted delivery of drugs to specific tissues while minimizing systemic exposure. Once the NPs are accumulated at the target site, an external magnetic field can be used to trigger drug release by inducing heat (magnetic hyperthermia) or through the application of magnetic fields to affect the structure or function of the NPs [1].

(II) Ultrasound-Responsive Drug Delivery: Ultrasound can be used as an external stimulus to trigger the release of drugs from NPs. When ultrasound waves are applied to gas-filled microbubbles or nanocarriers, they cause cavitation, which enhances the permeability of cell membranes or tumor vasculature, allowing for drug release. This approach is particularly beneficial in drug delivery to tumors and BBB crossing.

(III) Light-Responsive Drug Delivery: Light-sensitive NPs can be designed to release their drug payload in response to specific wavelengths of light, such as ultraviolet (UV) or near-infrared (NIR) light. This approach allows for spatial control over drug release, as light can be applied to a specific region of the body, such as a tumor or infected tissue, without affecting surrounding healthy tissues [18].

Stimuli-responsive drug delivery using NPs offers several key advantages. Stimuli-responsive NPs enable the spatial and temporal control of drug release, ensuring that the drug is delivered precisely to the site of disease. By releasing the drug specifically at the site of action, these systems reduce systemic exposure and toxicity, enhancing the safety profile of therapeutic agents. NPs can be designed to release drugs in a controlled manner over a prolonged period, reducing the need for frequent dosing and improving patient compliance. Stimuli-responsive NPs can be designed to respond to a wide range of internal and external triggers, providing customizable drug delivery systems for various diseases. Despite their promising potential, stimuli-responsive drug delivery systems face several challenges. Designing NPs that respond to specific stimuli requires careful selection of materials and precise engineering, which can increase the complexity and cost of production. While NPs offer controlled drug release, their biocompatibility and potential long-term effects in the body need further investigation to avoid toxic reactions. The large-scale production of stimuli-responsive NPs with consistent quality and functionality remains a challenge, limiting their clinical translation [19].

Stimuli-responsive drug delivery systems using NPs offer significant promise in improving the efficacy, precision, and safety of therapeutic agents. By responding to specific internal or external stimuli, these systems allow for targeted drug delivery, controlled release, and minimized side effects, making them particularly suitable for the treatment of cancer, infectious diseases, and chronic conditions. Despite challenges related to biocompatibility, scalability, and complexity, ongoing research in nanoparticle design and stimuli-responsive materials will likely overcome these barriers, paving the way for more effective and personalized therapies in the future.

3 Nanoparticle Design for Drug Delivery

NPs have revolutionized drug delivery systems by offering targeted, controlled, and sustained release of therapeutic agents. Their ability to encapsulate a wide variety of drugs, such as small molecules, proteins, and nucleic acids, while overcoming biological barriers, makes them an essential tool in modern pharmacology.

3.1 Size and Shape of Nanoparticles

The size and shape of NPs play a crucial role in determining their biodistribution, cellular uptake, and pharmacokinetics. NPs typically range from 1 to 200 nm in size, with the optimal size depending on the intended application. Smaller NPs (under 10 nm) may be rapidly cleared by the renal system, whereas larger NPs (over 200 nm) are more likely to be phagocytosed by the RES. 10–200 nm particles have ideal properties for tumor accumulation via the EPR effect. Spherical NPs are the most commonly used, but rod-shaped, disk-shaped, and core–shell structures offer unique advantages. Rod-shaped particles tend to have better cellular uptake and penetration through tissue than spherical ones. Multifunctional shapes can improve targeting efficacy by allowing functionalization with different ligands or imaging agents on different surfaces [20].

3.2 Surface Charge and Hydrophobicity

The surface charge and hydrophobicity of NPs significantly affect their stability, interactions with biological systems, and drug loading capacity. NPs can be positively charged, negatively charged, or neutral, each having distinct effects on cellular uptake and circulation time. Cationic NPs tend to be taken up more efficiently by cells but can be toxic due to interaction with cellular membranes. Anionic NPs are less toxic and offer longer circulation times but have lower cellular uptake. NPs with hydrophobic surfaces are often used for encapsulating hydrophobic drugs, such as chemotherapy agents. Surface modification with hydrophilic polymers like PEG can prevent aggregation, reduce protein adsorption, and prolong the circulation half-life of NPs [20].

3.3 Biodegradable and Biocompatible Materials

Materials used in nanoparticle design must be biocompatible, ensuring minimal toxicity and immunogenicity. Biodegradable materials are often preferred, as they degrade into non-toxic by-products after drug delivery, reducing long-term accumulation in the body. Poly(lactic-co-glycolic acid) (PLGA) and polylactic acid (PLA) are widely used due to their biodegradability and ability to encapsulate a variety of drugs, from small molecules to proteins. Polymeric micelles made from amphiphilic copolymers are particularly useful for delivering poorly water-soluble drugs. Liposomes and solid lipid NPs (SLNs) are popular lipid-based NPs used for drug encapsulation and targeted delivery. Liposomes, in particular, are versatile carriers with both hydrophilic and hydrophobic drug-loading capacities. Silica NPs and gold NPs offer unique properties, including optical properties for imaging and stability under physiological conditions. However, their biocompatibility must be carefully evaluated [21].

4 Nanoparticle Functionalization for Drug Delivery

4.1 Surface Functionalization with Targeting Ligands

Surface functionalization involves modifying the nanoparticle surface with targeting ligands that specifically bind to receptors overexpressed in diseased tissues, ensuring targeted drug delivery. Monoclonal antibodies or single-chain variable fragments (scFv) can be attached to NPs to target tumor-associated antigens (e.g., HER2 for breast cancer). Short peptides that bind to cell surface receptors, such as RGD peptides targeting integrins, are commonly used in nanoparticle functionalization for tumor targeting. DNA/RNA aptamers are small oligonucleotide sequences that bind to specific protein targets and can be conjugated to NPs to deliver drugs selectively to cancer cells. NPs functionalized with carbohydrate moieties can target bacterial infections by binding to specific lectins or surface glycoproteins on pathogens [22].

Ligand-functionalized NPs are internalized by cells via receptor-mediated endocytosis, enabling intracellular drug release. However, certain NPs target markers associated with the epithelial-mesenchymal transition, a process where tumor cells gain migratory properties, commonly seen in metastasis.

4.2 *Surface Coating with Stealth Polymers*

To prolong the circulation time of NPs in the bloodstream and prevent rapid clearance by the immune system, NPs are often coated with stealth polymers like PEG. PEGylation involves attaching PEG chains to the surface of NPs to mask the immune recognition and reduce protein adsorption, a phenomenon known as the "protein corona". PEGylation results in longer circulation times and reduces phagocytosis by macrophages, ensuring that NPs remain in the bloodstream longer. The "PEG dilemma" arises when long-term PEGylation leads to the development of anti-PEG antibodies, which can limit the efficacy of repeated treatments [23]. Polyvinyl alcohol (PVA) and polysaccharides such as chitosan are used for coating NPs to improve their biodegradability and biocompatibility.

4.3 *Multifunctional Nanoparticles*

To enhance therapeutic outcomes, NPs are increasingly designed to be multifunctional, integrating several functions, such as drug delivery and imaging (theranostics), gene delivery and immunotherapy, and temperature or pH-sensitive release. Theranostics NPs are those that combine therapeutic and diagnostic capabilities enable real-time monitoring of drug delivery and treatment efficacy. Gold NPs are used for both drug delivery and imaging, as they possess optical properties that allow for detection via surface-enhanced Raman spectroscopy or photoacoustic imaging [24]. Multifunctional NPs also can deliver multiple drugs simultaneously to treat complex diseases like cancer, which may require chemotherapy, immunotherapy, and gene therapy. For example, NPs loaded with both chemotherapy agents and immune checkpoint inhibitors are being investigated to target tumor cells and simultaneously modulate the immune system [25].

5 Applications of Targeting Strategies

5.1 Cancer Therapy

Cancer is one of the leading causes of morbidity and mortality worldwide. The traditional approaches to cancer therapy—such as surgery, chemotherapy, and radiation—have been effective in many cases but still face significant challenges. These therapies are often associated with non-specific cytotoxicity, leading to severe side effects and suboptimal therapeutic outcomes. Furthermore, tumors are inherently heterogeneous, and their microenvironment plays a critical role in both tumor progression and response to treatment. Therefore, innovative strategies targeting tumor cells directly as well as the tumor microenvironment have gained significant attention in recent years. Nanoparticle-based drug delivery systems offer a promising approach to overcome the limitations of conventional therapies by providing targeted delivery and sustained release of chemotherapeutic agents and therapeutic biomolecules to cancerous tissues.

5.1.1 The Challenges of Targeting Tumor Cells

Tumor cells exhibit several unique features that make them attractive targets for drug delivery. However, targeting these cells is not without challenges. Tumor cells are often genetically heterogeneous, meaning that the same type of tumor can have different characteristics depending on the location and the individual patient. Moreover, the cell surface markers that are often used for targeting may vary between tumors, limiting the effectiveness of certain targeting strategies. To address these challenges, NPs have been employed to selectively target tumor cells. Their size and surface properties can be finely tuned to recognize and bind to specific receptors overexpressed on tumor cells, ensuring that therapeutic agents are delivered precisely to the desired site. There are two main strategies for targeting tumor cells: passive targeting through the EPR effect and active targeting via specific ligand-receptor interactions.

5.1.2 Passive Targeting for Cancer Therapy

The EPR effect is one of the most well-known mechanisms used to target tumors passively. Tumor tissues are characterized by abnormal blood vessels that are more permeable than those in normal tissues. These blood vessels have larger endothelial gaps (ranging from 100 to 600 nm) that allow NPs to extravasate from the bloodstream into the tumor tissue. Additionally, the impaired lymphatic drainage in tumors prevents the rapid removal of NPs, causing them to accumulate in the tumor [26]. NPs that fall within the optimal size range (10–200 nm) can take advantage of the EPR effect and preferentially accumulate in tumors, reducing the risk of affecting healthy tissues. This strategy has been successfully employed in various types of nanoparticle formulations, including liposomes, polymeric micelles, and nanogels, which encapsulate chemotherapeutic agents for efficient delivery to tumor cells. Doxil®, a liposomal formulation of doxorubicin, exploits the EPR effect to deliver chemotherapy directly to the tumor site, resulting in reduced cardiotoxicity compared to free doxorubicin. The liposomal drug is taken up by tumor cells, where the chemotherapeutic agent is slowly released in a controlled manner [27].

5.1.3 Active Targeting for Cancer Therapy

In contrast to passive targeting, active targeting involves modifying the nanoparticle surface with specific ligands that bind to tumor-associated receptors overexpressed on the surface of tumor cells. These ligands can be antibodies, peptides, small molecules, or aptamers that specifically recognize tumor markers, thereby improving the selectivity and efficacy of drug delivery. Monoclonal antibodies (mAbs), such as trastuzumab (Herceptin®), target specific cell surface receptors like HER2 that are overexpressed in certain types of cancer, including breast and gastric cancer. Small peptides, like RGD peptides, target integrins on tumor vasculature, facilitating nanoparticle uptake by endothelial cells and increasing tumor perfusion. Aptamers, short DNA or RNA oligonucleotides can bind to overexpressed tumor cell markers like nucleolin, making them an excellent choice for targeted therapy [28].

The ligand-receptor binding triggers receptor-mediated endocytosis, allowing the NPs to enter the cell and release their drug payload inside the tumor cell, where it can exert its therapeutic effect. This strategy ensures that drugs are concentrated in tumor tissues, minimizing systemic side effects. For example, Trastuzumab-functionalized NPs specifically target HER2-positive breast cancer cells, improving therapeutic efficacy while reducing side effects associated with chemotherapy [29].

5.1.4 Targeting the Tumor Microenvironment

While targeting tumor cells is important, the tumor microenvironment also plays a crucial role in tumor progression and therapeutic resistance. The tumor microenvironment consists of various components, including the extracellular matrix, fibroblasts, immune cells, and blood vessels, all of which can influence the effectiveness of cancer therapies. The tumor microenvironment is often characterized by hypoxia, acidic pH, increased interstitial pressure, and immune suppression, all of which contribute to the complexity of cancer treatment [2]. Targeting the tumor microenvironment is a promising strategy for overcoming resistance to therapy and enhancing the effectiveness of drug delivery systems.

- **Hypoxia-Responsive NPs**

The hypoxic nature of many tumors contributes to their resistance to radiation and chemotherapy. Hypoxia induces the expression of genes that promote angiogenesis, survival, and drug resistance. Targeting hypoxia within the tumor by hypoxia-responsive NPs is an emerging strategy for improving treatment efficacy. These NPs are designed to release their drug payloads in response to the low oxygen conditions typically found in tumors. For example, NPs with hypoxia-sensitive linkers are stable in normal tissues but degrade under hypoxic conditions, thereby releasing their drugs selectively in the tumor. Hypoxia-responsive polymeric NPs have been designed to deliver chemotherapy agents selectively to tumor sites, reducing the risk of normal tissue toxicity and overcoming the challenge of tumor hypoxia [30].

- **pH-Sensitive Nanoparticles**

The tumor microenvironment is also typically acidic (pH ~ 6.5–7.0), compared to the neutral pH of healthy tissues (pH ~ 7.4). This difference in pH can be exploited to deliver drugs selectively to tumors via pH-sensitive NPs [31]. These NPs are designed to release their drug payload when exposed to the acidic conditions of the tumor microenvironment. pH-sensitive NPs are usually made from materials like polymeric micelles or liposomes, which have pH-sensitive bonds that degrade under acidic conditions, allowing for the controlled release of encapsulated drugs. pH-sensitive NPs have been used to deliver gefitinib to cancer cells, with the drug being released when the NPs reach the acidic tumor environment, thus reducing the potential for systemic toxicity [4].

- **Targeting the Tumor Vasculature**

The blood vessels that supply tumors are often abnormal and leaky, leading to increased vascular permeability. This feature has been exploited for drug delivery, particularly using NPs that target the tumor vasculature. Tumor-specific angiogenesis markers such as vascular endothelial growth factor (VEGF) and αvβ3 integrins are often overexpressed on the endothelial cells of tumor blood vessels, making them ideal targets for nanoparticle functionalization. By functionalizing NPs with ligands targeting these markers, drug delivery can be localized to the tumor vasculature, improving tumor penetration and reducing systemic toxicity. RGD-functionalized NPs target αvβ3 integrins on tumor blood vessels, allowing NPs to accumulate at the site of angiogenesis and release their payload directly into the tumor vasculature [32].

5.2 Neurological Disorders

The treatment of neurological disorders has always posed significant challenges due to the unique properties of the BBB. This highly selective semipermeable membrane restricts the entry of many therapeutic agents, including small molecules, proteins, and nucleic acids, into the brain, thus limiting the effectiveness of conventional therapies for diseases such as Alzheimer's, Parkinson's disease, multiple sclerosis, and brain cancers [7, 33, 34]. Although several strategies have been proposed to circumvent the BBB, nanoparticle-based drug delivery has emerged as a promising approach due to its ability to cross the BBB and deliver drugs directly to the central nervous system (CNS).

5.2.1 The Blood–Brain Barrier (BBB): Challenges and Mechanisms

The BBB is a specialized structure that regulates the exchange of materials between the bloodstream and the brain. It is composed of endothelial cells forming the walls of blood vessels in the brain, which are tightly joined by tight junctions. These tight junctions prevent the paracellular passage of most molecules, including drugs, from the bloodstream into the brain. In addition to the endothelial cells, the BBB also includes astrocytes, pericytes, and the basement membrane, which provide structural support and influence the function of the barrier. The BBB serves as a critical protective mechanism, preventing harmful substances, pathogens, and toxins from entering the brain while allowing essential nutrients like glucose and amino acids to pass through. However, this protective barrier also limits the delivery of therapeutic agents, posing a major challenge for treating neurological diseases.

Several mechanisms have been proposed to facilitate the crossing of the BBB by therapeutic agents including paracellular transport, transcellular transport, receptor-mediated transcytosis, and adsorptive-mediated endocytosis [35]. Paracellular transport occurs when molecules pass between endothelial cells. However, due to the tight junctions between the cells, paracellular transport is limited to small hydrophilic molecules and generally does not allow large molecules or NPs to cross the BBB. Transcellular transport process involves molecules passing through endothelial cells. Lipophilic drugs can cross the BBB through passive diffusion, but most therapeutics are hydrophilic and require active transport mechanisms. Certain molecules, including transferrin and insulin, can cross the BBB by binding to receptors on the endothelial cell surface, followed by internalization and transport across the cell via endocytosis (receptor-mediated transcytosis). This mechanism is particularly important for drug delivery strategies involving NPs functionalized with targeting ligands. Similarly, some NPs can cross the BBB through electrostatic interactions with endothelial cells, leading to their internalization (adsorptive-mediated endocytosis). This method is typically used for cationic NPs that interact with negatively charged cell membranes.

5.2.2 Nanoparticles as a Strategy for BBB Penetration

NPs offer several advantages in overcoming the BBB. NPs can be engineered to a size range (typically between 10 and 200 nm) that allows them to pass through endothelial cell junctions and utilize receptor-mediated transport mechanisms. NPs can be functionalized with biomolecules (e.g., antibodies, peptides, or ligands) that target specific receptors on the endothelial cells of the BBB. This enhances the ability of NPs to cross the BBB by receptor-mediated endocytosis or transcytosis. NPs can be engineered to possess an appropriate balance of hydrophobicity and hydrophilicity, allowing for passive diffusion across the BBB or interaction with transport proteins at the endothelial cell membrane. These properties enable NPs to overcome the selective permeability of the BBB and deliver therapeutic agents to the brain effectively.

Several strategies have been developed to enhance the ability of NPs to cross the BBB:

- **Surface Functionalization with Targeting Ligands**

Surface modification of NPs with targeting ligands is one of the most widely studied approaches for improving BBB penetration. By attaching ligands such as transferrin, rabies virus glycoproteins (RVG), or insulin, NPs can exploit receptor-mediated transcytosis to cross the BBB. These ligands bind to specific receptors on the endothelial cells, triggering the internalization of the nanoparticle and its transport across the endothelial cell layer. For example, transferrin-conjugated NPs have been shown to cross the BBB by binding to the transferrin receptor, which is overexpressed on the endothelial cells of the BBB [35]. This approach has been successfully used to deliver chemotherapeutic agents and gene therapy to the brain.

- **Use of Stimuli-Responsive NPs**

Stimuli-responsive NPs are designed to release their drug payload in response to specific environmental triggers. This can include changes in pH, temperature, or the presence of enzymes that are abundant in the brain or in the pathological tissue. pH-sensitive NPs have been developed to release their drug payload when exposed to the acidic environment of certain pathological brain conditions, such as gliomas or brain metastases. The NPs can be designed to degrade in response to the slightly acidic environment of the tumor, facilitating localized drug release [36].

- **Ultrasound-Mediated Drug Delivery**

Ultrasound-mediated drug delivery involves the use of microbubbles and ultrasound waves to temporarily open the BBB and facilitate the passage of NPs. Ultrasound waves generate mechanical forces that increase the permeability of endothelial cells, allowing NPs to enter the brain more efficiently. Ultrasound-mediated BBB disruption has been used to deliver lipid-based NPs loaded with chemotherapeutic agents to brain tumors. The ultrasound waves induce a transient opening of the BBB, allowing the NPs to pass through and deliver their payload [37].

- **Intranasal Delivery of NPs**

Intranasal delivery offers a non-invasive route for delivering NPs directly to the brain. NPs administered via the nasal cavity can travel through the olfactory bulb and the trigeminal nerve, bypassing the BBB and reaching the brain through neural pathways. Polymeric NPs have been delivered intranasally to treat Alzheimer's disease by targeting the brain's cognitive centers. The NPs deliver neuroprotective agents to the brain without the need for systemic circulation [38].

5.3 Infectious Diseases

Infectious diseases remain one of the leading causes of morbidity and mortality worldwide, despite advances in medicine and public health. The rise of antimicrobial resistance and emerging viral infections such as COVID-19 have exacerbated the challenges in treating infections [39]. Traditional antimicrobial agents, including antibiotics and antivirals, are becoming less effective due to the ability of pathogens to develop resistance, and the emergence of novel viruses has underscored the need for innovative treatment approaches. NPs are rapidly emerging as a promising alternative for the treatment of infectious diseases. Their unique physicochemical properties, including their small size, large surface area, and tunable surface functionalities, make them excellent candidates for targeting both bacterial infections and viral diseases. NPs cannot only act as direct antimicrobial agents but can also be used to enhance the efficacy of conventional antibiotics and antivirals, offer targeted delivery, and reduce side effects.

5.3.1 Nanoparticles in Antimicrobial Therapy

NPs exhibit several mechanisms of action against bacteria, making them potent antimicrobial agents. These include, physical disruption of the bacterial membrane, generation of ROS, antimicrobial peptide mimicry, and targeting bacterial DNA and proteins [40]. NPs, especially metallic NPs like silver (AgNPs) and copper (CuNPs), can physically interact with the bacterial cell wall or membrane. The sharp edges of NPs cause damage to the cell wall, leading to membrane disruption, leakage of cellular contents, and eventually bacterial death. Many NPs, including gold (AuNPs) and zinc oxide (ZnO) NPs, can generate ROS under physiological conditions. These ROS induce oxidative stress, leading to damage to bacterial DNA, lipids, and proteins, effectively killing the bacteria. NPs can mimic antimicrobial peptides (AMPs) by interacting with the bacterial membrane. Some lipid-based NPs or polymers exhibit AMP-like activity by disrupting the bacterial membrane and causing membrane potential loss, which results in bacterial death. Some NPs are designed to interact directly with bacterial DNA, inhibiting replication or transcription processes. NPs may also target bacterial enzymes or proteins, interfering with their normal functions and leading to bacterial cell death.

Various types of NPs have been developed for antimicrobial applications, each offering unique advantages:

- **Metallic NPs (AgNPs, CuNPs, ZnONPs)**

Metallic NPs, particularly AgNPs, have shown broad-spectrum activity against a range of bacterial pathogens. AgNPs disrupt bacterial cell membranes and generate ROS, leading to cell death. They are also less prone to the development of resistance compared to traditional antibiotics. For example, AgNPs have been used to treat drug-resistant bacterial infections, such as methicillin-resistant Staphylococcus aureus. Their ability to disrupt cell membranes and generate ROS makes them an effective tool in combating infections that are resistant to conventional antibiotics [41].

- **Polymeric NPs**

Polymeric NPs made from biodegradable polymers like PLGA or chitosan can encapsulate antimicrobial agents and deliver them directly to the site of infection. These NPs can be engineered to release drugs in a controlled manner, prolonging the therapeutic effect and reducing the need for frequent dosing. Chitosan NPs have been shown to possess antimicrobial properties and are used to deliver antibiotics such as gentamicin and ciprofloxacin to infected tissues. Chitosan's natural biocompatibility and antibacterial properties make it a promising material for developing antimicrobial nanoparticle-based drug delivery systems [42].

- **Lipid-Based NPs (Liposomes, SLNs)**

Lipid-based NPs, such as liposomes and SLNs, can encapsulate hydrophobic antimicrobial agents and enhance their solubility and stability. These lipid-based carriers also protect the encapsulated drugs from degradation and provide targeted delivery to infected tissues. Liposome-encapsulated antibiotics like doxorubicin have been studied for their ability to target tuberculosis infections, providing a more controlled and localized release of the drug, thereby improving therapeutic outcomes.

The use of NPs in antimicrobial therapy provides some advantages. The use of NPs can help mitigate the growing problem of antibiotic resistance, as they do not induce the same resistance mechanisms as traditional antibiotics. Also, NPs can be engineered to target specific bacterial strains, ensuring that the antimicrobial agents reach the infection site with minimal systemic exposure. Furthermore, NPs can provide synergistic effects when combined with conventional antibiotics, enhancing their antimicrobial activity. However, there are some challenges remain in this approach such as nanoparticles toxicity. Some NPs, particularly metallic NPs, can exhibit cytotoxicity at high concentrations. Careful optimization of nanoparticle size and dosage is necessary to ensure safety. In addition, the production of NPs at a commercial scale while maintaining consistent quality remains a challenge.

5.3.2 Nanoparticles in Antiviral Therapy

NPs offer several mechanisms for inhibiting viral infections. These include blocking viral entry, inhibition of viral replication, and immunomodulation [43]. NPs can bind to the viral surface proteins (e.g., spike proteins of coronaviruses) and prevent the virus from attaching to and entering host cells. This prevents the initiation of the viral lifecycle. Some NPs can interfere with the viral replication machinery, either by inhibiting viral enzymes or by directly interacting with the viral genome. NPs can enhance the body's immune response by promoting the release of interferons or activating T cells (immunomodulation). This helps the body fight the infection more effectively.

Various types of NPs have been developed for antiviral applications, each offering unique advantages:

- **Metallic NPs (Silver, Gold)**

AgNPs have demonstrated potent antiviral activity against a wide range of viruses, including HIV, influenza, and herpes simplex virus (HSV). The antiviral activity of AgNPs is thought to result from their ability to interact with viral proteins and disrupt the virus structure. For example, AgNPs have been shown to inhibit HSV infection by binding to the viral envelope and preventing the virus from attaching to host cells [44].

- **Polymeric NPs**

Polymeric NPs can be used to deliver antiviral drugs like acyclovir for herpes virus infections or oseltamivir for influenza. The controlled release of these drugs from polymeric NPs can improve their pharmacokinetics and reduce side effects. Polymeric NPs have been used to deliver antiviral agents to the respiratory tract for the treatment of influenza, improving drug bioavailability and reducing the need for frequent dosing [45].

- **Liposomes and SLNs**

Liposomes and SLNs can encapsulate antiviral drugs, protecting them from enzymatic degradation and improving their delivery to infected tissues. Additionally, liposomes can be functionalized with viral surface protein-specific ligands for targeted delivery. Liposome-encapsulated antiviral agents, such as zidovudine, are being investigated for their potential in HIV therapy, allowing for prolonged circulation times and increased drug accumulation at the site of infection [46].

The use of NPs in Antiviral therapy provides some advantages. NPs can be engineered to target a wide range of viruses, including both DNA and RNA viruses. Similar to antimicrobial therapy, NPs are less likely to induce viral resistance compared to conventional antiviral drugs. NPs can be designed to deliver antiviral agents directly to the site of infection, increasing efficacy and reducing systemic side effects. However, there are some challenges remain in this approach such as toxicity and regulatory issues. Some NPs, particularly metallic NPs, can exhibit toxicity at high concentrations. The safety profile of NPs must be carefully evaluated before clinical use. The approval process for nanoparticle-based drugs is complex and requires thorough preclinical and clinical testing to ensure safety and efficacy.

5.4 *Cardiovascular Diseases*

Chronic diseases, including cardiovascular diseases and inflammatory diseases, are among the leading causes of global morbidity and mortality. Traditional treatments often involve medications with significant side effects, long-term use, and limited efficacy in specific populations. Over the past few decades, the development of nanoparticle-based therapies has emerged as a promising solution for enhancing the delivery, efficacy, and targeting of therapeutic agents. NPs, with their unique physicochemical properties, such as size, surface charge, surface functionalization, and biodegradability, offer significant advantages over conventional drug delivery systems. This chapter explores the applications of NPs in the treatment of cardiovascular diseases and inflammatory diseases, focusing on how NPs are used to target specific tissues, enhance bioavailability, and provide controlled release for long-term therapeutic benefits [47].

Cardiovascular diseases, including atherosclerosis, myocardial infarction, heart failure, and stroke, are the leading causes of death worldwide. Despite significant advances in pharmacology, many current treatments for cardiovascular diseases are limited by issues such as poor bioavailability, side effects, and limited specificity to the diseased tissues. NPs offer several advantages for overcoming these challenges, including targeted drug delivery, improved solubility, and prolonged circulation.

5.4.1 Mechanisms of Nanoparticle Action in Cardiovascular Diseases

NPs can improve cardiovascular therapy through a variety of mechanisms. NPs can be engineered to deliver therapeutic agents directly to atherosclerotic plaques, the endothelial lining, or infarcted myocardial tissues, improving the effectiveness of the drug and reducing off-target effects. For example, NPs can be functionalized with ligands targeting vascular cell adhesion molecules (VCAM-1), which are overexpressed in the inflamed endothelial cells of atherosclerotic lesions. Liposomes and polymeric NPs can increase the permeability of endothelial cell membranes, allowing for better drug penetration into the targeted tissues, such as atherosclerotic plaques or ischemic heart tissue. NPs can be designed to sustainably release drugs over extended periods, reducing the frequency of dosing and improving therapeutic adherence. This controlled release is particularly useful for anti-inflammatory agents and antioxidants, which are commonly used to treat cardiovascular diseases. NPs can serve as both therapeutic and diagnostic agents, allowing for real-time monitoring of drug delivery and therapeutic efficacy. AuNPs or iron oxide NPs are frequently used in cardiovascular imaging, providing high-resolution contrast for magnetic resonance imaging and photoacoustic imaging [47].

5.4.2 Applications of Nanoparticles in Cardiovascular Diseases

Atherosclerosis Disease: Atherosclerosis is a chronic inflammatory disease of the arteries, characterized by the accumulation of lipids, foam cells, and extracellular matrix components in the arterial walls, forming plaques that obstruct blood flow. NPs are used to target and treat atherosclerotic plaques more effectively. NPs loaded with statins or anti-inflammatory agents can specifically accumulate in atherosclerotic plaques due to the EPR effect. NPs can reduce inflammation, stabilize plaques, and even reverse some aspects of the disease. Polymeric NPs can be functionalized with VCAM-1 antibodies or peptides to target the inflammatory endothelial cells at the site of plaques, improving the specificity of drug delivery [48].

Myocardial Infarction (Heart Attack): After a heart attack, cardiac tissue undergoes necrosis, and its ability to regenerate is limited. NPs can aid in cardiac tissue regeneration by delivering growth factors, stem cells, or genetic material to promote healing and reduce scarring. NPs loaded with stem cells or cardioprotective drugs like angiotensin-converting enzyme inhibitors can be injected into the myocardium to promote tissue regeneration and prevent post-infarction remodeling [49].

Heart Failure: In heart failure, the heart's ability to pump blood is impaired, leading to fluid retention and decreased oxygen supply to tissues. NPs can help deliver heart failure medications, such as beta-blockers, diuretics, and anti-hypertensive agents, directly to the heart. Liposomes loaded with beta-blockers can specifically target cardiac tissue, improving heart function without affecting other systems.

Stroke: Stroke involves the interruption of blood flow to the brain, leading to neuronal damage. NPs can be used to deliver neuroprotective drugs, such as NMDA receptor antagonists, growth factors, and antioxidants, to the brain after stroke. Liposomes containing neurotrophic factors such as brain-derived neurotrophic factor can be used to promote neuronal repair and reduce ischemic damage.

5.5 Nanoparticles in Inflammatory Diseases

Inflammatory diseases such as rheumatoid arthritis, inflammatory bowel disease (IBD), and systemic lupus erythematosus (SLE) are characterized by chronic inflammation that leads to tissue damage and organ dysfunction. The conventional therapies for these diseases, including nonsteroidal anti-inflammatory drugs (NSAIDs) and glucocorticoids, often have significant side effects, including gastric ulcers, immune suppression, and osteoporosis. NPs offer several advantages in treating these conditions by enabling targeted drug delivery and localized treatment, thus reducing side effects and improving therapeutic efficacy [50].

5.5.1 Mechanisms of Nanoparticle Action in Inflammatory Diseases

NPs can be engineered to selectively target the sites of inflammation. By functionalizing NPs with specific ligands such as integrin peptides or lectins, drugs can be delivered directly to the inflamed tissues, such as the synovium in rheumatoid arthritis or the intestinal mucosa in IBD. They can also deliver anti-inflammatory agents like corticosteroids and NSAIDs directly to the inflamed tissues, providing more localized effects and reducing the risk of systemic side effects. In addition, some NPs, especially those derived from lipids or polymers, can be used to modulate the immune response. For example, NPs can deliver immunomodulatory agents to regulate the activity of T-cells or macrophages, which are often overactive in chronic inflammation.

5.5.2 Applications of Nanoparticles in Inflammatory Diseases

Rheumatoid Arthritis (RA): RA is an autoimmune disease characterized by chronic inflammation of the joints. The delivery of anti-inflammatory agents or disease-modifying antirheumatic drugs using NPs has shown promise in reducing joint damage and improving patient outcomes. Polymeric NPs have been used to deliver methotrexate to inflamed joints, where it can be released in a controlled manner, reducing the need for oral dosing and minimizing systemic toxicity [51].

Inflammatory Bowel Disease: IBD, which includes Crohn's disease and ulcerative colitis, is characterized by inflammation of the gastrointestinal tract. NPs can be used to deliver biologics like TNF-α inhibitors and immune modulators directly to the site of inflammation in the gut, improving local treatment efficacy. Liposome-encapsulated corticosteroids have been studied for the treatment of ulcerative colitis, offering targeted delivery and reducing the side effects associated with systemic steroid use.

Systemic Lupus Erythematosus (SLE): SLE is a systemic autoimmune disease characterized by widespread inflammation affecting multiple organs. NPs can be used to deliver immunosuppressive agents and anti-inflammatory drugs to specific tissues, reducing disease activity and preventing organ damage. Dendrimers loaded with hydrocortisone have been used to modulate the immune response in lupus, targeting specific immune cells and tissues involved in the disease [52].

NPs offer a promising approach to the treatment of cardiovascular and inflammatory diseases by enhancing drug targeting, providing controlled release, and reducing systemic toxicity. The ability to deliver therapeutic agents directly to the site of disease allows for more effective treatments with fewer side effects. While challenges related to toxicity, scalability, and regulation remain, ongoing advancements in nanoparticle design, functionalization, and manufacturing are expected to overcome these hurdles, making nanoparticle-based therapies a key tool in the future of chronic disease management.

6 Future Perspectives and Conclusions

Environmental sustainability is becoming increasingly important in nanomedicine. Green synthesis of NPs using plant extracts or bio-based materials offers an eco-friendly and cost-effective alternative to traditional nanoparticle synthesis methods, with reduced toxicity and environmental impact. There are also some concerns about the biocompatibility and toxicity of NPs. Despite their potential, NPs may accumulate in organs like the liver and spleen, leading to toxicity. Furthermore, NPs could induce immune responses or cause unwanted inflammatory reactions. Extensive studies are needed to evaluate the long-term safety of NPs for clinical use. Another challenge is immune system interactions and clearance. NPs are often recognized by the immune system as foreign particles, leading to rapid clearance through the liver and spleen. Strategies such as PEGylation are employed to reduce immune recognition and extend circulation time, but challenges remain in ensuring that NPs reach their target site before being cleared [27].

The scale-up of nanoparticle production is challenging due to the need for reproducible formulations, precise control over size and shape, and regulatory compliance. Advances in manufacturing techniques, such as microfluidic devices, are being explored to overcome these challenges [31, 53, 54]. The regulatory approval process for nanoparticle-based therapies is complex, requiring extensive preclinical and clinical testing to demonstrate safety and efficacy. Regulatory frameworks are still evolving to address the unique challenges posed by nanomedicine.

Recent advancements have focused on the development of multifunctional NPs that combine drug delivery with diagnostic capabilities, known as theranostics. These NPs allow for real time monitoring of therapeutic efficacy while simultaneously delivering the therapeutic agent, making them an ideal choice for personalized medicine [55]. Personalized and precision nanomedicine is also a trending approach. Personalized medicine involves tailoring medical treatment to the individual characteristics of each patient. By combining targeting strategies with personalized therapeutic agents based on the patient's genetic profile, nanoparticle-based therapies can improve treatment outcomes and reduce adverse effects [56]. Integration of artificial intelligence (AI) and machine learning in nanoparticle design in another revolutionary approach. The integration of AI and machine learning is revolutionizing nanoparticle drug delivery by enabling better prediction of nanoparticle behavior, optimization of drug release profiles, and design of NPs tailored to specific disease conditions.

The future success of nanoparticle-based drug delivery will depend on the integration of knowledge across disciplines, including chemistry, biology, engineering, and medicine. Collaboration among scientists, clinicians, and regulatory bodies will be crucial for overcoming current challenges and bringing nanoparticle-based therapies to market.

References

1. Ahmadi, M.: Iron oxide nanoparticles for delivery purposes. In: Nanoengineered Biomaterials for Advanced Drug Delivery, pp. 373–393. Elsevier (2020)
2. Ahmadi, M., et al.: Stimuli-sensitive drug delivery systems. In: Nanoengineered Biomaterials for Advanced Drug Delivery, pp. 37–59. Elsevier (2020)
3. Dastghaib, S., et al.: Smart nanocarriers in glucose transporters-targeted delivery of anticancer drugs. In: Smart Nanocontainers, pp. 251–269. Elsevier (2020)
4. Khademzadeh, N., et al.: Microfluidic aerosol-assisted synthesis of gefitinib anticancer drug nanocarrier based on chitosan natural polymer. J. Drug Deliv. Sci. Technol. **99**, 105992 (2024)
5. Mahdian, M., et al.: Dual stimuli-responsive gelatin-based hydrogel for pH and temperature-sensitive delivery of curcumin anticancer drug. J. Drug Deliv. Sci. Technol. **84**, 104537 (2023)
6. Seyfoori, A., et al.: Emerging advances of nanotechnology in drug and vaccine delivery against viral associated respiratory infectious diseases (VARID). Int. J. Mol. Sci. **22**(13), 6937 (2021)
7. Weng, X., et al.: Lipidomics-driven drug discovery and delivery strategies in glioblastoma. Biochim. Biophys. Acta (BBA)-Mol. Basis Dis. 167637 (2024)
8. Wu, J.: The enhanced permeability and retention (EPR) effect: the significance of the concept and methods to enhance its application. J. Pers. Med. **11**(8), 771 (2021)
9. Nooreen, R., et al.: Polymer nanotherapeutics: a versatile platform for effective rheumatoid arthritis therapy. J. Control. Release **348**, 397–419 (2022)
10. Sultana, S., et al.: Nanoparticles-mediated drug delivery approaches for cancer targeting: a review. J. Drug Target. **21**(2), 107–125 (2013)
11. Yoo, J., et al.: Active targeting strategies using biological ligands for nanoparticle drug delivery systems. Cancers **11**(5), 640 (2019)
12. Sitia, L., et al.: HER-2-targeted nanoparticles for breast cancer diagnosis and treatment. Cancers **14** (2022). https://doi.org/10.3390/cancers14102424
13. Sanati, M., et al.: RGD-engineered nanoparticles as an innovative drug delivery system in cancer therapy. J. Drug Deliv. Sci. Technol. **84**, 104562 (2023)
14. Kemp, J.A., et al.: "Combo" nanomedicine: co-delivery of multi-modal therapeutics for efficient, targeted, and safe cancer therapy. Adv. Drug Deliv. Rev. **98**, 3–18 (2016)
15. Karimi, M., et al.: Temperature-responsive smart nanocarriers for delivery of therapeutic agents: applications and recent advances. ACS Appl. Mater. Interfaces **8**(33), 21107–21133 (2016)
16. Yang, Y., Sun, W.: Recent advances in redox-responsive nanoparticles for combined cancer therapy. Nanoscale Adv. **4**(17), 3504–3516 (2022)
17. Li, M., et al.: Enzyme-responsive nanoparticles for anti-tumor drug delivery. Front. Chem. **8** (2020)
18. Tao, Y., et al.: Light: a magical tool for controlled drug delivery. Adv. Func. Mater. **30**(49), 2005029 (2020)
19. Al-Eryani, Y., et al.: Toxicity, therapeutic applicability, and safe handling of magnetic nanomaterials. In: Magnetic Nanomaterials in Analytical Chemistry, pp. 61–83. Elsevier (2021)
20. Mitchell, M.J., et al.: Engineering precision nanoparticles for drug delivery. Nat. Rev. Drug Discov. **20**(2), 101–124 (2021)
21. Opriş, O., et al.: An overview of biopolymers for drug delivery applications. Appl. Sci. **14**(4), 1383 (2024)
22. Seidu, T.A., Kutoka, P.T.: Functionalization of nanoparticulate drug delivery systems and its influence in cancer therapy. **14**(5) (2022)
23. Zalba, S., et al.: Stealth nanoparticles in oncology: facing the PEG dilemma. J. Control. Release **351**, 22–36 (2022)
24. Liu, Y., Crawford, B.M., Vo-Dinh, T.: Gold nanoparticles-mediated photothermal therapy and immunotherapy. Immunotherapy **10**(13), 1175–1188 (2018)
25. Yu, M.K., Park, J., Jon, S.: Targeting strategies for multifunctional nanoparticles in cancer imaging and therapy. Theranostics **2**(1), 3 (2012)
26. Maeda, H., et al.: Tumor vascular permeability and the EPR effect in macromolecular therapeutics: a review. J. Control. Release **65**(1), 271–284 (2000)

27. Allen, T.M., Cullis, P.R.: Liposomal drug delivery systems: from concept to clinical applications. Adv. Drug Deliv. Rev. **65**(1), 36–48 (2013)
28. Shi, P., et al.: Active targeting schemes for nano-drug delivery systems in osteosarcoma therapeutics. J. Nanobiotechnol. **21**(1), 103 (2023)
29. Untch, M., et al.: Nab-paclitaxel versus solvent-based paclitaxel in neoadjuvant chemotherapy for early breast cancer (GeparSepto—GBG 69): a randomised, phase 3 trial. Lancet Oncol. **17**(3), 345–356 (2016)
30. Li, Y., Jeon, J., Park, J.H.: Hypoxia-responsive nanoparticles for tumor-targeted drug delivery. Cancer Lett. **490**, 31–43 (2020)
31. Alizadeh, H., Ahmadi, M., Heydari Shayesteh, O.: On chip synthesis of a pH sensitive gefitinib anticancer drug nanocarrier based on chitosan/alginate natural polymers. Sci. Rep. **14**(1), 772 (2024)
32. Liu, W., et al.: RGD peptide-conjugated selenium nanocomposite inhibits human glioma growth by triggering mitochondrial dysfunction and ROS-dependent MAPKs activation. Front. Bioeng. Biotechnol. **9**, 781608 (2021)
33. Dabiri, S.M.H., et al.: Potential theranostic targets in glioblastoma. In: New Insights Into Glioblastoma, pp. 631–665. Academic Press (2023)
34. Vitorino, C., Balana, C., Cabral, C.: New Insights into Glioblastoma: Diagnosis, Therapeutics and Theranostics. Elsevier (2023)
35. Pardridge, W.M.: The blood-brain barrier: bottleneck in brain drug development. NeuroRx **2**, 3–14 (2005)
36. Pinheiro, R.G., et al.: Nanoparticles for targeted brain drug delivery: what do we know? Int. J. Mol. Sci. **22**(21), 11654 (2021)
37. Aryal, M., et al.: Ultrasound-mediated blood–brain barrier disruption for targeted drug delivery in the central nervous system. Adv. Drug Deliv. Rev. **72**, 94–109 (2014)
38. Chen, Y., et al.: Intranasal drug delivery: the interaction between nanoparticles and the nose-to-brain pathway. Adv. Drug Deliv. Rev. **207**, 115196 (2024)
39. Peymani, P., et al.: Statins in patients with COVID-19: a retrospective cohort study in Iranian COVID-19 patients. Transl. Med. Commun. **6**, 1–14 (2021)
40. Mondal, S.K., et al.: Antimicrobial nanoparticles: current landscape and future challenges. RSC Pharmaceut. (2024)
41. Deshmukh, S.P., et al.: Silver nanoparticles as an effective disinfectant: a review. Mater. Sci. Eng. C **97**, 954–965 (2019)
42. Rashki, S., et al.: Chitosan-based nanoparticles against bacterial infections. Carbohyd. Polym. **251**, 117108 (2021)
43. Rios-Ibarra, C.P., et al.: Nanoparticle-based antiviral strategies to combat the influenza virus. Biomed. Rep. **20**(4), 65 (2024)
44. Zhang, X.-F., et al.: Silver nanoparticles: synthesis, characterization, properties, applications, and therapeutic approaches. Int. J. Mol. Sci. **17**(9), 1534 (2016)
45. Delshadi, R., et al.: Development of nanoparticle-delivery systems for antiviral agents: a review. J. Control. Release **331**, 30–44 (2021)
46. Reshna, K.R., Balakrishnan, P., Gopi, S.: Systematic review on activity of liposomal encapsulated antioxidant, antibiotics, and antiviral agents. J. Liposome Res. **32**(4), 340–353 (2022)
47. Ullah, A., Ullah, M., Lim, S.I.: Recent advancement in nanotechnology based drug delivery for the management of cardiovascular disease. Curr. Probl. Cardiol. 102396 (2024)
48. Chen, W., et al.: Macrophage-targeted nanomedicine for the diagnosis and treatment of atherosclerosis. Nat. Rev. Cardiol. **19**(4), 228–249 (2022)
49. Wei, Q., et al.: Advances in nanoparticles in the prevention and treatment of myocardial infarction. Molecules **29**(11), 2415 (2024)
50. Brusini, R., Varna, M., Couvreur, P.: Advanced nanomedicines for the treatment of inflammatory diseases. Adv. Drug Deliv. Rev. **157**, 161–178 (2020)
51. Siddique, R., et al.: Promising role of polymeric nanoparticles in the treatment of rheumatoid arthritis. Inflammopharmacology **30**(4), 1207–1218 (2022)

52. Rawding, P.A., et al.: Dendrimers for cancer immunotherapy: avidity-based drug delivery vehicles for effective anti-tumor immune response. Wiley Interdiscip. Rev. Nanomed. Nanobiotechnol. **14**(2), e1752 (2022)
53. Ghoorchian, A., et al.: Lab-on-a-chip miniaturized analytical devices. In: Micro- and Nanotechnology Enabled Applications for Portable Miniaturized Analytical Systems, pp. 261–284 (2021)
54. Pattanayak, P., et al.: Microfluidic chips: recent advances, critical strategies in design, applications and future perspectives. Microfluid. Nanofluid. **25**(12), 99 (2021)
55. Bauri, S., et al.: Nanomaterials as theranostic agents for cancer therapy. ACS Appl. Nano Mater. **6**(23), 21462–21495 (2023)
56. Mahmood, A.A.R., Jha, A.M., Manivannan, K.: Precision medicine: personalizing the fight against cancer. Int. J. Trends OncoSci. 10–18 (2024)

Nanocarriers for Chemotherapy and Drug Delivery

Ayesha Latif, Laila Maryam, Adil Abid Khan, and Muhammad Abdul Basit

Abstract The development of nanocarriers for targeted drug delivery has advanced significantly in the field of nanomedicine recently, and they assure to enhance efficacy and reducing side effects related to traditional chemotherapy. Gold, magnetic, polymeric, carbon-based, and silver nanoparticles are most commonly used as nanocarriers for chemotherapy. They exhibit different properties according to their shape, size and surface functionality for specific applications. These nanocarriers have wide the range of applications including targeted drug delivery, CNS drug delivery, pharmaceutical industry, tissue regeneration and gene therapy. Although they have some challenges and limitations, ongoing research hopes to eradicate them. This chapter presents a summary of mechanism, design and characterization, toxicity, applications and future advancement of nanocarriers in targeted drug delivery and chemotherapy.

Keywords Drug delivery · Nanocarriers · Chemotherapy · Nanomedicine · Tumors

1 Introduction

Unregulated cell growth and avoidance of cell death are features common to a group of disorders known as cancer. Chemotherapy is a treatment used for several types of cancers specifically lungs cancer, breast cancer, leukemia, lymphoma, colon cancer,

A. Latif · L. Maryam · A. A. Khan · M. A. Basit (✉)
Department of Materials Science & Engineering, Institute of Space Technology, Islamabad, Pakistan
e-mail: ab_saim@hotmail.com; a.basit@ist.edu.pk

A. Latif
e-mail: ayeshalatif323@gmail.com

L. Maryam
e-mail: laila.maryam14@gmail.com

A. A. Khan
e-mail: adilabidkhan02@gmail.com

N. J. Joshi and G. Patel (eds.), *Nanoparticle Delivery in Cancer Care*,
Series in BioEngineering, https://doi.org/10.1007/978-981-95-0457-2_5

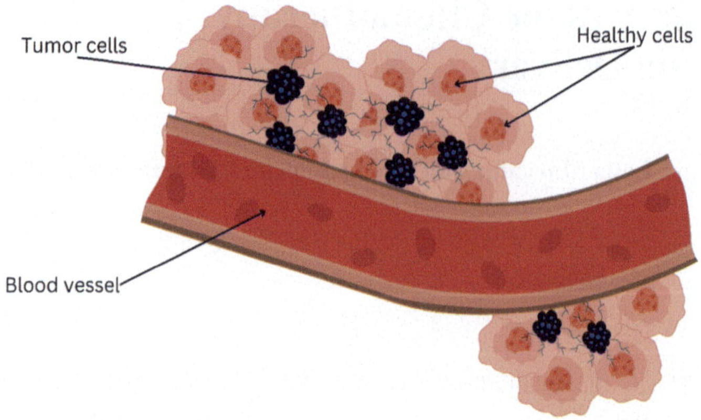

Fig. 1 Schematic diagram of tumor cells

ovarian cancer, and prostate cancer. Tumors are abnormal cell masses formed when cancer cells multiply. A brain tumor is an abnormal development of cells, either benign or malignant (cancerous) [1]. Malignant brain tumors are known for their complexity and unsatisfactory prediction as shown in Fig. 1. The variations of these tumors are due to the presence of cancer stem cells (CSC), which initiate tumor growth [2].

One of the therapies for brain tumors is chemotherapy. Over the past 15 years, it has become essential for palliative care [3]. The primary aim of chemotherapy is to kill cancer cells and limit their production, by which they can prevent further growth and spread. Types of chemotherapy involved monotherapy and combination chemotherapy. Combination chemotherapy and having multiple drugs are mostly prescribed [4]. The advanced field focused on materials with dimensions in the nanoscale range is called nanotechnology. A carrier system with dimensions ranging from 1 to 1000 nm is referred to as a "nanoparticle." Recently, the field of nanomedicine has received significant improvements in the evolution of nanocarriers and lower harmful effects associated with conventional chemotherapy. There are several nanocarriers having improved efficacy and better targeting due to their reduced size in nanometers. Nanocarriers are advantageous in terms of therapeutics regardless of production cost and stability issues [5].

2 Types of Nanocarriers

There are a variety of nanocarriers available for chemotherapy as indicated in Fig. 2.

Table 1 presents various types of nanocarriers, along with their shapes, properties, and applications. Due to their distinct forms and characteristics, these nanocarriers are used for therapeutic, diagnostic, and targeted drug delivery.

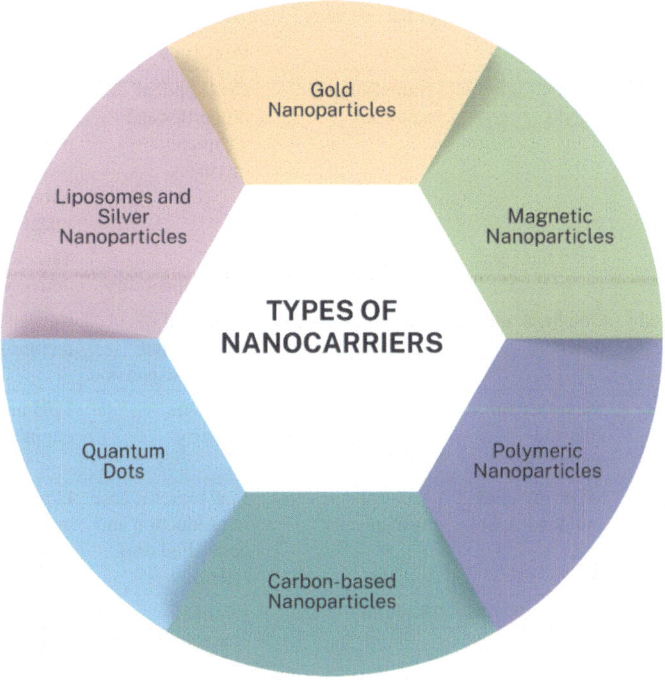

Fig. 2 Nanocarriers for chemotherapy and drug delivery

2.1 Gold Nanoparticles

Gold nanoparticles have exceptional chemical inertness and optical properties, facilitating their surface altering very easily, which increases their application in various fields, not only targeted drug delivery, but also in photothermal therapy and photoacoustic imaging. They are highly biocompatible and stable nanocarriers with good permeability inside living cells, making them suitable for applications in intelligent diagnosis and treatment [6]. Biological stimuli or light activation can be used to transport and control drug release from gold nanoparticle surfaces. Drugs can also attach themselves to surfaces by physical absorption and ionic or covalent bonding [7].

ZD-CAR nanoparticles were produced through the use of Au–S covalent bonds to bind mercaptopropionic acid to the surface of ZD-CA nanoparticles. Figure 3 shows the SEM and TEM characterization of ZD-CAR nanoparticles.

Table 1 Overview of different types of nanocarriers for chemotherapy and drug delivery [6–14]

Types	Material/drug	Shape	Properties	Applications
Metals nanoparticles	AuNPs AgNPs	Nanoshell and nanorod	Good optical properties and chemical inertness	Targeted drug delivery, photothermal therapy and photoacoustic imaging
Liposomes	Lipid bilayers	Nanospheres	Good biocompatibility	Targeted drug delivery
Carbon-based nanoparticles	Graphene and fullerene	Carbon nanotube and nanographite	Structural stiffness and thermal expansion	Cancer theranostics
Magnetic nanoparticles	Nickel, cobalt, zinc and magnesium	Cube and octahedron	Magnetic property	MRI and tissue engineering
Polymeric nanoparticles	MOF, micelles and dendrimers	Nanocapsules and nanospheres	Low toxicity, ease in production and controlled drug release,	Drug targeting and immunotherapy
Quantum dots	Semiconductors	Bowl-like shape	Luminous characteristics	Therapeutic and diagnostic applications, in vivo imaging and tissue engineering

2.2 *Magnetic Nanoparticles*

These anticancer drug carriers, which also serve as contrast agents in magnetic resonance imaging (MRI) and exhibit greater magnetic susceptibility than traditional contrast agents, are prime examples of theragnostic nanosystems. Their inorganic cores consist of iron oxide, manganese oxide, or other magnetic materials [9]. The functionality of these magnetic particles can be enhanced by adding fluorescent dyes. Once the anticancer drugs are delivered and accumulate in tumors (chemotherapy), the magnetic nanoparticles can be subjected to alternating magnetic fields, generating heat to perform thermal ablation on cancers located in deep or difficult-to-access areas [10].

2.3 *Polymeric Nanoparticles*

Biodegradable polymers make up the nanoscale colloidal particles known as polymeric nanoparticles. Two types of polymeric nanoparticles are nanocapsules and

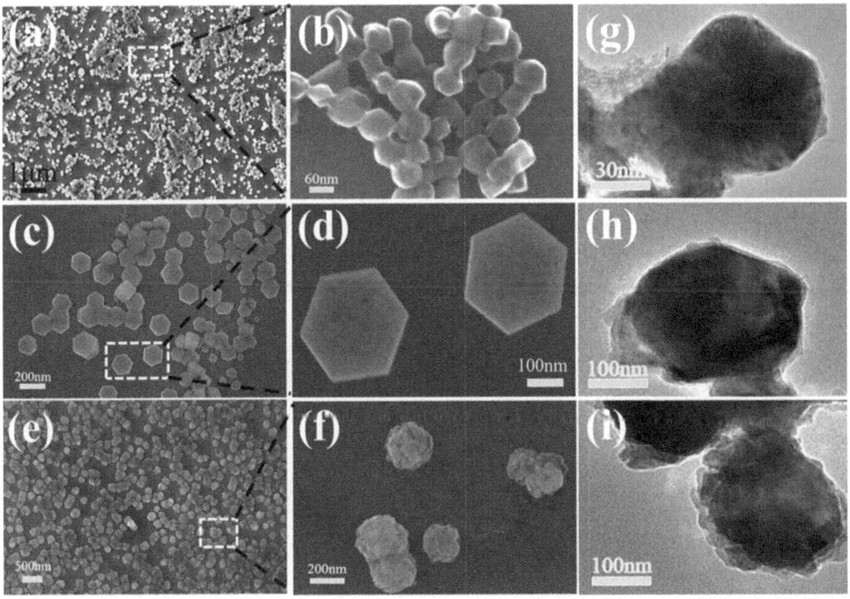

Fig. 3 SEM images of Zn-Co ZIF are shown in **a**, **b**, ZD-C in **c**, **d**, and ZD-CAR nanoparticles in **e**, **f**. TEM images display Zn-Co ZIF in **g**, ZD-C in **h**, and ZD-CAR in **i** [8] (copyrights 2024 frontiers)

nanospheres based on their structure. A range of synthetic and natural biocompatible polymers are used in their creation. Their fundamental qualities that make them an excellent choice for brain targeted therapy are their low toxicity, low cost, ease in production, controlled drug release, low immunogenic response, pharmacotechnological stability, and biodegradability. Uncertain potential toxicity and slow degradability are some disadvantages of polymeric nanoparticles [11].

2.4 *Carbon-Based Nanoparticles*

Since carbon-based nanocarriers may change their structure, their structural stiffness, thermal expansion, and high surface-to-volume ratio make them useful for biological research. Carbon nanotubes (CNTs), have seen widespread application in cancer theranostics. NIR fluorescent imaging tools and Raman probes are some applications of CNTs due to their scattering characteristics'. Based on their sizes and shapes, carbonaceous nanomaterials can also be further classified as fullerene (C60), nanotubes, nanodots, nanodiamonds, and graphene derivatives [12].

2.5 *Quantum Dots*

These are the colloidal semiconductor nanocrystals with diameters of 2–10 nm. They exhibit distinct luminous characteristics that mainly include cadmium compounds that contain sulfide, selenide, and telluride. It can be created from different kinds of semiconductor material by colloidal synthesis or electrochemistry. Quantum dots can be used for both diagnostic and therapeutic applications, including chemotherapy and drug delivery. The careful management of electron size, shape, and amount allows for the bonding of therapeutic molecules for medication administration, in vivo imaging, tissue engineering, and other applications [13].

2.6 *Liposomes*

Numerous hydrophilic anticancer medications can be trapped in liposomes, which are self-assembling colloidal vesicles with an aqueous core and phospholipid bilayer membrane. Liposomes contain natural phospholipids, making them biocompatible and slightly cytotoxic [14]. Liposomes' surface functionalization allows them to effectively and uniformly target tumor cells. Liposomes have a long circulation time in the blood. They also have an easily modified surface. It is possible to alter the physiochemical characteristics of liposomes by combining several commercial lipid compounds.

2.7 *Silver Nanoparticles*

Silver nanoparticles are commonly used as antibacterial agents in both medicine and various other sectors. Their effectiveness in treating multidrug-resistant (MDR) cancer is closely linked to their size. Metal nanoparticles, particularly silver and gold, have been widely studied for cancer treatment. Bioactive molecules, such as antibodies, can be attached to the surface of these nanoparticles to target cancer cells. The performance of metal nanoparticles is primarily influenced by factors like their size, shape, and surface properties.

3 Drug Delivery Mechanisms Using Nanocarriers

It has been discovered that nanodelivery equipment enhances biodistribution and administration routes for medications with little immunogenicity and negative effects. Drugs are either actively or passively delivered to the location of illness using nanocarriers. Nanocarriers can minimize the side effects of medical therapy by administering drugs to specific locations in a more regulated and decreased dose manner.

3.1 Mechanism of Size-Shrinkage

Nanocarriers with smaller sizes (e.g., less than 50 nm) penetrate deeper into tumor tissues, but the high interstitial fluid pressure inside the tumor makes it easy to pump them back into the bloodstream. Larger nanocarriers, on the other hand, have difficulty penetrating yet benefit from retention at tumor sites [15]. The extracellular matrix of tumor is complicated and can block larger nanocarriers, whereas smaller particles can easily pass through it. This shows a correlation between nanocarrier retention and penetration, which is based on optimizing nanocarrier size for improved dispersion in tumor tissues. To achieve equilibrium between the retention and penetration effectiveness of nanocarriers in tumor sites, researchers created a series of size-deformable nanocarriers that improve both retention and penetration, resulting in better anti-tumor efficacy. Acidic pH-triggered size shrinkage and light and redox-responsive induced size shrinkage are some categories of size-shrinkage mechanisms. Chemotherapy remains limited due to inefficient drug delivery caused by the tumor's heterogeneous nature. To address this, an intelligent nanoparticle (IDDHN) was developed, featuring hyaluronic acid shells that shrink in response to hyaluronidase (HAase). As illustrated in Fig. 4, this nanoparticle was combined with IDD containing DOX as a chemotherapy drug, ICG as a photothermal agent, and a nitric oxide donor (HN) that responds to near-infrared (NIR) laser light [16].

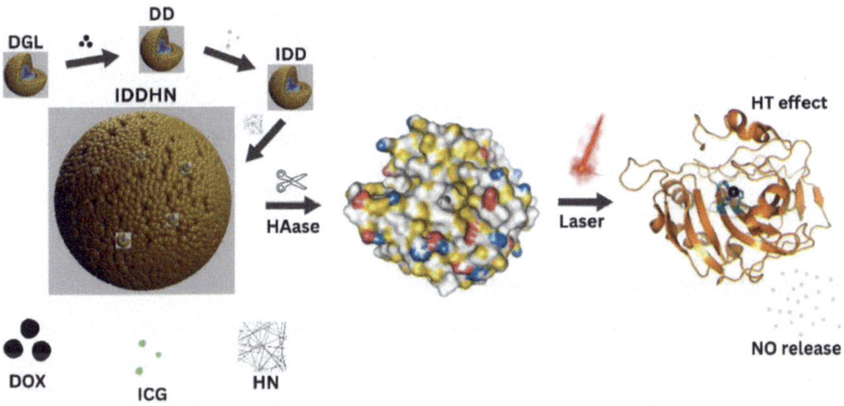

Fig. 4 Schematic diagram of IDDHN

3.2 Aggregative Properties

Statistical data covering the previous few decades in the field of nanocarrier-based drug delivery suggests that while almost seventy percent of nanocarriers can reach the tumor site following systemic the procedure, only a small number (roughly 0.7%, median) settle and concentrate there. Nanomedicines reduced developing efficiency decreases their therapeutic effectiveness when given systemically [17]. Large-sized nanocarriers are often kept for an extended period of time in tumor tissues diverse vascular environments. Researchers have developed an aggregation technique to craft tumor stimulus-responsive nanocarriers for medicine delivery. This technique enables nanocarriers to shift from small to larger sizes at the tumor site, improving retention. Aggregative properties of drug delivery include salt-induced aggregation, redox-responsive aggregation, light-triggered aggregation, enzyme-induced aggregating nanocarriers, and pH-responsive crosslinking nanocarriers.

3.3 Stimuli-Triggered Nanocarriers

It seems that stimuli-responsive nanoparticles will provide excellent targets for anticancer drugs [18]. Platinum-based anticancer medicines are essential for treating a variety of cancers, but their usage is restricted due to severe adverse effects and the growth of drug resistance. The development of stimuli-responsive nanocarrier systems offers a practical approach for increasing the efficacy of platinum-based medications while minimizing negative effects. The use of ultrasound provides spatiotemporal control to ensure that medications are released at the appropriate spot, preventing adverse effects on healthy organs. The non-invasive nature of ultrasounds, their lack of ionizing radiation, and their ease of control over the depth of tissue penetration with simple adjustments to frequency, duty cycle, and exposure duration make them a desirable tool. Light irradiation has another application besides inducing or increasing medication release: it can be an antidote. Higher band gap molecular orbitals will result in more light-induced photon energy [19].

3.4 Changing Shapes Capabilities

Therapeutic efficacy can be increased by using nanocarriers that exhibit tumor-specific morphological changes, such as fibrotic transformation, particle swelling, and phase transitions. These changes can improve tumor penetration, retention, immune evasion, and tumor cell internalization. Specific nanocarrier morphologies, such as fibrous structures, allow for increased penetration across tumor interstitial gaps due to their elongated shapes, resulting in more accurate treatment distribution. Following morphological transformation, these nanocarriers may remain within the tumor tissue for an prolonged duration, allowing for controlled drug release [20]. Additionally, certain morphological changes can improve the [21] interaction of nanocarriers with cellular membranes, leading to increased effectiveness. A new approach to cancer treatment involves the strategic design of nanocarriers that undergo shape changes at tumor sites, enhancing drug delivery efficiency and improving therapeutic results.

3.5 *Personalized Treatment Approaches*

This method is crucial for advancing toward precision medicine, in which health products are personalized for individual patient traits. By identifying accurate biomarkers, healthcare providers can select the most effective treatments using nanomedicine tailored to the patient's needs. Other important personalized therapy approach proposed is the use of nanomaterials to deliver biologically active compounds targeted to individual patient needs. The potential of novel therapeutic classes as well as small peptides, which can be customize to the distinct traits of each patient. Biodegradable and biocompatible materials, including as lipids and polymers, are the basis of particles utilized for therapeutic administration. These materials can be further functionalized into more complicated forms. The development of particular nanocarrier formulations that might minimize undesirable side effects while achieving optimal delivery of active compounds to targeted cells is currently of great interest [22].

4 Design and Characterization

4.1 *Shape and Size*

There are several shapes and sizes of nanocarriers as depicted in Fig. 5. Sphere, rod, nanoshell, nanocage, cube, and nanostar are the most common shapes of nanocarriers for the treatment of cancer [23]. The size of these nanocarriers ranges from 1 to 150 nm, with different morphologies having unique chemical properties and physical properties for the specific applications [24].

4.2 *Surface Functionality*

Surface functionalization of nanoparticles involves integrating various organic and inorganic compounds at the nanoscale by utilizing covalent and non-covalent interactions, including hydrogen bonds, electrostatic forces, and van der Waals interactions.

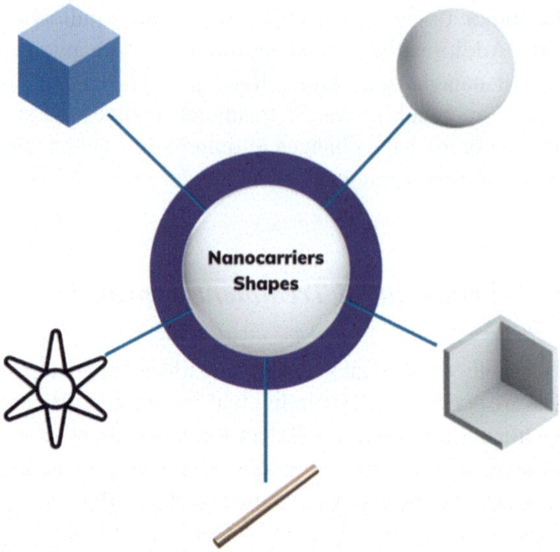

Fig. 5 Different shapes of nanocarriers for targeted drug delivery

For ligands and NP surfaces to create covalent connections, many linker molecules are usually used. One advantage of non-covalent is their simplicity; they do not alter the target biological material's interactions or the structure of the molecules used. Nonetheless, a variety of variables, including ionic strength and pH, can readily affect non-covalent changes [25]. Enhancing the qualities that will make NPs more advantageous for use in medical applications is the aim of NP surface functionalization. Since different types of nanomaterials have different chemical characteristics and exposed functional groups, they can be applied in different functionalization stages [26].

4.3 In Vitro Characterization

In vitro characterization is the process of testing NPs in controlled laboratory environments, such as a test tube or petri dish. Accurate and reliable assessment of nanoparticles in environmental and biological systems is crucial for ensuring their desired optical, electrical, thermal, and chemical characteristics and functionality. Accurate control of such characteristics is critical for developing safe biological applications. Most critically, a number of physicochemical parameters contribute to NPs' cytotoxicity. Nanoscale particles may be more harmful than larger ones. Size, composition, particle size distribution, morphology, surface properties, aggregation, purity, and hydrophobicity are some nanoparticle characterization parameters. Techniques for physiochemical characterization are atomic force microscopy (AFM), Fourier transform infrared spectroscopy (FTIR), nuclear magnetic resonance (NMR), mass spectroscopy, and scanning electron microscopy. In theory, characterization appears to be straightforward. However, this isn't always the case. Methodological and technical challenges might lead to biased results and incorrect characterizations. Inaccuracies highlight the need for a deeper understanding of the instrument's function and potential factors influencing outcomes. It is critical that the researcher keeps working for the most accurate, highest quality measurements and moves away from just collecting data.

4.4 In Vivo Characterization

In vitro characterization refers to the tests, experiments, and procedures that researchers conduct on a whole living organism, such as a human, laboratory animal, or plant. Numerous in vitro assays are available to effectively establish the effects of the nanomedicine; nevertheless, they require patient or animal samples. Analyzing the cellular intake of therapeutic NPs is made easier with the use of in vitro tests [27]. In vitro monitoring and detection approaches that improve the efficacy and resolution of conventional procedures while reducing their invasiveness and toxicity have been created because of nanotechnology. It is now possible to monitor NPs as they move through the human body. A single nanocarrier containing both therapeutic and imaging materials is referred to as a theranostic nanocarrier because of their capacity for codelivery; these NPs enable both imaging and treatment to occur simultaneously. Luminescent fluorescence generated by quantum dots is a common imaging agent. The emission properties of these dots can be adjusted to modify their sensitivity and depth penetration. Some imaging techniques are optical imaging, computed tomography, MRI, ultrasound imaging, PET, and single-photon-emission CT. In vivo can be classified into nanotherapeutics, nanocarriers, nanoimaging, and contrast agents [28].

5 Clinical Applications

5.1 Pharmaceutical Industry

There are various applications of targeted drug delivery as illustrated in Fig. 6. An effective drug delivery system meets several pharmaceutical requirements, such as enhanced therapeutic outcomes, good bioresorbability, targeted accumulation, and controlled drug release at the desired site [29]. The use of nanocarriers has emerged as a successful method for improving the delivery of both small and large molecules, including genes and peptides [30]. For the release of small and big molecules fot therapy, a variety of nanocarriers have been investigated, including quantum dots, liposomes, dendrimers, viruses, and virus-like nanoparticles. Therapeutic liposomes range in size from 50 to 200 nm, and their phospholipid (hydrophobic) plasma

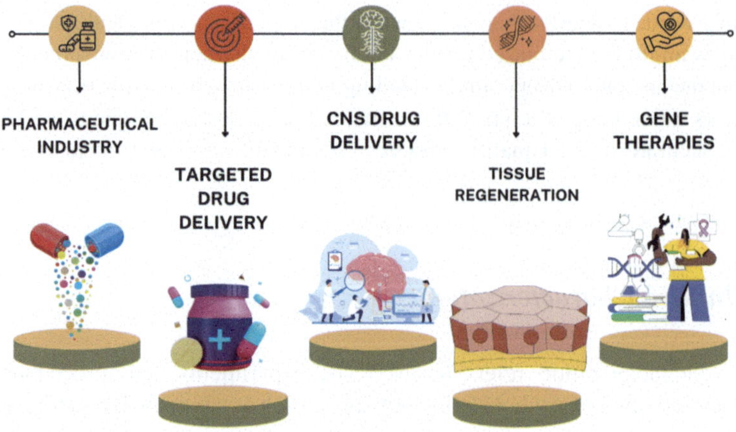

Fig. 6 Schematic illustration of diverse applications

membrane can include water-insoluble or water-soluble therapeutic medicines. Micelles are colloidal matter dispersions with sizes ranging from 5 to 100 nm [31]. Micelles can enhance the solubility and bioavailability of medications with poor solubility. When therapeutic agents and nanocarriers work together, the carrier-loaded medicines have improved pharmacokinetic qualities, which enhances therapeutic action. Compared to other drug delivery techniques, these technologies offer easier manage over composition, size, control, and in vivo stability [32]. Liposomes and micelles are pharmaceutical nanocarrier will extend the lifespan of particular medications and can open up new revenue streams for the biotech and pharmaceutical sectors [33].

5.2 *Targeted Drug Delivery*

Large-sized materials have various challenges in drug delivery, such as poor bioavailability, in vivo stability, and issues with target-specific delivery. Delivered drugs have also many side effects. Targeted drug delivery using nanocarriers can promote stability through enhancing the bioavailability and accumulation of medicines, especially in the targeted disease region. These colloidal systems can shield drugs from biological breakdown, reduce toxicity, improve therapeutic efficacy, and enable temporal and spatial management of medicines in the precise site of an illness by delivering medications to targeted locations [34]. A novel strategy has been presented, which involves conjugation methods or magnetic fields to improve drug delivery to target areas while actively targeting through the addition of particular ligands. Chemotherapy drug delivery methods and formulations could be innovative as a result of nanotechnology as shown in Fig. 7. An effective and site-specific distribution of chemicals makes it easier to achieve a therapeutic outcome that can resist

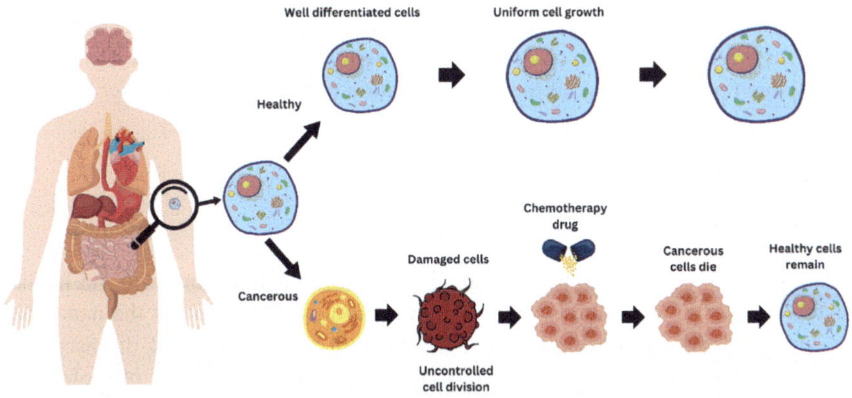

Fig. 7 Difference between healthy cell and cancerous cell and the use of chemotherapy drug

immunological disorders, tumoral diseases, or neurodegenerative disorders [35]. The protein known as hepatocyte growth factor receptor, or MET, accumulates in certain cancer cells. By varying the density of MET-nanobodies, the targeting efficacy of liposomes targeted to MET can be controlled. MET-targeted liposomes bind to tumor cells that express MET in a density-dependent manner; however, a decrease in cell selectivity is observed on extensively conjugated surfaces [36].

5.3 CNS Drug Delivery

One of the most widespread and catastrophic diseases is cancer of the central nervous system (CNS). The primary cause of the illnesses' low treatment effectiveness is the blood–brain tumor barrier (BBTB) and other barriers that prevent targeted delivery of anticancer medicines to the place of interest. Cytotoxic chemicals and the bulk of anticancer medications are actively and selectively prevented from entering the brain by the blood–brain barrier, a strong biological barrier that divides brain extracellular fluid from circulating blood. The main barriers to targeted drug delivery into CNS cancers are the BBB and BBTB [37]. Chemobrain, or chemotherapy-induced cognitive impairment, is becoming more common, and this is an urgent concern given the rise in the worldwide cancer burden, particularly breast cancer. The field of nanomedicine is advancing in its efforts to prevent chemobrain by creating innovative therapeutic delivery methods capable of crossing the blood–brain barrier (BBB). Polymer- and lipid-based nanoparticles allow for targeted drug release, enhancing therapeutic effectiveness. Administering drugs via the intranasal route may improve drug delivery to the central nervous system (CNS) by bypassing first-pass metabolism [38]. Despite challenges, nanoformulations hold promise for enhancing brain drug delivery. Ongoing research into chemical processes and formulation strategies is essential for developing effective treatments. Peposertib may be an effective radiosensitizer for brain tumors with an open BBB, though its limited diffusion into healthy brain tissue reduces the risk of increased radiation exposure [39].

5.4 *Tissue Regeneration*

For tissue engineering and medication administration, nanotechnology and nanostructured materials constitute new and growing orders. Chemicals at the nanoscale are employed. Major innovations and advances in the treatment of diseases in a range of fields have been made recently, including the development of biological, immunotherapy, and chemotherapeutic treatments. By developing biological replacements that protect, restore, or enhance the function of injured tissues and organs, tissue engineering aims to regenerate and repair damaged organs and tissues. 3D nanoparticles, which can be made of natural or synthetic biomaterials, are seeded using patient cells that have been isolated [40]. Bacterial infections that are resistant to drugs limit tissue regeneration and present difficult therapeutic issues. Antibiotic-free disease treatment is achieved through photodynamic therapy utilizing metal–organic frameworks (MOFs). However, under visible light, the MOF-based PDT often experiences little photon absorbance. By adding silver nanoparticles (AgNPs) to MOFs, AgNPs-MOFs with improved visible light sensitivity and charge carrier separation is created to increase PDT efficiency as shown in Fig. 8. AgNPs-MOFs that exhibit enhanced photodynamic efficiency in the presence of visible light. Furthermore, the addition of AgNPs gives nanosystems the potential to perform chemotherapy, which damages the membrane of bacteria cells [41].

5.5 *Gene Therapies*

Gene therapy is becoming a practical substitute for treating a range of human diseases because of the advancement of genome editing equipment. One of the biggest challenges still facing genome editing technologies is getting them to target cells efficiently. The scalability, chemical and thermal stability, and simplicity of functionalization of nanoparticles enable their practical uses. According to recent studies, some kinds of nanomaterials, such as porosity, lipid-based, polymer, and gold nanoparticles, can efficiently enable gene editing while also improving DNA stability and safeguarding the target gene [42]. Advancements in gene therapy offer the potential for long-lasting effects in tissue regeneration, such as improved bone healing, cartilage repair, and regeneration after cancer removal, compared to traditional approaches like surgery. Gene therapy can enhance the localized production of therapeutic factors by reducing gene complex loss or diffusion and allowing for precise control over the timing and location of gene product release for tissue regeneration. Advanced gene/drug nanocarriers are especially appealing among gene delivery vectors due to their

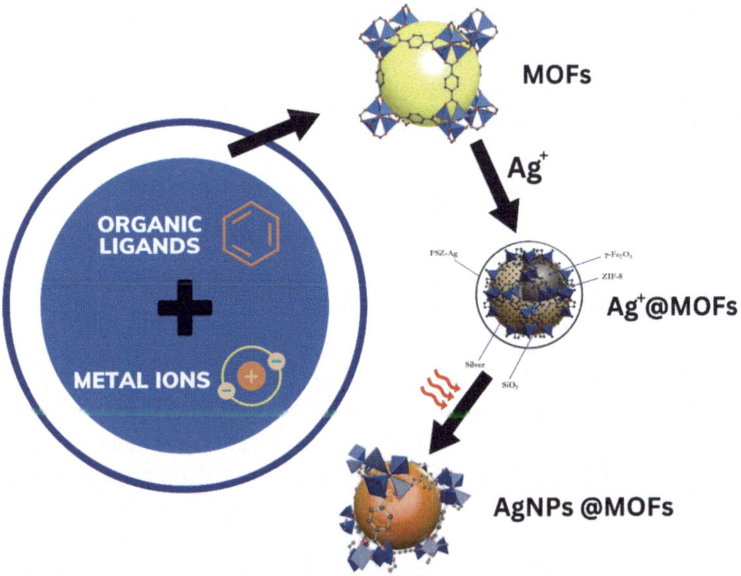

Fig. 8 Schematic of photodynamic therapy on metal–organic frameworks

adaptable properties, making them highly effective for regenerating bone, cartilage, blood vessels, nerves, and tissue damaged by cancer resection [43].

6 Toxicology and Safety of Nanocarriers

The fascinating qualities of nanoparticles, such as their target site specificity, level of solubility, and systemic stability, have transformed biomedicine, particularly in the field of drug administration. The need to carry out nanotoxicology research before to using it on humans is made clear by the fact that it can have both positive and negative effects depending on its properties in different circumstances. According to studies, CNTs can be harmful to the immune system, the lungs, the heart, the embryo, the genus, the embryo, and the nervous system. Studies on the risks and safety of using drugs indicate infusion-related reactions, which can be minimized by taking vitamin A supplements and taking some medications, such as corticosteroids and antihistamines, prior. Headache, altered blood pressure, and respiratory problems are a few of the frequent side effects related to drug use [44].

7 Future Perspectives and Challenges

The ongoing advancements in materials science, nanoscience, cancer biology, biotechnology, bioinformatics, data science, instrumentation techniques, computational chemistry, and analytical chemistry are revolutionizing the field of nanoformulations for cancer treatment.

7.1 Advancements in Nanotechnology

Nanotechnology has the potential to change medical treatment by improving chemotherapy's effectiveness against hydatid cysts. In order to target hydatid cysts more effectively, distribute therapeutic drugs with more accuracy and efficiency, and minimize tissue interference, nanotechnology is essential. However, there are issues with toxicity, resistance, and biodistribution related to nanotherapeutics. Hydatid cysts can be found early, non-invasively, quickly, and precisely with a nanobiosensor. Nanotechnology may enhance therapeutic options and methods for identifying hydatid cysts by offering more accurate and tailored alternatives for treatment [45].

7.2 Overcoming Multidrug Resistance (MDR)

Significant progress has been made in the fight against MDR malignancies, largely due to the growth of nanocarriers that acknowledge to specific stimuli and nanotherapeutics. By using nanotechnology, it has become possible to overcome the limitations of traditional medication delivery methods. Due to tumor microenvironment (TME) substantial variability, a few issues must be resolved. The causes of multidrug resistance (MDR) are likewise quite complex and may vary between individuals and tumor cells. Despite the fact that non-targeted nanocarriers can gather by utilizing the EPR effect, it is important to remember that the majority of research is conducted on in vivo animal models, which do not perfectly replicate how tumors progress inside of human bodies. Consequently, the development of precisely targeted nanomedicines with precise drug release and distribution is needed. In the near future, cancer patients may be able to expect personalized nanotherapies to treat MDR. A significant obstacle in the creation of novel nanomaterials for the treatment of chemoresistant malignancies is attaining high anticancer efficacy while preserving the drug delivery system's safe systemic biocompatibility in vivo. Systems must allow for the use of additional cancer-killing procedures, including photodynamic treatment (PDT) or photothermal treatment (PTT), which can boost the therapeutic effect by increasing cell sensitivity. If this technology could also see cancer cells directly in the body, that would be amazing [46].

7.3 *Combination Therapies*

Nanomedicines are becoming more and more widely accepted due to their many advantageous applications, especially in the treatment of cancer. However, the transfer of nanomedicines from experimental to real-world settings has proven difficult. Nanotechnology offers increased adaptability and targeting. They have all been studied as drug delivery systems, malignant gene transfer vehicles, and diagnostic contrast chemicals. Nanoparticles made from synthetic polymers combined with organic, inorganic, lipid, or glycan materials have proven advantageous in oncology treatments. However, the therapeutic effectiveness of cancer-targeting nanomedicines is hindered by inadequate drug delivery into cancerous cells. Nanotechnology produces particles between 1 and 100 nm in size, which are utilized as drug nanocarriers due to their unique properties, including optimal size, solubility, hydrophilicity, high specificity, and favorable drug release profiles [47]. Chemotherapeutic strategies for cancer may have adverse impacts, including chemoresistance and chronic toxicity, which can lead to treatment failure and complicate the development of clinical cancer therapies.

8 Conclusion

Over the past few years, nanocarrier technology has advanced significantly, especially in chemotherapy and drug delivery applications. Nevertheless, there are several challenges, including toxicity, optimizing drug loading, biocompatibility issues, and precise drug targeting. Recent studies have concentrated on the development of deformable nanocarriers for chemotherapy as an effort to overcome these challenges. To enhance drug delivery efficiency and therapeutic results, these nanocarriers are engineered to undergo structural and functional changes in response to specific stimuli present in the tumor microenvironment. These nanocarriers have distinctive properties that not only limit the challenges but also increase the effectiveness and targeted delivery of drugs. There is ongoing research on the advancements of nanotechnology and nanocarriers that will increase their applications in various fields. Going forward, the integration of biology, medicine, and nanotechnology is likely to propel the advancement of nanocarriers toward more complex and tailored designs to meet demanding biomedical needs. At the same time, we should anticipate a wave of new developments in nanocarriers that move on to clinical trials and eventually, commercial availability.

References

1. Pichaivel, M.: An overview of brain tumor. In: Brain Tumor, p. 33. (2022)
2. Lin, S.: Cancer stem cells in brain tumors: from origin to clinical implications. **4**(4), Wiley. https://doi.org/10.1002/mco2.341
3. Galton, D.A.: Chemotherapy in malignant disease. Introduction. Postgrad. Med. J. **37**(428), 338–341 (1961). https://doi.org/10.1136/pgmj.37.428.338
4. Behranvand, N.: Chemotherapy: a double-edged sword in cancer treatment. **71**, 526 (2021)
5. Cao, Z.: Deformable nanocarriers for enhanced drug delivery and cancer therapy. Wiley (2024). https://doi.org/10.1002/EXP.20230037
6. Her, S.: Gold nanoparticles for applications in cancer radiotherapy: mechanisms and recent advancements. **109**, 101 (2017). https://doi.org/10.1016/j.addr.2015.12.012
7. Kong, F.-Y.K.: Unique roles of gold nanoparticles in drug delivery, targeting and imaging applications. **22**(9). https://doi.org/10.3390/molecules22091445
8. Yang, C.: Zn-Co metal organic frameworks coated with chitosand and Au nanoparticles for chemo-photothermal-targeted combination therapy of liver cancer. **13** (2023). https://doi.org/10.3389/fonc.2023.1110909
9. Bae, K.H., Lee, K., Kim, C., Park, T.G.: Surface functionalized hollow manganese oxide nanoparticles for cancer targeted siRNA delivery and magnetic resonance imaging. Biomaterials **32**(1), 176–184 (2011). https://doi.org/10.1016/j.biomaterials.2010.09.039
10. Park, J.W., et al.: Anti-HER2 immunoliposomes: enhanced efficacy attributable to targeted delivery. Clin. Cancer Res. **8**(4), 1172–1181 (2002)
11. Prasanth, Y.: Nanocarriers and their types for targeted drug delivery. 28 (2022). https://doi.org/10.47583/ijpsrr.2022.v77i01.004
12. Gong, H.: Carbon nanotubes for biomedical imaging: the recent advances. Adv. Drug Deliv. Rev. **65**(15), 1963 (2013). https://doi.org/10.1016/j.addr.2013.10.002
13. Larson, D.R., et al.: Water-soluble quantum dots for multiphoton fluorescence imaging in vivo. Science **300**(5624), 1434–1436 (2003). https://doi.org/10.1126/science.1083780
14. Sercombe, L., Veerati, T., Moheimani, F., Wu, S.Y., Sood, A.K., Hua, S.: Advances and challenges of liposome assisted drug delivery. Front. Pharmacol. **6**, 286 (2015). https://doi.org/10.3389/fphar.2015.00286
15. Hansen, A.E., et al.: Positron emission tomography based elucidation of the enhanced permeability and retention effect in dogs with cancer using copper-64 liposomes. ACS Nano **9**(7), 6985–6995 (2015). https://doi.org/10.1021/acsnano.5b01324
16. Hu, C., et al.: Enzyme-triggered size shrink and laser-enhanced NO release nanoparticles for deep tumor penetration and combination therapy. Biomaterials **168**, 64–75 (2018). https://doi.org/10.1016/j.biomaterials.2018.03.046
17. Price, L.S.L., Stern, S.T., Deal, A.M., Kabanov, A.V., Zamboni, W.C.: A reanalysis of nanoparticle tumor delivery using classical pharmacokinetic metrics. Sci. Adv. **6**(29), eaay9249 (2020). https://doi.org/10.1126/sciadv.aay9249
18. Yoon, H.-M., et al.: Stimuli-responsive drug delivery of doxorubicin using magnetic nanoparticle conjugated poly(ethylene glycol)-g-chitosan copolymer. Int. J. Mol. Sci. **22**(23), 13169 (2021). https://doi.org/10.3390/ijms222313169
19. Wang, T.: Stimuli-responsive nanocarrier delivery systems for Pt-based antitumor complexes: a review. RSC Adv. (24) (2023). [Online]. https://pubs.rsc.org/en/content/articlelanding/2023/ra/d3ra00866e
20. Gong, Z., et al.: Enzyme-induced transformable peptide nanocarriers with enhanced drug permeability and retention to improve tumor nanotherapy efficacy. ACS Appl. Mater. Interfaces **13**(47), 55913–55927 (2021). https://doi.org/10.1021/acsami.1c17917
21. Buljan, M., Wick, P.: Tailoring design of nanomaterials and systems to individualize patient treatments. Chimia (Aarau) **76**(3), 236–241 (2022). https://doi.org/10.2533/chimia.2022.236
22. Lohse, S.E.: The quest for shape control: a history of gold nanorod synthesis. (2013). https://doi.org/10.1021/cm303708p

23. Personick, M.L., Mirkin, C.A.: Making sense of the mayhem behind shape control in the synthesis of gold nanoparticles. J. Am. Chem. Soc. **135**(49), 18238–18247 (2013). https://doi.org/10.1021/ja408645b
24. Ahmad, F.: Unique properties of surface-functionalized nanoparticles for bio-application: functionalization mechanisms and importance in application. (2022). https://doi.org/10.3390/nano12081333
25. Nanoparticle surface functionalization: how to improve biocompatibility and cellular internalization. Frontiers (2020). https://doi.org/10.3389/fmolb.2020.587012
26. Louie, A.Y., et al.: In vivo visualization of gene expression using magnetic resonance imaging. Nat. Biotechnol. **18**(3), 321–325 (2000). https://doi.org/10.1038/73780
27. Singer, A.: Nanoscale drug-delivery systems: in vitro and in vivo characterization, p. 419. (2019). [Online]. https://www.sciencedirect.com/science/article/abs/pii/B9780128140338000138
28. Carstens, M.G.: Pharmaceutical Micelles: Combining Longevity, Stability, and Stimuli Sensitivity, p. 308. Springer Link. [Online]. https://link.springer.com/chapter/10.1007/978-0-387-76554-9_9
29. Magadala, P.: Multifunctional polymeric nanosystems for tumor-targeted delivery. In: Fundamental Biomedical Technologies, vol. 4. [Online]. https://link.springer.com/chapter/10.1007/978-0-387-76554-9_2
30. Pharmaceutical nanocarriers (liposomes and micelles) in cancer therapy. J. Nanomed. Nanotechnol. (2016). https://doi.org/10.4172/2157-7439.1000e138
31. Lipid-based parenteral drug delivery systems: biological implications. (2006). [Online]. https://www.researchgate.net/publication/229658464_Lipid-Based_Parenteral_Drug_Delivery_Systems_Biological_Implications
32. Brown, P.D., Patel, P.R.: Nanomedicine: a pharma perspective. Wiley Interdiscip. Rev. Nanomed. Nanobiotechnol. **7**(2), 125–130 (2015). https://doi.org/10.1002/wnan.1288
33. Marinelli, L.: Solid lipid nanoparticles for efficient delivery of capsaicin-rich extract: potential neuroprotective effects in Parkinson's disease. **91** (2024). https://doi.org/10.1016/j.jddst.2023.105097
34. Di Stefano, A.: Nanotechnology in targeted drug delivery. Int. J. Mol. Sci. **24**(9), 8194 (2023). https://doi.org/10.3390/ijms24098194
35. Mesquita, B.S.: The impact of nanobody density on the targeting efficiency of PEGylated liposomes. **23**(23) (2022). https://doi.org/10.3390/ijms232314974
36. Yu, H., Yang, Z., Li, F., Xu, L., Sun, Y.: Cell-mediated targeting drugs delivery systems. Drug Deliv. **27**(1), 1425–1437 (2020). https://doi.org/10.1080/10717544.2020.1831103
37. Alexander, A., Saraf, S.: Nose-to-brain drug delivery approach: a key to easily accessing the brain for the treatment of Alzheimer's disease. Neural Regen. Res. **13**(12), 2102–2104 (2018). https://doi.org/10.4103/1673-5374.241458
38. Talele, S., et al.: Central nervous system delivery of the catalytic subunit of DNA-dependent protein kinase inhibitor peposertib as radiosensitizer for brain metastases. J. Pharmacol Exp. Ther. **381**(3), 217–228 (2022). https://doi.org/10.1124/jpet.121.001069
39. Matic, A.: Nanostructured materials for drug delivery and tissue engineering applications. (2023). [Online]. https://link.springer.com/article/10.1007/s12033-023-00784-1
40. Xie, W.: Multi-mechanism antibacterial strategies enabled by synergistic activity of metal–organic framework-based nanosystem for infected tissue regeneration. **19**(14) (2023). https://doi.org/10.1002/smll.202205941
41. Wei Hu, S.: Nanobiomaterial vectors for improving gene editing and gene therapy. **66**, 136 (2023). https://doi.org/10.1016/j.mattod.2023.04.011
42. Zhang, W.: Advanced gene nanocarriers/scaffolds in nonviral-mediated delivery system for tissue regeneration and repair. (2024). [Online]. https://jnanobiotechnology.biomedcentral.com/articles/10.1186/s12951-024-02580-8
43. Sharma, S., Parveen, R., Chatterji, B.P.: Toxicology of nanoparticles in drug delivery. Curr. Pathobiol. Rep. **9**(4), 133–144 (2021). https://doi.org/10.1007/s40139-021-00227-z

44. Sadr, S., et al.: Challenges and prospective of enhancing hydatid cyst chemotherapy by nanotechnology and the future of nanobiosensors for diagnosis. Trop. Med. Infect. Dis. **8**(11), 494 (2023). https://doi.org/10.3390/tropicalmed8110494
45. Ashique, S.: Nanodelivery systems: an efficient and target-specific approach for drug-resistant cancers. **12**(18). https://doi.org/10.1002/cam4.6502
46. Bhandari, K.: Nanocarriers in combined medication delivery: issues and prospects. (2024). [Online]. https://www.researchgate.net/publication/383338900_Chapter_17_Nanocarriers_in_combined_medi
47. Eljack, S., David, S., Faggad, A., Chourpa, I., Allard-Vannier, E.: Nanoparticles design considerations to co-deliver nucleic acids and anti-cancer drugs for chemoresistance reversal. Int J Pharm X **4**, 100126 (2022). https://doi.org/10.1016/j.ijpx.2022.100126

Cancer Immunotherapy Advancements Using Nanoparticles

Preeti Rajesh, Harsimran Kaur, Anshu Kumar, Sarvjeet Das, Juhi Saxena, and Anupam Jyoti

Abstract Cancer is the leading cause of mortality worldwide with more than 10 million deaths in 2020, hence pose a significant health burden. Several treatment options are available for the treatment of cancer, among which cancer immunotherapy has gained attention in past few years. During immune-therapy, the host immune responses activated against cancer cells. However, the success rate of cancer immunotherapy is limited due to poor therapeutic targeting and undesirable side effects. To confer the specificity, nanoparticles (NPs) mediated immunotherapy has been emerged as promising platforms for revolutionizing cancer treatment. The current chapter explores the diverse applications of NPs in cancer immunotherapy as carriers for immunotherapeutic agents, immune-modulators for natural killer cells, T cells, and dendritic cells. Various types of NPs are explored for immune-stimulation, with strategies aimed at optimizing their immune-stimulatory effects. In conclusion, the clinical translation of nano-immunotherapy holds immense potential in revolutionizing cancer treatment paradigms.

Preeti Rajesh and Harsimran Kaur: Both authors contributed equally.

P. Rajesh
Department of Biotechnology, Brainware University, Kolkata, India

H. Kaur
Department of Biotechnology, University Institute of Biotechnology, Chandigarh University, Mohali, India

A. Kumar · S. Das · A. Jyoti (✉)
Department of Life Sciences, Parul Institute of Applied Science, Parul University, Vadodara, India
e-mail: anupamjyoti03@gmail.com

J. Saxena
Department of Biotechnology, Parul Institute of Technology, Parul University, Vadodara, India

N. J. Joshi and G. Patel (eds.), *Nanoparticle Delivery in Cancer Care*,
Series in BioEngineering, https://doi.org/10.1007/978-981-95-0457-2_6

1 Introduction

Cancer has been one of the most challenging diseases worldwide in terms of cellular complexities, and abnormal irregular metabolism and it has been a daunting task for scientific communities and researchers to study the physiology, possible therapies, treatments, prevention and control. Most of the cancer cases lead to death because of uncontrolled malignancy shown by the tumor cells [1]. According to Centers for Disease Control and Prevention (CDC), cancer cells are abnormal and capable of uncontrolled cell division, invasion, and migrating to other tissues or organs through blood and lymphatic systems showing metastasis properties [2]. Invasiveness, ability to survive stress, smartly avoiding death by apoptosis, unregulated and unchecked cell division cycles, genetic instability and accumulation of mutations, anchorage-independent growth, loss of density-dependent and contact inhibition exhibited by normal cells, angiogenesis, ability to maintain their telomere lengths by telomerase activity elongating their chromosome ends (avoiding replicative cell senescence) are some of the remarkable properties exhibited by cancerous cells [3].

Cancer cells tend to shift their metabolism from oxidative phosphorylation to aerobic glycolysis; even in the presence of oxygen, they produce more than 85% Lactate [4]. Cancerous cells show the Warburg effect (a phenomenon showing abnormally high uptake of glucose) which makes them selectively visible in whole-body scans as well as helps in monitoring the tumor development and responses to treatments [5]. Benign tumors can be fully cured by removal or destruction of the cell mass locally by conventional therapies, chemotherapy, or radiation therapy [6]. Malignant tumors can be called cancers as they metastasize to distant organs and invade tissues located distant from the site of origin [7]. The tumor microenvironment (TME) is distinguished by the aberrant tumor vasculature, which results in hypoxia that can alter the function of the normal cells, enhance tumor growth, as well as resist therapeutic drugs. TME has been shown to influence targeted drug delivery and is also associated with drug resistance and alleviated rates of response [8].

1.1 Cancer Epidemiology

According to IARC (International Agency for Research on Cancer), WHO GLOBOCAN (Global Cancer Observatory) 2022 data, incidence of diseases leading to all kinds of cancers was close to 20 million, out of which approximately 9.75 million deaths were reported of all ages making it a major health concern. American Cancer Society estimated about 2 million new cancer cases in United States for the year 2024 [9]. Cancers associated with lung, breasts, stomach, liver, cervix-uteri, bladder, pancreas, thyroid and lymphatic systems are the major types diagnosed predominantly, out of which lung cancers are the most prevalent (18.7% of total cancer deaths) being the leading cause of deaths among other types. Lung and bronchiol (18.7%), colorectal (9.3%), liver (7.8%), breast (6.9%), stomach (6.8%) cancers are the leading cancers in terms of mortality rate [10]. Global Burden of Disease (GBD) study 2019 comparative risk assessment framework considered behavioural, environmental and occupational, metabolic risk attributes leading to major types of cancers globally [11]. Mortality was high relative to corresponding incidence rates in Asian and African regions because of respective distribution of cancer types, low to medium Human Development Index (HDI) and in part because of late-stage and poor diagnoses, while there has been very high incidence and respective mortality in European, North-American and Australian countries despite high to very high HDI [12]. Among males, Asia (49.4%), America (21.4%), and Europe (22.9%) contribute to the high incidence rates out of 10.3 million new cases worldwide; Asia (58.9%), America (14.0%), Europe (20.1%), Africa (6.3%) deaths out of 5.4 million deaths worldwide (GLOBOCAN 2022 data). Among females, breast cancer is the most common and leading cause of death followed by lung and colorectal cancers in terms of both incidence and mortality rates; whereas lung cancer is the most frequent and leading cause of death among males followed by prostate and colorectal cancers in terms of incidence; liver and colorectal in terms of mortality rates [13].

1.2 Current Therapy

Cancer treatment mainly depends on cancer type, stage and severity; whether it's benign (restricted to the tumor site of origin with confined boundaries) or malignant, timely screening tests, age related factors, behavioural risk factors, perceived lifestyle as well as immune strength of the individual suffering from the disease [6]. Some cancers might respond to a single kind of treatment while others might require joint therapy which combines surgery, chemotherapy, radiation therapy, thermal ablation and immunotherapy [14]. Cancer therapies currently are associated with numerous side effects such as alopecia (hair fall out), neutropenia (decrease in WBC count), thrombocytopenia (low platelet count, bleeding and bruising), lymphedema (by radiation therapy or surgical removal of lymph nodes), peripheral neuropathy, arrythmia, blood clot (Deep vein thrombosis), cancer pain (arthralgia and myalgia), delirium, cognitive impairment (referred as "chemo brain") [15]. Nanoparticles in cancer therapy show enhanced permeability and retention effect (EPR) and have been shown to prolong retention time of therapeutic drugs in the tumor microenvironment (Table 1) [16].

1.3 Cancer Cells Resistance to Multi-faceted Therapeutic Strategies

Tumor cells exhibit resistance to the current therapies, mainly targeted therapy due to gene instability and accumulation of secondary mutations post-treatment [26].

Table 1 Types of cancer therapy, overview and associated side effects

Currently available therapies	Description and listed gold standard therapies	Side effects, adverse outcomes and key factors	References
Radiation therapy	• External beam (3D-CRT, IMRT, Proton beam, IGRT, SRT) and internal radiation therapy (permanent implants, temporary internal) • Dosage and timespan of therapy needs to be tightly regulated	• Acute radiation toxicity, consequential damage and latency to post-mitotic cells • Nausea, vomiting, anorexia, radiation pneumonitis, fibrosis, necrosis, hematuria, mucositis, cardiotoxicity, vascular damage, infertility in females, testicular dysfunction in males • Radiation induced secondary malignancies	[17]
Surgery	• Works best for solid tumors localized to a particular region as well as depends on the stage of cancer such as low risk early-stage breast cancers • Radical gastrectomy has been the gold standard treatment for gastric cancer • Mohs surgery has been the gold standard for skin cancer treatment • Surgery in pancreatic adenocarcinoma patients have been effective • Surgical resection has been the gold standard for early-stage non-squamous cell Lung cancer (NSCLC)	• Side effects vary on location of tumor mass and how advanced the particular cancer is; surgery usually avoided if near a major blood vessel • Before surgery, neoadjuvant therapy allows shrinkage of tumor or adjuvant therapy to destroy leftover tumor cells post-surgery	[18]
Targeted therapy	• Inhibitor therapies to treat solid cancers; PARP, angiogenesis, Histone deacetylase, mTOR, MDM2 inhibitors, Hedgehog pathway blockers, Receptor tyrosine kinase inhibitors, proteasome inhibitors	• Extravasation (drug escape and leak into surrounding healthy tissues) • Drug dosage and timespan of therapy needs to be tightly regulated, controlled activation/inhibition of receptors	[19]

(continued)

Table 1 (continued)

Currently available therapies	Description and listed gold standard therapies	Side effects, adverse outcomes and key factors	References
Chemotherapy	• Neoadjuvant (shrinks tumor before surgery) traditional chemotherapy is the gold standard for locally advanced, inoperable HER-2 positive breast cancer over targeted therapies, IV or injectable chemotherapy (chemo infusion) • Dosage and timespan of therapy needs to be tightly regulated	• Nausea and vomiting are the most common side effects • Alopecia (hair loss), dizziness, loss of appetite, fatigue, sore mouth (stomatitis or oral mucositis), dry mouth Discomfort, swelling or pain by IV injection, inflamed vein-Phlebitis etc.	[20]
Vaccine against certain cancers	• HPV, prostate, bladder cancers	• Soreness or swelling at the injection site, moderate fever, or weariness	[21]
Antibody drug-conjugate therapy	• Cell-based immunotherapy (cytokine-induced killer therapy) • "Biological missile" for targeted drug delivery conjugated with monoclonal antibody to achieve high accuracy, specificity, elimination of tumor cells	• Side effects may include infusion responses, fever, chills, tiredness, and an increased risk of infection • In certain situations, individuals may develop autoimmune responses or cytokine release syndrome (CRS), which can be fatal if not treated	[22]
Oncolytic virus therapy (gene immunotherapy)	• Modified HSV in order to target tumor cell destruction; T-VEC got FDA approval in 2015 for patients with advanced melanoma	• The most common adverse effects include flu-like symptoms (fever, chills, weariness), injection site discomfort, and local inflammation • There is a danger of immune-related side effects	[23]

(continued)

Table 1 (continued)

Currently available therapies	Description and listed gold standard therapies	Side effects, adverse outcomes and key factors	References
Combination therapy	• Chemoradiotherapy (CRT), Thermal Chemotherapy, Combination therapy for Acute Lymphocytic Leukemia (ALL) and Non-Hodgkin's Lymphoma • Systemic triple therapy for mHSPC has shown positive results in patients with resistance to monotherapy	• Curative chemotherapy combined with surgery/radiation makes treatment more effective reducing symptoms • Cryoablation combined with surgery for invasive breast cancer	[24]
Ablation therapy	• Thermal ablation, Hyperthermia-Radiofrequency ablation (RFA), Microwave Ablation (MWA) for hepatocellular carcinoma • Cryoablation (CYA), Laser ablation, High-intensity focused Ultrasound (MRI-HIFU) Nanomaterial mediated Ablation Therapy (NMAT)-photothermal and magnetic hyperthermia	• Most common adverse effects include skin or muscle burns • Minimally invasive alternative therapy for early-stage breast cancer to lower associated risks and morbidity	[25]

Distorted blood vessels and rapid growth of tumor cells deprive the tissues of oxygen resulting in its hypoxic state as well as further rapid formation of new blood vessels (Angiogenesis) contribute more to the resistance to therapeutic drug delivery [27]. Hypoxia-induced tissues might not allow immune cells to function properly resulting in the weakening of immune response against tumor microenvironment [28] (Fig. 1).

1.4 Cancer Immunotherapy

Prior to the development of chemotherapy or radiation therapy for cancer, William B. Coley first came up with the idea of injecting bacteria directly into the inoperable tumor of one of his patients in 1891 [29]. Coley and other physicians practiced this immunotherapy method for many years successfully, using bacterial products that became known as "Coley's Toxins" to treat certain aggressive cancers. This treatment was later replaced with chemo- and radiotherapy. Nonetheless, today we realise that Coley's observations led to the beginning of a novel approach to fight cancer [30].

Cancer cells tend to show heterogeneity and are therefore, more likely to develop resistance to certain chemotherapeutic drugs or show relapse after certain kinds of

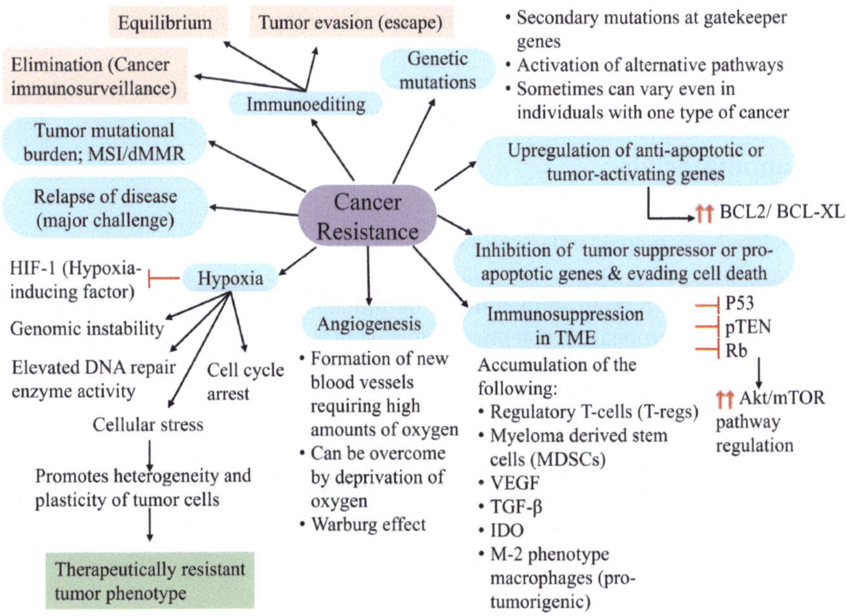

Fig. 1 Mechanism of drug resistance to targeted cancer therapy

treatments. There have been enormous research and clinical trials performed with a wide variety of therapy options for the destruction of cancerous cells localized to a particular region or as post-surgical interventions or targeted delivery methods; each depending on the type of cancer or associated side effects to a particular therapy [31].

In 2015, the U.S. Food and Drug Administration (FDA) approved the first oncolytic virus immunotherapy for the treatment of cancer—T-VEC (Talimogene laherparepvec) sold under brand name Imlygic® for advanced melanoma skin cancer patients [32]. This treatment involving a herpes virus has been genetically engineered in a way to not harm healthy cells and besides lead infected cancer cells to produce the immune-stimulating GM-CSF (Granulocyte Monocyte-Colony Stimulation Factor) protein enhancing local and systematic anti-tumor responses. Cancer Research Institute (CRI) has been undergoing clinical trials with combination of oncolytic virus and immune checkpoint inhibitor in patients with advanced appendicial cancer, colorectal cancer and ovarian cancer [33]. Treating triple-negative breast cancer patients with oncolytic virus therapy prior to surgery led to increase in their response to checkpoint immunotherapy (Reviewed by John C. Bell; CRI) [34].

James P. Allison and Tasuko Honjo shared 2018 Nobel Prize in Physiology or Medicine for their discovery of cancer therapy through releasing the 'brakes' on the immune system triggering the immune defense by accelerating the effect of T-cells in order to kill the tumor cells besides a tight control on regulating excessive immune activation that might lead to destruction of body's own healthy cells. James P. Allison had developed an antibody that could bind to a immune checkpoint regulatory

T-cell protein CTLA-4, blocking it's function and thereby stimulating a hyped T-cell response [35].

2 Nanomaterials

Nanomaterials encompass a size of less than 100 nm and find a role in diverse aspects of diseases including diagnostics, and therapeutics. Nanoparticle-based delivery methods have shown promise in stimulating immune cells such as T cells, natural killer cells, and macrophages, potentially increasing the effectiveness of immunotherapies [36]. Researchers have overcome specificity, bioavailability, and adverse effects by encapsulating immunostimulatory agents in nanoparticles, making nanoparticle-mediated immunotherapy a promising approach for enhancing antitumor immune responses and improving clinical outcomes. Various types of nanomaterials including NP, CNT, Dendrimers, Liposomes, and Micelles, Polymer nanoparticles, Nanoemulsion, have been utilized in the targeted drug delivery against various diseases [37] shown in Fig. 2.

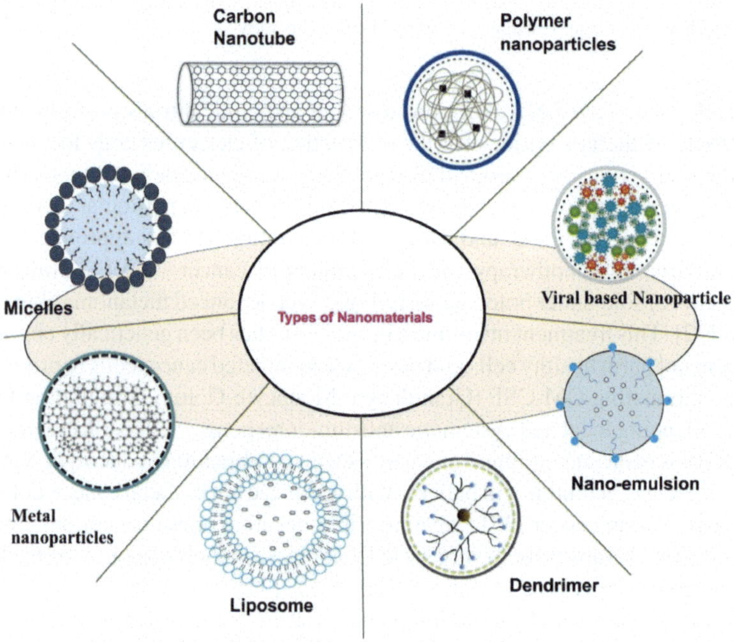

Fig. 2 Types of nanomaterial for drug delivery

2.1 Types of NM for Drug Delivery

These nanomaterials encompass unique physio-chemical properties that help in achieving specific applications in drug delivery. Nanocarriers specifically nanoparticles, offer a platform for targeted drug delivery, enhancing therapeutic efficacy and mitigating systemic side effects [38]. These therapeutic agents have small and large molecules they consist of a porous structure that can offer a high surface area and can hold enzymes, creating a better microenvironment than flat surfaces and facilitating better interaction bonds between the enzyme and Carrier surface that would be helpful for drug delivery [39]. The detailed features of these NMs have been mentioned in Table 2.

2.2 NM-Anti Cancer Drug Conjugate

Nanomaterials from past few decades have proven to significantly enhance the safety and efficacy of drug delivery system. In cancer theranostics, nanomaterials have proven to improve pharmacokinetics, accurate targeting of cancerous cells, minimized side effects, and reduced drug resistance. Nanomaterials are chosen based on the pathophysiological development of tumours. Various physiochemical properties of nanoparticles viz., size, shape, load carrying capacity, tethering with homing ligands, ability to encapsulate both hydrophobic and hydrophilic drug molecules makes them a new class of vehicles in nanomedicine [50, 51]. The emergence of organic and inorganic nanomaterials as a drug conjugated nanoparticles has opened up new possibilities in the field of theranostics by reducing the toxic side effect and increasing efficacy of drug. Various types of organic nanoparticles (polymers, lipids, dendrimers) and inorganic nanoparticles (Table 3) (iron, silver, gold) [52, 53].

The concept of Polymer–drug conjugates (PDCs) was originated in 1955 by Horst Jatzkewitz who developed antianxiety polymer-drug conjugate of polyvinylpyrrolidone (PVP) conjugated with glycyl-L-leucyl-mescaline. PDCs are categorised as synthetic (poly N-(2hydroxypropyl)-methacrylamide-(poly-HPMA), polyethylene glycol (PEG), polylactic acid (PLA), poly(lactic-co-glycolicacid) (PLGA), polyphosphazene, polyphosphoesters (PPEs), and polyglycerol) and natural polymers (Polyglutamic acid, albumin, gelatin, alginate, chitosan, hyaluronic acid, and hydroxyethyl starch) [66–68]. The broad range application of PDCs is ascribable to their numerous benefits of sustainable, controlled and targeted release of therapeutic molecules. In various formulations of PDCs are under clinical/preclinical trials and categorised as next generation chemotherapeutics for clinical cancer treatment [69]. PDCs have become the thrust area of research due to its wide application potentials and many researchers have proposed PDCs using Polyphosphazenes, Polyphosphoester, Polyglycerol, chitosan, alginate. Covalently linked dextran and paclitaxel based PDCs was reported as a reduction sensitive PDC for sustainable delivery of drug against cancer. Spherical nanoparticle of 110 nm size was synthesized with

Table 2 Physiochemical properties of NMs

Types of NP's	Classification	Physiochemical property	References
Dendrimer	• 1–100 nm diameter • Branched synthetic polymer	• Tunable with various functional groups and high monodispersity, and ethylene diamine core	[40]
Liposome	• 30–1000 nm diameter • Phospholipid bilayer-containing it can be modified with polymers or ligands	• Water soluble and contains acyl chain and phospholipids like, (di steroyl phosphatidylcholine), etc.	[41, 42]
Micelles	• The size is around 10–100 nm and amphiphilic molecule as well as it consists of a hydrophilic shell and a hydrophobic core	• The core materials consist of amphiphilic copolymers and represented as hydrophobic and hydrophilic parts they most able for drug delivery	[43]
Nano-emulsion	• The size is around 10–1000 nm mostly they are oil and water stabilized by surfactants which can be modified with emulsifiers	• Polyethylene, and a high solubilization capacity • Such as homogenous structure and mostly they have low surfactant consumption and protect encapsulated active compounds • Depending on the oil and water phases	[44]
Carbon nanotube	• 10–100 nm Diameter • Shape cylindrical • low solubility • polysaccharides, polymers	• Electrical and mechanical properties • Long carbon chain with graphite and high aspect ratio	[45]
Polymer nanoparticles	• 10–1000 nm • Amphillic blocks • Co-polymers	• Naturally polymers polyacrylamide, polystyrene, polymethylmethacrylate, etc. • Tunable properties and lower polydispersity	[46]

(continued)

Table 2 (continued)

Types of NP's	Classification	Physiochemical property	References
Viral based nanoparticle	• 20–200 nm diameter • Viral capsid protein shell and can mimic natural viral targeting variables	• Biocompatible • Inherent targeting ability • Double antigen loading • Carriers are amino acid chains, capsid protein and lipid	[47, 48]
Metal nanoparticles	• 1–100 nm diameter • Various metals (e.g., gold, silver), etc. • It can be functionalized with variable ligands	• Unique optical and electronic properties	[49]

Table 3 List of nano formulation for cancer treatment

Nanocomposite	Drug	Cancer	References
Dextran	Paclitaxel	Breast	[54]
Poly (ethylene oxide)	Curcumin	Lung	[55]
Polycaprolactone	Docetaxel	Breast	[56]
Poly (lactic-co-glycolic acid)	Losmapimod	Multiple Myeloma	[57]
Cholesterol SLNs	Curcumin	Breast	[58]
Compritol 888 ATO-pluronic F127	Docetaxel	Lung	[59]
Compritol 888 ATO	Pterostilbene	Breast	[60]
Carboxylated carbon nanotubes	Ferulic acid and diosgenin	Liver, lung	[61]
PEGylated carbon nanotubes	Cisplatin	Breast	[62]
ZIF-8-on-ZIF-8	Doxorubicin	Breast	[63]
Europium-MOF	Fucoidan	Lung	[64]
Gelatin	Epigallocatechin gallate-cisplatin	Lung	[65]

drug loading capacity of 39%. Synthesized PDCs were found to release the paclitaxel in response to the reducing environment of the tumour cells in BT-549 and MCF-7 cell lines [54]. In another approach a colloidal nanosphere of poly (ethylene oxide)–poly (propylene oxide)–poly (ethylene oxide) encapsulated with curcumin was demonstrated as nanocomposite for lung cancer under light irradiations. The loading efficiency of proposed was reported to be 14.6%. In addition to cytotoxicity, proposed nanocomposite has also shown biological impact of DNA breaks on cancer cell lines when exposed to irradiation in-turn highlighting the photodynamic therapy application of the PDCs [55].

Polycaprolactone-Polyethylene glycol based promising alternative PDCs triblock copolymer Bi(mPEG-SeSe)-PCL, was formulated for the delivery of docetaxel for treatment of tumor. Bi(mPEG-SeSe)-PCL nanoparticles were able to carry 49% of docetaxel. The cell viability studies revealed that DOX-loaded Bi(mPEG-SeSe)-PCL nanoparticles have antitumor properties against HeLa cell lines. Furthermore, the prepared copolymer has the unique ability to self-assemble into nanoparticles and disassemble into under aqueous and redox conditions due to diselenide bond reduced into selenols (RSeH) groups [56]. The fabrication of poly(lactic-co-glycolic acid) nanoparticles encapsulated with Losmapimod (LOS@PLGA-NPs) was directed for multiple myeloma treatment. LOS@PLGA-NPs showed considerable increase in the apoptosis activity in cancer cells along with antimetastatic efficacy [57].

Solid lipid nanoparticles (SLNs) are considered as a promising candidate in NDDS as anticancer drug delivery carrier. Because of great potential of SLNs in the theranostics many researches have proposed formulation of SLNs using palmitostearate, glyceryl mono-stearate, stearic acid, oleic acid. For instance, curcumin loaded cholesterol SLNs were formulated with a size of 166 nm and 76% encapsulation. Curcumin loaded SLNs lead to increase in the cytotoxicity of human breast cancer cell due to increase in the apoptosis in the tumor cells [70]. Likewise, highly stable docetaxel encapsulated SLNs (SLN-DTX) based on compritol 888 ATO-pluronic F127 were explicated. During the cellular uptake and cell cytotoxicity assay analysis, it was established from increased apoptosis and microtubule damage that SLN-DTX has a superior antitumor property as compared to bare docetaxel. Furthermore, histological studies revealed that SLN-DTX has a remarkable impact on preventing lung metastasis in 4T1 tumor-bearing mice [59].

Inorganic based nanomaterials viz., gold, silver, zinc, quantum dots and various carbon-based nanomaterials including graphene, graphene oxide, carbon nanotubes and fullerenes are extensively used in nanomedicine. Carbon nanotubes are predominantly use in cancer theranostics due to their ability to get internalized into the intracellular compartments of tumor cells and causes therapeutic effect on the growth of tumor cells [71]. Anticancer drug loaded in CNTs leads to detrimental effect by interfering in the DNA replication of tumor cells. In addition, CNTs have remarkable properties of optical, mechanical, and electrical characteristics, making it a promising candidate in nanomedicine for cellular imaging along with drug delivery carrier. In view of the importance of CNTs in nanomedicine, carboxylated multi-wall carbon nanotube (CNTCOOH) was reported as a nanocarrier of anticancer agents: ferulic acid (FUA) and diosgenin (DGN). CNTs were found to elevate the anticancer effect of the loaded pro-drugs on the hepatocellular carcinoma cells and lung cancer under in-vitro conditions. The surfacing of synthesized CNTCOOH nanoparticle with chitosan and stearic acid rendered synergistic effect on hepatocellular carcinoma cell lines (HepG2) by causing apoptosis and targeted delivery of drug [72, 73]. In view of the importance of solubility of the chemotherapeutic drugs, magnetized PEGylated carbon nanotube was fabricated for cisplatin a poorly water-soluble chemotherapeutic drug. The fabricated carbon nanotubes/polyethyleneglycol@Fe_3O_4/CPTMS@Cis-dichloro-bis(5-(4-pyridyl)-5-phenylhydantoin)platinum(II): (CNTs/PEG@M/CPTMS@CisPt(II) nanoparticle was conjugated with Platinum(II) complex to enhance the solubility of cisplatin. Notably, CNTs are have remarkable property to get across the cell membrane through endocytosis and conglomerate in the intracellular compartments of the cells. Increased ability of intracellular penetration for delivery of cisplatin have been demonstrated by magnetized PEGylated-CNTs. Surface modification of the synthesized CNTs with platinum has enhanced the solubility of the loaded cisplatin drug molecule, thus reducing tumor cell viability [62].

The distinctive characteristics of nanomaterials have rendered them a promising candidate in the biomedical field for several decades. Improved absorption and targeted delivery of pharmaceuticals can be attained by an innovative drug nanocarriers known as Metal–organic frameworks (MOFs). MOFs have gained recognition owing to their diverse physiochemical characteristics: high load carrying capacity, ability to infiltrate intracellular compartments, which facilitate drug delivery and minimizing the adverse effects of the chemotherapeutic drugs. MOFs are typically referred as porous coordination polymers formed from the synergistic interaction of organic ligands and metal ions leading to the formation of structured framework [74, 75]. Recently, researchers have explicated nanostructured ZIF-8-on-ZIF-8 based MOFs loaded with doxorubicin. In the tumor microenvironment, the DOX@ZIF-8-on-ZIF-8 were able to release doxorubicin due to change in pH and increase concentration of glutathione. ZIF-on-ZIF nanostructured MOFs represents a promising strategy for targeted drug delivery in the treatment of breast cancer with high loading efficiency [76].

In an innovative approach, europium metal–organic framework (EuMOF) was developed for the pH responsive delivery of a sea weed polysaccharide fucoidan with antitumor activity. The fabricated FU@Eu-MOF exhibited high drug loading efficiency (22.15%) along with sustainable release of the encapsulated drug molecules (85.3% in 48 h). The primary mechanisms underlying the anticancer effect of the synthesized FU@Eu-MOF have been identified as mitochondrial dysfunction and DNA damage, which ultimately result in apoptosis [64]. Chemoresistance is a significant limiting factor in the recurrence of tumor, and has become a considerable challenge in treating tumor effectively. Numerous studies have been reported to create nanomaterials-based formulations to reduce the early regrowth of tumor. In related studies, epigallocatechin gallate-cisplatin loaded gelatin nanoparticles (GE-Pt NP) were developed to address the issue of drug resistance associated with cisplatin for lung cancer treatment. The proposed self-assembled nano-formulation of EGCG/cisplatin-loaded gelatin nanoparticle showed significant cytotoxic effect on human lung adenocarcinoma cell line (A549). The combination therapy has improved the intracellular infiltration of cisplatin, thus elevating the drug concentration in the tumor cell [65]. This nanoformulation can be considered as a potential solution for the drug resistance challenge encountered during treatment of lung cancer.

3 Nanoparticle-Based Cancer Immunotherapy

Cancer is a global concern that has a significant influence on human health. Cancer immunotherapy can regulate the autoimmune system, restoring normal anti-cancer immune response, and showing remarkable therapeutic promise against cancer cells [77]. NPs have unique physical, chemical, and biological features because of their tiny size and high surface area-to-volume ratio. Nanoparticles are currently attracting a lot of interest in the field of immunotherapy due to their ability to increase medication administration, immunomodulation, and imaging applications and have shown promise in immunotherapy by addressing challenges such as poor pharmacokinetics, nuclease degradation, and inefficient delivery to target cells [78]. Nanoparticles may stimulate the immune system to boost its anti-cancer response and immune cells like dendritic cells and T cells by conjugating immunostimulatory molecules to their surfaces. These immune cells are essential in beginning and sustaining anti-cancer immune responses. NPs can also be designed to resemble infections, activating the immune system's identification and response system [79].

NPs play a wide role in vaccination platforms, delivering tumor antigens or antigens from cancer cells. This strategy prepares the immune system to recognize and target cancer cells selectively. The combination of immune activation and antigen delivery allows for a more targeted and effective immune response against cancer cells [80]. They can be used as delivery systems for small molecule agonists, enhancing their bioavailability and specificity. Nanoparticles can also be utilized in adoptive cell therapy, targeting specific cells and compensating for the limitations of CAR-T therapy [81]. Polymeric nanoparticles have been particularly effective in enhancing the adjuvanticity of cancer vaccines, improving the potency, quality, and durability of the vaccine response. Additionally, nanoparticle-based vaccine delivery has been shown to activate both myeloid dendritic cells (mDCs) and plasmacytoid dendritic cells (pDCs), highlighting the importance of activating both cell types for effective cancer vaccines. Overall, nanoparticles offer a promising approach to immunotherapy, addressing various challenges and enhancing the efficacy of different immunotherapeutic modals [82].

3.1 Nanomaterial-Based Drug Delivery to Antigen Presenting Cells (APCs)

Tumor blood vessels show uncontrolled angiogenesis, hyper-vascular permeability and defective lymphatic drainage which could lead to ineffective anti-tumor responses by immune cells. Nanoparticles in drug delivery and nano-formulated pharmaceutical agents have gained recent importance due to the active targeting of tumor cells, enhancing efficacy of the adjuvant administered, prolong the retention time (shielded from proteolytic enzymes) and minimizing off-target dose dependent systemic toxicity to normal healthy cells [83].

Nanoparticle-based drug delivery systems that target specifically tumor-infiltrating immune cells in circulation rather than direct tumor cell targeting is a preferred way to effectively localize immunotherapeutic agents in tumors. Dendritic cells are the most effective Antigen Presenting cells (APCs) and if properly modulated, can as well play a crucial role in determining anti-tumor responses. DCs maturation with exogenous secondary activation stimuli and further manipulation help in priming helper T cells [84].

3.2 Antigen and Adjuvant Delivery to Tumor Microenvironment

Within in the tumor microenvironment (TME) a diverse array of immune cells plays critical role in mediating both innate and adaptive immune responses. The innate and adaptive responses are maintained by variety of cells including the dendritic cells (DCs), macrophages, T-cells, B-cells, cytokines, fibroblasts, endothelial cells, inflammatory cells, and lymphocytes. Extracellular matrix (ECM), Vascularization of tumor and lymphatic drainage system are the key factors in TME. Innate immune cells (macrophages and dendritic cells) and chemokines of TME stimulates the eradication of tumor by exhibiting pro/anti-tumor events [65, 85, 86]. Furthermore, adaptive immunity specifically identifies the tumours and initiate its elimination by the participation of various immune system mechanisms. The framework of the TME is critically sustained by associative production of extracellular matrix (ECM) by activated fibroblasts for the regulation of tumor progression. Hypoxic conditions are prevalent within the TME due to aberrant blood vessels and increased proliferation in the tumor cells which in-turn causes accumulation of immunosuppressive cells within the TME. This change in the microenvironment favours tumour advancement and poses significant challenges to tumour therapy due to drug resistance and low response rates. Owing to this, researchers are concentrated on cancer immunotherapy by developing nanoplatform which are engineered to address the existing challenges associated with TME [87–89]. Nanomaterials have been surface decorated with various modulating agents (tumour-specific CD8, cytotoxic lymphocyte antigen 4, Toll-like receptor, interleukin, integrins) of TME to elevate the therapeutic potentials. The concept of integrating chemotherapy with immunotherapy have headed to the development of vaccination-induced antitumor immunity [90, 91]. The conductive state of development of Chemotherapy serves as an adjunct treatment that transforms the tumor microenvironment into a hi^{+}ghly conducive state for the development of vaccination-induced antitumor immunity. This integrated therapy, chemotherapy transforms the tumour microenvironment and immune components elevated the immune response against tumour cells [92]. A pivotal mechanism was proposed by a research group, where they successfully developed a cancer vaccine by surface engineering the Iron nanoparticles (nano-Fe_3O_4) into tumor-derived antigenic microparticle (T-MPs) via CpG-loaded liposomes (CpG/Lipo). CpG when administered with vaccines is known to act as an immunostimulatory (antigen-specific immunity) by interacting with Toll-like receptor 9 (TLR9) on antigen presenting cells (APCs). CpG has reported to be diminish the activation of antigen-specific immunity under immunosuppressive tumor microenvironment thereby causing chemoresistance. This immunostimulant mechanism of Fe_3O_4/T-MPs-CpG/Lipo has demonstrated a notable decrease in the tumour growth due to elevated level of cytokine (TNF-α, IL-12, IFN-γ), cytotoxic T lymphocytes [93]. The advance approach of immunomodulating macrophages for the treatment of cancer has become an emerging research area in therapeutics. Various challenges are still to be taken into consideration for effective activation of macrophages for antitumor immunotherapy. Major obstacles are to inhibit the interaction of signal regulatory protein alpha (SIRPα) on macrophages and CD47 to enhance the phagocytic activity of cancer cells; secretion of colony stimulating factors by cancer cells has the ability to divert tumour-associated macrophages (TAMs) towards a tumour-promoting M2 phenotype [94]. In an innovative approach, genetically engineered cell-membrane-coated magnetic nanoparticles (gCM-MNs) was constructed to diminish the interaction of SIRPα and CD47. gCM-MNs were overexpressed with high affinity SIRPα variants, thereby, effectively obstructs the CD47-SIRPα interaction. Magnetic nanoparticle was facilitating the repolarization of M2 TAMs, thereby enhancing macrophage immune responses [95].

Delivery of tumor specific antigen peptide into the dendritic cells has been reported by Liu et al. [96] through lipid-coated calcium phosphate nanocarriers. The coating of nanocarrier with lipid bilayer was found to overcome the physical barrier along with inhibited the degradation of cargo loaded drug. Moreover, an elevated antigen-specific T cell response and interferon-γ (IFN-γ) production of was attained by the nano formulation [96]. The novel combinatorial adjuvant delivery platform for cancer vaccine was proposed by Li et al. [94] to improve the existing therapeutics. The formulation was designed to facilitate the photothermal ablation and immunotherapy for effective treatment of tumor. Liposomes decorated with a multivalent immunoadjuvant (HA-CpG) and encapsulated fluorophores IR-7 nanocarrier (IR-7-lipo/HA-CpGmediated) was fabricated. IR-7 is known for its high phototherapeutic properties. Anti-tumor of the fabricated IR-7-lipo/HA-CpG mediated was analysed on a mouse colon cancer model. An exposure to NIR irradiation was found to have an impact on the population of myeloid derived suppressor cells (MDSCs) along with increase in the $CD8^+$ effector T cell count [97].

Diverse array of organic and inorganic adjuvants is categorised as an activator to immune response. In a special reference, nanoparticles prepared from the inorganic materials (Silica, iron, carbon) were observed to act as an adjuvant and further enhances the immune responses. In a report, an albumin-binding nano-vaccines (AlbiVax) conjugated with Evans blue (EB), were explicated for its novel feature to self-assemble into albumin nanocomplex (13 nm in size) under in-vivo conditions. These nanocomplex also acted as an efficient delivery vehicle for CpG and antigens into lymph nodes. These albumin nanocomplexes stimulated the release of the antigen-specific $CD8^+$ cytotoxic T lymphocytes and thereby showed promising results in controlling colorectal tumor growth [98].

Cancer associated fibroblasts (CAFs) represents a factor in tumorigenesis as TME causes unavoidable interaction of tumor cells with CAFs. Nanoparticles due to their unique properties are considered as a potential candidate to inhibit the tumorigenesis during cancer. It has been corroborated that various nanoparticle like gold, Poly (lactic-co-glycolic acid) (PLGA) and albumin nanostructures can target fibroblast activation protein in CAFs and inhibits its interaction with cancer cells and delineating tumorigenesis. Thus, engineered nanoparticles coupled with immunologically active components can elicit anti-tumor responses in chemotherapy [99, 100].

3.3 Immune Checkpoint Inhibitors (ICIs) Delivery

The onset of autoimmunity can be attained by diminishing T cell activation by preventing excessive immune response through Immune checkpoint pathways. Immune checkpoint regulation plays a crucial role in regulating various functions related to immune response viz., inflammatory response, immune tolerance and other activities. These co-inhibitory immune checkpoints are advantageous for the tumor cells as under tumor microenvironment (TME), these checkpoints cause immunosuppressive conditions. Immunosuppression, enhances the immune evasion mechanism of tumor cells by getting resistant to T-Cells [101, 102]. As a result, advance therapeutic approaches can be manifested by targeting co-inhibitory immune checkpoints through antibodies, drugs and ligands. The concept of immune checkpoint blockade therapy (ICBT) has emerged as a promising strategy in cancer immunotherapy. The main participant of ICBT is Cytotoxic T lymphocyte-associated antigen-4 (CTLA-4) and programmed cell death-1 (PD-1), ICIs are designed to target these immune regulatory receptors [103]. In addition, various metabolic checkpoints, like indoleamine-2,3 dioxygenase (IDO1, adenosine 2A receptor (A2AR), B and T lymphocyte attenuator (BTLA), B7 homolog 3 protein (B7-H3), CD47, CD200R, and indoleamine 2,3-dioxygenase 1 (IDO-1) are currently under exploration for their potential in exhibiting anti-tumor effects [104–106].

In a study conducted by Tanaka et al., intradermal administration of ICIs based lymphatic drug delivery was proposed. The study was conducted on mouse and non-human primate model and show different expression level of PD-L1 in tumor. Anti-PD-L1 antibody was effectively delivered into tumor-draining lymph nodes by internal administration inducing significant tumor growth inhibition [107].

ICIs being the most considerable approach for tumor therapeutics, still have various disadvantages. It is noteworthy that in case of ICIs, immune-related adverse effect has been reported to cause severe impact on distinct organs. The above-mentioned limitation, can be overcome by combining ICIs and nanostructures, further improving the efficacy of the therapy model. Likewise, nanohybrid (AA@PN) of hollow gold and poly(lactic-co-glycolic acid) (PLGA) nanoparticles was tagged with anti-PD-1 peptide (APP). These fabricated nanohybrids were observed to increase the photothermal ablations of tumor cells. Plasmon absorption gold nanoshell causes destruction of PLGA and further stimulates the release of APP. APP promoted the infiltration of functional cytotoxic T lymphocytes (CTLs) into the tumor [108].

In similar approach, a compound effect of ICIs and photothermal therapy and chemotherapy was demonstrated by synthesizing aPD-L1 and doxorubicin (DOX)-loaded hollow gold nanoshell (T-HGNS-DOX). Hyperthermia led to release of DOX, and aPD-L1 release in TME and manifesting observable apoptosis of tumor cells inhibiting angiogenesis in tumor [109]. In another report, change in the TME from hypoxia to normoxia was noticed after the systemic administration of biodegradable indocyanine green (ICG) and PD-L1-specific siRNA-loaded $CaCO_3/MnO_2$ nanoplatform (Mn@$CaCO_3$/ICG@siRNA) [110].

Research in the development of hybrid nanocarriers: Metal–organic frameworks (nMOFs) is gaining interest due to its exceptionally high surface area presents opportunities for the stable loading of immune checkpoint inhibitors (ICIs). Recently, controlled delivery of Nivolumab through zeolitic imidazolate frameworks (NV-ZIFs) was proposed in combination with ICIs. The nanohybrid offered a promising delivery system as it was found to sustainably release ICIs and activation of T-cells under low pH of TME [111]. A novel strategy, has been delineated, integrating the chemotherapy agent paclitaxel (PTX), the anti-cancer stem cell (CSC) agent thioridazine (THZ), and the PD-L1 inhibitor HY19991 (HY) into a dual-responsive nanoparticle formulation. The formulation comprises a double layer of MMP-9 peptide tagged micelle–liposome, wherein, PTX is encapsulated responsive pol-yethylene glycol-block-poly[(1,4-butanediol)-diacrylate-β-N,N-diisopropylethylenediamine] (PDB), along with HY and THZ as a load. Due to enhance permeability of TME, the fabricated nanoparticles aggregates into the tumor. MMP responsive peptides will interact with MMP of TME, causing the bursting of liposome and releasing HY, THZ and PTX [112].

The application of nanocarrier-mediated immune checkpoint inhibitor (ICI) cancer immunotherapy offers significant potential for enhancing ICI treatments, especially by enabling the delivery of higher therapeutic doses directly within the tumor microenvironment. However, substantial biological and technological challenges persist, making the effective targeting of ICI-loaded nanocarriers to the designated tumor sites uncertain. Nonetheless, despite the potential of ICIs in treating metastatic cancer, these therapies are often expensive and have been associated with low rates of clinical response and immune-related adverse effects.

3.4 Nanomaterials Modulator as in Innate Immune System

Nanoparticles have been found to improve the delivery of immunomodulators, to cells enhancing the effectiveness of immunotherapy. Most studies indicate that targeting cells is a promising approach for immunotherapy. The body's innate immune system serves as the defense against pathogens like bacteria, viruses, and fungi [113]. In contrast, the adaptive immune system focuses on antigens and Develops memory. The innate immune system provides species-conserved protection. Nanomaterial modulators have the potential to revolutionize disease management by strengthening the response [114]. Recent advancements in nanotechnology have created opportunities for immunomodulatory strategies. Nanomaterials can stimulate responses within the immune system leading to the release of cytokines such as tumor necrosis factor alpha (TNF α) and interleukin 6 (IL 6) which can enhance reactions. Conversely nanomaterials may also trigger the production of inflammatory cytokines like interleukin 10 (IL 10) which help suppress immune responses and promote tolerance [115]. Researchers are investigating how nanomaterials can modulate pathways, in the system for therapeutic purposes. In the field of cancer immunotherapy vaccines utilizing nanomaterials show potential in enhancing responses, against cancer cells through assisting with antigen presentation and boosting cell function [116].

3.5 Nanomaterials for Modulating Adaptive Immune System

Nanomaterials have displayed the potential to impact the system. Their unique characteristics, including their size and extensive surface area, allow interaction, with immune cells and regulation of immune responses. Depending on the desired outcome these nanomaterials can. Stimulate or suppress the immune system. Additionally, modifications to nanomaterials in terms of composition, size, shape, and surface chemistry can enhance drug delivery, and life within the body control release rates, and dictate distribution patterns [117]. This customization enables the administration of medications and enhances the effectiveness of immunotherapy. Nanotechnology has shown promise in cancer immunotherapy by leveraging nanomaterials to modulate the immune system. Functionalizing nanoparticles with tumor antigens or immune stimulants can boost the presentation and activation of cells [118]. These nanomaterials can also transport drugs directly to tumor sites to increase the infiltration of cells and induce death in cancer cells. Overall utilizing nanomaterials to alter the system holds significant potential in realms such as immunotherapy and cancer treatment. T lymphocytes and B lymphocytes play roles as components of the adaptive immune system (AIS). These lymphocytes possess antigen receptors that recognize a range of antigens, from malignant cells or other infectious agents. Importantly these white blood cells can transform into memory cells, which store information from encounters with substances and trigger a swift immune reaction when they come across those same substances again [119]. The part of the response involving antibodies (B cells) relies on cells that target specific antigens as well as antigen-presenting cells. B lymphocytes are crucial, in eliminating pathogens from the body's spaces and they detect and react to specific structural characteristics [120]. The antigens are arranged in a repeated pattern they link up with B cell receptors (BCRs) prompting signalling through these receptors. While single protein antigens can initiate some signalling through BCRs clustered antigens lead to signalling, internalization of the antigen and preparation for presentation to $CD4^+$ T cells. This process is vital for B cells to receive assistance from T cells in producing antibodies. Activation of Toll receptors (TLRs) in B cells is essential for generating antibody responses that depend on T cells and are specific to certain antigens. T cells play a role in the body's ability to combat growths and intracellular invaders. Treatments such, as transferring T cells or using checkpoint blockade antibodies rely heavily on the effectiveness of $CD8^+$ T cells that destroy cancer cells and enter tumours resulting in improvements seen during medical treatment [121].

T cells that target tumor antigens can naturally multiply in patients be boosted by a cancer vaccine or be utilized in cell therapy to eliminate cancer cells. The key focus lies in implementing methods to foster the growth of these T cells within the body and thwarting signals that hinder the effectiveness. The researchers have explored nanoparticle delivery systems to enhance T-cell responses through approaches [122].

3.6 Targeting/Stimulation of Dendritic Cells (DCs)

The tumor microenvironment is an extremely complex milieu with many cellular and chemical components influencing the growth and survival of the tumor cells. TDACs or Tumor-associated dendritic cells are one of the major components of this heterogeneous environment. These cells are professional antigen-presenting cells (APCs) with a pivotal role in mounting the anti-tumor immune response via priming of naïve T-cells. Apart from initiating the T cell response DCs also produce proinflammatory cytokines and can exhibit direct cytotoxicity against tumor cells. Immunogenic DCs can potentially enhance the antitumoral properties of NK cells and T lymphocytes and promote tumor elimination. Thus, contributing to the clinical efficacy of immunotherapeutic strategies, and improving patient survival. Many immunosuppressive factors in the tumor microenvironment can lead to DC dysfunction which abrogates the anti-tumor immune response, promotes tumor progression, and is linked to poor prognosis.

Nanoparticle targeting systems are increasingly being tested to regulate dendritic cells' response to generate therapeutic immunity against cancer as well as to improve the effectiveness of immunotherapies for cancer treatment. NP-mRNA antitumor vaccines hold great promise in inducing antigen-specific immune responses through dendritic cell stimulation. NPs can potentially overcome the challenge of transfecting dendritic cells. Passive or active targeting strategies can be used to stimulate DCs with antigen-loaded nanoparticles. Typically, subcutaneous delivery is used to provide these nano vaccines, which then drain into lymph nodes to present antigens. The size of the NPs is an important factor to take into consideration as small NPs with size <100 nm can readily drain into lymph nodes while the large NPs with size >500 nm are primarily retained at the administration site and can only be absorbed by the skin resident DCs. The response of the intermediate-sized NPs (100–500 nm) is heterogeneous, exhibiting both free and cell-based drainage to the lymph nodes. Apart from size, surface charge, hydrophobicity, and NP shape are significant aspects to take into consideration [123]. For the active targeting of NPs, multiple strategies have been employed to initiate the stimulation of DCs. Ligands that specifically target surface receptors on DCs including mannose receptor, DCIR2, CLEC9A, DEC205, and C-type lectin receptors can be grafted onto the surfaces of formed nanoparticles.

These ligands are identified by the uptake mechanisms particular to DCs enabling the functionalized nanoparticles to be taken up only by DCs. This strategy lowers the effective dosages of vaccine required minimizing undesirable side effects. Several nanoparticle delivery technologies and ways for ligand conjugation are being tested to stimulate DCs for cancer immunotherapy successfully. Functionalizing the NPs can also improve their efficiency in addition to homing the formulation to the DCs. To activate DCs and T-cells, for instance, the ligands specific for intracellular TLRs (TLR 3, 7, 8, and 9) have been used as vaccine adjuvants. Overall, actively targeted nanoparticles specific to DCs have been proven to elicit better immune responses than non-specific delivery.

Both organic and inorganic materials have been investigated to synthesize nanoparticles for targeting the DCs. Poly (D, L-lactic-co-glycolic acid) (PLGA) copolymers are organic compounds with excellent biocompatibility, biodegradability, and unique physical and chemical properties, making them one of the most popular and efficient polymers for drug delivery systems [124]. PLGA-based nanoparticles can efficiently deliver antigens to DCs and even act as an adjuvant resulting in potent T-cell responses [125]. Table 4 summarizes a few NP-based delivery systems for the stimulation of the DCs. Lin et al. reported the synthesis of the PLGA-ICG-R848 NPs, with a mean diameter of 157.7 nm, co-loaded with a fluorescent dye, indocyanine green (ICG), and a TLR7/8 agonist resiquimod (R848). The model was tested in the subcutaneous Xenograft Model of RM9-Luc PCa Cells, showing significant maturation of BMDCs with significantly elevated proportions of $CD11c^{+}$ $CD86^{+}$ and $CD11c^{+}$ $CD80^{+}$ cells and enhanced immune response against prostate cancer [126]. A study by Ben-Akiva et al. reports the synthesis of lipophilic poly (beta-amino ester) (PBAE) NPs with diacrylate backbone, amine side-chain, and endcap monomers to form the final structure. PBAEs are synthetic biodegradable cationic polymers mainly studied for their potential as gene delivery vehicles as they can spontaneously self-assemble with anionic nucleic acids into NPs in aqueous solutions. In addition to the biodegradable ester bonds in the backbone structure, disulfide bonds are incorporated to create bio-reducible (R) polymers to enable environmentally triggered rapid degradation in the reducing environment of the cytosol, increasing the safety of the polymers. These nanoparticles with a diameter of approximately 100 nm were loaded with the ovalbumin (OVA) tumor antigen mRNA and TLR3 agonist polyinosinic-polycytidylic acid (poly (I: C)) or the TLR9 agonist CpG oligodeoxynucleotide (CpG ODN). The antitumor response was evaluated in the B16-F10-OVA mouse melanoma model and MC38-OVA mouse colon carcinoma model demonstrating enhanced antigen-specific $CD8^{+}$ T cell response, reduced tumor burden, and prolonged median survival [127].

Twilhaar et al. have reported the use of ganglioside GM3-containing liposome vectors to induce DC maturation and T-cell priming. The liposomes with a size of 160–190 nm and the TLR4L monophosphoryl lipid A (MPLA) were demonstrated to be potent inducers of moDC maturation in a GM3-dependent manner [128]. Recently, Cao et al. explored the potential of calcium nanoparticles as a vehicle to deliver calcium into the cytosol of DCs. PEGylated $Ca(OH)_2/SiO_2$ core–shell nanoparticles coupled with an anti-CD205 antibody (AnCHNPs) and size of 295.3

Table 4 Examples of nanoparticles targeting dendritic cells

NP	Diameter	Package/load	Model	Response	References
Poly(lactic-co-glycolic acid) PLGA-ICG-R848 NPs	Mean diameter 157.7 nm	Co-loading of a fluorescent dye, indocyanine green (ICG), and a toll-like receptor 7/8 (TLR7/8) agonist resiquimod (R848)	Subcutaneous Xenograft Model of RM9-Luc PCa cells	Significant maturation of BMDCs with significantly elevated proportions of $CD11c^+$ $CD86^+$ and $CD11c^+$ $CD80^+$ cells	[126]
Lipophilic poly (beta-amino ester) PBAE NPs	Approximately 100 nm	Ovalbumin (OVA) as a tumor antigen and TLR3 agonist polyinosinic-polycytidylic acid (poly(I:C)) or the TLR9 agonist CpG oligodeoxynucleotide (CpG ODN)	B16-F10-OVA mouse melanoma model and MC38-OVA mouse colon carcinoma model	Antigen-specific $CD8^+$ T cell response, reduced tumor burden, and prolonged median survival	[127]
Ganglioside GM3 containing liposomes	160–190 nm	TLR4 agonist: monophosphoryl lipid A (MPLA)	Monocyte-derived dendritic cells and mouse splenocytes	Maturation of Dendritic Cells (DCs) and $CD4^+$ and $CD8^+$ T cell priming	[128]
PEGylated $Ca(OH)_2/SiO_2$ core–shell nanoparticles coupled an anti-CD205 antibody (AnCHNPs)	295.3 ± 46.7 nm	$Ca(OH)_2$	Mouse bone marrow-derived dendritic cells (BMDCs) and B16F10-OVA tumor-bearing C57BL/6 mice	Enhanced antitumor immune response and augmented efficacy of both radiotherapy and chemotherapy	[129]
α-AP-fmNPs [Phospholipid coated oleic acid-stabilized iron oxide nanocrystals]	Approximately 25 nm	α-APgp100 antigen peptide, iron oxide nanocrystals, and indocyanine green	BMDCs and C57BL/6 mice	Enhanced in vivo migration efficiency under magnetic pull force (MPF), resulting in a significantly improved antitumor efficacy	[130]

$\pm$ 46.7 nm were tested on mouse bone marrow-derived dendritic cells (BMDCs) and B16F10-OVA tumor-bearing C57BL/6 mice, showing enhanced antitumor immune response and augmented efficacy of both radiotherapy and chemotherapy [129]. α-AP-fmNPs (phospholipid-coated oleic acid-stabilized iron oxide nanocrystals) with a size of approximately 25 nm were used with α-APgp100 antigen peptide and indocyanine green. These NPs were tested on BMDCs and C57BL/6 mice, showing enhanced in vivo migration efficiency under magnetic pull force (MPF), resulting in significantly improved antitumor efficacy [130].

3.7 Targeting/Stimulation of T-Cells

In the middle of the twentieth century, Sir Macfarlane Burnet and Lewis Thomas introduced the concept of cancer immunosurveillance suggesting that lymphocytes could act as mediators of anti-tumor response [131, 132]. However, T cells were not recognized as the primary adaptive immunity mediators until much later and these days, the field of study on T cells' role in cancer immunotherapy is expanding swiftly [133]. Over the years, the discovery of antitumor T cells infiltrating the tumors has opened doors for T cell immunotherapy targeting tumor antigens. Many different immunotherapeutic approaches are being developed for the treatment of cancer. These immunotherapeutic seek to skew the development of naïve T cells into one of the effector subsets, deplete the particular pro- or anti-inflammatory subsets of T cells, or employ T cells as drug carriers to the site of disease. In addition, there is a tremendous amount of interest in modulating the immune checkpoint inhibition as a tactic to increase T-cell responses to sick or infected cells. Recent developments in genome editing and gene transfer technology have led to breakthrough in chimeric antigen receptor (CAR) T cell treatment, which involves genetically modifying T cells outside of the body. These cells can seek out and penetrate the tumor microenvironment or target autoantibody-producing or infected cells after reinfusion (adoptive transfer). Furthermore, T cells directly suffer from disease in the form of T cell lymphoma and T cell lymphocytic leukaemia.

Even though T-cell immunotherapy has advanced, significant challenges exist in the targeted delivery of therapeutic agents to T cells and inadequate therapeutic efficacy for solid tumors. Emerging nanomaterials could enhance the delivery of immunomodulatory agents specifically to the T-cells amplifying the immune response against cancer cells [134, 135]. The utilization of advanced nanoparticle formulations presents a promising strategy for drug delivery. However, it remains a challenge to maintain the stability and biocompatibility of these formulations while also mitigating potential cytotoxic effects [136, 137]. Researchers worldwide are putting a lot of effort into the development of nanoparticles that can effectively navigate the tumor microenvironment and deliver therapeutic agents to T-cells without inducing systemic toxicity to ensure safe and effective targeting of T-cells in the clinical setting [138–140].

Over the years, many different research studies have reported the use of nanoparticles composed of a wide variety of material classes, including polymers like polyethyleneimine, polyesters, chitosan, alginate, hyaluronan, polyamidoamine, lipids, and inorganic materials like gold, iron oxide, and silica for both in vitro and in vivo activation of T cells. Guasch et al. reported the development of polyethylene glycol (PEG) hydrogel decorated with a quasi-hexagonal array of gold nanoparticles (AuNPs). These nanogels were further cross-linked with two fibronectin-derived peptides, cyclic Arg-Gly-Asp (cRGD) and cyclic Leu-Asp-Val (cLDV), that are known to activate different integrins and functionalized with the activating antibody CD3 to activate primary T cells [141]. Many such strategies are being evaluated. Table 5 summarizes a few such nanoparticle-based approaches for enhancing T cell-mediated cancer immunotherapy. Many of the evaluated formulations employ a targeted approach that allows for the delivery of therapeutic agents directly to T cells, minimizing off-target effects and enhancing the therapeutic index. Amphiphilic organic ligand-protected gold nanoparticles (2.5–3.5 nm) conjugated with anti-CD8 antibodies demonstrated enhanced uptake of TGF-β inhibitors in T cells and increased cytokine production by antigen-specific $CD8^+$ T cells [142]. Liposomes (approximately 100 nm) composed of DPPC, cholesterol, and PEG2000 PE exhibited improved T-cell activation and delayed tumor progression in a multiple myeloma mouse model [143]. Silica nanocapsules (231 ± 111 nm) stabilized with CTAC effectively delivered siRNA to downregulate PD-L1 in $CD8^+$ T-cells, thereby enhancing T-cell function [144]. Dual-binding magnetic nanoparticles (330 ± 42 nm) comprising PEG-maleimide, hyaluronic acid, and Fe3O4 enhanced anti-solid tumor efficacy in adoptive cell therapy [145]. In addition, anti-CD4 conjugated polymeric nanogels (20–30 nm) exhibited cell growth inhibition in primary $CD4^+$ T cells and a CD4 high T cell lymphoma model [146]. The development of engineered nanoparticle systems has significantly progressed T cell-mediated cancer immunotherapies by optimizing drug delivery mechanisms, enhancing T cell activation, and refining tumor targeting strategies. These nano-formulations have demonstrated improved specificity and augmented T cell targeting, thereby strengthening T cell responses to tumors and enhancing the efficacy of cancer treatment methodologies. The integration of such advanced technologies into clinical practice holds the potential for more effective and personalized therapeutic strategies.

3.8 Targeting/Stimulation of Natural Killer (NK) Cells

Anticancer immunotherapy aims to bolster the immune response and counteract the immunosuppressive tumor microenvironment (TME). Multiple nanoparticle-based tumor immunotherapy strategies targeting B cells, T cells, DCs, and NK cells have been reported (Fig. 3). The immune system can effectively remove tumor cells through the activation of innate and adaptive immune responses in the initial stages of cancer development and progression. In the elimination phase of cancer immunoediting, a combination of both adaptive and innate immune cells provides systemic protection against cancer [147]. Innate immune cells including the natural killer (NK) cells, dendritic cells, macrophages, neutrophils, basophils, and eosinophils form the

Table 5 Examples of nanoparticles targeting T-cells

NP	Diameter	Package/load	Model	Response	References
Amphiphilic organic ligand-protected gold nanoparticles (amph-NPs) with anti-CD8 antibodies	2.5–3.5 nm	TGF-β inhibitor SB525334	Primary $CD8^+$ T cells and melanoma murine model	Enhanced uptake of a small molecule TGF-β inhibitor in T cells and increased cytokine production by antigen-specific $CD8^+$ T cells	[142]
DPPC, cholesterol, and 16:0 PEG2000 PE Liposomes	Approx. 100 nm	Phytohemagglutinin (PHA)	Immunocompetent multiple myeloma mouse model	Improved T-cell activation, significant delayed tumor progression and prolonged survival of cancer-bearing mice	[143]
Silica Nanocapsules (SiNC) stabilized with cationic surfactant cetyltrimethylammonium chloride (CTAC)	231 ± 111 nm	siRNA to Knockdown PD-L1	$CD8^+$ T-cells	Knockdown of PD-L1 and improved T-cell function	[144]
Dual-binding magnetic nanoparticles (DBMN) comprised of PEG-maleimide (Mal), hyaluronic acid (HA) and Fe_3O_4	330 ± 42 nm	$CD8^+$ T cells [magnetic T cells (DBMN-T)]	Bilateral subcutaneous E.G7-OVA tumor model	Improved anti-solid tumor efficacy in adoptive cell therapy (ACT)	[145]
Anti-CD4 conjugate Polyethylene glycol and pyridyl disulfide polymeric p(PDS-*co*-PEG) nanogels	20–30 nm (micelles)	Anti-cancer drug mertansine	Primary $CD4^+$ T cells and a $CD4^{high}$ T cell lymphoma	Cell growth inhibition	[146]

first line of host defense. Adaptive immune responses mediated by the T lymphocytes and B lymphocytes form the second line of defense against cancer. Recognition and presentation of the tumor-associated antigens by the antigen-presenting cells trigger the activation and clonal selection of T cells and B cells and the generation of cytotoxic and antibody-based responses [148]. NK cells are cytotoxic lymphocytes. These professional killer cells can recognize and rapidly eliminate infected or transformed cells and contribute towards cancer immune surveillance, particularly cancer metastasis. NK cells lack an antigen-specific cell surface receptor and have a broader reactivity. Many cancers suppress the surface expression of MHC-I expression making them vulnerable to NK cell-mediated assault. These cells do not require prior exposure to the malignant cell to respond. In addition to the direct cytotoxic effect, these cells can potentially amplify the B cell and T cell responses thus playing a crucial part in immune responses to malignancies [147, 149, 150].

The strategies devised for NK cell-based immunotherapies include increased cell proliferation, expression levels of cytotoxic agents, infiltration capability, blockage of inhibitory signaling pathways, and prevention of cell exhaustion. Thus the primary approach to developing NK cell-based immunotherapies would be to upregulate or activate NK cell-activating receptors such as NKG2D, DNAM1, NCR, CD16, NCAM/CD56, CD96, CXCR1, CX3CR1, IL-2/IL-15, CD11, and CCR7; while inhibit or downregulate the NK cell-inhibitory receptors such as CD94/NKG2A, TIM-3, TIGIT, KIRs, LAG-3, and PD-1 [151]. Multiple studies demonstrate the potential of nanoparticle-based approaches in enhancing NK cell-mediated anti-tumor immunity. Different polymers, organics, and inorganic nanomaterials as well as multiple functionalization strategies have been employed to enhance the homing and infiltration capabilities of NK cells (Table 6). Immunogene TUSC2 (Tumor

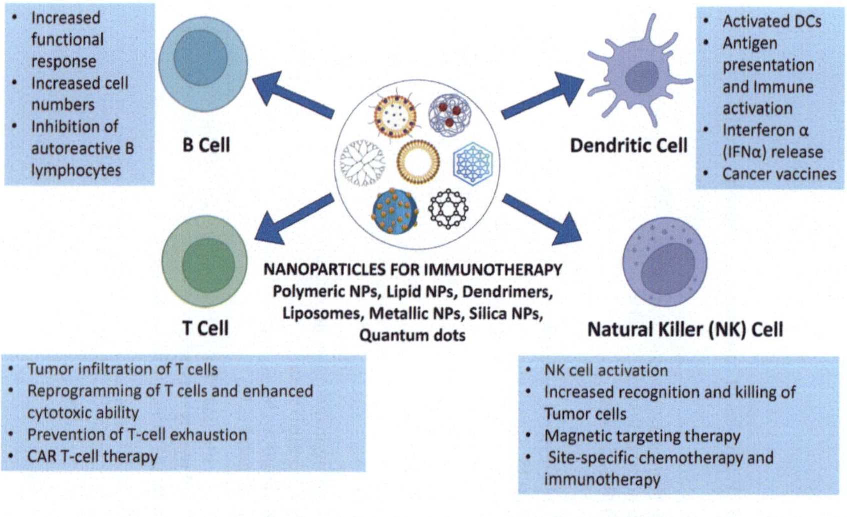

Fig. 3 Schematic illustration of the interaction between NPs and different immune cells in cancer immunotherapy

Suppressor Candidate 2)-loaded liposomes or IFN-γ coated nano-composite microspheres demonstrate a significant increase in NK cell infiltration [152], 153]. In addition, novel lipid-based platforms based on imidazole lipids produced by microfluidic mixing and a size of approximately 100 nm can transfect NK-92 cells, KHYG-1 cells, and primary NK cells with high efficiency and without cytotoxicity. The lipoplexes and lipid nanoparticles reported in this study showcase lipid-based platforms for mRNA engineering of NK cells [154]. The use of magnetic nanoparticles synthesized using different strategies has been evaluated in multiple studies. Magnetic nanoparticles with iron-oxide core coated with (3-aminopropyl) triethoxysilane (APS-MNP) and cationic magnetic nanoparticles (cNPs) coated with polydopamine (PDA) and then conjugated with the cationic polymer polyethylenimine (PEI) were evaluated in independent research studies. APS-MNPs are attached to the surface of the NK cells without impairing their functions. Specific in vitro retention of the functional NK cells was achieved using the electromagnetic field confirming the potential of this approach to promote the specific accumulation of these cytolytic cells in vivo [155]. The NK cells treated with the cNPs were effectively tracked in vivo by magnetic resonance imaging and exhibited tumor growth inhibition in an in vivo tumor animal model of triple-negative breast cancer [156]. These kinds of therapeutic strategies can enhance the delivery of fully functional NK cells to the tumor site and result in improved treatment efficacy and superior clinical outcomes.

3.9 Regulating Immune-Suppressive Tumour Microenvironment (TME)

Cancer immunoediting is the term assigned to the immune-suppressive conditions in the tumour microenvironment (TME). The immune system has a dual property of acting as a pro- or anti tumor in the context of carcinogenesis. Immune system is reported to contribute significantly to tumor progression and metastasis by modulating tumor immunity, chronic inflammation, and further suppressing antitumor immunity. Tumor cells constitute of a complex mechanism which undermine standard immune regulation and facilitates various immune evasion strategies, inducing an immunosuppressive tumor microenvironment [159, 160]. The tumor microenvironment (TME) is made up of tumor cells, lymphocytes, lymphatic cells, myeloid-derived suppressor cells (MDSCs), endothelial cells, fibroblasts tumor associated macrophages (TAMs), dendritic cells (DCs), bone marrow-derived cells (BMDCs), and various cytokines/chemokines [161], 162].

The anti-tumor response of immune system involves series of effects, initiated by specifically recognizing the tumor antigens via tumor associated antigens (TAAs). Antigen presenting cells (APCs) processes the antigens and activates the elimination of tumor cells by programmed cell death (apoptosis), thus controlling the tumor growth. In the second phase of immune response, an immunological equilibrium is generated due to slow elimination of tumor cells by the immune system as tumor cells grows at a very high rate. Under this situation, various antitumor activities

Table 6 Examples of nanoparticles targeting NK cells

NP	Diameter	Package/load	Model	Response	References
iCD-Lx (imidazole cationic delivery lipoplexes) and iD-LNP (imidazole delivery lipid nanoparticles)	60–130 nm	IL-2 mRNA	NK-92 and primary NK cells	Enhancing natural killer cells proliferation and cytotoxicity	[154]
APS-MNP (3-aminopropyl-triethoxysilane magnetic nanoparticles) with iron-oxide cores	82 nm	–	NK-92MI cells and primary murine NK cells	No significant effect on the functionality and successful magnetic retention	[155]
cNPs (Cationic Nanoparticles) [Magnetic nanoparticles coated with polydopamine (PDA); cationic polymer polyethylenimine (PEI) immobilized with the PDA layer]	41 nm	–	Primary NK and NK-92MI cells	Effective inhibition of tumor growth in an in vivo tumor animal model of triple-negative breast cancer by cNP-treated NK cells	[156]
Trispecific PEG-PLGA NPs [poly(ethylene glycol)–block-poly(lactide-co-glycolide)] functionalized with cetuximab (anti-human EGFR antibody, α-EGFR) and anti-CD16 (α-CD16) and anti–4-1BB (α-4-1BB) antibodies	112 ± 7 nm	Epirubicin	A431 and MB468 tumor models in T cell-deficient Nu mice	Effective inhibition of the EGFR-overexpressed cancer growth in vivo	[157]
$pMnO_2$-NPs [Manganese dioxide NPs conjugated with poly (ethylene glycol)]	13.86 ± 5.05 nm	siRNA against TGFBR2	NK cell line NK-92	Increased NK cell function and infiltration of natural killer (NK)-92 cells into H1299 lung cancer cell spheroids	[158]

of immune system are dysfunctional, resulting in immunosuppressive TME and promoting tumor escape [163, 164]. Progressively, recognition of tumor antigens by the immune cells is reduced to the surface modification in the MHC-I complex of tumor cells. MHC-I complex play crucial role in identification and activation of tumor cell lysis by binding to the cytotoxic $CD8^+$ tumor-infiltrating lymphocytes (TILs) [165]. During the tumor growth, various types of cytokines are released by the stromal cells, immune cells and cancer cells. Excessive secretion of immunosuppression-associated cytokines induces tremendous change in the TME, making it more resistant towards existing immune system [166]. As the growth of tumor cells progresses, additional requirement of nutrients and oxygen is taken care by overexpression of pro-angiogenic factor: vascular endothelial growth factor (VEGF). In addition to this, VEGF causes high leakage, enhanced permeability and increased interstitial fluid. VEGF has been reported to repair the damaged tumor cells along with suppressing immune response by inhibiting T-cell proliferation. IL-2 has proven to be an important participant in generating immunosuppressive TME by directly regulating the activation of T-cells [167].

Another prominent player for inducing immunosuppressive TME if Tumor associated macrophages (TAM). In TME, immature macrophages differentiate into M1 (pro-inflammatory) and M2 (anti-inflammatory) phenotypes. Further, the infiltrating macrophages into the tumor cells are converted into M2 phenotype resulting in secretion of large number of immunosuppressive cytokines, chemokines and supports tumor immune evasion mechanism [168, 169]. Myeloid derived suppressor cells (MDSCs) are also known to facilitate the release of cytokines and generating ROS under TME. Similarly, Tregs, CTLA-4, cancer-associated fibroblasts (CAFs) and programmed death receptor-1 (PD-1) are considered to play crucial role in inhibition of T-cell activation, thereby, promoting cancer growth. Cancer-associated fibroblasts (CAFs) are known for its ability to determine the fate of anti-tumor immune response by the action of chemokines, active proteins of extracellular matrix and anti-inflammatory cytokines [170, 171]. Secretion of cytokines via CAFs has led to a condition in number of regulatory T cells (Tregs) is elevated along with inhibition of effector T cells and cytotoxic T cells ($CD8^+$ T cells) activity [172].

Tumor microenvironment (TME) represents a cocktail of complex and dynamic immune cell networking with tumor cells. Effective advancement has been made in the cancer treatment based on anti-cytotoxic T-lymphocyte associated protein (CTLA)-4 and anti-programmed cell death receptor (PD)-(L)1 antibodies by targeting immune checkpoint inhibitors (ICIs) [173]. Despite the potential of ICIs based immunotherapy, they are characterized by certain disadvantages like low response rates and high rates of adverse effects on multiple organs during immunotherapy. Consequently, the development of advanced therapeutic strategies to promote anti-tumor responses and reversing immune suppression within the TME is predominantly important. A comprehensive understanding of the TME and implementation of targeted therapeutic strategies through nanomedicine can be considered as an effective approach to address immunoresistance. Likewise, numerous nanoformulation have been reported targeting various components of immunosuppressive network persisting in TME. In one of the report, dendritic cell (DC) membrane coated nanoaggregates (DC@AIEdots) with photosensitizers (AIEgens) was fabricated. The prepared nanoaggregates were found to initiate T cell proliferation and activation by artificially processing and presenting antigens. Along with this, DC@AIEdots elevated the levels of tumor necrosis factor (TNF-a) and interferon (IFN-g) in by many folds. The activation of anti-tumor immune response further led to the reduction in the size of tumor [174]. Similarly, fucoidan functionalized DOX-loaded micelle (FD/DOX) was formulated to retain the antitumor immune response in TME. The interaction of fucoidan with P-selectin has been recognized as a method to monitor tumor cells and enhance adaptive immunity, thereby counteracting the immunosuppressive tumor microenvironment (TME) [175]. Furthermore, docetaxel demonstrated anti-tumor efficacy in both in vitro and in vivo studies, offering a straightforward yet significant approach to cancer therapy.

3.10 Clinical Translation of Nano-immunotherapy

Emerging Nano-immunotherapeutic approaches can be related to pre-existing immunoregulatory circuits with in TME. Tumor cells possess capabilities to outlay the immune system to ensure their survival. Novel and innovative approaches based on immuno-oncology can be achieved by addressing the various limitations present in existing tumor therapies. Consequently, new therapeutic approaches have been developed by exploring nano/biomaterials in immunotherapy to address cancer progression. Diverse array of nanomedicines has undergone the clinical trials and were approved as an active chemotherapeutics [176]. The Cortellis Drug Discovery Intelligence (CDDI) database serves as a comprehensive platform for accessing information regarding nanomedicine that have received market approval or are currently undergoing clinical trials. The clinical translation of different types of nanomaterials viz., lipid-based nanoparticle, polymer-drug/protein nanohybrids, polymer-antibody-drug conjugates and dendrimer has been documented [177]. PEGylated liposomal formulation of doxorubicin namely, Doxil with various immunomodulatory properties was the first cancer nanomedicine approved by FDA in 1995 (76). Later the years various chemotherapeutic drugs were approved by FDA (Table 7). Nanomedicines and chemotherapeutics focused on biomaterials have attained considerable clinical significance in recent years (Table 8).

The treatment of lymphoma was clinically studied for the polymer-protein-drug conjugate of PEG-SN38 loaded with drug for cancer treatment. Enzon has proposed this formulation which underwent phase I and II clinical trials [178]. An advance version of paclitaxel: ANG-1005, was developed by AngioChem for the treatment of malignant tumor is currently in phase II clinical development [179]. Similarly, Mayo clinic has fabricated paclitaxel albumin stabilized nanoparticle as AR-160. Ar-160 was tagged with a chimeric monoclonal antibody against CD20 (rituximab). AR-160 is under clinical trial for non-Hodgkin lymphoma [180]. Paclitaxel have been extensively explored for wide range of nanoformulation. Likewise, Abraxane is a nanoformulation of albumin-paclitaxel nanoparticles. Abraxane under phase III clinical trial

Table 7 List of nanoformulation under clinical trials for cancer treatment

Cancer	Nanoformulation	Clinical trial phase/NCT number	Status
Esophageal	NBTXR3 capecitabine, carboplatin, paclitaxel	Phase 1 NCT04615013	Recruiting
Sarcoma	NBTXR3	Phase 2 and 3 NCT02379845	Completed
Breast	Nab-paclitaxel, nivolumab and ipilimumab	Phase 1 NCT04132817	Active, not recruiting
Lung	Nab-paclitaxel, carboplatin, ipilimumab, nivolumab	Phase 2 and 3 NCT04929041	Recruiting
Breast	Nab-paclitaxel atezolizumab	Phase 3, NCT041489113	Active, not recruiting
Breast	Nab-paclitaxel atezolizumab and carboplatin	Phase 2 NCT05266937	Recruiting
Lung	Nab-paclitaxel atezolizumab and carboplatin	Phase 2 NCT04865250	Recruiting
Breast	Nab-paclitaxel and liposomal, doxorubicin	Phase 3 NCT04895358	Recruiting
Ovarian	Liposomal, doxorubicin, Pembrolizumab, paclitaxel	Phase 2 NCT03539328	Unknown
Glioblastoma	Ferumoxytol ascorbate, temozolomide	Phase 1 NCT04900792	Not yet recruiting
Lung	Nab-paclitaxel CC-486 and duravalumab	Phase 2 NCT02250326	Active, not recruiting

has proven to be a potential immunomodulatory (for dendritic cells, lymphocytes, natural killer (NK) cells) in breast cancer patients along with reducing the number of regulatory T cells (Tregs) and regulating the M1 phenotype. FDA approval has been given to Abraxane and atezolizumab for metastatic triple-negative breast cancer [181]. PEGylated liposomal formulation of doxorubicin namely, Doxil with various immunomodulatory properties was the first cancer nanomedicine approved by FDA in 1995 [182]. Currently, clinical investigations of innovative formulations such as targeted nanomedicine have demonstrated substantiate clinical promise for providing more effective immunotherapeutic options for cancer patients.

Table 8 List of FDA approved Immunotherapy based nanoformulation for cancer treatment

FDA (year)	Drug	Targeted receptors	Mode of action
2019	Polatuzumab vedotin	Humanised anti-CD79b	Neoantigen recognition Target therapy
2018	Cemiplimab	Human anti-PD1	Checkpoint inhibition
2017	Tisagenlecleucel	CAR-T cell anti-CD19	Neoantigen recognition
2017	Durvalumab	Human anti-PDL1	Checkpoint inhibition
2017	Durvalumab	Human anti-PDL1	Checkpoint inhibition
2017	Avelumab	Human anti-PDL1	Checkpoint inhibition
2017	Axicabtagene ciloleucel	CAR-T cell anti-CD19	Neoantigen recognition
2016	Atezolizumab	Humanised Anti-PDL1	Checkpoint inhibition
2015	Tremelimumab	Human anti-CTLA4	Checkpoint inhibition
2015	Necitumumab	Human anti-EGFR	Neoantigen recognition target therapy
2015	Daratumumab	Human anti-CD38	Neoantigen recognition immunomodulation
2014	Nivolumab human	Anti-PD1	Checkpoint inhibition
2014	Ramucirumab	Human anti-VEGFR-2	Target therapy: anti-angiogenic effect
2014	Blinatumomab	Murine anti-CD19 and anti-CD3	Neoantigen recognition

4 Conclusion

Cancer immunotherapy holds promise as it confers safer approach towards treatment. Stimulation of immune cells for recognizing and eliminating cancer cells at early stage is an effective and endogenous way to manage cancer within body. With the advent of nanotechnology, the specific and targeted stimulation of immune cells lead the better treatment options. However, few challenges including correct choice of safer nanomaterial, and optimum stimulation of immune cells need to be addressed and researched properly so that such therapy may reach to the clinics.

References

1. Seyfried, T.N., Shelton, L.M.: Cancer as a metabolic disease. Nutr. Metab. **7**, 1–22 (2010)
2. Pina-Sanchez, P., Chavez-Gonzalez, A., Ruiz-Tachiquin, M., Vadillo, E., Monroy-Garcia, A., Montesinos, J.J., et al.: Cancer biology, epidemiology, and treatment in the 21st century: current status and future challenges from a biomedical perspective. Cancer Control **28** (2021)
3. Pfeffer, C.M., Singh, A.T.: Apoptosis: a target for anticancer therapy. Int. J. Mol. Sci. **19**(2), 448 (2018)
4. Vander Heiden, M.G., Cantley, L.C., Thompson, C.B.: Understanding the Warburg effect: the metabolic requirements of cell proliferation. Science **324**(5930), 1029–1033 (2009)

5. Liao, M., Yao, D., Wu, L., Luo, C., Wang, Z., Zhang, J., Liu, B.: Targeting the Warburg effect: a revisited perspective from molecular mechanisms to traditional and innovative therapeutic strategies in cancer. Acta Pharmaceutica Sinica B. (2023)
6. Debela, D.T., Muzazu, S.G., Heraro, K.D., Ndalama, M.T., Mesele, B.W., Haile, D.C., et al.: New approaches and procedures for cancer treatment: current perspectives. SAGE Open Med. **9** (2021)
7. Fares, J., Fares, M.Y., Khachfe, H.H., Salhab, H.A., Fares, Y.: Molecular principles of metastasis: a hallmark of cancer revisited. Signal Transduct. Target. Ther. **5**(1), 28 (2020)
8. Chen, Z., Han, F., Du, Y., Shi, H., Zhou, W.: Hypoxic microenvironment in cancer: molecular mechanisms and therapeutic interventions. Signal Transduct. Target. Ther. **8**(1), 70 (2023)
9. Sung, H., Ferlay, J., Siegel, R.L., Laversanne, M., Soerjomataram, I., Jemal, A., Bray, F.: Global cancer statistics 2020: GLOBOCAN estimates of incidence and mortality worldwide for 36 cancers in 185 countries. CA Cancer J. Clin. **71**(3), 209–249 (2021)
10. Bray, F., Laversanne, M., Sung, H., Ferlay, J., Siegel, R.L., Soerjomataram, I., Jemal, A.: Global cancer statistics 2022: GLOBOCAN estimates of incidence and mortality worldwide for 36 cancers in 185 countries. CA Cancer J. Clin. **74**(3) (2024)
11. Collaborators, G.B.D., Tran, K.B., Lang, J.J., Compton, K., Xu, R., Acheson, A.R., et al.: The global burden of cancer attributable to risk factors, 2010–19: a systematic analysis for the Global Burden of Disease Study 2019. Lancet **400**(10352) (2022)
12. Soerjomataram, I., Cabasag, C., Bardot, A., Fidler-Benaoudia, M.M., Miranda-Filho, A., Ferlay, J., et al.: Cancer survival in Africa, central and south America, and Asia (SURVCAN-3): a population-based benchmarking study in 32 countries. Lancet Oncol. **24**(1) (2023)
13. Cao, W., Qin, K., Li, F., Chen, W.: Comparative study of cancer profiles between 2020 and 2022 using global cancer statistics (GLOBOCAN). J. Natl. Cancer Center (2024)
14. Dai, Q., Cao, B., Zhao, S., Zhang, A.: Synergetic thermal therapy for cancer: state-of-the-art and the future. Bioengineering **9**(9) (2022)
15. Jameus, A., Kennedy, A.E., Thome, C.: Hematological changes following low dose radiation therapy and comparison to current standard of care cancer treatments. Dose-Response **19**(4) (2021)
16. Wu, J.: The enhanced permeability and retention (EPR) effect: the significance of the concept and methods to enhance its application. J. Pers. Med. **11**(8) (2021)
17. Majeed, H., Gupta, V.: Adverse effects of radiation therapy. (2020)
18. Rosa, F., Schena, C.A., Laterza, V., Quero, G., Fiorillo, C., Strippoli, A., et al.: The role of surgery in the management of gastric cancer: state of the art. Cancers **14**(22), 5542 (2022)
19. Nguyen, N.M., Cho, J.: Hedgehog pathway inhibitors as targeted cancer therapy and strategies to overcome drug resistance. Int. J. Mol. Sci. **23**(3) (2022)
20. Wankhade, D., Gharde, P., Dutta, S.: The current role of neoadjuvant chemotherapy in the management of HER2-positive, triple-negative, and micropapillary breast cancer: a narrative review. Cureus **15**(11) (2023)
21. Ashique, S., Hussain, A., Fatima, N., Altamimi, M.A.: HPV pathogenesis, various types of vaccines, safety concern, prophylactic and therapeutic applications to control cervical cancer, and future perspective. VirusDisease **34**(2) (2023)
22. Yu, P., Zhu, C., You, X., Gu, W., Wang, X., Wang, Y., et al.: The combination of immune checkpoint inhibitors and antibody-drug conjugates in the treatment of urogenital tumors: a review insights from phase 2 and 3 studies. Cell Death Dis. **15**(6) (2024)
23. Nassief, G., Anaeme, A., Moussa, K., Chen, D., Ansstas, G.: Where are we now with oncolytic viruses in melanoma and nonmelanoma skin malignancies? Pharmaceuticals **17**(7) (2024)
24. Thomas, C., Baunacke, M., Erb, H.H., Füssel, S., Erdmann, K., Putz, J., Borkowetz, A.: Systemic triple therapy in metastatic hormone-sensitive prostate cancer (mHSPC): ready for prime time or still to be explored? Cancers **14**(1) (2021)
25. Bo, X.W., Sun, L.P., Yu, S.Y., Xu, H.X.: Thermal ablation and immunotherapy for hepatocellular carcinoma: recent advances and future directions. World J. Gastrointest. Oncol. **13**(10) (2021)

26. Lei, Z.N., Tian, Q., Teng, Q.X., Wurpel, J.N., Zeng, L., Pan, Y., Chen, Z.S.: Understanding and targeting resistance mechanisms in cancer. MedComm 4(3) (2023)
27. Lugano, R., Ramachandran, M., Dimberg, A.: Tumor angiogenesis: causes, consequences, challenges and opportunities. Cell. Mol. Life Sci. **77** (2020)
28. Emami Nejad, A., Najafgholian, S., Rostami, A., Sistani, A., Shojaeifar, S., Esparvarinha, M., et al.: The role of hypoxia in the tumor microenvironment and development of cancer stem cell: a novel approach to developing treatment. Cancer Cell Int. **21**, 1–26 (2021)
29. McCarthy, E.F.: The toxins of William B. Coley and the treatment of bone and soft-tissue sarcomas. Iowa Orthopaed. J. **26** (2006)
30. Carlson, R.D., Flickinger Jr, J.C., Snook, A.E.: Talkin'toxins: from Coley's to modern cancer immunotherapy. Toxins **12**(4) (2020)
31. Maleki, E.H., Bahrami, A.R., Matin, M.M.: Cancer cell cycle heterogeneity as a critical determinant of therapeutic resistance. Genes Di. **11**(1) (2024)
32. Pol, J., Kroemer, G., Galluzzi, L.: First oncolytic virus approved for melanoma immunotherapy. Oncoimmunology **5**, e1115641 (2016)
33. Aldrak, N., Alsaab, S., Algethami, A., Bhere, D., Wakimoto, H., Shah, K., et al.: Oncolytic herpes simplex virus-based therapies for cancer. Cells **10**(6) (2021)
34. Bell, J.C.: How Oncolytic Virus Therapy is Changing Cancer Treatment. (2020). Cancer Research Institute. Retrieved February, 19, 2022
35. Allison, J.P., Honjo, T.: Discovery of cancer therapy by inhibition of negative immune regulation. The Nobel Assembly at Karolinska Institute (2018)
36. Yetisgin, A.A., Cetinel, S., Zuvin, M., Kosar, A., Kutlu, O.: Therapeutic nanoparticles and their targeted delivery applications. Molecules **25**(9) (2020)
37. Chehelgerdi, M., Chehelgerdi, M., Allela, O.Q.B., Pecho, R.D.C., Jayasankar, N., Rao, D.P., et al.: Progressing nanotechnology to improve targeted cancer treatment: overcoming hurdles in its clinical implementation. Mol. Cancer **22**(1) (2023)
38. Singh, A.K.: Structure, synthesis, and application of nanoparticles. In: Singh, A.K. (ed.) Engineered Nanoparticles-Structure, Properties and Mechanisms of Toxicity, pp. 19–76. (2016)
39. Al-Thani, A.N., Jan, A.G., Abbas, M., Geetha, M., Sadasivuni, K.K.: Nanoparticles in cancer theragnostic and drug delivery: a comprehensive review. Life Sci. (2024)
40. Pérez-Ferreiro, M., Abelairas, M., Criado, A., Gómez, I.J., Mosquera, J.: Dendrimers: exploring their wide structural variety and applications. Polymers **15**(22) (2023)
41. Nsairat, H., Khater, D., Sayed, U., Odeh, F., Al Bawab, A., Alshaer, W.: Liposomes: structure, composition, types, and clinical applications. Heliyon **8**(5) (2022)
42. Juszkiewicz, K., Sikorski, A.F., Czogalla, A.: Building blocks to design liposomal delivery systems. Int. J. Mol. Sci. **21**(24) (2020)
43. Bose, A., Roy Burman, D., Sikdar, B., Patra, P.: Nanomicelles: types, properties and applications in drug delivery. IET Nanobiotechnol. **15**(1) (2021)
44. Preeti, Sambhakar, S., Malik, R., Bhatia, S., Al Harrasi, A., Rani, C., et al.: Nanoemulsion: an emerging novel technology for improving the bioavailability of drugs. Scientifica **2023**(1) (2023)
45. Mahor, A., Singh, P.P., Bharadwaj, P., Sharma, N., Yadav, S., Rosenholm, J.M., Bansal, K.K.: Carbon-based nanomaterials for delivery of biologicals and therapeutics: a cutting-edge technology. C **7**(1), 19 (2021)
46. Adeyemi, S.B., Akere, A.M., Orege, J.I., Ejeromeghene, O., Orege, O.B., Akolade, J.O.: Polymeric nanoparticles for enhanced delivery and improved bioactivity of essential oils. Heliyon **9**(6) (2023)
47. Mellid-Carballal, R., Gutierrez-Gutierrez, S., Rivas, C., Garcia-Fuentes, M.: Viral protein nanoparticles (Part 1): pharmaceutical characteristics. Eur. J. Pharmaceut. Sci. **187** (2023)
48. Chung, Y.H., Cai, H., Steinmetz, N.F.: Viral nanoparticles for drug delivery, imaging, immunotherapy, and theranostic applications. Adv. Drug Deliv. Rev. **156** (2020)
49. Burlec, A.F., Corciova, A., Boev, M., Batir-Marin, D., Mircea, C., Cioanca, O., et al.: Current overview of metal nanoparticles' synthesis, characterization, and biomedical applications, with a focus on silver and gold nanoparticles. Pharmaceuticals **16**(10) (2023)

50. Jahan, S.T., Sadat, S.M.A., Walliser, M., Haddadi, A.: Targeted therapeutic nanoparticles: an immense promise to fight against cancer. J. Drug Deliv. **2017** (2017). https://doi.org/10.1155/2017/9090325
51. Kadam, R.S., Bourne, D.W.A., Kompella, U.B.: Nano-advantage in enhanced drug delivery with biodegradable nanoparticles: contribution of reduced clearance. Drug Metabol. Dispos. **40**(7) (2012). https://doi.org/10.1124/dmd.112.044925
52. Liu, D., Yang, F., Xiong, F., Gu, N.: The smart drug delivery system and its clinical potential. Theranostics **6**(9) (2016). https://doi.org/10.7150/thno.14858
53. Qiu, L.Y., Bae, Y.H.: Polymer architecture and drug delivery. Pharmaceut. Res. **23**(1) (2006). https://doi.org/10.1007/s11095-005-9046-2
54. Kanwal, S., Naveed, M., Arshad, A., Arshad, A., Firdous, F., Faisal, A., Yameen, B.: Reduction-sensitive dextran-paclitaxel polymer–drug conjugate: synthesis, self-assembly into nanoparticles, and in vitro anticancer efficacy. Bioconjug. Chem. **32**(12), 2516–2529 (2021). https://doi.org/10.1021/acs.bioconjchem.1c00492
55. Bechnak, L., Khalil, C., El Kurdi, R., Khnayzer, R.S., Patra, D.: Curcumin encapsulated colloidal amphiphilic block co-polymeric nanocapsules: colloidal nanocapsules enhance photodynamic and anticancer activities of curcumin. Photochem. Photobiol. Sci. **19**(8) (2020). https://doi.org/10.1039/d0pp00032a
56. Hailemeskel, B.Z., Hsu, W.H., Addisu, K.D., Andrgie, A.T., Chou, H.Y., Lai, J.Y., Tsai, H.C.: Diselenide linkage containing triblock copolymer nanoparticles based on bi(methoxyl poly(ethylene glycol))-poly(ε-carprolactone): selective intracellular drug delivery in cancer cells. Mater. Sci. Eng. C **103** (2019). https://doi.org/10.1016/j.msec.2019.109803
57. Ye, S., Hou, Z., Su, Y., Ye, M., Liu, T., Ying, W., Su, M.: Rationally engineered Losmapimod encapsulating polymeric nanoparticles for treatment of human multiple myeloma cells. Appl. Nanosci. (Switz.) **11**(9) (2021). https://doi.org/10.1007/s13204-021-02010-3
58. Rompicharla, S.V.K., Bhatt, H., Shah, A., Komanduri, N., Vijayasarathy, D., Ghosh, B., Biswas, S.: Formulation optimization, characterization, and evaluation of in vitro cytotoxic potential of curcumin loaded solid lipid nanoparticles for improved anticancer activity. Chem. Phys. Lipid. **208**, 10–18 (2017). https://doi.org/10.1016/j.chemphyslip.2017.08.009
59. Da Rocha, M.C.O., Da Silva, P.B., Radicchi, M.A., Andrade, B.Y.G., De Oliveira, J.V., Venus, T., Merker, C., Estrela-Lopis, I., Longo, J.P.F., Báo, S.N.: Docetaxel-loaded solid lipid nanoparticles prevent tumor growth and lung metastasis of 4T1 murine mammary carcinoma cells. J. Nanobiotechnol. **18**(1) (2020). https://doi.org/10.1186/s12951-020-00604-7
60. Aly, S., El-Kamel, A.H., Sheta, E., El-Habashy, S.E.: Chondroitin/Lactoferrin-dual functionalized pterostilbene-solid lipid nanoparticles as targeted breast cancer therapy. Int. J. Pharm. **642**, 123163 (2023). https://doi.org/10.1016/j.ijpharm.2023.123163
61. AbouAitah, K., Abdelaziz, A.M., Higazy, I.M., Swiderska-Sroda, A., Hassan, A.M.E., Shaker, O.G., Szałaj, U., Stobinski, L., Malolepszy, A., Lojkowski, W.: Functionalized carbon nanotubes for delivery of ferulic acid and diosgenin anticancer natural agents. ACS Appl. Bio Mater. **7**(2), 791–811 (2024). https://doi.org/10.1021/acsabm.3c00700
62. Moniriyan, F., Jamal Tabatabaei Rezaei, S., Javad Sabounchei, S.: Pegylated magnetic carbon nanotubes for efficient and safe delivery of poorly water-soluble platinum anticancer drugs. Res. Chem. **7** (2024). https://doi.org/10.1016/j.rechem.2023.101275
63. Li, Y., Sun, H., Yang, C., Yan, Z., Ge, J.-Y., Wang, J., Chen, Z., Qiu, L., Bai, Y.: Metal-organic frameworks-based nanoplatform for treatment of breast cancer. ACS Appl. Nano Mater. **7**(9), 10532–10542 (2024). https://doi.org/10.1021/acsanm.4c00984
64. Raju, P., Balakrishnan, K., Mishra, M., Ramasamy, T., Natarajan, S.: Fabrication of pH responsive FU@Eu-MOF nanoscale metal organic frameworks for lung cancer therapy. J. Drug Deliv. Sci. Technol. **70** (2022). https://doi.org/10.1016/j.jddst.2022.103223
65. Chen, Y.J., Wang, Z.W., Lu, T.L., Gomez, C.B., Fang, H.W., Wei, Y., Tseng, C.L.: The synergistic anticancer effect of dual drug-(cisplatin/epigallocatechin gallate) loaded gelatin nanoparticles for lung cancer treatment.J. Nanomater. **2020** (2020).https://doi.org/10.1155/2020/9181549

66. Chaudhuri, A., Ramesh, K., Kumar, D.N., Dehari, D., Singh, S., Kumar, D., Agrawal, A.K.: Polymeric micelles: a novel drug delivery system for the treatment of breast cancer. J. Drug Deliv. Sci. Technol. **77** (2022). https://doi.org/10.1016/j.jddst.2022.103886
67. Duncan, R.: Polymer therapeutics as nanomedicines: new perspectives. Curr. Opin. Biotechnol. **22**(4) (2011). https://doi.org/10.1016/j.copbio.2011.05.507
68. Li, C., Wallace, S.: Polymer-drug conjugates: recent development in clinical oncology. Adv. Drug Deliv. Rev. **60**(8) (2008). https://doi.org/10.1016/j.addr.2007.11.009
69. Feng, Q., Tong, R.: Anticancer nanoparticulate polymer-drug conjugate. Bioeng. Transl. Med. **1**(3) (2016). https://doi.org/10.1002/btm2.10033
70. Wang, W., Chen, T., Xu, H., Ren, B., Cheng, X., Qi, R., Liu, H., Wang, Y., Yan, L., Chen, S., Yang, Q., Chen, C.: Curcumin-loaded solid lipid nanoparticles enhanced anticancer efficiency in breast cancer. Molecules **23**(7) (2018). https://doi.org/10.3390/molecules23071578
71. Gupta, N., Gupta, S.M., Sharma, S.K.: Carbon nanotubes: synthesis, properties and engineering applications. Carbon Lett. **29**(5) (2019). https://doi.org/10.1007/s42823-019-00068-2
72. Naser Mohammed, S., Mishaal Mohammed, A., Al-Rawi, K.F.: Novel combination of multiwalled carbon nanotubes and gold nanocomposite for photothermal therapy in human breast cancer model. Steroids **186** (2022). https://doi.org/10.1016/j.steroids.2022.109091
73. Zhang, Y., Bai, Y., Yan, B.: Functionalized carbon nanotubes for potential medicinal applications. Drug Discov. Today **15**(11–12) (2010). https://doi.org/10.1016/j.drudis.2010.04.005
74. Kumar, R., Mondal, K., Panda, P.K., Kaushik, A., Abolhassani, R., Ahuja, R., Rubahn, H.G., Mishra, Y.K.: Core-shell nanostructures: perspectives towards drug delivery applications. J. Mater. Chem. B **8**(39) (2020). https://doi.org/10.1039/d0tb01559h
75. Liu, J., Huang, J., Zhang, L., Lei, J.: Multifunctional metal-organic framework heterostructures for enhanced cancer therapy. Chem. Soc. Rev. **50**(2) (2021). https://doi.org/10.1039/d0cs00178c
76. Tran, V.A., Lee, S.W.: pH-triggered degradation and release of doxorubicin from zeolitic imidazolate framework-8 (ZIF8) decorated with polyacrylic acid. RSC Adv. **11**(16) (2021). https://doi.org/10.1039/d0ra10423j
77. Kciuk, M., Yahya, E.B., Mohamed Ibrahim Mohamed, M., Rashid, S., Iqbal, M.O., Kontek, R., et al.: Recent advances in molecular mechanisms of cancer immunotherapy. Cancers **15**(10) (2023)
78. Xuan, Y., Zhang, W., Zhu, X., Zhang, S.: An updated overview of some factors that influence the biological effects of nanoparticles. Front. Bioeng. Biotechnol. **11** (2023)
79. Li, Q., Liu, Y., Huang, Z., Guo, Y., Li, Q.: Triggering immune system with nanomaterials for cancer immunotherapy. Front. Bioeng. Biotechnol. **10** (2022)
80. Bezbaruah, R., Chavda, V.P., Nongrang, L., Alom, S., Deka, K., Kalita, T., et al.: Nanoparticle-based delivery systems for vaccines. Vaccines **10**(11) (2022)
81. Balakrishnan, P.B., Sweeney, E.E.: Nanoparticles for enhanced adoptive T cell therapies and future perspectives for CNS tumors. Front. Immunol. **12** (2021)
82. Grego, E.A., Siddoway, A.C., Uz, M., Liu, L., Christiansen, J.C., Ross, K.A., et al.: Polymeric nanoparticle-based vaccine adjuvants and delivery vehicles. In: Nanoparticles for Rational Vaccine Design, pp. 29–76. (2020)
83. Zhu, D., Li, Y., Zhang, Z., Xue, Z., Hua, Z., Luo, X., et al.: Recent advances of nanotechnology-based tumor vessel-targeting strategies. J. Nanobiotechnol. **19**, 1–17 (2021)
84. Wu, P., Han, J., Gong, Y., Liu, C., Yu, H., Xie, N.: Nanoparticle-based drug delivery systems targeting tumor microenvironment for cancer immunotherapy resistance: current advances and applications. Pharmaceutics **14**(10) (2022)
85. Denton, A.E., Roberts, E.W., Fearon, D.T.: Stromal cells in the tumor microenvironment. Adv. Exp. Med. Biol. **1060**, 99–114 (2018). https://doi.org/10.1007/978-3-319-78127-3_6
86. Wu, T., Dai, Y.: Tumor microenvironment and therapeutic response. Cancer Lett. **387** (2017). https://doi.org/10.1016/j.canlet.2016.01.043

87. Guo, S., Deng, C.X.: Effect of stromal cells in tumor microenvironment on metastasis initiation. Int. J. Biol. Sci. **14**(14) (2018). https://doi.org/10.7150/ijbs.25720
88. Hirata, E., Sahai, E.: Tumor microenvironment and differential responses to therapy. Cold Spring Harbor Perspect. Med. **7**(7) (2017). https://doi.org/10.1101/cshperspect.a026781
89. Musetti, S., Huang, L.: Nanoparticle-mediated remodeling of the tumor microenvironment to enhance immunotherapy. ACS Nano **12**(12) (2018). https://doi.org/10.1021/acsnano.8b05893
90. Meneveau, M.O., Kumar, P., Lynch, K.T., Patel, S.P., Slingluff, C.L.: The vaccine-site microenvironment: Impacts of antigen, adjuvant, and same-site vaccination on antigen presentation and immune signaling. J. Immuno Ther. Cancer **10**(3) (2022). https://doi.org/10.1136/jitc-2021-003533
91. Ren, D., Xiong, S., Ren, Y., Yang, X., Zhao, X., Jin, J., Xu, M., Liang, T., Guo, L., Weng, L.: Advances in therapeutic cancer vaccines: harnessing immune adjuvants for enhanced efficacy and future perspectives. Comput. Struct. Biotechnol. J. **23**, 1833–1843 (2024). https://doi.org/10.1016/j.csbj.2024.04.054
92. Neeson, P., Paterson, Y.: Effects of the tumor microenvironment on the efficacy of tumor immunotherapy. Immunol. Invest. **35**(3–4), 359–394 (2006). https://doi.org/10.1080/08820130600755009
93. Zhao, H., Zhao, B., Wu, L., Xiao, H., Ding, K., Zheng, C., Song, Q., Sun, L., Wang, L., Zhang, Z.: Amplified cancer immunotherapy of a surface-engineered antigenic microparticle vaccine by synergistically modulating tumor microenvironment. ACS Nano **13**(11) (2019). https://doi.org/10.1021/acsnano.9b03288
94. Li, M., Zhang, F., Su, Y., Zhou, J., Wang, W.: Nanoparticles designed to regulate tumor microenvironment for cancer therapy. Life Sci. **201** (2018). https://doi.org/10.1016/j.lfs.2018.03.044
95. Rao, L., Zhao, S.K., Wen, C., Tian, R., Lin, L., Cai, B., Sun, Y., Kang, F., Yang, Z., He, L., Mu, J., Meng, Q.F., Yao, G., Xie, N., Chen, X.: Activating macrophage-mediated cancer immunotherapy by genetically edited nanoparticles. Adv. Mater. **32**(47) (2020). https://doi.org/10.1002/adma.202004853
96. Liu, Q., Zhu, H., Liu, Y., Musetti, S., Huang, L.: BRAF peptide vaccine facilitates therapy of murine BRAF-mutant melanoma. Cancer Immunol. Immunother. **67**(2) (2018). https://doi.org/10.1007/s00262-017-2079-7
97. Li, L., Yang, S., Song, L., Zeng, Y., He, T., Wang, N., Yu, C., Yin, T., Liu, L., Wei, X., Wu, Q., Wei, Y., Yang, L., Gong, C.: An endogenous vaccine based on fluorophores and multivalent immunoadjuvants regulates tumor micro-environment for synergistic photothermal and immunotherapy. Theranostics **8**(3) (2018). https://doi.org/10.7150/THNO.19826
98. Zhu, G., Lynn, G.M., Jacobson, O., Chen, K., Liu, Y., Zhang, H., Ma, Y., Zhang, F., Tian, R., Ni, Q., Cheng, S., Wang, Z., Lu, N., Yung, B.C., Wang, Z., Lang, L., Fu, X., Jin, A., Weiss, I.D., et al.: Albumin/vaccine nanocomplexes that assemble in vivo for combination cancer immunotherapy. Nat. Commun. **8**(1) (2017). https://doi.org/10.1038/s41467-017-02191-y
99. Jin, J., Krishnamachary, B., Barnett, J.D., Chatterjee, S., Chang, D., Mironchik, Y., Wildes, F., Jaffee, E.M., Nimmagadda, S., Bhujwalla, Z.M.: Human cancer cell membrane-coated biomimetic nanoparticles reduce fibroblast-mediated invasion and metastasis and induce T-cells. ACS Appl. Mater. Interfaces **11**(8) (2019). https://doi.org/10.1021/acsami.8b22309
100. Yu, Q., Qiu, Y., Li, J., Tang, X., Wang, X., Cun, X., Xu, S., Liu, Y., Li, M., Zhang, Z., He, Q.: Targeting cancer-associated fibroblasts by dual-responsive lipid-albumin nanoparticles to enhance drug perfusion for pancreatic tumor therapy. J. Control. Rel. **321** (2020). https://doi.org/10.1016/j.jconrel.2020.02.040
101. Mahoney, K.M., Rennert, P.D., Freeman, G.J.: Combination cancer immunotherapy and new immunomodulatory targets. Nat. Rev. Drug Discov. **14**(8) (2015). https://doi.org/10.1038/nrd4591
102. Webster, R.M.: The immune checkpoint inhibitors: where are we now? Nat. Rev. Drug Discov. **13**(12) (2014). https://doi.org/10.1038/nrd4476

103. Dougan, M., Dougan, S.K.: Targeting immunotherapy to the tumor microenvironment. J. Cell. Biochem. **118**(10) (2017). https://doi.org/10.1002/jcb.26005
104. Chiappori, A.A., Creelan, B., Tanvetyanon, T., Gray, J.E., Haura, E.B., Thapa, R., Barlow, M.L., Chen, Z., Chen, D.T., Beg, A.A., Boyle, T.A., Castro, J., Morgan, L., Morris, E., Aregay, M., Hurtado, F.K., Manenti, L., Antonia, S.: Phase I study of Taminadenant (PBF509/NIR178), an Adenosine 2A receptor antagonist, with or without Spartalizumab (PDR001), in patients with advanced non-small cell lung cancer. Clin. Cancer Res. **28**(11) (2022). https://doi.org/10.1158/1078-0432.CCR-21-2742
105. Choe, D., Choi, D.: Cancel cancer: the immunotherapeutic potential of CD200/CD200R blockade. Front. Oncol. **13** (2023). https://doi.org/10.3389/fonc.2023.1088038
106. Wang, Y., Zhang, H., Liu, C., Wang, Z., Wu, W., Zhang, N., Zhang, L., Hu, J., Luo, P., Zhang, J., Liu, Z., Peng, Y., Liu, Z., Tang, L., Cheng, Q.: Immune checkpoint modulators in cancer immunotherapy: recent advances and emerging concepts. J. Hematol. Oncol. **15**(1) (2022). https://doi.org/10.1186/s13045-022-01325-0
107. Tanaka, R., Hiramitsu, M., Shimizu, S., Kawashima, S., Sato, A., Iwase, Y.: Efficient drug delivery to lymph nodes by intradermal administration and enhancement of anti-tumor effects of immune checkpoint inhibitors. Cancer Treat. Res. Commun. **36** (2023). https://doi.org/10.1016/j.ctarc.2023.100740
108. Luo, L., Yang, J., Zhu, C., Jiang, M., Guo, X., Li, W., Yin, X., Yin, H., Qin, B., Yuan, X., Li, Q., Du, Y., You, J.: Sustained release of anti-PD-1 peptide for perdurable immunotherapy together with photothermal ablation against primary and distant tumors. J. Control. Rel. **278** (2018). https://doi.org/10.1016/j.jconrel.2018.04.002
109. Banstola, A., Poudel, K., Emami, F., Ku, S.K., Jeong, J.H., Kim, J.O., Yook, S.: Localized therapy using anti-PD-L1 anchored and NIR-responsive hollow gold nanoshell (HGNS) loaded with doxorubicin (DOX) for the treatment of locally advanced melanoma. Nanomed. Nanotechnol. Biol. Med. **33** (2021). https://doi.org/10.1016/j.nano.2020.102349
110. Liu, Y., Pan, Y., Cao, W., Xia, F., Liu, B., Niu, J., Alfranca, G., Sun, X., Ma, L., De La Fuente, J.M., Song, J., Ni, J., Cui, D.: A tumor microenvironment responsive biodegradable $CaCO_3$/MnO_2-based nanoplatform for the enhanced photodynamic therapy and improved PD-L1 immunotherapy. Theranostics **9**(23) (2019). https://doi.org/10.7150/thno.37586
111. Ni, K., Luo, T., Nash, G.T., Lin, W.: Nanoscale metal-organic frameworks for cancer immunotherapy. Acc. Chem. Res. **53**(9) (2020). https://doi.org/10.1021/acs.accounts.0c00313
112. Lang, T., Liu, Y., Zheng, Z., Ran, W., Zhai, Y., Yin, Q., Zhang, P., Li, Y.: Cocktail strategy based on spatio-temporally controlled nano device improves therapy of breast cancer. Adv. Mater. **31**(5) (2019). https://doi.org/10.1002/adma.201806202
113. Jang, H., Kim, E.H., Chi, S.G., Kim, S.H., Yang, Y.: Nanoparticles targeting innate immune cells in tumor microenvironment. Int. J. Mol. Sci. **22**(18) (2021)
114. Aljabali, A.A., Obeid, M.A., Gammoh, O., El-Tanani, M., Mishra, V., Mishra, Y., et al.: Nanomaterial-driven precision immunomodulation: a new paradigm in therapeutic interventions. Cancers **16**(11) (2024)
115. Huang, Y., Guo, X., Wu, Y., Chen, X., Feng, L., Xie, N., Shen, G.: Nanotechnology's frontier in combatting infectious and inflammatory diseases: prevention and treatment. Signal Transduct. Target. Ther. **9**(1), 34 (2024)
116. Luo, Z., Liu, X.: Nanomaterials for cancer immunotherapy, what is the next? Next Nanotechnol. **1** (2023)
117. Aljabali, A.A., Obeid, M.A., Bashatwah, R.M., Serrano-Aroca, Á., Mishra, V., Mishra, Y., et al.: Nanomaterials and their impact on the immune system. Int. J. Mol. Sci. **24**(3) (2023)
118. Zhang, P., Meng, J., Li, Y., Yang, C., Hou, Y., Tang, W., et al.: Nanotechnology-enhanced immunotherapy for metastatic cancer. Innovation **2**(4) (2021)
119. Yao, Y., Zhou, Y., Liu, L., Xu, Y., Chen, Q., Wang, Y., et al.: Nanoparticle-based drug delivery in cancer therapy and its role in overcoming drug resistance. Front. Mol. Biosci. **7**, 193 (2020)
120. Rastogi, I., Jeon, D., Moseman, J. E., Muralidhar, A., Potluri, H. K., McNeel, D.G.: Role of B cells as antigen presenting cells. Front. Immunol. **13** (2022)

121. Ferapontov, A., Omer, M., Baudrexel, I., Nielsen, J.S., Dupont, D.M., Juul-Madsen, K., et al.: Antigen footprint governs activation of the B cell receptor. Nat. Commun. **14**(1) (2023)
122. Ahmed, H., Mahmud, A.R., Faijanur-Rob-Siddiquee, M., Shahriar, A., Biswas, P., Shimul, M.E.K., et al.: Role of T cells in cancer immunotherapy: opportunities and challenges. Cancer Pathogen. Ther. **1**(02) (2023)
123. Tran, T.H., Tran, T.T.P., Nguyen, H.T., Phung, C.D., Jeong, J.H., Stenzel, M.H., Jin, S.G., Yong, C.S., Truong, D.H., Kim, J.O.: Nanoparticles for dendritic cell-based immunotherapy. Int. J. Pharmaceut. **542**(1–2) (2018). https://doi.org/10.1016/j.ijpharm.2018.03.029
124. Guo, X., Zuo, X., Zhou, Z., Gu, Y., Zheng, H., Wang, X., Wang, G., Xu, C., Wang, F.: PLGA-based micro/nanoparticles: an overview of their applications in respiratory diseases. Int. J. Mol. Sci. **24**(5) (2023). https://doi.org/10.3390/ijms24054333
125. Horvath, D., Basler, M.: PLGA particles in immunotherapy. Pharmaceutics **15**(2) (2023). https://doi.org/10.3390/pharmaceutics15020615
126. Lin, W., Li, C., Xu, N., Watanabe, M., Xue, R., Xu, A., Araki, M., Sun, R., Liu, C., Nasu, Y., Huang, P.: Dual-functional PLGA nanoparticles co-loaded with indocyanine green and resiquimod for prostate cancer treatment. Int. J. Nanomed. **16** (2021). https://doi.org/10.2147/IJN.S301552
127. Ben-Akiva, E., Karlsson, J., Hemmati, S., Yu, H., Tzeng, S.Y., Pardoll, D.M., Green, J.J.: Biodegradable lipophilic polymeric mRNA nanoparticles for ligand-free targeting of splenic dendritic cells for cancer vaccination. Proc. Natl. Acad. Sci. USA **120**(26). https://doi.org/10.1073/pnas.2301606120
128. Nijen Twilhaar, M.K., Czentner, L., Bouma, R.G., Olesek, K., Grabowska, J., Wang, A.Z., Affandi, A.J., Belt, S.C., Kalay, H., van Nostrum, C.F., van Kooyk, Y., Storm, G., den Haan, J.M.M.: Incorporation of toll-like receptor ligands and inflammasome stimuli in GM3 liposomes to induce dendritic cell maturation and T cell responses. Front. Immunol. **13** (2022). https://doi.org/10.3389/fimmu.2022.842241
129. Cao, Z., Yang, X., Yang, W., Chen, F., Jiang, W., Zhan, S., Jiang, F., Li, J., Ye, C., Lang, L., Zhang, S., Feng, Z., Lai, X., Liu, Y., Mao, L., Cai, H., Teng, Y., Xie, J.: Modulation of dendritic cell function via nanoparticle-induced cytosolic calcium changes. ACS Nano **18**(10) (2024). https://doi.org/10.1021/acsnano.4c00550
130. Jin, H., Qian, Y., Dai, Y., Qiao, S., Huang, C., Lu, L., Luo, Q., Chen, J., Zhang, Z.: Magnetic enrichment of dendritic cell vaccine in lymph node with fluorescent-magnetic nanoparticles enhanced cancer immunotherapy. Theranostics **6**(11) (2016). https://doi.org/10.7150/thno.15102
131. Burnet, M.: Cancer—a biological approach: III. Viruses associated with neoplastic conditions. IV. Practical applications. BMJ **1**(5023), 841–847 (1957). https://doi.org/10.1136/bmj.1.5023.841
132. Thomas, L.: On immunosurveillance in human cancer. Yale J. Biol. Med. **55**(3–4), 329–333 (1982)
133. Want, M.Y., Bashir, Z., Najar, R.A.: T cell based immunotherapy for cancer: approaches and strategies. Vaccines **11**(4) (2023). https://doi.org/10.3390/vaccines11040835
134. Hristova-Panusheva, K., Xenodochidis, C., Georgieva, M., Krasteva, N.: Nanoparticle-mediated drug delivery systems for precision targeting in oncology. Pharmaceuticals **17**(6), 677 (2024). https://doi.org/10.3390/ph17060677
135. Zhou, L., Zou, M., Xu, Y., Lin, P., Lei, C., Xia, X.: Nano drug delivery system for tumor immunotherapy: next-generation therapeutics. Front. Oncol. **12** (2022). https://doi.org/10.3389/fonc.2022.864301
136. Ezike, T.C., Okpala, U.S., Onoja, U.L., Nwike, C.P., Ezeako, E.C., Okpara, O.J., Okoroafor, C.C., Eze, S.C., Kalu, O.L., Odoh, E.C., Nwadike, U.G., Ogbodo, J.O., Umeh, B.U., Ossai, E.C., Nwanguma, B.C.: Advances in drug delivery systems, challenges and future directions. Heliyon **9**(6) (2023). https://doi.org/10.1016/j.heliyon.2023.e17488
137. Jiang, L., Ding, L., Liu, G.: Nanoparticle formulations for therapeutic delivery, pathogen imaging and theranostic applications in bacterial infections. Theranostics **13**(5) (2023). https://doi.org/10.7150/thno.82790

138. Fang, L.R., Wang, Y.H., Xiong, Z.Z., Wang, Y.M.: Research progress of nanomaterials in tumor-targeted drug delivery and imaging therapy. OpenNano **14** (2023). https://doi.org/10.1016/j.onano.2023.100184
139. Schmid, D., Park, C.G., Hartl, C.A., Subedi, N., Cartwright, A.N., Puerto, R.B., Zheng, Y., Maiarana, J., Freeman, G.J., Wucherpfennig, K.W., Irvine, D.J., Goldberg, M.S.: T cell-targeting nanoparticles focus delivery of immunotherapy to improve antitumor immunity. Nat. Commun. **8**(1). https://doi.org/10.1038/s41467-017-01830-8
140. Zhu, X., Li, S.: Nanomaterials in tumor immunotherapy: new strategies and challenges. Mol. Cancer **22**(1) (2023). https://doi.org/10.1186/s12943-023-01797-9
141. Guasch, J., Muth, C.A., Diemer, J., Riahinezhad, H., Spatz, J.P.: Integrin-assisted T-cell activation on nanostructured hydrogels. Nano Lett. **17**(10) (2017). https://doi.org/10.1021/acs.nanolett.7b02636
142. Yang, Y.S.S., Moynihan, K.D., Bekdemir, A., Dichwalkar, T.M., Noh, M.M., Watson, N., Melo, M., Ingram, J., Suh, H., Ploegh, H., Stellacci, F.R., Irvine, D.J.: Targeting small molecule drugs to T cells with antibody-directed cell-penetrating gold nanoparticles. Biomater. Sci. **7**(1) (2019). https://doi.org/10.1039/c8bm01208c
143. Alhallak, K., Sun, J., Muz, B., Jeske, A., O'Neal, J., Ritchey, J.K., Achilefu, S., DiPersio, J.F., Azab, A.K.: Liposomal phytohemagglutinin: in vivo T-cell activator as a novel pan-cancer immunotherapy. J. Cell. Mol. Med. **26**(3) (2022). https://doi.org/10.1111/jcmm.16885
144. Thiramanas, R., Li, M., Jiang, S., Landfester, K., Mailänder, V.: Cellular uptake of sirna-loaded nanocarriers to knockdown pd-l1: strategies to improve t-cell functions. Cells **9**(9) (2020). https://doi.org/10.3390/cells9092043
145. Luo, Z., Luo, L., Lu, Y., Zhu, C., Qin, B., Jiang, M., Li, X., Shi, Y., Zhang, J., Liu, Y., Shan, X., Yin, H., Guan, G., Du, Y., Cheng, N., You, J.: Dual-binding nanoparticles improve the killing effect of T cells on solid tumor. J. Nanobiotechnol. **20**(1) (2022). https://doi.org/10.1186/s12951-022-01480-z
146. Canakci, M., Singh, K., Munkhbat, O., Shanthalingam, S., Mitra, A., Gordon, M., Osborne, B.A., Thayumanavan, S.: Targeting CD4+ cells with anti-CD4 conjugated mertansine-loaded nanogels. Biomacromolecules **21**(6) (2020). https://doi.org/10.1021/acs.biomac.0c00442
147. Garner, H., de Visser, K.E.: Immune crosstalk in cancer progression and metastatic spread: a complex conversation. Nat. Rev. Immunol. **20**(8) (2020). https://doi.org/10.1038/s41577-019-0271-z
148. Wang, R., Lan, C., Benlagha, K., Camara, N.O.S., Miller, H., Kubo, M., Heegaard, S., Lee, P., Yang, L., Forsman, H., Li, X., Zhai, Z., Liu, C.: The interaction of innate immune and adaptive immune system. MedComm **5**(10) (2024). https://doi.org/10.1002/mco2.714
149. Bald, T., Krummel, M.F., Smyth, M.J., Barry, K.C.: The NK cell–cancer cycle: advances and new challenges in NK cell–based immunotherapies. Nat. Immunol. **21**(8) (2020). https://doi.org/10.1038/s41590-020-0728-z
150. Shimasaki, N., Jain, A., Campana, D.: NK cells for cancer immunotherapy. Nat. Rev. Drug Discov. **19**(3), 200–218 (2020). https://doi.org/10.1038/s41573-019-0052-1
151. Murugan, D., Murugesan, V., Panchapakesan, B., Rangasamy, L.: Nanoparticle enhancement of natural killer (NK) cell-based immunotherapy. Cancers **14**(21) (2022). https://doi.org/10.3390/cancers14215438
152. Meraz, I.M., Majidi, M., Cao, X., Lin, H., Li, L., Wang, J., Baladandayuthapani, V., Rice, D., Sepesi, B., Ji, L., Roth, J.A.: TUSC2 immunogene therapy synergizes with anti–PD-1 through enhanced proliferation and infiltration of natural killer cells in syngeneic *Kras*-mutant mouse lung cancer models. Cancer Immunol. Res. **6**(2), 163–177 (2018). https://doi.org/10.1158/2326-6066.CIR-17-0273
153. Park, W., Gordon, A.C., Cho, S., Huang, X., Harris, K.R., Larson, A.C., Kim, D.H.: Immunomodulatory magnetic microspheres for augmenting tumor-specific infiltration of natural killer (NK) cells. ACS Appl. Mater. Interfaces **9**(16) (2017). https://doi.org/10.1021/acsami.7b02258
154. Delehedde, C., Ciganek, I., Bernard, P.L., Laroui, N., Da Silva, C.C., Gonçalves, C., Nunes, J., Bennaceur-Griscelli, A.-L., Imeri, J., Huyghe, M., Even, L., Midoux, P., Rameix, N., Guittard,

G., Pichon, C.: Enhancing natural killer cells proliferation and cytotoxicity using imidazole-based lipid nanoparticles encapsulating interleukin-2 mRNA. Mol. Ther. Nucleic Acids **35**(3), 102263 (2024). https://doi.org/10.1016/j.omtn.2024.102263
155. Sanz-Ortega, L., Rojas, J.M., Portilla, Y., Pérez-Yagüe, S., Barber, D.F.: Magnetic nanoparticles attached to the NK cell surface for tumor targeting in adoptive transfer therapies does not affect cellular effector functions. Front. Immunol. **10**(AUG) (2019). https://doi.org/10.3389/fimmu.2019.02073
156. Kim, K.S., Han, J.H., Choi, S.H., Jung, H.Y., Park, J.D., An, H.J., Kim, S.E., Kim, D.H., Doh, J., Han, D.K., Kim, I.H., Park, W., Park, K.S.: Cationic nanoparticle-mediated activation of natural killer cells for effective cancer immunotherapy. ACS Appl. Mater. Interfaces **12**(51) (2020). https://doi.org/10.1021/acsami.0c16357
157. Au, K.M., Par, S.I., Wa, A.Z.: Trispecific natural killer cell nanoengagers for targeted chemoimmunotherapy. Sci. Adv. **6**(27) (2020). https://doi.org/10.1126/sciadv.aba8564
158. Adjei, I.M., Jordan, J., Tu, N., Trinh, T. Le, Kandell, W., Wei, S., Sharma, B.: Functional recovery of natural killer cell activity by nanoparticle-mediated delivery of transforming growth factor beta 2 small interfering RNA. J. Interdiscip. Nanomed. **4**(4) (2019). https://doi.org/10.1002/jin2.63
159. Munn, D.H., Bronte, V.: Immune suppressive mechanisms in the tumor microenvironment. Curr. Opin. Immunol. **39** (2016). https://doi.org/10.1016/j.coi.2015.10.009
160. Pansy, K., Uhl, B., Krstic, J., Szmyra, M., Fechter, K., Santiso, A., Thüminger, L., Greinix, H., Kargl, J., Prochazka, K., Feichtinger, J., Deutsch, A.J.: Immune regulatory processes of the tumor microenvironment under malignant conditions. Int. J. Mol. Sci. **22**(24) (2021). https://doi.org/10.3390/ijms222413311
161. Park, K., Veena, M.S., Shin, D.S.: Key players of the immunosuppressive tumor microenvironment and emerging therapeutic strategies. Front. Cell Dev. Biol. **10** (2022). https://doi.org/10.3389/fcell.2022.830208
162. Xu, Y., Xiong, J., Sun, X., Gao, H.: Targeted nanomedicines remodeling immunosuppressive tumor microenvironment for enhanced cancer immunotherapy. Acta Pharmaceutica Sinica B **12**(12) (2022). https://doi.org/10.1016/j.apsb.2022.11.001
163. Arner, E.N., Rathmell, J.C.: Metabolic programming and immune suppression in the tumor microenvironment. Cancer Cell **41**(3) (2023). https://doi.org/10.1016/j.ccell.2023.01.009
164. Finn, O.J.: Immuno-oncology: understanding the function and dysfunction of the immune system in cancer. Ann. Oncol. **23**(SUPPL.8) (2012). https://doi.org/10.1093/annonc/mds256
165. Dunn, G.P., Bruce, A.T., Ikeda, H., Old, L.J., Schreiber, R.D.: Cancer immunoediting: from immunosurveillance to tumor escape. Nat. Immunol. **3**(11) (2002). https://doi.org/10.1038/ni1102-991
166. Munn, D.H., Mellor, A.L.: IDO in the tumor microenvironment: inflammation, counter-regulation, and tolerance. Trends Immunol. **37**(3) (2016). https://doi.org/10.1016/j.it.2016.01.002
167. Park, K.S., Nam, J., Son, S., Moon, J.J.: Personalized combination nano-immunotherapy for robust induction and tumor infiltration of CD8+ T cells. Biomaterials **274** (2021). https://doi.org/10.1016/j.biomaterials.2021.120844
168. Franklin, R.A., Li, M.O.: Ontogeny of tumor-associated macrophages and its implication in cancer regulation. Trends Cancer **2**(1), 20–34 (2016). https://doi.org/10.1016/j.trecan.2015.11.004
169. Haist, M., Stege, H., Grabbe, S., Bros, M.: Review the functional crosstalk between myeloid-derived suppressor cells and regulatory t cells within the immunosuppressive tumor microenvironment. Cancers **13**(2) (2021). https://doi.org/10.3390/cancers13020210
170. Melincovici, C.S., Boşca, A.B., Şuşman, S., Mărginean, M., Mihu, C., Istrate, M., Moldovan, I.M., Roman, A.L., Mihu, C.M.: Vascular endothelial growth factor (VEGF)—key factor in normal and pathological angiogenesis. Romanian J. Morphol. Embryol. **59**(2) (2018)
171. Monteran, L., Erez, N.: The dark side of fibroblasts: cancer-associated fibroblasts as mediators of immunosuppression in the tumor microenvironment. Front. Immunol. **10**(AUG) (2019). https://doi.org/10.3389/fimmu.2019.01835

172. DeNardo, D.G., Ruffell, B.: Macrophages as regulators of tumour immunity and immunotherapy. Nat. Rev. Immunol. **19**(6) (2019). https://doi.org/10.1038/s41577-019-0127-6
173. Zhu, Y., Yu, X., Thamphiwatana, S.D., Zheng, Y., Pang, Z.: Nanomedicines modulating tumor immunosuppressive cells to enhance cancer immunotherapy. Acta Pharmaceutica Sinica B **10**(11) (2020). https://doi.org/10.1016/j.apsb.2020.08.010
174. Xu, X., Deng, G., Sun, Z., Luo, Y., Liu, J., Yu, X., Zhao, Y., Gong, P., Liu, G., Zhang, P., Pan, F., Cai, L., Tang, B.Z.: A biomimetic aggregation-induced emission photosensitizer with antigen-presenting and hitchhiking function for lipid droplet targeted photodynamic immunotherapy. Adv. Mater. **33**(33) (2021). https://doi.org/10.1002/adma.202102322
175. Guo, R., Deng, M., He, X., Li, M., Li, J., He, P., Liu, H., Li, M., Zhang, Z., He, Q.: Fucoidan-functionalized activated platelet-hitchhiking micelles simultaneously track tumor cells and remodel the immunosuppressive microenvironment for efficient metastatic cancer treatment. Acta Pharmaceutica Sinica B **12**(1) (2022). https://doi.org/10.1016/j.apsb.2021.05.012
176. Shan, X., Gong, X., Li, J., Wen, J., Li, Y., Zhang, Z.: Current approaches of nanomedicines in the market and various stage of clinical translation. Acta Pharmaceutica Sinica B **12**(7) (2022). https://doi.org/10.1016/j.apsb.2022.02.025
177. Mundekkad, D., Cho, W.C.: Nanoparticles in clinical translation for cancer therapy. Int. J. Mol. Sci. **23**(3) (2022). https://doi.org/10.3390/ijms23031685
178. Zhao, H., Rubio, B., Sapra, P., Wu, D., Reddy, P., Sai, P., Martinez, A., Gao, Y., Lozanguiez, Y., Longley, C., Greenberger, L.M., Horak, I.D.: Novel prodrugs of SN38 using multiarm poly(ethylene glycol) linkers. Bioconjugate Chem. **19**(4) (2008). https://doi.org/10.1021/bc700333s
179. Amiri-Kordestani, L., Mena, E., Lindenberg, M.L., Kurdziel, K., Choyke, P., Patronas, N., Frye, R., Lin, N., Bala, S., Fojo, T., Bates, S.: 18F-FLT-PET/CT for the prediction of response to ANG-1005 therapy in patients with brain metastases from breast cancer. Cancer Res. **73**(24) (2013)
180. Nevala, W.K., Butterfield, J.T., Sutor, S.L., Knauer, D.J., Markovic, S.N.: Antibody-targeted paclitaxel loaded nanoparticles for the treatment of CD20 + B-cell lymphoma. Sci. Rep. **7** (2017). https://doi.org/10.1038/srep45682
181. Schmid, P., Adams, S., Rugo, H.S., Schneeweiss, A., Barrios, C.H., Iwata, H., Diéras, V., Hegg, R., Im, S.-A., Shaw Wright, G., Henschel, V., Molinero, L., Chui, S.Y., Funke, R., Husain, A., Winer, E.P., Loi, S., Emens, L.A.: Atezolizumab and nab-paclitaxel in advanced triple-negative breast cancer. New Engl. J. Med. **379**(22) (2018). https://doi.org/10.1056/nejmoa1809615
182. Little, R.F., Aleman, K., Kumar, P., Wyvill, K.M., Pluda, J.M., Read-Connole, E., Wang, V., Pittaluga, S., Catanzaro, A.T., Steinberg, S.M., Yarchoan, R.: Phase 2 study of pegylated liposomal doxorubicin in combination with interleukin-12 for AIDS-related Kaposi sarcoma. Blood **110**(13) (2007). https://doi.org/10.1182/blood-2007-06-097568

Nanotechnology for RNA and DNA Delivery in Cancer Therapeutics

Kwasi Kantanka Safo, Matthew Ogoe, Noble George, Anjali Gupta, Komal Janiyani, Harjeet Singh, K. Gireesh Babu, and Shivani R. Pandya

Abstract Nanotechnology is transforming cancer therapeutics by enhancing the precision and efficacy of RNA and DNA delivery. This chapter outlines the history and significance of innovative delivery systems in overcoming biological barriers, focusing on RNA interference (RNAi) and gene therapy to target cancer cells while minimizing off-target effects. It discusses various nanoparticle platforms, including liposomes, polymers, dendrimers, and viral vectors, along with engineering strategies like surface modifications and stimuli-responsive release mechanisms. Specific delivery methods for RNA, such as siRNA, miRNA, and mRNA, and DNA approaches like plasmid delivery for CRISPR/Cas9 and DNA vaccines are examined. The chapter emphasizes targeted delivery mechanisms, characterizes nanoparticles for optimization, and navigates clinical translation, addressing regulatory and safety concerns. Emerging trends include next-generation platforms, personalized nanomedicine, and the integration of AI and machine learning in precision oncology, providing a comprehensive guide to harnessing nanotechnology in revolutionizing cancer treatment.

Keywords RNA delivery systems · CRISPR/Cas9 · Gene editing · Personalized nanomedicine · mRNA vaccines · Oncology

K. K. Safo · M. Ogoe · N. George · A. Gupta · K. Janiyani
Research and Development Cell, Parul University, Vadodara, Gujarat, India

K. G. Babu
Parul Institute of Applied Sciences, Parul University, Vadodara, Gujarat, India

H. Singh · S. R. Pandya (✉)
Narnarayan Shastri Institute of Technology-Institute of Forensic Science and Cyber Security, National Forensic Science University, Jetalpur, Ahmedabad, Gujarat, India
e-mail: shivpan02@gmail.com

N. J. Joshi and G. Patel (eds.), *Nanoparticle Delivery in Cancer Care*, Series in BioEngineering, https://doi.org/10.1007/978-981-95-0457-2_7

1 Introduction

1.1 Historical Overview of Nanomedicine

The field of nanomedicine refers to the collection of nanotechnological studies conducted in medicine [1]. The year 1995 marked the beginning of the era of nanomedicine when the U.S. Food and Drug Administration (FDA) officially approved the nanotherapeutic product Doxil. It is believed that science fiction films from the 1950s and 1960s and Richard Feynman's remarks, which are frequently credited as the source of all things "nano," are the origins of nanomedicine [2]. Through nanotechnology, academia and industry have made remarkable advances; in pharmaceutics and pharmacology, the generic property proposed by "nano" innovation has generated great anticipation in the field of nanomedicine for the treatment of cancer [3]. Consequently, there has been an exponential expansion in the field of nanomedicine, which has promoted interdisciplinary connections and collaborations. Using molecular information and maintaining control over cellular processes, nanomedicine creates molecular instruments for early diagnosis, therapy, and defence.

1.2 Significance of RNA and DNA Therapeutics in Cancer Treatment

With their ability to precisely modulate gene expression and their direct targeting of the genetic defects that propel cancer growth, RNA and DNA therapies guarantee a revolutionary approach to cancer treatment [4]. By targeting the underlying causes of cancer, RNA and DNA treatments enable more focused molecular interventions [5]. In contrast to standard medications that frequently show systemic toxicity and non-specific effect. For malignancies caused by particular mutations, DNA-based treatments such as antisense oligonucleotides and CRISPR-Cas9 gene editing enable accurate genetic alterations or transcript silencing, offering long-lasting and perhaps curative treatments [6].

2 Fundamentals of Nanoparticle Design

2.1 Types of Nanoparticles for Drug Delivery

2.1.1 Liposomic Nanoparticles

Liposomes are a type of lipid spherical vesicles that consist of an aqueous core surrounded by one or more phospholipid bilayers [7]. Because of their size, hydrophobic and hydrophilic properties, and capacity to encapsulate drug molecules in the lipophilic membrane or the aqueous interior of vesicles, liposomes are thought to have great promise for application as drug delivery vehicles [8]. Structurally, liposomes are either single-layered or multilayered spherical vesicles formed by the self-organization of diacyl-chain phospholipids like soybean phosphatidylcholine or synthetic dialkyl and trialkyl lipids in aqueous environments [9]. Several liposomal-based formulations have been effectively utilized in clinical settings for antitumor, antifungal therapies, and pain management. Doxil® was the first liposomal anticancer drug approved for the chemotherapy drug Doxorubicin [10].

2.1.2 Polymeric Nanomaterials

Polymeric nanocarriers containing conjugated or encapsulated medications, exhibit diverse structural morphologies such as polymeric micelles, nanospheres, nanogels, vesicles, dendrimers, and polymer–drug conjugates. Brigatinib, an inhibitor of the epidermal growth factor receptor, has been effectively loaded onto a PLGA nanocarrier to increase its anticancer efficacy against the lung carcinoma cell line A549 [11]. The co-administration of navitoclax, a Bcl-2 inhibitor, and decitabine, a nucleic acid synthesis inhibitor, using PLA-based copolymeric nanocarriers demonstrated strong in vitro synergistic cytotoxicity against acute myeloid leukemia (AML) and breast cancer cell lines, as well as potent tumor progression inhibition on AML xenografts [12].

2.1.3 Dendrimer Nanoparticles

Dendrimers are extensively branching polymeric which is very useful for chemical modification because it can provide vast and variable amounts of surface fictionalization. Dendritic polymers are composed of three main parts: an outer shell characterized by terminating functional groups; a middle or inner component composed of repeating branches; and a centered core [12]. Dendrimers are appropriate for a variety of applications due to their wide internal space for drug encapsulation, small molecule packaging, solubility in water, monodispersity, and varying surface activity [13]. Doxorubicin, administered using dendrimers, have been linked to lower cardiac toxicity, greater penetration, controlled release, and enhanced bioavailability

[14]. Gemcitabine encapsulated in amphiphilic dendrimer demonstrated a 2.5-fold decrease in tumor volume in mouse models of pancreatic cancer with minimal side effects, demonstrating the effectiveness of dendrimer-based gemcitabine delivery [13].

2.1.4 Micelles

Micelles are formed by self-assembly of amphiphilic block copolymers in aqueous solution and are characterized by a core–shell structure of hydrophobic core, sterically stabilized by a hydrophilic shell [15]. Polymeric micelles, in comparison to other nanocarriers, often exhibit smaller sizes, simpler manufacturing and sterilizing procedures, and strong solubilization capabilities. Dox and irinotecan have been incorporated in a phenylboronic acid micelle of about 30–40 nm to reduce tumor size appreciably devoid of systemic toxicity [16]. Cisplatin and paclitaxel have been administered together in polymeric micelles for a combined effects are more potent and higher anti-tumor activity compared to free form of respective chemo-drugs [16].

2.2 *Engineering Considerations, Surface Modifications and Functionalization of NPs for Targeted Delivery*

Engineering nanoparticles offers great potential to overcome traditional barriers to free medicines and enhance diagnosis and therapeutics appreciably [17]. The overall efficacy of medication delivery for precision applications can be enhanced by smart nanoparticle design, leading to an improvement in patient outcomes [18]. The unit focuses on different engineering considerations involved in nanoparticle design for drug delivery purposes. The first stage in a circulating NP's journey to the target tissue is extravasation [19]. Size is one of the NP features that can change extravasation; small NPs often pass through capillary walls with greater ease than large NPs [17, 20]. When it comes to size, for instance, studies have usually demonstrated that NPs smaller than 10 nm are quickly removed by the kidneys, while NPs bigger than 200 nm, unless specifically designed otherwise, run the danger of triggering the complement system [21]. The proportion of administered NPs reaching the target treatment site is limited by physiological obstacles found in the body, such as shear forces, protein adsorption, and fast clearance [22]. Numerous nanoparticle formulations use PEG as a covert coating to prevent quick excretion due to surface characteristics. This protects the nanoparticle surface from digestive enzymes and antibodies that could cause breakdown, secretion, and elimination [23]. Drug release and NP disintegration has been induced by endogenous stimuli, for instance the acidic as well as hypoxic environment of a tumor [21] With the concerns of improving the

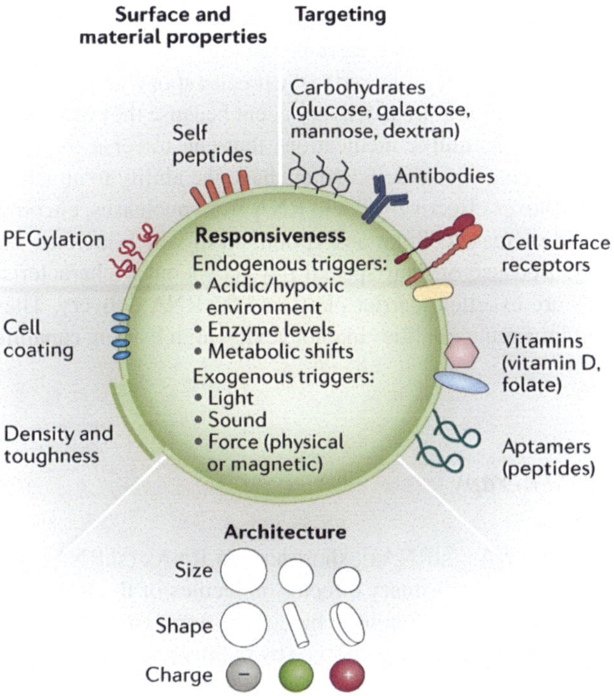

Fig. 1 Overview of NP engineering, including design, modification, and functionalization strategies aimed at enhancing drug delivery efficacy and overcoming traditional barriers in therapeutic applications. Reproduced from open-source [23]

stability of nanoparticles, polymer NPs have been stabilized by cross-linking techniques, numerous excipients, including helper lipids, PEGylated lipids, and cholesterol, have been synthesized with lipid-based NPs to improve their stability [24]. The different factors that are taken into account when creating nanoparticles for medication delivery with the goal of overcoming traditional obstacles are summarized in Fig. 1.

3 Nanoparticle Formulation for RNA Delivery

RNA therapy has a high selectivity compared to normal chemotherapy and can be utilized to target numerous important oncogenic drivers, decrease tumor cell resistance to drugs, and stop the progression of advanced-stage cancers [25]. However, the main barriers to RNA-based cancer therapeutic development and application are thought to be poor stability, immunogenicity, low cellular absorption rate, and problems in endosomal evacuation [26] These factors significantly restrict the development and clinical translation of RNA-based approaches in cancer therapy. In order for RNA to be transported into cancer cells in a safe and effective manner, it must typically be packaged in suitable carriers that allow the RNA to be absorbed by the target cells and subsequently released to the precise spot where it is needed [27]. Most RNA-based treatments require a delivery agent because they are either too big or too charged to enter cells, unlike medications that can traverse the lipid bilayer [28]. Polymers, in particular, cationic polymers have the ability to attach to nucleic acids and create polyplexes effectively shield RNA from nucleases, encourage endosomal emigration, and increase cellular absorption, all of which increase the efficiency of RNA delivery. Because of their special physicochemical characteristics, inorganic nanomaterials are excellent carrier platforms for RNA delivery. These characteristics give them intriguing qualities like longevity, high loading capability, and optical flexibility [29].

3.1 siRNA Therapy

Small interfering RNAs (siRNAs), short hairpin RNAs (shRNAs) and microRNAs (miRNAs) comprises the primary effector molecules of the RNA interference therapies (RNAi) [27]. RNAi technology blocks the expression of particular genes and produces a target gene deletion phenotype by rapidly and selectively degrading intracellular homologous messenger RNA [30]. siRNA, (a 20–25 base pair dsRNA), binds to the polymerase complex (RNA-induced silencing complex, RISC) and carries out the RNA interference effector function of mRNA degradation, which stops transcriptional expression of genes and eventually results in a decrease in gene expression [31].

3.2 miRNA Therapy

The small non-coding molecules of RNA known as miRNAs control the expression of genes by preventing mRNA from being translated while triggering mRNA degradation [29]. When miRNAs bind to the 3′ untranslated region of a target mRNA, they combine with siRNAs and the RISC to utilize the RNase component of RISC to degrade the targeted mRNA [32].

3.3 mRNA Therapy

With mRNA therapy, a particular RNA messenger is injected into the patient's body to encourage protein synthesis in cells as a variant to DNA therapy [33]. Due to its lower stability compared to DNA, mRNA is more easily targeted by endogenous

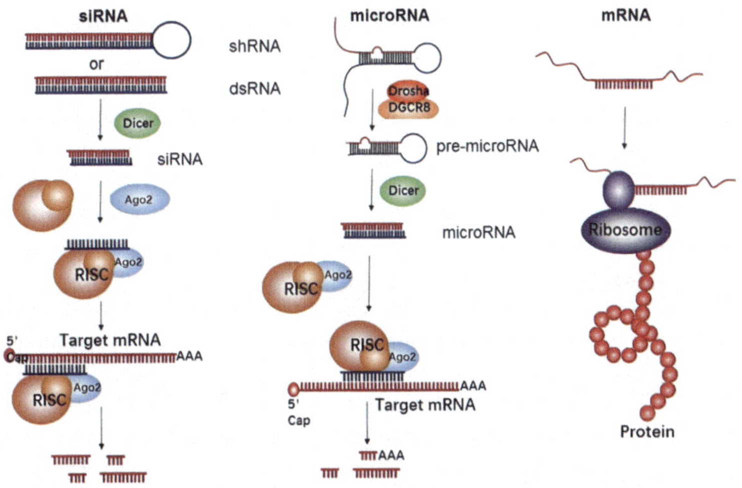

Fig. 2 The biological process by which target protein expressions are inhibited or elevated by siRNA, microRNA, and mRNA. Reproduced from open-source [35]

nucleases and may elicit immunological responses due to large quantities of neoantigens. The development of an immune-evading "pseudouracil" through alteration of the nucleoside part of uracil ribose has been investigated as a means of reducing these occurrences [34]. The biological process by which target protein expressions are inhibited or elevated by siRNA, microRNA, and mRNA is illustrated in Fig. 2.

4 Nanoparticle Formulation for DNA Delivery

4.1 Plasmid DNA Delivery Systems

Recombinant therapeutic genes can be successfully delivered to cells and tissues via plasmid DNAs [36]. Additionally, after being inserted into target cells, pDNAs contain extra sequences called enhancer and promoter elements essential for regulating the transcription and expression of the encoded protein. In clinical trials, pDNA containing therapeutic sequences utilizes cellular machinery to synthesize the necessary RNA, thereby modulating protein expression [37]. Various viral and non-viral gene delivery techniques have been developed for plasmid DNA-based therapeutic applications in cells [38].

4.1.1 Nanoplatforms Designed for the Delivery of Spherical Nucleic Acids Facilitate the Introduction of pDNA

Dizaji et al. developed a core/shell nanoplatform for delivering pDNA using $Fe_3O_4@SiO_2$ nanoparticles with a mean diameter of 50 nm, synthesized via a microemulsion technique. The nanoparticles were functionalized with negatively charged pDNA encoding enhanced green fluorescent protein and modified with poly(allylamine hydrochloride) (PAH) to impart a positive charge. This nanoplatform demonstrated enhanced cellular uptake and gene expression efficiency while protecting pDNA from lysosomal degradation through electrostatic interactions. Additionally, the system exhibited minimal cytotoxicity at high concentrations, positioning it as a promising gene delivery system for cancer therapy [39]. Wang et al. introduced Alkyl-polyethylenimine 2k-carbon dots as a low cytotoxicity nanocomposite for in vivo gene delivery. Utilizing Alkyl-PEI2k-Cdots/siRNA complexes, they effectively targeted and delivered luciferase pDNA to fluc-4T1/4T1 xenograft tumor tissues, resulting in efficient gene expression. Additionally, these complexes significantly suppressed luciferase expression, demonstrating their potential in gene therapy applications [40].

4.2 Gene Editing Tools (CRISPR/Cas9, TALENs) in Nanocarriers

Identifying key genes involved in tumor development through high-throughput screening can guide cancer treatment strategies. Various gene editing technologies, such as zinc-finger nucleases (ZFN), transcription activator-like effector nucleases (TALEN), and CRISPR/Cas9, have been developed for this purpose [41]. TALEN technology improves upon ZFNs by fusing TALENs with the Fok I cleavage domain, allowing direct DNA binding. Despite its advantages, TALENs face challenges due to their large size and repetitive sequences, which complicate their packaging and assembly [42]. CRISPR/Cas9, an adaptive immune system originally found in bacteria, is a third-generation gene editing technology that offers ease of construction and precision. The Type II CRISPR/Cas9 system is particularly well-researched and used in tumor treatment research [43]. However, its clinical application is limited by off-target effects and delivery challenges. Advances are being made to enhance the precision and reduce the side effects of CRISPR/Cas9 through improved control mechanisms [44]. Designing delivery vectors and selecting appropriate biomaterials is essential for utilizing CRISPR/Cas9 successfully (Fig. 3A).

Zhen et al. developed pH-responsive liposomes by incorporating pH-responsive peptides or polymers into traditional liposomes. These liposomes specifically release the CRISPR/Cas9 system at the tumor site due to the acidic environment of tumors [45]. Following this, Hu et al. created a hybrid nanoparticle with pH-responsive lipids and poly acid, which was rapidly taken up by dendritic cells and released

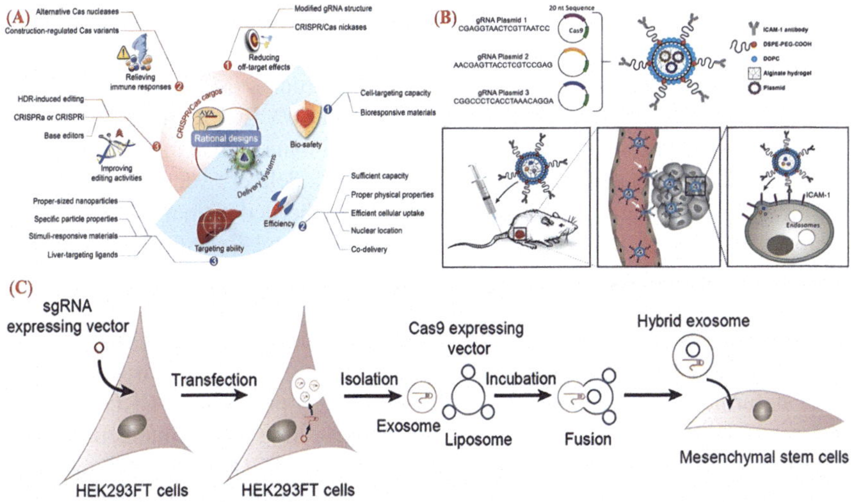

Fig. 3 **A** Devise methodologies for the targeted delivery of biomaterials leveraging CRISPR/Cas9 gene-editing systems [55]. **B** The composition of a deformable, noncationic nanolipogel and how CRISPR/Cas9 is delivered to cells for gene editing [47]. **C** Exosome-liposome hybrid nanoparticles allow for successful delivery of the CRISPR/Cas9 gene-editing system [54]. Reproduced from open-sources

antigens in conditions like those in dendritic cell endosomes, targeting the acidic tumor environment for selective killing of tumor cells [46].

Guo et al. developed a tumor-targeted nanogel system, modified with intercellular cell adhesion molecule-1 antibodies, for delivering CRISPR/Cas9 components. This system demonstrated high stability and resistance to premature drug release, significantly enhancing tumor delivery. Systemic administration effectively knocked out lipocalin 2 in triple-negative breast cancer (TNBC) cells, achieving a 77% inhibition rate. The study highlighted the biocompatibility, biodegradability, and improved solubility of hydrophobic drugs offered by nanogels, as well as their potential to optimize drug and mRNA delivery, warranting further clinical investigation (Fig. 3B) [47].

Building on previous advancements, Mout et al. investigated arginine-conjugated gold nanoparticles for delivering Cas9 protein and single guide RNA (sgRNA) into target cells. They synthesized arginine-AuNPs and loaded them with Cas9/sgRNA, testing the system on human cancer cell lines. The results demonstrated over 90% delivery efficiency of the CRISPR components, with a gene editing efficiency of 30%, highlighting the potential of arginine-AuNPs as effective carriers for CRISPR applications [48].

Lee et al. introduced CRISPR-Gold, a gold nanoparticle vector for delivering Cas9/sgRNA into refractory primary and stem cells, achieving 5.4% correction of the dystrophy gene and 18 times greater effectiveness than traditional methods in vivo [49]. Yang et al. developed ZIF-90/Cas9 nanoparticles, demonstrating a 35% gene editing efficiency in HeLa cells while highlighting the need to optimize synthesis for better reproducibility and address challenges like aggregation and metal ion toxicity [50]. Zhang et al. created mesoporous silica nanoparticles (MSNs) as a CRISPR/Cas9 delivery vector, showing biocompatibility, over 60% EGFR editing efficiency, and 85% tumor suppression in a hepatocellular carcinoma mouse model [51].

Yang et al. investigated nanodiamonds (NDs) as a delivery vector for CRISPR/Cas9 systems targeting the RS1 gene involved in inherited retinal diseases. They developed two linear DNA constructs—RS1-sgRNA and Cas9-GFP—leveraging NDs' unique adsorbent properties for binding DNA, proteins, and small molecules. Their findings confirmed successful RS1 gene editing in target cells, highlighting NDs' potential in retinal gene therapy despite challenges like high surface energy and agglomeration, which necessitate functionalization for enhanced dispersion [52]. Sun et al. further explored DNA nanowires as a delivery platform for CRISPR/Cas9, synthesized via rolling cycle amplification, demonstrating robust interactions with Cas9/sgRNA complexes. In human osteosarcoma cells (U2OS), the DNA nanoparticles achieved a 36% gene editing efficiency and facilitated gene knockout in EGFP U2OS mice. This study emphasized the biocompatibility and lower toxicity of DNA nanowires compared to synthetic systems, though clinical applications remain in preliminary stages and require further investigation [53]. Lin et al. investigated hybrid exosomes as delivery vectors for the CRISPR/Cas9 system by combining tumor-derived exosomes with cationic liposomes (Lipofectamine 2000) via electroporation. This approach enhanced the loading capacity and stability of the exosomes, facilitating efficient delivery of CRISPR/Cas9 components to mesenchymal cells. The hybrid exosomes demonstrated improved chemotaxis, addressing limitations of traditional exosomal delivery methods, such as low loading capacity due to their small size (Fig. 3C) [54].

4.3 DNA Vaccines for Cancer Immunotherapy

The discovery that plasmid DNA can directly enter animal cells in vivo has prompted investigations into using DNA plasmids to elicit immune responses. This approach involves injecting plasmids encoding antigenic proteins to stimulate an immune reaction. However, the immunogenicity of naked plasmid DNA vaccines is limited due to degradation by nucleases, which hampers effective delivery to immune cells. Encapsulating DNA in nanoparticles presents a promising strategy for enhancing vaccine development (Table 1) [56]. The small size, specialized surface, versatility, and targeting capabilities of nanoparticles can significantly improve the immunogenicity of vaccines while enabling controlled release [57]. DNA vaccines stimulate antigen expression by engaging MHC class I and II molecules, activating both cellular and humoral immune responses [58]. Antigen-presenting cells (APCs), such as macrophages, enhance cellular uptake through mannose receptors [59]. Delivery systems for DNA vaccines can be categorized into viral and non-viral vectors. While viral vectors (e.g., retroviral, lentiviral, adenoviral) have limited genomic capacity

Table 1 DNA vaccines conjugated to polymeric nanoparticles for application in various cancer types

Formulation	Cancer type	Route of administration	Animal model	Source
Mannosylated chitosan	Colorectal	Subcutaneous	Mouse	[61]
Chitosan	Cervical (HPV)	Intramuscular	Mouse	[62]
Chitosan-linked PEI	Melanoma	–	–	[63]
Chitosan	Colorectal	Intramuscular	Mouse	[64]
Alginic acid-coated chitosan	Breast	Intramuscular	Mouse	[65]
Chitosan	Fibrosarcoma	Subcutaneous	Mouse	[66]
Chitosan	Breast	Subcutaneous	Mouse	[67]
Chitosan	Melanoma	Subcutaneous	Mouse	[68]
Mannosylated chitosan	Prostate	Intramuscular	Mouse	[69]
PLGA	Melanoma	Intranasal and intraperitoneal	Mouse	[70]
PLGA-PEI	Breast	–	Mouse	[71]
PLGA	Prostate	Intraperitoneal	Mouse	[70]
PLGA-PEI	Leukemia	Intranasal	Mouse	[72]
PLGA	Lymphoma	Subcutaneous	Mouse	[73]

and raise safety concerns, synthetic non-viral vectors, including lipids, inorganic molecules, and polymers, are considered safer and can accommodate larger DNA constructs [60]. Among these, polymeric nanoparticles offer advantages such as a larger surface area for functional modifications, small size for navigating capillaries, longer clearance times, and controllable DNA release, making them effective carriers for targeted tissue delivery under various physiological conditions.

5 Characterization Techniques for Nanoparticles

5.1 Advanced Physicochemical Characterization: Unraveling Material Properties

Nanopharmaceuticals necessitate comprehensive physicochemical characterization across various properties to facilitate their application in preclinical and clinical research [74]. It is essential to assess the same property using multiple techniques, as the inherent strengths and limitations of each method influence the selection of the most appropriate approach [75, 76]. Furthermore, a combinatorial characterization strategy is frequently required. Additionally, the reproducibility of these characterizations is critical to ensure the reliability of nanotherapeutics in biomedical applications [77].

5.1.1 XRD

X-ray diffraction (XRD) is a key technique for analyzing crystalline and polycrystalline materials, enabling the determination of their atomic-level structures. It involves the reflection of a collimated X-ray beam on crystalline planes, as described by Bragg's law [78]. XRD is primarily used to assess the crystallinity of pharmaceuticals, carriers, and excipients, and to identify crystalline phases of contaminants in drug synthesis. In contrast, Small-Angle X-ray Scattering (SAXS) provides insights into particle size, shape, surface-to-volume ratio, and distribution by examining both crystalline and amorphous materials, including polymers and nanomaterials. SAXS is recognized for its ability to investigate material structures across length scales of 1–100 nm [79].

Rana et al. studied octaarginine-functionalized graphene oxide as a gene delivery nano-carrier. XRD confirmed the oxidation of exfoliated graphite into GO sheets, indicated by a peak at $2\theta = 26.8$. Oxidation introduced hydroxyl, epoxy, and carboxyl groups, weakening van der Waals forces. The non-graphitic oxidized sample (NGOS) showed a peak at $2\theta = 12.3$, reflecting increased interlayer spacing from exfoliation. Further carboxylation enhanced dispersion, evidenced by a broader peak at $2\theta = 12.3$. The R8-functionalized sample exhibited a peak at $2\theta = 12.6$, indicating increased d-spacing from R8 functionalization [80].

Kmaran et al. synthesized a Fe_3O_4/chitosan biopolymer grafted with a novel organosilane-modified 5-amino-1H-tetrazole as a gene delivery carrier. XRD (Fig. 4A) confirmed the cubic morphology of the magnetic nanoparticles with characteristic peaks at $2\theta = 11°$ and $2\theta = 20°$ corresponding to the chitosan biopolymer. The diffraction peaks (220), (311), (400), (422), (511), and (440) aligned with the inverse cubic spinel structure of Fe_3O_4 MNPs, while increased absorption bands between 5° and 20° indicated successful incorporation of MNPs onto the biopolymer surface [81].

In a separate study, Viktor et al. developed niosomes as a nanocarrier for encapsulating Lifeguard (LFG)-specific siRNA, incorporating superparamagnetic iron oxide nanoparticles (Fe_xO_y-NPs) within the niosomal bilayer to enhance cellular uptake via magnetic manipulation. The XRD (Fig. 4B) analysis revealed a crystalline mixed phase of magnetite and maghemite, complicating precise composition determination. Using the Debye–Scherrer equation, a crystallite size of 5.8 nm was calculated from the highest intensity reflection at 36.6°, consistent with sizes obtained through transmission electron microscopy (TEM) and dynamic light scattering (DLS) [82].

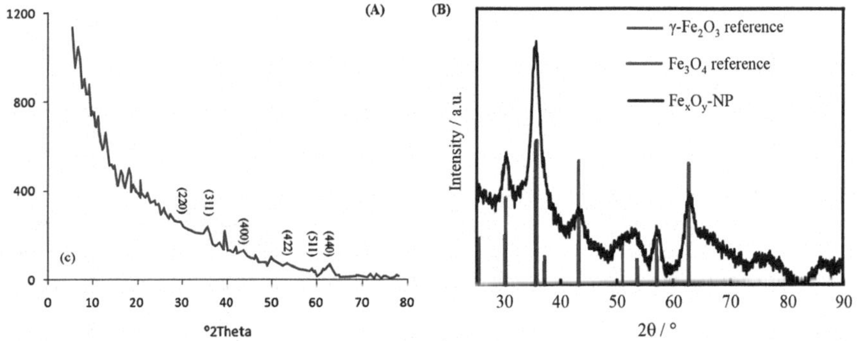

Fig. 4 XRD spectrum of **A** aminotetrazole-functionalized magnetic chitosan nanocomposite [81], reprinted with permission and **B** IONP entrapped niosomes, reprinted from an open source [82]

5.1.2 Microscopic Techniques

Gene delivery systems are mostly nanometric in size, therefore their imaging demands appropriate resolution microscopy techniques such as AFM, SEM and TEM.

SEM

The scanning electron microscope (SEM) is a vital analytical tool in cancer therapeutics, particularly for characterizing nanocarriers for DNA and RNA delivery [83]. SEM operates by bombarding a sample with a primary electron beam, which interacts with the surface to emit backscattered and secondary electrons. These emitted electrons are collected to generate detailed topographic images, revealing the morphology and structural integrity of nanocarriers [84]. A key advantage of SEM is its ability to produce high-resolution, three-dimensional images that highlight the surface characteristics of nanocarriers, which is crucial for effective drug delivery in cancer therapy. However, the high vacuum environment required for SEM necessitates that biological samples be dehydrated, freeze-dried, or fixed, complicating sample preparation but ultimately enhancing image quality [85].

The imaging process in SEM utilizes two primary types of electrons: backscattered electrons (BSEs) and secondary electrons (SEs). BSEs provide compositional contrast, as their generation is influenced by the atomic number of the sample's elements, allowing differentiation based on composition [86]. In contrast, SEs are sensitive to surface topography, making them ideal for revealing fine morphological details. The combination of BSE and SE imaging modes facilitates a comprehensive analysis of nanocarrier structures, including the shape and arrangement of oligonucleotide–polymer complexes, which can appear as spherical, porous, or rod-shaped forms [87].

Fig. 5 SEM representative micrographs of **A** PMMA sacrificial core, **B** NCs before and **C** after extraction of the core. Reproduced from an open source [88]

Elena et al. developed terpolymer-based nanocapsules through free radical polymerization of three monomers around a sacrificial core. These nanocapsules, loaded with DNA and functionalized with folic acid (FA) for active targeting of cancer cells, exhibited nanometric dimensions and a spherical morphology, as confirmed by SEM. Importantly, the nanocapsules maintained structural stability before and after core extraction (Fig. 5) [88].

TEM

Transmission electron microscopy (TEM) is essential for characterizing nanomaterials in cancer therapy. It allows electrons to pass through thin samples, providing high-resolution images of nanoparticles' internal structures, which are vital for assessing their morphology and functionality [89]. Unlike SEM, TEM reveals complex morphologies and size distributions critical for drug-loaded nanoparticles [90]. Though the vacuum environment facilitates imaging, it requires samples to be under 50 nm thick, making preparation complex and time-consuming. Despite this, TEM offers valuable insights into the structural integrity of nanoparticles post-drug integration. Techniques like negative staining further enhance visualization of gene delivery systems [91]. Overall, TEM's superior resolution and detailed structural information are crucial for optimizing nanocarrier designs in cancer therapeutics.

Rosa et al. developed a novel transporter for 5-fluorouracil (5-Fu) using biocompatible gold nanosystems, specifically a gold nanoparticle precursor (Au@16-Ph-16) stabilized by the gemini surfactant 16-Ph-16, and DNA/5-Fu nanocomplexes. TEM (Fig. 6) revealed that untreated cells had normal morphology, while those treated with formulations N2 and N3 displayed vacuoles and dense bodies from nanoparticle dispersion. Formulation N1 showed nanoparticle aggregates in large vesicles, indicating partial aggregation from biological media. Mitochondria remained normal but were more abundant in treated cells. Formulations N3 and C3 enhanced internalization and degradation in BJ-hTERT cells, supporting that smaller, spherical nanoparticles facilitate uptake [92].

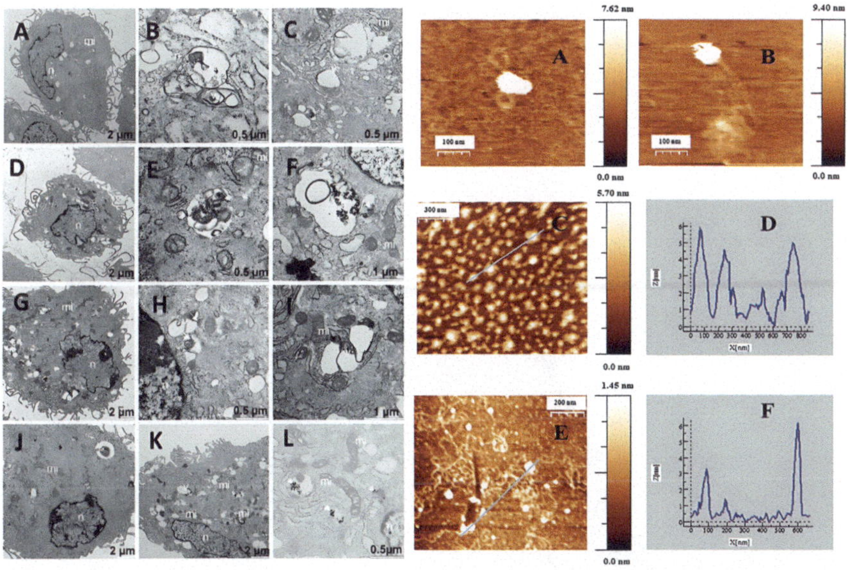

Fig. 6 TEM images (**Left hand side**) of BJ-hTERT cells and control cells (**A**–**C**); TEM images of cells treated with Au@16-pH-16, N1 (**D**–**F**), N2 (**G**–**I**), and N3 (**J**–**L**). AFM topography images (**Right hand side**) of CI complexes adsorbed on APTES modified mica surface in cacodylate buffer (I = 1.63 mM, pH = 7.4; CDNA = 0.3 μM, CDoxo = 7.5 nM), under different reaction conditions. (**A**, **B**) Intermediates of compaction, CAu@16-Ph-16 = 1.07 pM; **C** CI complexes at a compact stage, Cau@16-Ph-16 = 22.5 pM. **E** CI complexes at a decompacted stage, Cau@16-Ph-16 = 124.5 pM. **D**, **F** correspond to the cross section along the selected lines for (**C**, **E**). Reprinted from an open source [94]

Martha et al. developed graphene-based complexes for nucleic acid delivery into glioblastoma (GBM) and fibroblast cell lines. Ultrastructural analysis showed multi-layer formations in graphene oxide (GO) sheets and pronounced corrugation in reduced graphene oxide (rGO) sheets, which tended to agglomerate. Szczepaniak et al. noted similar agglomeration in TEM images (Fig. 7), indicating varying layer densities. In-situ functionalization with miRNA demonstrated a strong affinity of miRNA for GO, leading to a homogeneous distribution in GO + miRNA complexes, while rGO + miRNA complexes exhibited lower dispersion. TEM confirmed that GO and rGO flakes had diameters exceeding 100 nm and thicknesses below 1 nm [93].

AFM

Atomic force microscopy (AFM) is a high-resolution imaging technique that uses a sharp tip on a micro-machined cantilever to probe sample surfaces, enabling near-atomic resolution topographical mapping [95]. It assesses interactions between

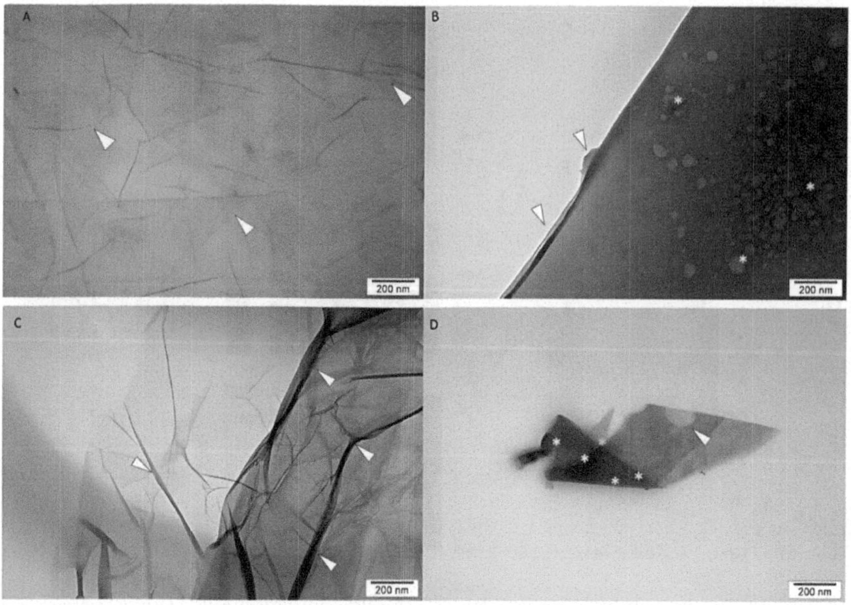

Fig. 7 TEM images of graphene oxide (**A**), graphene oxide functionalized with miRNA (**B**), reduced graphene oxide (**C**), and reduced graphene oxide functionalized with miRNA (**D**). Arrows: GO, rGO flakes corrugation, * miRNA, scale bars: 200 nm. Reprinted from an open source [93]

the tip and the sample, allowing observation of morphological changes in polyplexes and oligonucleotides, such as condensation and elongation with cationic polymers. Unlike other imaging methods, AFM does not require electrically conductive surfaces and can produce three-dimensional images of non-conductive biological samples. However, effective imaging necessitates immobilizing biological molecules on smooth substrates like highly oriented pyrolytic graphite (HOPG) to prevent displacement during scanning [96]. Overall, AFM's ability to provide detailed topographical insights makes it an essential tool for studying the structural properties of nanocarriers in biomedical research.

Rosa et al. studied compacted gold nanocomplexes (CAu@16-Ph-16) for anticancer drug delivery using DNA vehicles. AFM (Fig. 6) revealed that low concentrations ($R = 3.6 \times 10^{-6}$) induced DNA compaction through interactions between surfactant hydrophobic chains and DNA's phosphate groups. At higher concentrations ($R = 7.5 \times 10^{-6}$), compact globular structures averaging 67 ± 18 nm formed, demonstrating greater efficiency in DNA compaction than traditional cationic surfactants. At even higher ratios, circular dichroism (CD) spectra indicated partial DNA decompaction at $R = 4.2 \times 10^{-4}$. Thus, optimizing the R ratio is crucial for designing effective Au@16-Ph-16/DNA–Doxo complexes for targeted drug delivery [94].

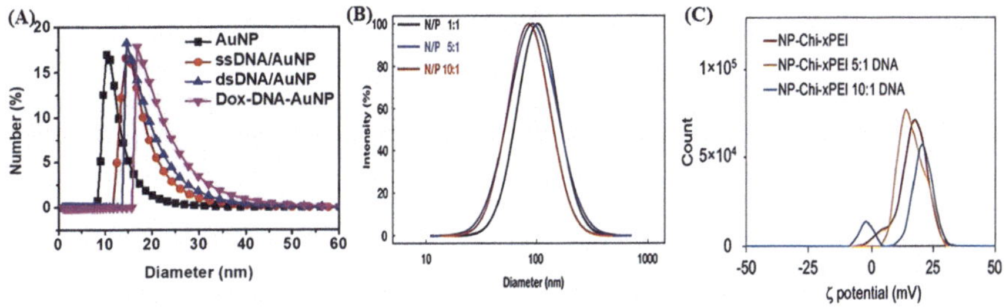

Fig. 8 Size distribution curves obtained by DLS of **A** polyplexes prepared in aqueous media at different N/P ratios: 1:1 (d = 122.6 ± 0.18 nm [99], **B** Dox–DNA–AuNP with 13 nm [100]. **C** Zeta potentials of NP-Chi-xPEI, NP-Chi-xPEI 5:1 DNA, and NP-Chi-xPEI 10:1 DNA [102]. Reprinted from an open sources

Dynamic Light Scattering

Dynamic light scattering (DLS) is essential for characterizing the hydrodynamic radius of particles in cancer therapeutics. It measures light scattering from particles in Brownian motion, providing insights into the size distribution and polydispersity of polyplexes used in RNA interference (RNAi) delivery systems [97]. Ensuring a homogeneous population is crucial, as aggregation can reduce efficacy and increase toxicity. While effective, DLS has limitations with heterogeneous solutions, requiring complementary microscopy techniques for complete characterization. An ideal particle size of 20–100 nm enhances blood circulation and tumor accumulation via the enhanced permeability and retention (EPR) effect. Overall, DLS is vital for optimizing nanoparticle-based drug delivery systems in cancer therapy [98].

Chang et al. demonstrated the in vitro and in vivo anticancer efficacy of a Dox-DNA-AuNP drug delivery system for ovarian cancer. DLS analysis (Fig. 8A) showed an increase in hydrodynamic diameter from 11 to 15 nm after binding single-stranded DNA to the AuNPs, confirming successful immobilization. The addition of capture DNA further increased the diameter to 16 nm, indicating the formation of double-stranded DNA on the nanoparticles. The subsequent loading of Dox also increased particle size, likely due to Dox intercalation into the G-C rich regions of the dsDNA on the AuNPs [99].

In a separate study, Kalinovva et al. synthesized a novel multi-amino-functional diblock copolymer for gene and drug delivery. DLS analysis (Fig. 8B) revealed average diameters of 122–86 nm with relatively narrow size distributions for the resulting polyplexes [100].

Zeta Potential

Zeta potential, the potential difference between a particle and its surrounding solution, is vital for RNAi delivery systems as it influences stability and cellular interactions [101]. A positive charge enhances internalization, but excessive charge can cause aggregation and immune activation; values above ±30 mV indicate stability. Guany et al. developed NP-Chi-xPEI, a DNA carrier that effectively transfects breast cancer cells in vitro. While a positive surface charge is crucial, excessive charge may harm organelles. NP-Chi-xPEI exhibited zeta potentials of 11.8, 17.0, and 18.3 mV, suitable for transfection (Fig. 8C). It showed superior hydrodynamic size and zeta potential profiles compared to non-viral carriers, attributed to its compact 10 nm NP core, demonstrating a size-conserving effect compared to Chi-xPEI-DNA without the NP [102].

Vibrational Spectroscopy

Raman and infrared (IR) spectroscopies are essential vibrational techniques for the structural characterization of nanoparticles, utilizing light-matter interactions to determine material composition. Raman spectroscopy relies on inelastic scattering of photons, providing insights into molecular vibrational states through shifts in frequency, while IR spectroscopy involves the absorption of light to promote vibrational transitions in molecules [103]. The resulting spectra are characteristic of specific substances, aiding in their identification. Raman spectroscopy is particularly advantageous for studying biological samples in aqueous solutions due to the weak Raman scattering of water, although it faces challenges such as fluorescence interference and limited spatial resolution [75]. Conversely, IR spectroscopy, including Fourier Transform Infrared (FTIR) and Attenuated Total Reflection FTIR (ATR-FTIR), effectively analyzes molecular structures and surface compositions of nanoparticles [104]. Both techniques complement each other, as the vibrational modes active in one are typically inactive in the other, enhancing the overall understanding of nanomaterials.

5.2 *Drug Loading and Release Kinetics*

Smart, stimuli-responsive nanoparticle-based drug delivery systems, including liposomes, dendrimers, micelles, and polymeric nanoparticles, offer a promising approach for sustained drug release. These systems respond to external triggers such as pH, temperature, or light, enabling controlled release of therapeutic agents and overcoming limitations of conventional drug delivery methods, thereby enhancing pharmacokinetics, targeted delivery, and therapeutic efficacy [105].

Liposomes, spherical lipid-based nanocarriers, can encapsulate both hydrophilic and hydrophobic compounds, with their lipid bilayer structure facilitating cellular uptake [106]. Advances in liposome technology include formulations with PEG for prolonged circulation and antibody-targeted ‘immunoliposomes’ for improved cancer targeting. Recent developments have led to smart liposomal systems capable of sustained drug release, such as hyaluronic acid-coated, peptide-modified liposomes loaded with curcumin and celecoxib, which demonstrated prolonged release for up to 72 h [107]. Liposomes are the most clinically utilized nanoparticle-based systems, with several approved formulations for cancer treatment, including the first thermo-sensitive, smart liposomes successfully used in clinical trials for enhanced local drug delivery.

Dendrimers are highly branched, three-dimensional macromolecules with a dense core and functional groups on the outer shell, making them versatile for drug delivery. Traditional dendrimers release encapsulated molecules within 24 h; however, recent advancements have produced smart, stimuli-responsive formulations, such as redox and pH-responsive dendrimer-heparin conjugates loaded with letrozole, which can release drugs over 100 h, modulated by glutathione concentration and pH [108].

Micelles, formed by the self-assembly of amphiphilic molecules, possess hydrophilic and hydrophobic domains that facilitate drug delivery. They can be engineered for specific functionalities, such as pH-dependent drug release. For instance, pH-sensitive micelles made from a four-arm star polymer demonstrated controlled doxorubicin release, with 24.5% release at pH 7.4 increasing to 51.8% at pH 5.0, mimicking the tumor microenvironment [109].

Zhang et al. developed smart nanocarriers integrating gold nanorods, thermo-sensitive polymers, and DNA nanorobots for targeted delivery of siRNA and doxorubicin. These nanocarriers enable stealthy circulation and tumor accumulation, with controlled release triggered by intracellular stimuli, resulting in synergistic tumor growth inhibition with minimal side effects. The advancement of DNA nanotechnology has facilitated the creation of multifunctional, stimulus-responsive nanodevices for personalized gene and drug therapies (Fig. 9) [110].

6 Targeting Strategies in Nanomedicine

Efficient nanoparticle-based drug delivery is vital for improving cancer treatment by enhancing therapeutic efficacy and reducing toxicity to normal cells. Anticancer agents should be delivered to tumors via nanoparticles that can traverse phagocytic barriers with minimal loss of blood activity. Once inside, the drug must selectively target cancer cells without affecting healthy tissue. Research on the interactions between nanomaterials and tumor cells focuses on two delivery mechanisms: passive targeting through non-ligand-mediated nanoparticles and active targeting via ligand-conjugated formulations (Fig. 10) [111].

6.1 Active Targeting Using Ligands and Antibodies

Active tumor targeting utilizes recognition motifs that specifically bind to receptors or other structural features found on cancer cells, tumor endothelial cells, and various cells within the tumor microenvironment. Additionally, components of the extracellular matrix (ECM), such as collagen and fibronectin, can also be employed as targets for active delivery [11].

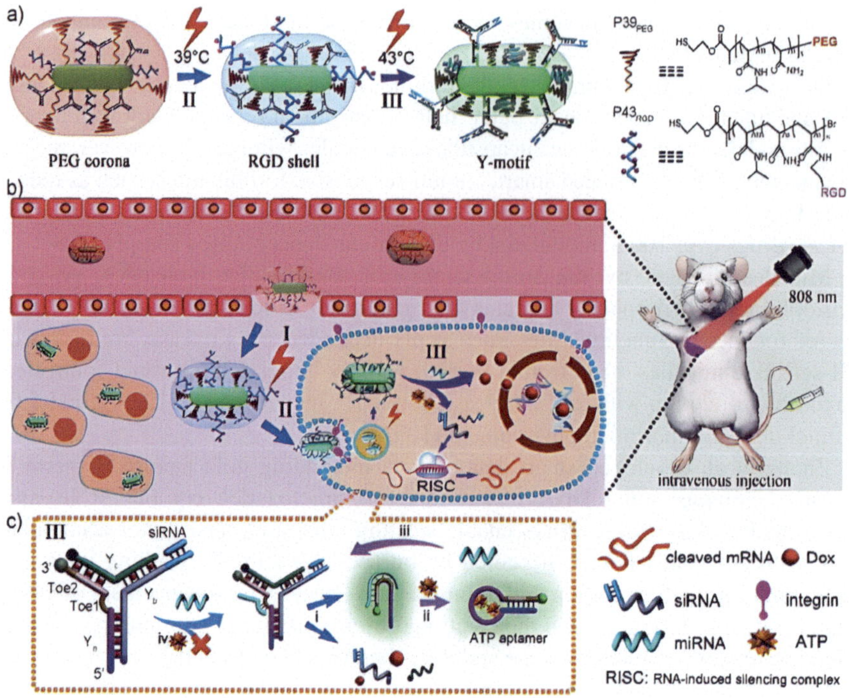

Fig. 9 Schematic design of the smart nanocarriers triggered by miRNA and fuelled by ATP. **a** Thermal-responsive regulation of the surface composition. **b** NIR-guided delivery and controlled release in vivo. (I) Passive accumulation at tumor sites via EPR effect; (II) specific uptake via receptor-mediated endocytosis; (III) controlled release by endogenous miRNA/ATP. **c** Amplified dissociation of Y-motifs based on TMSD cascade reactions with miRNA as trigger and ATP as fuel: (i) Toe1-TMSD triggered by miRNA; (ii) ATP (or H^+) recognition by aptamer; (iii) Toe2-TMSD for miRNA regeneration. Reprinted with permission [110]

6.1.1 Antibodies Mediated

Recent advancements in antibody research have led to antibodies with high specificity for targeted cancer therapy. Monoclonal antibodies (mAbs) are often used for nanoparticle (NP) delivery; however, whole antibodies can trigger immune responses, while antibody fragments reduce immunogenicity and enhance circulation half-life. Studies indicate effective targeting of B-cells in leukemia through cross-linking CD20 and CD38 [113].

Two main strategies for synthesizing antibody-nanoparticle conjugates are physical adsorption and covalent conjugation. Physisorption allows for non-covalent attachment of antibodies to NPs via interactions like hydrogen bonding, without chemical modifications. For example, ce6-conjugated hyaluronic acid and anti-PD-L1 antibodies can self-assemble into nanoparticles, facilitating tumor immunotherapy [114].

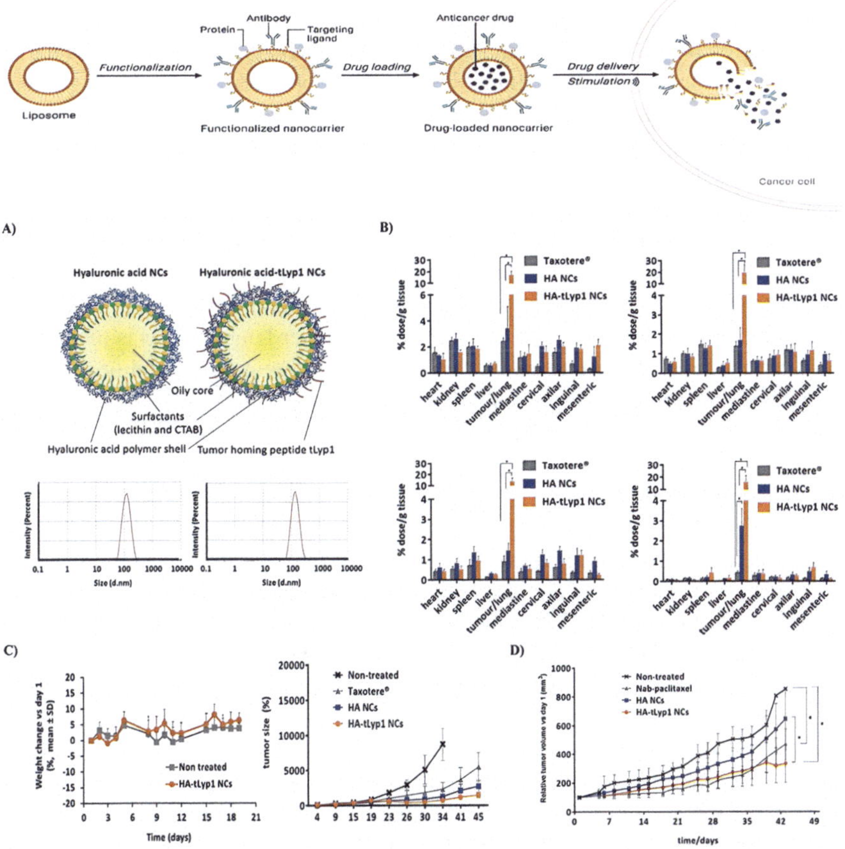

Fig. 10 (**Schematic representation**) Mechanism of targeted drug delivery using antibody and ligand-conjugated nanoparticles (NPs) for enhanced drug release in cancer cells. (**Down**) Tumor-homing peptide ligand tLyP-1-modified HA-based NCs show robust tumor accumulation. **A** Physicochemical characterization and schematic illustration of the HA and HA-tLyP1 NCs structure. **B** Biodistribution studies following intravenous injection of several therapies at 1, 4, 8, and 24 h in an orthotopic lung cancer model with lymphatic metastases. **c** In vivo toxicity and effectiveness tests using an orthotopic lung cancer tumor model. **d** The pancreatic cancer patient-derived xenograft model's relative tumor volume following treatment. Significant variations among the treatments (*) with $p < 0.001$. Reprinted with permission [112]

Covalent coupling is the predominant method for creating antibody-nanoparticle conjugates, utilizing techniques such as carbodiimide chemistry, maleimide binding, and click chemistry [115]. This approach offers stable and reproducible linkages compared to adsorption methods. For example, S. Jain et al. employed EDC/NHS chemistry with DSPE-PEG-COOH to conjugate VEGF antibodies to pH-sensitive docetaxel liposomes, enhancing therapeutic efficacy while minimizing side effects [116].

While various non-functionalized nanoparticles are available for cancer treatment, research on antibody-functionalized nanoparticles remains limited, primarily focused on formulation and preclinical studies. Few have advanced to clinical trials, particularly phases 1 and 2 [117]. One phase 1 study evaluated doxorubicin-loaded immunoliposomes targeting cetuximab Fab fragments, finding them well-tolerated at lower doses, with adverse events mainly due to tumor progression. Currently, a phase 2 trial is recruiting patients with advanced triple-negative breast cancer to assess the effectiveness of doxorubicin-loaded anti-EGFR immunoliposomes [118].

Antibody-nanoparticle conjugate systems enhance tumor targeting and reduce toxic side effects, making them valuable in cancer therapy. However, challenges persist in their preparation, particularly regarding linker length's effects on cellular uptake and toxicity. Future research should focus on optimizing modification and conjugation strategies for improved targeting efficiency.

6.1.2 Ligand Mediated

Diverse ligands, particularly peptides, are commonly used in designing active targeting nanomedicines (Table 2). Valino et al. identified tLyP-1 as an effective tumor-penetrating peptide (TPP) that enhances nanoparticle (NP) accumulation in tumors. Research shows that tLyP-1-functionalized hyaluronic acid nanoparticles (HA NCs) improve tumor retention and facilitate access to lymphatics in orthotopic lung cancer models. These targeted NPs delivered approximately 20% of the administered docetaxel dose per gram of tumor tissue, while docetaxel levels in other organs, such as the liver, remained below 5%. In contrast, Taxotere and its non-targeted formulations demonstrated limited targeting capacity [112].

Although hyaluronic acid (HA) has been widely used to improve tumor access, it is insufficient as a sole targeting ligand due to CD44's expression on non-tumor cells. Thus, dual targeting has been proposed, combining HA with additional ligands to enhance nanoparticle (NP) targeting and reduce off-target effects. NPs designed with monoclonal antibodies (mAb) anchored to HA enable targeting through CD44 and another receptor. This approach significantly improves siRNA delivery to blood cancer cells, as shown by marked BCL-2 downregulation in ex vivo analyses, indicating that multiple targeting ligands can enhance NP tumor accumulation [123].

In targeted nanomedicine design, various ligands enhance tumor accumulation. Anti-CD44 antibodies have shown significant nanoparticle (NP) accumulation in breast cancer models, effectively impeding metastasis. Transferrin-targeted PLGA nanoparticles carrying anti-BCR/ABL monoclonal antibodies improved targeting in chronic myeloid leukemia mouse models, reducing liver and kidney accumulation while enhancing bone targeting [124]. Tumor-penetrating peptides (TPPs) like iRGD and tLyP-1 further advance tumor cell targeting strategies, with potential optimization through coadministration of ligands and combinations with hyaluronic acid to

Table 2 An overview of activated nanosystems that target cancerous cells

Active ligand	Nanocarrier	Physico-chemical properties	Administration route	Tumor model	Outcome	Refs.
iRGD	PolyMTO-DSPE-PEG Micelles	92 nm, 40%	Intravenous (5 mg/mL)	Prostate cancer	Increased tumor accumulation and therapeutic efficacy	[119]
iRGD	Polymeric NPs (Survivin-siRNA)	70 nm	Intravenous (650 μg/kg siRNA)	Prostate cancer	Tumor accumulation is threefold greater with modified NPs	[120]
tLyP-1	HA NCs (DOX)	149 nm, −27 mV	Intravenous (5 mg/kg)	NSCLC (orthotopic)	Significant tumor accumulation and reduction in tumor load	[112]
HA	Lipid-polymer hybrid NPs	122 nm, −21 mV	Intravenous (10 mg/kg of erlotinib and/or bevacizumab)	NSCLC	Highest tumor accumulation and reduction in tumor volume	[121]
HA	LNPs (siPKL1 siRNA)	60 nm, +5 mV	Intraperitoneal (1 mg/kg siRNA)	Ovarian cancer	Enhanced tumor targeting relative to uncoated NPs	[122]
HA and CD20 mAb	HA-PLGA-PLA NPs (BCL2 siRNA)	135 nm, −39 mV	Intravenous (1 mg/kg siRNA)	Non-Hodgkin lymphoma	Improved siRNA delivery to tumor cells and enhanced survival	[123]
TfR	PLGA NPs (BCR/ABL mAb)	296 nm, −13 mV, 73%	Intravenous (2 mg per mouse)	Chronic myeloid leukemia	Extended survival	[124]
MPC	NCs (nimotuzumab or trastuzumab)	25 nm	Intravenous (10 mg/kg)	Glioblastoma (orthotopic)	Effective delivery of mAb to the tumor site and prolonged survival	[125]
CXCL13	PLA-PEG-PLA NCs (rituximab)	28 nm, −1 mV	Retro-orbital vein injection (4 mg/kg)	Non-Hodgkin lymphoma	Greater antitumor effect compared to nontargeted NCs	[126]

improve NP accumulation at tumor sites. Incorporating biomimetic materials in NP design holds promise for effective cancer therapeutic delivery [127].

6.2 Passive Targeting Through Enhanced Permeability and Retention (EPR) Effect

Passive targeting involves delivering drugs or drug-carrier systems to specific sites based on physicochemical and pharmacological properties. Solid tumors create an environment conducive to the accumulation of large drugs and colloidal delivery systems, such as micelles, liposomes, and polymeric nanoparticles. The increased permeability of blood vessels and poor lymphatic drainage in rapidly growing tumors allow these nanosystems to remain in the tumor longer, enhancing drug delivery efficacy. Many solid tumors, particularly those with a robust blood supply, exhibit an enhanced permeability and retention (EPR) effect, which facilitates the passive targeting of anti-tumor drugs. This effect arises from the permeable nature of tumor blood vessels and inadequate lymphatic drainage, leading to the accumulation of larger molecules and particles (approximately 20–500 nm in diameter) within tumor tissue [128].

Studies indicate that DSPE-PEG-modified nanoparticles exhibit significantly longer circulation times and higher tumor accumulation compared to other PEG-lipid modifications. The length of the alkyl chain in PEG-lipids (C14, C16, C18) directly influences the rate of detachment from the nanoparticle surface, with longer chains associated with slower dissociation rates and enhanced tumor targeting and gene silencing efficacy. Comparative efficacy studies consistently show that DSPE-PEG outperforms alternative PEG-lipids, making it a superior choice for optimizing the delivery of RNA and monoclonal antibodies (mAbs) [129]. Optimal PEGylation densities of 1.5–5% relative to total lipid components yield the best outcomes for passive targeting, maximizing nanoparticle stability and efficacy. While PEG2k is commonly used, preliminary findings suggest that higher molecular weights may further enhance nanoparticle properties, necessitating further research into the relationship between PEG molecular weight, density, and nanoparticle mechanics [26]. Achieving tumor accumulation levels of approximately 8% of the injected dose per gram of tissue is considered a significant benchmark for successful biological delivery in murine models. Although the type of biological payload does not significantly affect passive accumulation, its positioning may alter the protein corona, impacting overall efficacy. Additionally, PEGylated nanoparticles face challenges in being captured by the liver and spleen, limiting their effectiveness. Ongoing research aims to address these challenges, highlighting the importance of optimizing PEGylation strategies to enhance the delivery and efficacy of nanoparticle-based therapeutics [130].

6.3 Overcoming Biological Barriers (Blood–Brain Barrier, Tumor Microenvironment)

The blood–brain barrier (BBB) is typically a tightly regulated structure but can be compromised in conditions like trauma, neuroinflammation, or brain tumors, leading to the formation of the blood–brain tumor barrier (BBTB) in glioblastoma (GBM) [131]. This alteration results from the physical distortion of the neurovascular unit and loss of supportive signaling from pericytes and astrocytes, disrupting endothelial tight junctions. The high tumor mass induces hypoxia, promoting angiocrine factors like vascular endothelial growth factor (VEGF), which affect existing vasculature and encourage the formation of immature blood vessels [132].

Preclinical studies have identified several promising nanoparticle (NP) candidates that enhance brain accumulation and improve tumor responses in GBM [133]. However, only a few NPs have progressed to clinical evaluation. For instance, PEGylated liposomes encapsulating doxorubicin (Doxil-PEG-liposomes) have shown improved tumor concentration and reduced cardiotoxicity since the 1990s, yet a Phase II trial in GBM patients revealed no significant survival benefits. Despite preclinical evidence of enhanced tumor delivery and survival benefits from liposomal doxorubicin formulations, the effectiveness of chemotherapy for brain tumors remains limited [134].

Apolipoprotein E (ApoE) facilitates lipid transport and enhances NP transcytosis across the BBB when coated directly or via passive absorption [135]. Treatments combining ApoE peptide-coated polymers with immunoadjuvants have shown promise in inducing immunogenic cell death in tumors. Similarly, transferrin-coated NPs have been developed to enhance delivery by bypassing competition with serum transferrin, with peptide or antibody decorations improving NP transport across the BBB. Targeting solute carrier proteins, such as glucose transporter 1 (GLUT1), and glutathione (GSH) transporters has also been explored to facilitate NP accumulation in intracranial tumors [136]. Future advancements in NP design must integrate tumor-specific and BBB-targeted ligands to enhance brain tumor accumulation and minimize side effects, ultimately improving treatment options for GBM.

7 Emerging Trends and Future Directions

7.1 Next-Generation Nanoparticle Platforms

Advanced nanoparticle systems are addressing key challenges in nucleic acid delivery, including stability, cellular uptake, and targeted distribution, thereby enhancing cancer therapies [137]. These platforms are engineered to protect RNA and DNA from nuclease degradation, facilitate endosomal escape, and ensure precise targeting of cancer cells with therapeutic payloads.

Lipid nanoparticles (LNPs) are widely utilized for their ability to encapsulate nucleic acids within lipid bilayers, improving systemic circulation and promoting membrane fusion. Initially employed in mRNA vaccination technologies, LNPs have demonstrated significant efficacy. Polymeric nanoparticles, particularly those made from biocompatible and biodegradable materials like PEG and PLGA, provide regulated and sustained release, enhancing the therapeutic window [138].

Gold nanoparticles (AuNPs) offer strong surface chemistry for nucleic acid conjugation and unique optical properties for theranostic applications. Exosomes, as naturally occurring extracellular vesicles, present a biocompatible delivery system for RNA and DNA with minimal immune response. Additionally, mesoporous silica nanoparticles (MSNs) and peptide-based nanoparticles leverage customizable surface areas and self-assembly capabilities, allowing for precise delivery with reduced off-target effects [139].

Future advancements in precision-based gene therapies for cancer treatment will require optimizing targeting specificity, enhancing endosomal escape efficiency, and improving biocompatibility to overcome existing hurdles.

7.2 Personalized Nanomedicine Approaches

Custom-tailored nanomedicine for RNA and DNA delivery in cancer therapies aims to optimize nanoparticle platforms according to a patient's genetic and molecular tumor profile. By utilizing nanoparticles for targeted gene modification, these strategies seek to enhance therapeutic efficacy while minimizing off-target effects and systemic toxicity. Tumor-specific ligands can be integrated into lipid nanoparticles (LNPs) and polymeric nanoparticles to facilitate the targeted delivery of RNA-based therapies, such as messenger RNA (mRNA), small interfering RNA (siRNA), and DNA-based gene editing tools like CRISPR/Cas9 [140].

These nanoparticle platforms protect nucleic acids from degradation and promote cellular uptake. Surface modifications, such as polyethylene glycol (PEG) conjugation, reduce immune clearance and extend circulation time. Personalized approaches enhance selective accumulation in malignant tissues by designing nanoparticles based on patient-derived tumor markers, including specific receptors or antigens. Exosomes, being endogenous, can be genetically modified to carry patient-specific DNA or RNA sequences, offering high immunogenicity and biocompatibility [141]. Furthermore, nanomedicine platforms can be optimized for effective endosomal escape, controlled release, and precise geographic targeting, thereby improving the overall efficacy of cancer treatments.

8 Conclusions

Advanced nanomaterials and engineering techniques are enhancing the efficacy and safety of RNA and DNA delivery in cancer therapies. Optimized lipid nanoparticles (LNPs) incorporate surface ligands and improved endosomal escape, while polymeric nanoparticles like PEG and PLGA enable sustained release and minimize off-target effects. Gold nanoparticles provide stability and multifunctionality, and peptide-based and mesoporous silica nanoparticles offer versatile platforms for targeted treatments, including mRNA vaccinations and CRISPR gene editing. Exosomes deliver therapeutic RNA and DNA with high specificity and low immunogenicity. Hybrid nanoparticles combine inorganic and organic components for multifunctional applications. Additionally, advancements in machine learning and artificial intelligence are facilitating the design of optimal delivery systems. Collaborative efforts among materials scientists, oncologists, and bioengineers are crucial for translating these innovations into clinical practice, positioning nanotechnology to transform cancer therapies through enhanced nucleic acid delivery.

References

1. Yayın, G., Altunel, Y.E., Hazırlayan, Y., Kitaplığı, G.: A brief overview of developments in nanotechnology and nanomedicine. Gecekitapligi.Com (n.d.). https://www.gecekitapligi.com/Webkontrol/uploads/Fck/3-_Academic_Studies_In_Health_Sciences_DK_V2_4.pdf#page=31. Accessed 22 Sept 2024
2. Wang, J., Zhao, W., Zhang, Z., Liu, X., Xie, T., Wang, L., Xue, Y., Zhang, Y.: A journey of challenges and victories: a bibliometric worldview of nanomedicine since the 21st century. Adv. Mater. **36** (2024). https://doi.org/10.1002/ADMA.202308915
3. Kostarelos, K.: Establishing nanomedicine. Nanomedicine **1**, 259–260 (2006). https://doi.org/10.2217/17435889.1.3.259
4. Yahya, E.B., Alqadhi, A.M.: Recent trends in cancer therapy: a review on the current state of gene delivery. Life Sci. **269**, 119087 (2021). https://doi.org/10.1016/J.LFS.2021.119087
5. Manuela Drăgoi, C., Rodríguez-Gascón, A., Crengut, A., Dumitrescu, I.-B., Rakicevic, L.: DNA and RNA molecules as a foundation of therapy strategies for treatment of cardiovascular diseases. Pharmaceutics **15**, 2141 (2023). https://doi.org/10.3390/PHARMACEUTICS15082141
6. Limeres, M.J., Toscanini, M.A., Devoto, T.B., Ladetto, M.F., Castro, G.R., Cuestas, M.L.: Application in gene therapy and DNA/RNA vaccines. In: Smart Nanomaterials for Bioencapsulation, pp. 39–62 (2022). https://doi.org/10.1016/B978-0-323-91229-7.00001-5
7. Nsairat, H., Khater, D., Sayed, U., Odeh, F., Al Bawab, A., Alshaer, W.: Liposomes: structure, composition, types, and clinical applications. Heliyon **8**, e09394 (2022). https://doi.org/10.1016/j.heliyon.2022.e09394
8. Chen, T., Gong, T., Zhao, T., Fu, Y., Zhang, Z., Gong, T.: A comparison study between lycobetaine-loaded nanoemulsion and liposome using nRGD as therapeutic adjuvant for lung cancer therapy. Eur. J. Pharm. Sci. **111**, 293–302 (2018). https://doi.org/10.1016/J.EJPS.2017.09.041
9. Nakhaei, P., Margiana, R., Bokov, D.O., Abdelbasset, W.K., Jadidi Kouhbanani, M.A., Varma, R.S., Marofi, F., Jarahian, M., Beheshtkhoo, N.: Liposomes: structure, biomedical applications, and stability parameters with emphasis on cholesterol. Front. Bioeng. Biotechnol. **9**, 705886 (2021). https://doi.org/10.3389/fbioe.2021.705886

10. Cabanes, A., Even-Chen, S., Zimberoff, J., Barenholz, Y., Kedar, E., Gabizon, A.: Enhancement of antitumor activity of polyethylene glycol-coated liposomal doxorubicin with soluble and liposomal interleukin 2. Clin. Cancer Res. **5**, 687–693 (1999)
11. Lammers, T., Ferrari, M.: The success of nanomedicine. Nano Today **31**, 100853 (2020). https://doi.org/10.1016/j.nantod.2020.100853
12. Mehrotra, N., Anees, M., Tiwari, S., Kharbanda, S., Singh, H.: Polylactic acid based polymeric nanoparticle mediated co-delivery of navitoclax and decitabine for cancer therapy, Nanomedicine Nanotechnology. Biol. Med. **47**, 102627 (2023). https://doi.org/10.1016/J.NANO.2022.102627
13. Undre B., Pandya S.R., Kumar V., Singh, M.: Dendrimers as smart materials for developing the various applications in the field of biomedical sciences. Adv. Mat. Lett. **7**, 502–516 (2016). https://doi.org/10.5185/amlett.2016.6153
14. Singh, V., Kesharwani, P.: Dendrimer as a promising nanocarrier for the delivery of doxorubicin as an anticancer therapeutics. J. Biomater. Sci. Polym. Ed. **32**, 1882–1909 (2021). https://doi.org/10.1080/09205063.2021.1938859
15. Ghezzi, M., Pescina, S., Padula, C., Santi, P., Del Favero, E., Cantù, L., Nicoli, S.: Polymeric micelles in drug delivery: an insight of the techniques for their characterization and assessment in biorelevant conditions. J. Control. Release **332**, 312–336 (2021). https://doi.org/10.1016/j.jconrel.2021.02.031
16. Li, J., Li, Z., Zhou, T., Zhang, J., Xia, H., Li, H., He, J., He, S., Wang, L.: Positively charged micelles based on a triblock copolymer demonstrate enhanced corneal penetration. Int. J. Nanomedicine **10**, 6027–6037 (2015). https://doi.org/10.2147/IJN.S90347
17. Blanco, E., Shen, H., Ferrari, M.: Principles of nanoparticle design for overcoming biological barriers to drug delivery. Nat. Biotechnol. **33**, 941–951 (2015). https://doi.org/10.1038/NBT.3330
18. Mitragotri, S., Lammers, T., Bae, Y.H., Schwendeman, S., De Smedt, S., Leroux, J.C., Peer, D., Kwon, I.C., Harashima, H., Kikuchi, A., Oh, Y.K., Torchilin, V., Hennink, W., Hanes, J., Park, K.: Drug delivery research for the future: expanding the nano horizons and beyond. J. Control. Release **246**, 183–184 (2017). https://doi.org/10.1016/J.JCONREL.2017.01.011
19. Jarvis, M., Arnold, M., Ott, J., Krishnan, V., Pant, K., Prabhakarpandian, B., Mitragotri, S.: Detachment of ligands from nanoparticle surface under flow and endothelial cell contact: assessment using microfluidic devices. Bioeng. Transl. Med. **3**, 148 (2018). https://doi.org/10.1002/BTM2.10089
20. von Roemeling, C., Jiang, W., Chan, C.K., Weissman, I.L., Kim, B.Y.S.: Breaking down the barriers to precision cancer nanomedicine. Trends Biotechnol. **35**, 159–171 (2017). https://doi.org/10.1016/J.TIBTECH.2016.07.006
21. Li, H.J., Du, J.Z., Du, X.J., Xu, C.F., Sun, C.Y., Wang, H.X., Cao, Z.T., Yang, X.Z., Zhu, Y.H., Nie, S., Wang, J.: Stimuli-responsive clustered nanoparticles for improved tumor penetration and therapeutic efficacy. Proc. Natl. Acad. Sci. U. S. A. **113**, 4164–4169 (2016). https://doi.org/10.1073/PNAS.1522080113
22. Chetty, R., Pandya, S.R., Singh, M.: Physicochemical interaction of cerium oxide nanoparticles with simulated biofluids, hemoglobin, insulin, and ds-DNA at 310.15 K. New J. Chem. **44**, 1825–1845 (2020). https://doi.org/10.1039/C9NJ04155A
23. Mitchell, M.J., Billingsley, M.M., Haley, R.M., Wechsler, M.E., Peppas, N.A., Langer, R.: Engineering precision nanoparticles for drug delivery. Nat. Rev. Drug Discov. **20**, 101–124 (2021). https://doi.org/10.1038/s41573-020-0090-8
24. Cheng, X., Lee, R.J.: The role of helper lipids in lipid nanoparticles (LNPs) designed for oligonucleotide delivery. Adv. Drug Deliv. Rev. **99**, 129–137 (2016). https://doi.org/10.1016/J.ADDR.2016.01.022
25. Liang, X., Li, D., Leng, S., Zhu, X.: RNA-based pharmacotherapy for tumors: from bench to clinic and back. Biomed. Pharmacother. **125** (2020). https://doi.org/10.1016/J.BIOPHA.2020.109997

26. Zhang, B.C., Wu, P.Y., Zou, J.J., Jiang, J.L., Zhao, R.R., Luo, B.Y., Liao, Y.Q., Shao, J.W.: Efficient CRISPR/Cas9 gene-chemo synergistic cancer therapy via a stimuli-responsive chitosan-based nanocomplex elicits anti-tumorigenic pathway effect. Chem. Eng. J. **393** (2020). https://doi.org/10.1016/j.cej.2020.124688
27. Sayed, N., Allawadhi, P., Khurana, A., Singh, V., Navik, U., Pasumarthi, S.K., Khurana, I., Banothu, A.K., Weiskirchen, R., Bharani, K.K.: Gene therapy: comprehensive overview and therapeutic applications. Life Sci. **294**, 120375 (2022). https://doi.org/10.1016/J.LFS.2022.120375
28. Lin, Y.-X., Wang, Y., Blake, S., Yu, M., Mei, L., Wang, H., Shi, J.: RNA nanotechnology-mediated cancer immunotherapy. Theranostics **10**, 281–299 (2020). https://doi.org/10.7150/thno.35568
29. Ghidini, M., Silva, S.G., Evangelista, J., do Vale, M.L.C., Farooqi, A.A., Pinheiro, M.: Nanomedicine for the delivery of RNA in cancer. Cancers (Basel) **14**, 2677 (2022). https://doi.org/10.3390/cancers14112677
30. Elbashir, S.M., Harborth, J., Lendeckel, W., Yalcin, A., Weber, K., Tuschl, T.: Duplexes of 21-nucleotide RNAs mediate RNA interference in cultured mammalian cells. Nature **411**, 494–498 (2001). https://doi.org/10.1038/35078107
31. Chen, D., Bao, C., Zhao, F., Yu, H., Zhong, G., Xu, L., Yan, S.: Exploring specific miRNA-mRNA axes with relationship to taxanes-resistance in breast cancer. Front. Oncol. **10** (2020). https://doi.org/10.3389/FONC.2020.01397
32. Pinel, K., Diver, L.A., White, K., McDonald, R.A., Baker, A.H.: Substantial dysregulation of miRNA passenger strands underlies the vascular response to injury. Cells **8**, 83 (2019). https://doi.org/10.3390/CELLS8020083
33. Liu, C., Shi, Q., Huang, X., Koo, S., Kong, N., Tao, W.: mRNA-based cancer therapeutics. Nat. Rev. Cancer **238**(23), 526–543 (2023). https://doi.org/10.1038/s41568-023-00586-2
34. Hoecke, V., Med, R.J.T., Van Hoecke, L., Roose, K.: How mRNA therapeutics are entering the monoclonal antibody field. J. Transl. Med. **171**(17), 1–14. https://doi.org/10.1186/S12967-019-1804-8
35. Hayward, S.L., Francis, D.M., Kholmatov, P., Kidambi, S.: Targeted delivery of MicroRNA125a-5p by engineered lipid nanoparticles for the treatment of HER2 positive metastatic breast cancer. J. Biomed. Nanotechnol. **12**, 554–568 (2016). https://doi.org/10.1166/JBN.2016.2194
36. Bathula, S.R., Huang, L.: Gene therapy with plasmid DNA. In: Burger's Medicinal Chemistry, Drug Discovery and Development, pp. 457–499. Wiley (2010). https://doi.org/10.1002/0471266949.bmc073.pub2
37. Beals, N., Kasibhatla, N., Basu, S.: Efficient delivery of plasmid DNA using incorporated nucleotides for precise conjugation of targeted nanoparticles. ACS Appl. Bio Mater. **2**, 717–727 (2019). https://doi.org/10.1021/acsabm.8b00596
38. Liu, Y., Musetti, S., Huang, L.: Gene therapy with plasmid DNA. In: Burger's Medicinal Chemistry, Drug Discovery and Development, pp. 1–35. Wiley (2021). https://doi.org/10.1002/0471266949.bmc073.pub3
39. Shi, M., Liu, Y., Xu, M., Yang, H., Wu, C., Miyoshi, H.: Core/shell Fe_3O_4@SiO_2 nanoparticles modified with PAH as a vector for EGFP plasmid DNA delivery into HeLa cells. Macromol. Biosci. **11**, 1563–1569 (2011). https://doi.org/10.1002/mabi.201100150
40. Wang, L., Wang, X., Bhirde, A., Cao, J., Zeng, Y., Huang, X., Sun, Y., Liu, G., Chen, X.: Carbon-dot-based two-photon visible nanocarriers for safe and highly efficient delivery of siRNA and DNA. Adv. Healthc. Mater. **3**, 1203–1209 (2014). https://doi.org/10.1002/adhm.201300611
41. Cesur-Ergün, B., Demir-Dora, D.: Gene therapy in cancer. J. Gene Med. **25** (2023). https://doi.org/10.1002/jgm.3550
42. Kim, Y.G., Cha, J., Chandrasegaran, S.: Hybrid restriction enzymes: zinc finger fusions to Fok I cleavage domain. Proc. Natl. Acad. Sci. U. S. A. **93**, 1156–1160 (1996). https://doi.org/10.1073/pnas.93.3.1156

43. Zhang, M., Eshraghian, E.A., Al Jammal, O., Zhang, Z., Zhu, X.: CRISPR technology: the engine that drives cancer therapy. Biomed. Pharmacother. **133** (2021). https://doi.org/10.1016/j.biopha.2020.111007
44. Zhuo, C., Zhang, J., Lee, J.H., Jiao, J., Cheng, D., Liu, L., Kim, H.W., Tao, Y., Li, M.: Spatiotemporal control of CRISPR/Cas9 gene editing. Signal Transduct. Target. Ther. **6** (2021). https://doi.org/10.1038/s41392-021-00645-w
45. Zhen, S., Liu, Y., Lu, J., Tuo, X., Yang, X., Chen, H., Chen, W., Li, X.: Human papillomavirus oncogene manipulation using clustered regularly interspersed short palindromic repeats/Cas9 delivered by pH-sensitive cationic liposomes. Hum. Gene Ther. **31**, 309–324 (2020). https://doi.org/10.1089/hum.2019.312
46. Hu, Y., Zhao, Z., Ehrich, M., Fuhrman, K., Zhang, C.: In vitro controlled release of antigen in dendritic cells using pH-sensitive liposome-polymeric hybrid nanoparticles. Polymer (Guildford) **80**, 171–179 (2015). https://doi.org/10.1016/j.polymer.2015.10.048
47. Guo, P., Yang, J., Huang, J., Auguste, D.T., Moses, M.A.: Therapeutic genome editing of triple-negative breast tumors using a noncationic and deformable nanolipogel. Proc. Natl. Acad. Sci. U. S. A. **116**, 18295–18303 (2019). https://doi.org/10.1073/pnas.1904697116
48. Mout, R., Ray, M., Yesilbag Tonga, G., Lee, Y.W., Tay, T., Sasaki, K., Rotello, V.M.: Direct cytosolic delivery of CRISPR/Cas9-ribonucleoprotein for efficient gene editing. ACS Nano **11**, 2452–2458 (2017). https://doi.org/10.1021/acsnano.6b07600
49. Lee, K., Conboy, M., Park, H.M., Jiang, F., Kim, H.J., Dewitt, M.A., Mackley, V.A., Chang, K., Rao, A., Skinner, C., Shobha, T., Mehdipour, M., Liu, H., Huang, W.C., Lan, F., Bray, N.L., Li, S., Corn, J.E., Kataoka, K., Doudna, J.A., Conboy, I., Murthy, N.: Nanoparticle delivery of Cas9 ribonucleoprotein and donor DNA in vivo induces homology-directed DNA repair. Nat. Biomed. Eng. **1**, 889–901 (2017). https://doi.org/10.1038/s41551-017-0137-2
50. Ma, Y., Gao, W., Zhang, Y., Yang, M., Yan, X., Zhang, Y., Li, G., Liu, C., Xu, C., Zhang, M.: Biomimetic MOF nanoparticles delivery of C-dot nanozyme and CRISPR/Cas9 system for site-specific treatment of ulcerative colitis. ACS Appl. Mater. Interfaces **14**, 6358–6369 (2022). https://doi.org/10.1021/acsami.1c21700
51. Zhang, B.C., Luo, B.Y., Zou, J.J., Wu, P.Y., Jiang, J.L., Le, J.Q., Zhao, R.R., Chen, L., Shao, J.W.: Co-delivery of sorafenib and CRISPR/Cas9 based on targeted core-shell hollow mesoporous organosilica nanoparticles for synergistic HCC therapy. ACS Appl. Mater. Interfaces **12**, 57362–57372 (2020). https://doi.org/10.1021/acsami.0c17660
52. Yang, T.C., Chang, C.Y., Yarmishyn, A.A., Mao, Y.S., Yang, Y.P., Wang, M.L., Hsu, C.C., Yang, H.Y., Hwang, D.K., Chen, S.J., Tsai, M.L., Lai, Y.H., Tzeng, Y., Chang, C.C., Chiou, S.H.: Carboxylated nanodiamond-mediated CRISPR-Cas9 delivery of human retinoschisis mutation into human iPSCs and mouse retina. Acta Biomater. **101**, 484–494 (2020). https://doi.org/10.1016/j.actbio.2019.10.037
53. Sun, W., Ji, W., Hall, J.M., Hu, Q., Wang, C., Beisel, C.L., Gu, Z.: Self-assembled DNA nanoclews for the efficient delivery of CRISPR-Cas9 for genome editing. Angew. Chem. **127**, 12197–12201 (2015). https://doi.org/10.1002/ange.201506030
54. Lin, Y., Wu, J., Gu, W., Huang, Y., Tong, Z., Huang, L., Tan, J.: Exosome–liposome hybrid nanoparticles deliver CRISPR/Cas9 system in MSCs. Adv. Sci. **5** (2018). https://doi.org/10.1002/advs.201700611
55. Kong, H., Ju, E., Yi, K., Xu, W., Lao, Y.H., Cheng, D., Zhang, Q., Tao, Y., Li, M., Ding, J.: Advanced nanotheranostics of CRISPR/Cas for viral hepatitis and hepatocellular carcinoma. Adv. Sci. **8** (2021). https://doi.org/10.1002/advs.202102051
56. Farris, E., Brown, D.M., Ramer-Tait, A.E., Pannier, A.K.: Micro- and nanoparticulates for DNA vaccine delivery. Exp. Biol. Med. **241**, 919–929 (2016). https://doi.org/10.1177/1535370216643771
57. Xiao, Y., Shi, K., Qu, Y., Chu, B., Qian, Z.: Engineering nanoparticles for targeted delivery of nucleic acid therapeutics in tumor. Mol. Ther. Methods Clin. Dev. **12**, 1–18 (2019). https://doi.org/10.1016/j.omtm.2018.09.002
58. Lopes, A., Vandermeulen, G., Préat, V.: Cancer DNA vaccines: current preclinical and clinical developments and future perspectives. J. Exp. Clin. Cancer Res. **38** (2019). https://doi.org/10.1186/s13046-019-1154-7

59. Saade, F., Petrovsky, N.: Technologies for enhanced efficacy of DNA vaccines. Expert Rev. Vaccines **11**, 189–209 (2012). https://doi.org/10.1586/erv.11.188
60. Danaeifar, M.: Recent advances in gene therapy: genetic bullets to the root of the problem. Clin. Exp. Med. **23**, 1107–1121 (2023). https://doi.org/10.1007/s10238-022-00925-x
61. Kim, T.H., Jin, H., Kim, H.W., Cho, M.H., Cho, C.S.: Mannosylated chitosan nanoparticle-based cytokine gene therapy suppressed cancer growth in BALB/c mice bearing CT-26 carcinoma cells. Mol. Cancer Ther. **5**, 1723–1732 (2006). https://doi.org/10.1158/1535-7163.MCT-05-0540
62. Tahamtan, A., Barati, M., Tabarraei, A., Mohebbi, S.R., Shirian, S., Gorji, A., Ghaemi, A.: Antitumor immunity induced by genetic immunization with chitosan nanoparticle formulated adjuvanted for HPV-16 E7 DNA vaccine. Iran. J. Immunol. **15**, 269–280 (2018). https://iji.sums.ac.ir/article_40591_b34bb68a84a5c18288e4a402cfe3f9c6.pdf. Accessed 23 Sept 2024
63. Chen, Y.Z., Yao, X.L., Ruan, G.X., Zhao, Q.Q., Tang, G.P., Tabata, Y., Gao, J.Q.: Gene-carried chitosan-linked polyethylenimine induced high gene transfection efficiency on dendritic cells. Biotechnol. Appl. Biochem. **59**, 346–352 (2012). https://doi.org/10.1002/bab.1036
64. Tan, L., Han, S., Ding, S., Xiao, W., Ding, Y., Qian, L., Wang, C., Gong, W.: Chitosan nanoparticle-based delivery of fused NKG2D-IL-21 gene suppresses colon cancer growth in mice. Int. J. Nanomedicine **12**, 3095–3107 (2017). https://doi.org/10.2147/IJN.S128032
65. Liu, Z., Lv, D., Liu, S., Gong, J., Wang, D., Xiong, M., Chen, X., Xiang, R., Tan, X.: Alginic acid-coated chitosan nanoparticles loaded with legumain DNA vaccine: effect against breast cancer in mice. PLoS One **8** (2013). https://doi.org/10.1371/journal.pone.0060190
66. Razi Soofiyani, S., Hallaj-Nezhadi, S., Lotfipour, F., Mohammad Hosseini, A., Baradaran, B.: Gene therapy based on interleukin-12 loaded chitosan nanoparticles in a mouse model of fibrosarcoma. Iran. J. Basic Med. Sci. **19**, 1238–1244 (2016). https://doi.org/10.22038/ijbms.2016.7825
67. Behnaz, N., Karizi, S.Z., Nazarian, S., Kazemi, R., Motamedi, M.J., Fasihi-Ramandi, M., Amani, J.: Construction and structural assessment of nanocapsule containing HER2-MUC1 chimeric protein as a candidate for a vaccine against breast cancer. Int. J. Cancer Manag. **12**, 66671 (2019). https://doi.org/10.5812/ijcm.66671
68. Shi, G.N., Zhang, C.N., Xu, R., Niu, J.F., Song, H.J., Zhang, X.Y., Wang, W.W., Wang, Y.M., Li, C., Wei, X.Q., Kong, D.L.: Enhanced antitumor immunity by targeting dendritic cells with tumor cell lysate-loaded chitosan nanoparticles vaccine. Biomaterials **113**, 191–202 (2017). https://doi.org/10.1016/j.biomaterials.2016.10.047
69. Yao, W., Peng, Y., Du, M., Luo, J., Zong, L.: Preventative vaccine-loaded mannosylated chitosan nanoparticles intended for nasal mucosal delivery enhance immune responses and potent tumor immunity. Mol. Pharm. **10**, 2904–2914 (2013). https://doi.org/10.1021/mp4000053
70. Ma, W., Chen, M., Kaushal, S., McElroy, M., Zhang, Y., Ozkan, C., Bouvet, M., Kruse, C., Grotjahn, D., Ichim, T., Minev, B.: PLGA nanoparticle-mediated delivery of tumor antigenic peptides elicits effective immune responses. Int. J. Nanomedicine **7**, 1475–1487 (2012). https://doi.org/10.2147/IJN.S29506
71. Ebrahimian, M., Taghavi, S., Mokhtarzadeh, A., Ramezani, M., Hashemi, M.: Co-delivery of doxorubicin encapsulated PLGA nanoparticles and Bcl-xL shRNA using alkyl-modified PEI into breast cancer cells. Appl. Biochem. Biotechnol. **183**, 126–136 (2017). https://doi.org/10.1007/s12010-017-2434-3
72. Pai Kasturi, S., Qin, H., Thomson, K.S., El-Bereir, S., Chul Cha, S., Neelapu, S., Kwak, L.W., Roy, K.: Prophylactic anti-tumor effects in a B cell lymphoma model with DNA vaccines delivered on polyethylenimine (PEI) functionalized PLGA microparticles. J. Control. Release **113**, 261–270 (2006). https://doi.org/10.1016/j.jconrel.2006.04.006
73. Nikitczuk, K.P., Schloss, R.S., Yarmush, M.L., Lattime, E.C.: PLGA-polymer encapsulating tumor antigen and CpG DNA administered into the tumor microenvironment elicits a systemic antigen-specific IFN-γ response and enhances survival. J. Cancer Ther. **04**, 280–290 (2013). https://doi.org/10.4236/jct.2013.41035

74. Bisso, S., Leroux, J.C.: Nanopharmaceuticals: a focus on their clinical translatability. Int. J. Pharm. **578**, 119098 (2020). https://doi.org/10.1016/j.ijpharm.2020.119098
75. George, N., Singh, H., Jotaniya, R., Pandya, S.R.: Raman spectroscopy for the determination of forensically important bio-fluids. Forensic Sci. Int. **340**, 111441 (2022). https://doi.org/10.1016/j.forsciint.2022.111441
76. Attinti, D.D., Safo, K.K., Balakrishna, M., George, N., Pandya, S.: Nuclear magnetic resonance spectroscopy: a versatile tool for forensic sample analysis. In: Advances in Analytical Techniques for Forensic Investigation, pp. 113–147. John Wiley & Sons, Ltd (2024). https://doi.org/10.1002/9781394167340.ch5
77. Singh, H., Desimone, M.F., Pandya, S., Jasani, S., George, N., Adnan, M., Aldarhami, A., Bazaid, A.S., Alderhami, S.A.: Revisiting the green synthesis of nanoparticles: uncovering influences of plant extracts as reducing agents for enhanced synthesis efficiency and its biomedical applications. Int. J. Nanomedicine **18**, 4727–4750 (2023). https://doi.org/10.2147/IJN.S419369
78. Bolje, A., Gobec, S.: Analytical techniques for structural characterization of proteins in solid pharmaceutical forms: an overview. Pharmaceutics **13**, 534 (2021). https://doi.org/10.3390/pharmaceutics13040534
79. Li, T., Senesi, A.J., Lee, B.: Small angle X-ray scattering for nanoparticle research. Chem. Rev. **116**, 11128–11180 (2016). https://doi.org/10.1021/acs.chemrev.5b00690
80. Imani, R., Emami, S.H., Faghihi, S.: Synthesis and characterization of an octaarginine functionalized graphene oxide nano-carrier for gene delivery applications. Phys. Chem. Chem. Phys. **17**, 6328–6339 (2015). https://doi.org/10.1039/c4cp04301d
81. Iravani Kashkouli, K., Torkzadeh-Mahani, M., Mosaddegh, E.: Synthesis and characterization of aminotetrazole-functionalized magnetic chitosan nanocomposite as a novel nanocarrier for targeted gene delivery. Mater. Sci. Eng. C **89**, 166–174 (2018). https://doi.org/10.1016/j.msec.2018.03.032
82. Maurer, V., Altin, S., Seleci, D.A., Zarinwall, A., Temel, B., Vogt, P.M., Strauß, S., Stahl, F., Scheper, T., Bucan, V., Garnweitner, G.: In-vitro application of magnetic hybrid niosomes: targeted sirna-delivery for enhanced breast cancer therapy. Pharmaceutics **13**, 394 (2021). https://doi.org/10.3390/pharmaceutics13030394
83. Depalo, N., Fanizza, E., Vischio, F., Denora, N., Laquintana, V., Cutrignelli, A., Striccoli, M., Giannelli, G., Agostiano, A., Curri, M.L., Scavo, M.P.: Imaging modification of colon carcinoma cells exposed to lipid based nanovectors for drug delivery: a scanning electron microscopy investigation. RSC Adv. **9**, 21810–21825 (2019). https://doi.org/10.1039/c9ra02381j
84. Falsafi, S.R., Rostamabadi, H., Assadpour, E., Jafari, S.M.: Morphology and microstructural analysis of bioactive-loaded micro/nanocarriers via microscopy techniques; CLSM/SEM/TEM/AFM. Adv. Colloid Interface Sci. **280**, 102166 (2020). https://doi.org/10.1016/j.cis.2020.102166
85. Goldstein, J.I., Newbury, D.E., Echlin, P., Joy, D.C., Fiori, C., Lifshin, E.: Preparation of biological samples for scanning electron microscopy. In: Scanning Electron Microscopy and X-ray Microanalysis, pp. 495–539. Springer, Boston, MA (1981). https://doi.org/10.1007/978-1-4613-3273-2_11
86. Koga, D., Kusumi, S., Shibata, M., Watanabe, T.: Applications of scanning electron microscopy using secondary and backscattered electron signals in neural structure. Front. Neuroanat. **15**, 759804 (2021). https://doi.org/10.3389/fnana.2021.759804
87. Jones, C.G.: Scanning electron microscopy: preparation and imaging for SEM. Methods Mol. Biol. **915**, 1–20 (2012). https://doi.org/10.1007/978-1-61779-977-8_1
88. Bellotti, E., Cascone, M.G., Barbani, N., Rossin, D., Rastaldo, R., Giachino, C., Cristallini, C.: Targeting cancer cells overexpressing folate receptors with new terpolymer-based nanocapsules: toward a novel targeted DNA delivery system for cancer therapy. Biomedicines **9**, 1275 (2021). https://doi.org/10.3390/biomedicines9091275
89. Chen, C.C., Zhu, C., White, E.R., Chiu, C.Y., Scott, M.C., Regan, B.C., Marks, L.D., Huang, Y., Miao, J.: Three-dimensional imaging of dislocations in a nanoparticle at atomic resolution. Nature **496**, 74–77 (2013). https://doi.org/10.1038/nature12009

90. Wang, W.: Imaging the chemical activity of single nanoparticles with optical microscopy. Chem. Soc. Rev. **47**, 2485–2508 (2018). https://doi.org/10.1039/c7cs00451f
91. Dini, L., Panzarini, E., Mariano, S., Passeri, D., Reggente, M., Rossi, M., Vergallo, C.: Microscopies at the nanoscale for nano-scale drug delivery systems. Curr. Drug Targets **16**, 1512–1530 (2015). https://doi.org/10.2174/1389450116666150531160851
92. Giráldez-Pérez, R.M., Grueso, E., Domínguez, I., Pastor, N., Kuliszewska, E., Prado-Gotor, R., Requena-Domenech, F.: Biocompatible DNA/5-fluorouracil-gemini surfactant-functionalized gold nanoparticles as promising vectors in lung cancer therapy. Pharmaceutics **13**, 423 (2021). https://doi.org/10.3390/pharmaceutics13030423
93. Kutwin, M., Sosnowska, M.E., Strojny-Cieślak, B., Jaworski, S., Trzaskowski, M., Wierzbicki, M., Chwalibog, A., Sawosz, E.: MicroRNA delivery by graphene-based complexes into glioblastoma cells. Molecules **26**, 5804 (2021). https://doi.org/10.3390/molecules26195804
94. Giráldez-Pérez, R.M., Grueso, E., Montero-Hidalgo, A.J., Luque, R.M., Carnerero, J.M., Kuliszewska, E., Prado-Gotor, R.: Gold nanosystems covered with doxorubicin/DNA complexes: a therapeutic target for prostate and liver cancer. Int. J. Mol. Sci. **23**, 15575 (2022). https://doi.org/10.3390/IJMS232415575/S1
95. Johnson, D., Oatley-Radcliffe, D.L., Hilal, N.: Atomic force microscopy (AFM). In: Membrane Characterization, pp. 115–144. Elsevier (2017). https://doi.org/10.1016/B978-0-444-63776-5.00007-3
96. Oliveira Brett, A.M., Chiorcea, A.M.: Atomic force microscopy of DNA immobilized onto a highly oriented pyrolytic graphite electrode surface. Langmuir **19**, 3830–3839 (2003). https://doi.org/10.1021/la027047d
97. Nogueira, S.S., Samaridou, E., Simon, J., Frank, S., Beck-Broichsitter, M., Mehta, A.: Analytical techniques for the characterization of nanoparticles for mRNA delivery. Eur. J. Pharm. Biopharm. **198**, 114235 (2024). https://doi.org/10.1016/j.ejpb.2024.114235
98. Swain, S., Sahu, P., Beg, S., Babu, S.: Nanoparticles for cancer targeting: current and future directions. Curr. Drug Deliv. **13**, 1290–1302 (2016). https://doi.org/10.2174/1567201813666160713121122
99. Lee, C.S., Kim, T.W., Oh, D.E., Bae, S.O., Ryu, J., Kong, H., Jeon, H., Seo, H.K., Jeon, S., Kim, T.H.: In vivo and in vitro anticancer activity of doxorubicin-loaded DNA-AuNP nanocarrier for the ovarian cancer treatment. Cancers (Basel) **12**, 634 (2020). https://doi.org/10.3390/cancers12030634
100. Kalinova, R., Mladenova, K., Petrova, S., Doumanov, J., Dimitrov, I.: Solvent-free synthesis of multifunctional block copolymer and formation of DNA and drug nanocarriers. Nanomaterials **13**, 2936 (2023). https://doi.org/10.3390/nano13222936
101. Rudzinski, W.E., Aminabhavi, T.M.: Chitosan as a carrier for targeted delivery of small interfering RNA. Int. J. Pharm. **399**, 1–11 (2010). https://doi.org/10.1016/j.ijpharm.2010.08.022
102. Lin, G., Huang, J., Zhang, M., Chen, S., Zhang, M.: Chitosan-crosslinked low molecular weight PEI-conjugated iron oxide nanoparticle for safe and effective DNA delivery to breast cancer cells. Nanomaterials **12**, 584 (2022). https://doi.org/10.3390/NANO12040584/S1
103. Zielińska, A., Szalata, M., Gorczyński, A., Karczewski, J., Eder, P., Severino, P., Cabeda, J.M., Souto, E.B., Słomski, R.: Cancer nanopharmaceuticals: physicochemical characterization and in vitro/in vivo applications. Cancers (Basel) **13**, 1896 (2021). https://doi.org/10.3390/cancers13081896
104. Kalmodia, S., Parameswaran, S., Yang, W., Barrow, C.J., Krishnakumar, S.: Attenuated total reflectance Fourier transform infrared spectroscopy: an analytical technique to understand therapeutic responses at the molecular level. Sci. Rep. **5**, 1–14 (2015). https://doi.org/10.1038/srep16649
105. Siafaka, P.I., Üstündağ Okur, N., Karavas, E., Bikiaris, D.N.: Surface modified multifunctional and stimuli responsive nanoparticles for drug targeting: current status and uses. Int. J. Mol. Sci. **17**, 1440 (2016). https://doi.org/10.3390/ijms17091440

106. De Leo, V., Maurelli, A.M., Giotta, L., Catucci, L.: Liposomes containing nanoparticles: preparation and applications. Colloids Surf. B Biointerfaces **218**, 112737 (2022). https://doi.org/10.1016/j.colsurfb.2022.112737
107. Naziris, N., Saitta, F., Chrysostomou, V., Libera, M., Trzebicka, B., Fessas, D., Pispas, S., Demetzos, C.: PH-responsive chimeric liposomes: from nanotechnology to biological assessment. Int. J. Pharm. **574**, 118849 (2020). https://doi.org/10.1016/j.ijpharm.2019.118849
108. Nguyen, T.L., Nguyen, T.H., Nguyen, C.K., Nguyen, D.H.: Redox and pH responsive poly(amidoamine) dendrimer-heparin conjugates via disulfide linkages for letrozole delivery. Biomed. Res. Int. **2017**, 8589212 (2017). https://doi.org/10.1155/2017/8589212
109. Bai, X., Smith, Z.L., Wang, Y., Butterworth, S., Tirella, A.: Sustained drug release from smart nanoparticles in cancer therapy: a comprehensive review. Micromachines **13**, 1623 (2022). https://doi.org/10.3390/mi13101623
110. Zhang, P., Wang, C., Zhao, J., Xiao, A., Shen, Q., Li, L., Li, J., Zhang, J., Min, Q., Chen, J., Chen, H.Y., Zhu, J.J.: Near infrared-guided smart nanocarriers for MicroRNA-controlled release of doxorubicin/siRNA with intracellular ATP as fuel. ACS Nano **10**, 3637–3647 (2016). https://doi.org/10.1021/acsnano.5b08145
111. Tiwari, H., Rai, N., Singh, S., Gupta, P., Verma, A., Singh, A.K., Kajal, Salvi, P., Singh, S.K., Gautam, V.: Recent advances in nanomaterials-based targeted drug delivery for preclinical cancer diagnosis and therapeutics. Bioengineering **10**, 760 (2023). https://doi.org/10.3390/bioengineering10070760
112. Teijeiro-Valiño, C., Novoa-Carballal, R., Borrajo, E., Vidal, A., Alonso-Nocelo, M., de la Fuente Freire, M., Lopez-Casas, P.P., Hidalgo, M., Csaba, N., Alonso, M.J.: A multifunctional drug nanocarrier for efficient anticancer therapy. J. Control. Release **294**, 154–164 (2019). https://doi.org/10.1016/j.jconrel.2018.12.002
113. Kampel, L., Goldsmith, M., Ramishetti, S., Veiga, N., Rosenblum, D., Gutkin, A., Chatterjee, S., Penn, M., Lerman, G., Peer, D., Muhanna, N.: Therapeutic inhibitory RNA in head and neck cancer via functional targeted lipid nanoparticles. J. Control. Release **337**, 378–389 (2021). https://doi.org/10.1016/j.jconrel.2021.07.034
114. Li, Q., Zhang, D., Zhang, J., Jiang, Y., Song, A., Li, Z., Luan, Y.: A three-in-one immunotherapy nanoweapon via cascade-amplifying cancer-immunity cycle against tumor metastasis, relapse, and postsurgical regrowth. Nano Lett. **19**, 6647–6657 (2019). https://doi.org/10.1021/acs.nanolett.9b02923
115. Giuliani, C.: The flavonoid quercetin induces AP-1 activation in FRTL-5 thyroid cells. Antioxidants **8** (2019). https://doi.org/10.3390/antiox8050112
116. Jain, S., Deore, S.V., Ghadi, R., Chaudhari, D., Kuche, K., Katiyar, S.S.: Tumor microenvironment responsive VEGF-antibody functionalized pH sensitive liposomes of docetaxel for augmented breast cancer therapy. Mater. Sci. Eng. C **121**, 111832 (2021). https://doi.org/10.1016/j.msec.2020.111832
117. Eskandari, Z., Bahadori, F., Celik, B., Onyuksel, H.: Targeted nanomedicines for cancer therapy, from basics to clinical trials. J. Pharm. Pharm. Sci. **23**, 132–157 (2020). https://doi.org/10.18433/jpps30583
118. Mamot, C., Ritschard, R., Wicki, A., Stehle, G., Dieterle, T., Bubendorf, L., Hilker, C., Deuster, S., Herrmann, R., Rochlitz, C.: Tolerability, safety, pharmacokinetics, and efficacy of doxorubicin-loaded anti-EGFR immunoliposomes in advanced solid tumours: a phase 1 dose-escalation study. Lancet Oncol. **13**, 1234–1241 (2012). https://doi.org/10.1016/S1470-2045(12)70476-X
119. Xu, X., Saw, P.E., Tao, W., Li, Y., Ji, X., Bhasin, S., Liu, Y., Ayyash, D., Rasmussen, J., Huo, M., Shi, J., Farokhzad, O.C.: ROS-responsive polyprodrug nanoparticles for triggered drug delivery and effective cancer therapy. Adv. Mater. **29** (2017). https://doi.org/10.1002/adma.201700141
120. Xu, X., Wu, J., Liu, Y., Yu, M., Zhao, L., Zhu, X., Bhasin, S., Li, Q., Ha, E., Shi, J., Farokhzad, O.C.: Ultra-pH-responsive and tumor-penetrating nanoplatform for targeted siRNA delivery with robust anti-cancer efficacy. Angew. Chem. - Int. Ed. **55**, 7091–7094 (2016). https://doi.org/10.1002/anie.201601273

121. Pang, J., Xing, H., Sun, Y., Feng, S., Wang, S.: Non-small cell lung cancer combination therapy: Hyaluronic acid modified, epidermal growth factor receptor targeted, pH sensitive lipid-polymer hybrid nanoparticles for the delivery of erlotinib plus bevacizumab. Biomed. Pharmacother. **125** (2020). https://doi.org/10.1016/j.biopha.2020.109861
122. Singh, M.S., Ramishetti, S., Landesman-Milo, D., Goldsmith, M., Chatterjee, S., Palakuri, R., Peer, D.: Therapeutic gene silencing using targeted lipid nanoparticles in metastatic ovarian cancer. Small **17** (2021). https://doi.org/10.1002/smll.202100287
123. Choi, K.Y., Correa, S., Min, J., Li, J., Roy, S., Laccetti, K.H., Dreaden, E., Kong, S., Heo, R., Roh, Y.H., Lawson, E.C., Palmer, P.A., Hammond, P.T.: Binary targeting of siRNA to hematologic cancer cells in vivo using layer-by-layer nanoparticles. Adv. Funct. Mater. **29** (2019). https://doi.org/10.1002/adfm.201900018
124. Jiang, G., Huang, Z., Yuan, Y., Tao, K., Feng, W.: Intracellular delivery of anti-BCR/ABL antibody by PLGA nanoparticles suppresses the oncogenesis of chronic myeloid leukemia cells. J. Hematol. Oncol. **14** (2021). https://doi.org/10.1186/s13045-021-01150-x
125. Han, L., Liu, C., Qi, H., Zhou, J., Wen, J., Wu, D., Xu, D., Qin, M., Ren, J., Wang, Q., Long, L., Liu, Y., Chen, I., Yuan, X., Lu, Y., Kang, C.: Systemic delivery of monoclonal antibodies to the central nervous system for brain tumor therapy. Adv. Mater. **31** (2019). https://doi.org/10.1002/adma.201805697
126. Wen, J., Wu, D., Qin, M., Liu, C., Wang, L., Xu, D., Vinters, H.V., Liu, Y., Kranz, E., Guan, X., Sun, G., Sun, X., Lee, Y.J., Martinez-Maza, O., Widney, D., Lu, Y., Chen, I.S.Y., Kamata, M.: Sustained delivery and molecular targeting of a therapeutic monoclonal antibody to metastases in the central nervous system of mice. Nat. Biomed. Eng. **3**, 706–716 (2019). https://doi.org/10.1038/s41551-019-0434-z
127. Cai, Y., Xu, Z., Shuai, Q., Zhu, F., Xu, J., Gao, X., Sun, X.: Tumor-targeting peptide functionalized PEG-PLA micelles for efficient drug delivery. Biomater. Sci. **8**, 2274–2282 (2020). https://doi.org/10.1039/c9bm02036e
128. Maeda, H., Wu, J., Sawa, T., Matsumura, Y., Hori, K.: Tumor vascular permeability and the EPR effect in macromolecular therapeutics: a review. J. Control. Release **65**, 271–284 (2000). https://doi.org/10.1016/S0168-3659(99)00248-5
129. Zhu, X., Tao, W., Liu, D., Wu, J., Guo, Z., Ji, X., Bharwani, Z., Zhao, L., Zhao, X., Farokhzad, O.C., Shi, J.: Surface de-PEGylation controls nanoparticle-mediated siRNA delivery in vitro and in vivo. Theranostics **7**, 1990–2002 (2017). https://doi.org/10.7150/thno.18136
130. Nikpoor, A.R., Tavakkol-Afshari, J., Sadri, K., Jalali, S.A., Jaafari, M.R.: Improved tumor accumulation and therapeutic efficacy of CTLA-4-blocking antibody using liposome-encapsulated antibody: in vitro and in vivo studies. Nanomedicine Nanotechnol. Biol. Med. **13**, 2671–2682 (2017). https://doi.org/10.1016/j.nano.2017.08.010
131. Greene, C., Connolly, R., Brennan, D., Laffan, A., O'Keeffe, E., Zaporojan, L., O'Callaghan, J., Thomson, B., Connolly, E., Argue, R., Martin-Loeches, I., Long, A., Cheallaigh, C.N., Conlon, N., Doherty, C.P., Campbell, M.: Blood–brain barrier disruption and sustained systemic inflammation in individuals with long COVID-associated cognitive impairment. Nat. Neurosci. **27**, 421–432 (2024). https://doi.org/10.1038/s41593-024-01576-9
132. Ahir, B.K., Engelhard, H.H., Lakka, S.S.: Tumor development and angiogenesis in adult brain tumor: glioblastoma. Mol. Neurobiol. **57**, 2461–2478 (2020). https://doi.org/10.1007/s12035-020-01892-8
133. Ak, G., Ünal, A., Karakayalı, T., Özel, B., Selvi Günel, N., Hamarat Şanlıer, Ş.: Brain-targeted, drug-loaded solid lipid nanoparticles against glioblastoma cells in culture. Colloids Surf. B Biointerfaces **206** (2021). https://doi.org/10.1016/j.colsurfb.2021.111946
134. Ananda, S., Nowak, A.K., Cher, L., Dowling, A., Brown, C., Simes, J., Rosenthal, M.A.: Phase 2 trial of temozolomide and PEGylated liposomal doxorubicin in the treatment of patients with glioblastoma multiforme following concurrent radiotherapy and chemotherapy. J. Clin. Neurosci. **18**, 1444–1448 (2011). https://doi.org/10.1016/j.jocn.2011.02.026
135. Hartl, N., Adams, F., Merkel, O.M.: From adsorption to covalent bonding: apolipoprotein e functionalization of polymeric nanoparticles for drug delivery across the blood–brain barrier. Adv. Ther. **4** (2021). https://doi.org/10.1002/adtp.202000092

136. Puris, E., Fricker, G., Gynther, M.: Targeting transporters for drug delivery to the brain: can we do better? Pharm. Res. **39**, 1415–1455 (2022). https://doi.org/10.1007/s11095-022-03241-x
137. Mendes, B.B., Conniot, J., Avital, A., Yao, D., Jiang, X., Zhou, X., Sharf-Pauker, N., Xiao, Y., Adir, O., Liang, H., Shi, J., Schroeder, A., Conde, J.: Nanodelivery of nucleic acids. Nat. Rev. Methods Prim. **21**(2), 1–21 (2022). https://doi.org/10.1038/s43586-022-00104-y
138. Pilkington, E.H., Suys, E.J.A., Trevaskis, N.L., Wheatley, A.K., Zukancic, D., Algarni, A., Al-Wassiti, H., Davis, T.P., Pouton, C.W., Kent, S.J., Truong, N.P.: From influenza to COVID-19: lipid nanoparticle mRNA vaccines at the frontiers of infectious diseases. Acta Biomater. **131**, 16–40 (2021). https://doi.org/10.1016/J.ACTBIO.2021.06.023
139. Ferreira, D., Fontinha, D., Martins, C., Pires, D., Fernandes, A.R., Baptista, P.V.: Gold nanoparticles for vectorization of nucleic acids for cancer therapeutics. Molecules **25**, 3489 (2020). https://doi.org/10.3390/MOLECULES25153489
140. Beach, M.A., Nayanathara, U., Gao, Y., Zhang, C., Xiong, Y., Wang, Y., Such, G.K.: Polymeric nanoparticles for drug delivery. Chem. Rev. **124**, 5505–5616 (2024). https://doi.org/10.1021/ACS.CHEMREV.3C00705
141. Fan, Z., Zhu, P., Zhu, Y., Wu, K., Li, C.Y., Cheng, H.: Engineering long-circulating nanomaterial delivery systems. Curr. Opin. Biotechnol. **66**, 131–139 (2020). https://doi.org/10.1016/J.COPBIO.2020.07.006

Nanoparticles in Cancer Imaging and Treatment Monitoring: Challenges and Perspectives

Elsa M. Materón, Juliana Martins Dias, Desirée Tamara Scheitd, Daiane Dias, Emanuel Carrilho, and Cecilia de Carvalho Castro Silva

Abstract Biosynthesized nanoparticles have increasingly drawn attention in the scientific community, especially in biomedical fields. This growing interest stem from a wide range of studies highlighting their remarkable advantages, including biocompatibility, eco-friendly production, and functional versatility. These nanoparticles show great promise in cancer-related applications; from controlled drug delivery to acting as contrast agents in imaging techniques, as well as enabling disease monitoring through electrochemical biosensors. With their natural origin and adaptable properties, biosynthesis stands out as an innovative and accessible path for developing safer, more personalized, and less invasive therapies in the future of medicine.

The impacts of cancer extend across social, economic, and public health dimensions, prompting a scientific focus on improving both diagnostic and therapeutic approaches [1, 2]. Traditional diagnostic methods, such as surgical biopsies and tumor removal, are invasive and come with significant risks. Likewise, radiation, chemotherapy, and immunotherapy treatments often lead to side effects due to their lack of active targeting mechanisms. As a result, higher drug doses are frequently required, increasing the risk of drug resistance [1, 3]. To overcome these limitations, nanomedicine, as illustrated in Fig. 1, offers considerable promise for cancer diagnosis and treatment. Custom-engineered nanoparticles can serve as contrast agents

E. M. Materón (✉) · C. de Carvalho Castro Silva
MackGraphe – Instituto Mackenzie de Pesquisa em Grafeno e Nanotecnologia, Instituto Presbiteriano, São Paulo, SP, Brazil
e-mail: elsa_materon@yahoo.com

Escola de Engenharia, Universidade Presbiteriana Mackenzie, São Paulo, SP, Brazil

J. M. Dias · D. Dias
Escola de Química e Alimentos, Universidade Federal do Rio Grande, Rio Grande, RS, Brazil

D. T. Scheitd · E. Carrilho
Instituto de Química de São Carlos, Universidade de São Paulo, São Carlos, SP, Brazil

E. Carrilho
Instituto Nacional de Ciência e Tecnologia de Bioanalítica – Lauro Kubota, Campinas, SP, Brazil

N. J. Joshi and G. Patel (eds.), *Nanoparticle Delivery in Cancer Care*,
Series in BioEngineering, https://doi.org/10.1007/978-981-95-0457-2_8

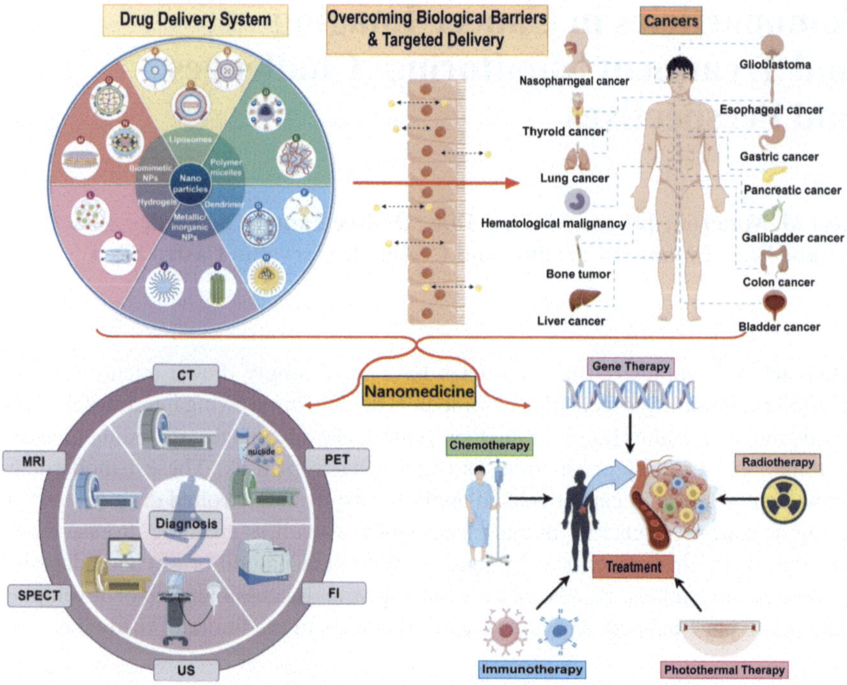

Fig. 1 Nanomedicine in cancer diagnosis and treatment. CT computed tomography, MRI magnetic resonance imaging, PET positron emission tomography, SPECT single-photon emission tomography, US ultrasound, FI fluorescence imaging. Copyright permission from Ref. [4]

in cancer diagnostics, facilitating high sensitivity and resolution in tumor detection through imaging techniques.

Additionally, the application of nanoprobes and nanobiosensors enables innovative strategies for tumor labeling and identification [4, 5]. Nanoparticles (NPs) present numerous advantages over other materials. A key benefit is the enhanced permeability and retention (EPR) effect, which enables NPs to accumulate at tumor sites. Furthermore, the surface of NPs can be engineered to minimize nonspecific uptake by the reticuloendothelial system [4]. The large surface-area-to-volume ratio of NPs also enables the efficient entrapment of therapeutic agents, protecting them from degradation by the tumor microenvironment [6]. Nanomaterials are also being used to develop more controlled and precise drug-release technologies [7]. Given these advantages, incorporating NPs into the design of multimodal synergistic therapies may yield greater therapeutic efficacy compared to monotherapy [8]. This chapter examines the application of nanoparticles for cancer imaging and treatment monitoring using electrochemical biosensors. It highlights innovative approaches to enhance early detection and personalized treatment strategies, with a focus on monitoring cancer biomarkers. Furthermore, the chapter addresses challenges in nanoparticle integration, their unique properties in medical applications, and the potential of biosynthesis and sustainable nanoparticle production as biocompatible, eco-friendly alternatives.

1 Nanoparticle-Integrated Electrochemical Biosensors for Cancer Treatment Monitoring

Nanotechnology is a rapidly advancing field that is profoundly impacting biosensors, thereby improving the diagnosis, prognosis, and monitoring of cancer. In the context of treatment monitoring, there has been extensive exploration into the development of biosensors capable of detecting biomarkers associated with various cancer types and evaluating drug efficacy at different target sites [4, 9]. Currently, a wide range of immunoassay methods is crucial for identifying tumor biomarkers (Fig. 2). Techniques such as molecular hybridization, gene amplification, DNA sequencing, gene-editing technologies, and immunohistochemistry (IHC) have significantly accelerated the discovery of prognostic and predictive cancer biomarkers. In addition, emerging approaches such as liquid biopsy, advanced microscopy, and single-cell sequencing analysis have further enhanced the precision and convenience of cancer diagnosis and treatment, providing more accurate insights into disease progression and therapeutic response [10].

However, certain limitations—such as high costs, the need for skilled supervision, complex data analysis, and invasive procedures like biopsies—limit the widespread use of these techniques despite their promising results. In this context, point-of-care (POC) biosensors have been proposed as a more cost-effective alternative for cancer diagnosis. These devices are fast, sensitive, portable, easy to use, and highly accurate, offering rapid response times due to their ability to directly analyze physiological fluids such as saliva, urine, blood, milk, and serum. Additionally, they can detect cancer cells even before symptoms manifest [11]. As illustrated in Fig. 3, a biosensor is a fully integrated analytical device that combines a biological recognition element—such as enzymes, antibodies, antigens, peptides, DNA, aptamers, or living cells—with a transduction element, which may be electrochemical, optical, or mechanical [12–14].

Among the various biosensing platforms, electrochemical biosensors—which integrate a biorecognition element with an electrochemical transducer such as an electrode or field-effect transistor (Fig. 4)—are particularly advantageous [14, 15]. Their design enables easy miniaturization, batch fabrication, and seamless integration with an electronic acquisition module on a single chip. Moreover, electrochemical signals, including electrical current and potential, can be captured using simple, portable, and cost-effective peripheral instruments that consume minimal power [16]. Techniques such as cyclic voltammetry (CV), differential pulse voltammetry (DPV), and chronoamperometry (i-t) can be employed to collect these signals [13]. In field-effect transistor (FET) biosensors, the solid–liquid interface plays a crucial role, as these sensors detect variations in surface potential ($\psi 0$) caused by the charges

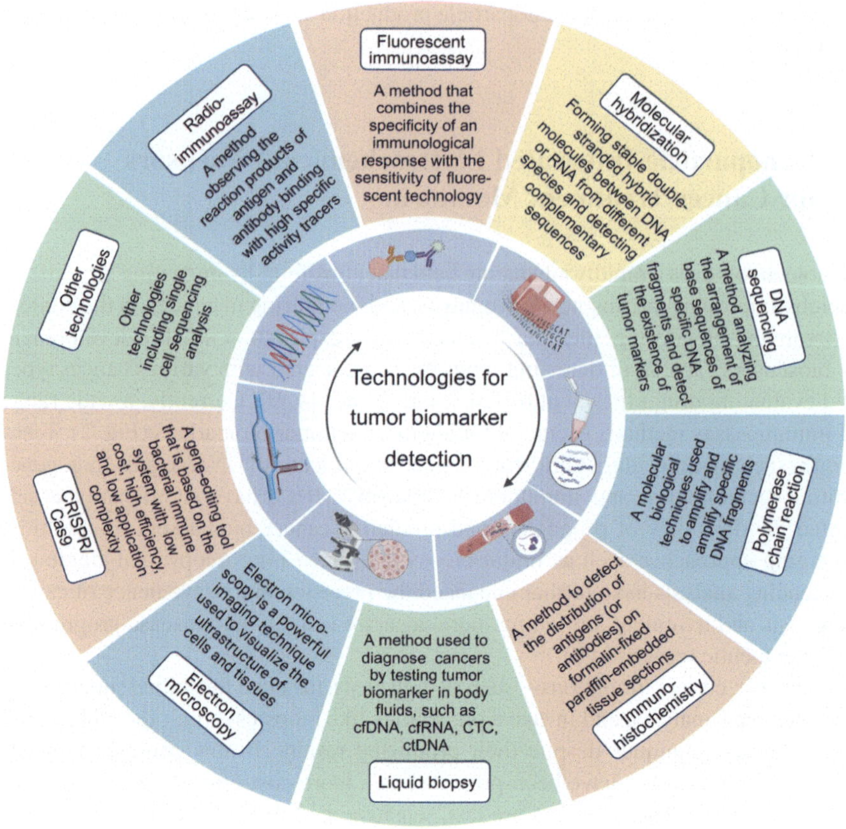

Fig. 2 Technologies for the detection of tumor biomarkers. Copyright permission from Ref. [10]

of analytes [15]. A FET consists of key components, including the gate, source, and drain [17]. When a voltage is applied to the gate, it alters the concentration of charge carriers in the semiconductor channel between the source and drain, thereby regulating the drain-source current (IDS). This ionic-electronic interaction induces a significant change in the channel's conductivity, which in turn results in a substantial modulation of the electronic current flowing from the source to the drain through the transistor channel [18].

Nanoparticles, typically with sizes ranging from 1 to 100 nm, exhibit unique physicochemical properties that make them promising tools for early-stage cancer diagnosis and therapy [5, 19, 20]. In recent years, nanoparticles have been extensively studied due to their wide range of potential applications, including nanosensors, data storage devices, nanoelectronics, optoelectronics, catalysts, magnetic recording media, drug delivery systems, and microelectronics. Additionally, nanotechnology research has introduced new methods for early cancer detection and treatment. Materials such as carbon nanotubes (CNTs), graphene, and gold nanoparticles (GNPs)

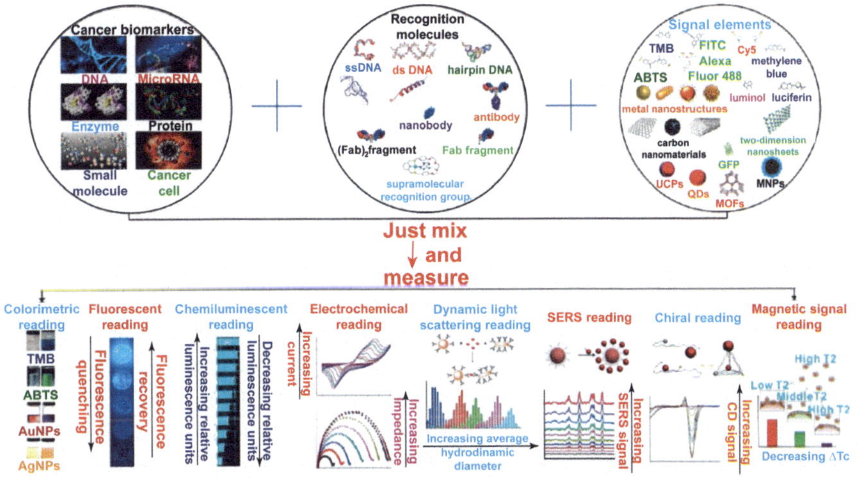

Fig. 3 Examples of possible systems applied to biosensing for cancer diagnosis. Used with permission from Copyright 2017 American Chemical Society [14]

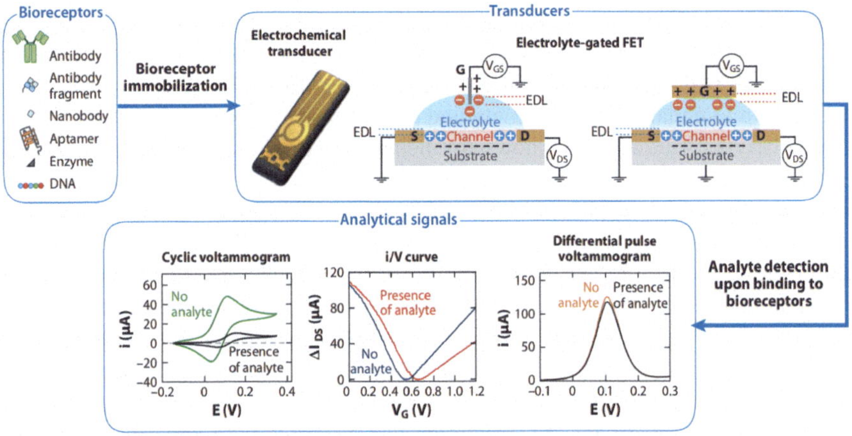

Fig. 4 Schematic design of an electrode-based electrochemical biosensor and a field-effect transistor (FET) electrochemical biosensor, along with their corresponding electrical responses. Adapted with permission from Ref. [18]

have garnered significant attention due to their applications in electrical biosensors [21]. Various devices have been explored by incorporating nanostructures into both synthetic and natural hydrogels. For instance, Mn_2O-based nanoparticles have been investigated in combination with carbonized polymer dot (PD)-encapsulated mesoporous MnO_2 (MnO_2@PD) incorporated into hydrogels for the electrochemical and fluorescence detection of glutathione (GSH) and hydrogen peroxide (H_2O_2). Elevated levels of GSH and H_2O_2 in the cancer microenvironment trigger the breakdown of

MnO_2 into Mn^{2+}, resulting in changes in conductivity and the release of PDs, which restore fluorescence. This mechanism allows for the differentiation between tumor and non-tumor cells upon contact with the sensor [22]. Additionally, the authors incorporated a wireless sensing system into the hydrogel, enabling real-time monitoring via a smartphone. Another approach involves embedding metal nanoparticles to develop conducting polymer hydrogels. One example is a hydrogel based on polypyrrole (PPy), nanocrystalline cellulose (CNC), and sodium benzenesulfonate (BSNa), coated with gold nanoparticles (GNPs) electrodeposited on the surface of an electrochemical immunosensor. This system (BSNa-CNC-PPy) was constructed on a glassy carbon electrode (GCE) for the detection of carcinoembryonic antigen (CEA) [23]. CEA is a biomarker associated with various types of cancer, and the sensor developed combines characteristics such as reliability, selectivity, stability, and reproducibility [25]. To evaluate early symptoms of non-small cell lung cancer (NSCLC) [24], a dual-modal aptasensor was proposed by integrating electrochemical and fluorescence detection within a single device. Gold nanoparticles (GNPs) were functionalized with exosome-capturing probes consisting of thiol-modified CD63 aptamers and immobilized onto screen-printed gold electrodes. Simultaneously, carboxylated CD63 aptamers were immobilized on the surface of PB-modified graphitic carbon nitride (g-C3N4) loaded with cobalt single-atom nanoparticles (Co-SANs) to form Co@g-C3N4@PB. This design created a sandwich structure (GNPs/Apt1/Exo/Apt2-Co@g-C3N4@PB), enabling the specific recognition of exosomes for the detection of non-small cell lung cancer (NSCLC). Such nanoparticle-based approaches have been shown to enhance the sensitivity and ease of biomolecule modification [25].

Similarly, FET biosensors, known for their exceptional electrical and mechanical properties, are emerging as a promising tool for early cancer detection. These biosensors have been widely studied for their ability to detect cancer biomarkers, including genes, antigens, and proteins, providing a more sensitive alternative to traditional diagnostic methods [17, 26]. Nanoparticles functionalized onto the graphene-based channel enhance the performance of biosensors by facilitating electron transfer to the channel and increasing the density of bioreceptor attachment, thereby improving the electrical signal [27]. In this context, field-effect transistors have been utilized for cancer detection, incorporating gold nanoparticles [28–30], quantum dots [31, 32], magnetic nanoparticles [33], platinum nanoparticles [34], and other materials [35]. Yu et al. developed an electrical and label-free method for directly detecting exosomes with high sensitivity based on a reduced graphene oxide (RGO) field-effect transistor (FET) biosensor. An RGO FET biosensor modified with the specific antibody CD63 in the sensing area was fabricated and used for the electrical and label-free quantification of exosomes [36]. Similarly, Liang et al. [37] proposed a carbon nanotube field-effect transistor (CN FET) biosensor that utilizes a floating gate (FG) composed of a 6 nm Y_2O_3 insulating layer. The Y_2O_3 layer is decorated with gold nanoparticles (Au NPs), which serve as effective linkers to facilitate the connection of biological probes.

2 Electrochemistry: Innovating Nanoparticle Synthesis and Future Detection of Cancer Biomarkers Using Laser-Induced Graphene

The use of lasers to modify the surface of polymeric materials has been a topic of research since the early 1990s. At that time, researchers were focused on developing methods to enhance the electrical conductivity of these materials without compromising their mechanical and thermal properties. Building on this goal, Schumann et al. [38] used a KrF ultraviolet laser (248 nm wavelength) to irradiate a polyimide film. Their experiments revealed the formation of dark, electrically conductive regions on the material's surface, with a conductivity approximately 15 times higher than that of the unirradiated polymer substrate [38].

These early findings spurred further research into laser-based techniques aimed at enhancing the conductivity of polymeric materials. Among these, the CO_2 laser has proven particularly effective in transforming natural carbon sources into highly conductive materials. The research group led by Tour et al. [39] was the first to pioneer this approach. By irradiating a polyimide film with a CO_2 laser, they successfully created conductive graphitic patterns in porous 3D layers on the surface of the insulating polymer. The materials produced by this method have been referred to as "laser-induced graphene" (LIG), "laser-scribed graphene" (LSG), or "laser-derived graphene" (LDG). Since this groundbreaking work, laser induction has become a widely adopted technique for producing graphene-based materials, with applications in supercapacitors [39–41], electrocatalysts [42], sensors [43–45], and biosensors [46, 47].

With the continued advancements in laser-based techniques, increasingly innovative applications have emerged. Researchers are now harnessing the precision and versatility of CO_2 lasers to create hybrid materials that combine graphene-based structures with the formation of nanoparticles. This synergistic approach significantly enhances the physical and chemical properties of the resulting materials, opening new possibilities for a wide range of applications.

Once again, the research group led by James Tour [48] was the first to demonstrate this significant advancement in laser techniques. In this pioneering work, the authors synthesized their polyimide (PI) film, allowing them to incorporate nanocrystal precursors (metal complexes) directly into the film's precursor, polyamic acid (PAA) (Fig. 5a). By varying the metal complex precursor mixed with the PI film, different metal oxide nanocrystals, such as Co_3O_4, MoO_2, and Fe_3O_4, could be selectively formed. When the CO_2 laser was scribed onto the PI film containing these metal complexes, crystalline nanoparticles were generated and embedded within a porous graphene matrix [48]. This process successfully produced metal oxide nanoparticle/graphene composites, with the average diameter of the nanocrystals ranging from 5.3 ± 1.3 to 14 ± 5.5 nm (Fig. 5b). After laser scribing, these printable films were annealed under argon and proved to be effective electrocatalysts in converting O_2 to OH^-.

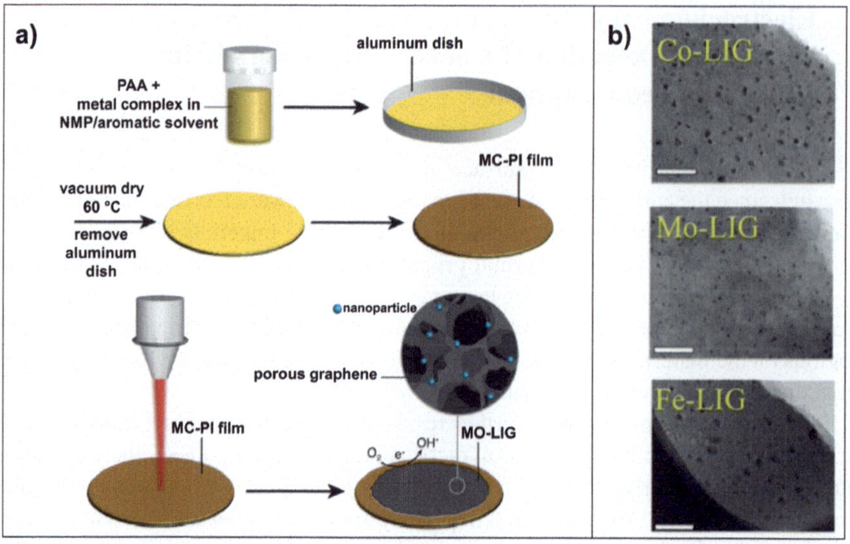

Fig. 5 **a** Schematic illustration of the formation of Metal oxide-Laser induced graphene (MO-LG) from metal-complex containing polyimide and **b** Transmission electron microscopy images of metal oxide nanocrystals of Co_3O_4, MoO_2, and Fe_3O_4. Imprinted with permission from [48]

The primary challenge with the strategy proposed by the Tour group lies in the necessity of a custom polymer formulation, which introduces additional complexity and costs. To address this issue, Clereci et al. [49] proposed an alternative approach to simplify the fabrication of these hybrid materials further. Their method involved spin-coating a mixture of 2 mL of a commercial dispersion of molybdenum disulfide (MoS_2) flakes (18 mg/L in an ethanol/water solution) with 1 mL of nonionic surfactant (Triton X-100) onto the surface of a commercial polyimide (PI) foil. When the CO_2 laser irradiated the modified surface, it simultaneously induced the graphitization of the polymer and facilitated the decoration of the resulting porous graphene network with MoS_2 nanoflakes, streamlining the overall process while maintaining the desired material properties. With this material, they created a high-performance electrode for flexible supercapacitors.

Building on these pioneering studies, a variety of substrates and applications have been explored for the fabrication of hybrid materials, revealing numerous potential uses. In this context, particular attention has been given to the development of sensors and biosensors, where these hybrid materials offer significant electrochemical enhancements. The following discussion will focus on how these materials enhance the performance of bio/sensing devices, highlighting the optimal conditions for forming the hybrid materials and the electrochemical signal amplification that can be achieved.

In 2020, You et al. [50] explored the use of graphene hybrid nanocomposites to develop an impedimetric immunosensor for detecting *Escherichia coli* O157. A PI film was employed as the substrate for this process. To begin, the hydrophilicity of the PI film was enhanced using a hydrophilic reagent, facilitating the subsequent deposition of nanoparticle precursors. The hydrogel ink was then applied to the prepared surface using the blade coating method. This hydrogel was formulated by mixing 5 mL of a 40 mg/mL chitosan solution (in 1.0% acetic acid) with 2 mL of an aqueous $HAuCl_4$ solution, where the concentration of $HAuCl_4$ was varied at 25, 50, 75, and 100 mM to assess the influence of precursor concentration on the average diameter of the resulting gold nanoparticles (AuNPs). These mixtures were stirred for 1 h to produce a metal precursor-chitosan hydrogel ink, which was then coated onto the PI film surface and dried at 50 °C for 10 min. The hybrid materials, consisting of LIG and AuNPs, were subsequently generated by directly laser-scribing the PI surface containing the hydrogel ink. During this process, the laser power was varied between 2.3, 2.75, 3.30, and 3.85 W, with a scan rate of 2.0 cm s^{-1}. The results indicated that the size of the AuNPs increased with higher laser intensities.

Additionally, when varying the scan rate at 3.0, 2.0, and 1.0 cm s^{-1} while maintaining a laser power of 2.23 W, the size of the AuNPs increased as the scan rate decreased. These findings demonstrate that the laser cutter parameters directly influence the size of the formed nanoparticles. The concentration of the $HAuCl_4$ solution also affected the AuNP size, with the smallest nanoparticles obtained at the lowest concentration. After optimizing these parameters, high-magnification SEM images revealed that the nanoparticles were uniformly and densely distributed on the LIG surface, with an average diameter of 10.1 nm (Fig. 6a). The authors subsequently prepared an AuNPs-LIG interdigitated array electrode, which was utilized to fabricate a biosensor (immunosensor) for detecting foodborne pathogens. Impedimetric measurements revealed that the electron transfer resistance (Ret) decreased as the AuNPs grew on the surface of the LIG electrodes, indicating that these hybrid materials have the potential to enhance the sensitivity of the immunosensor.

In a similar study, Zhao et al. [51] utilized a commercial PI film as the substrate to develop a hybrid material. Unlike the previous work that employed a hydrophilic reagent to enhance the wettability of the PI film, they achieved the same effect by roughening the surface with an ultraviolet laser. Following this, 1.5 mL of varying concentrations of silver nitrate ($AgNO_3$ in ultrapure water) was drop-cast onto the film and air-dried. The LIG-silver nanoparticle (LIG-AgNP) electrode was then created by laser engraving the modified PI surface using a laser power of 4.2 W, a speed of 50 mm s^{-1}, and a defocus distance of 10 mm. A control electrode was also fabricated by laser engraving the PI film without the $AgNO_3$ solution. The morphology of the AgNP-containing electrode was examined by SEM (Fig. 6b), revealing a clear dispersion of nanoparticles across the graphene nanosheets. As observed in previous work, the concentration of the $AgNO_3$ solution has a significant impact on nanoparticle formation. As the concentration increased, the AgNPs transitioned from uniform particles to coarse, rod-shaped structures, further demonstrating that a lower concentration is optimal for producing uniform particles. The authors explained that the formation of AgNPs under laser irradiation occurs through chemical reactions and particle nucleation. When the $AgNO_3$ layer on the PI film is exposed to laser irradiation, the high temperature causes the silver nitrate to dissolve and decompose, leading to the formation of silver clusters. These clusters then

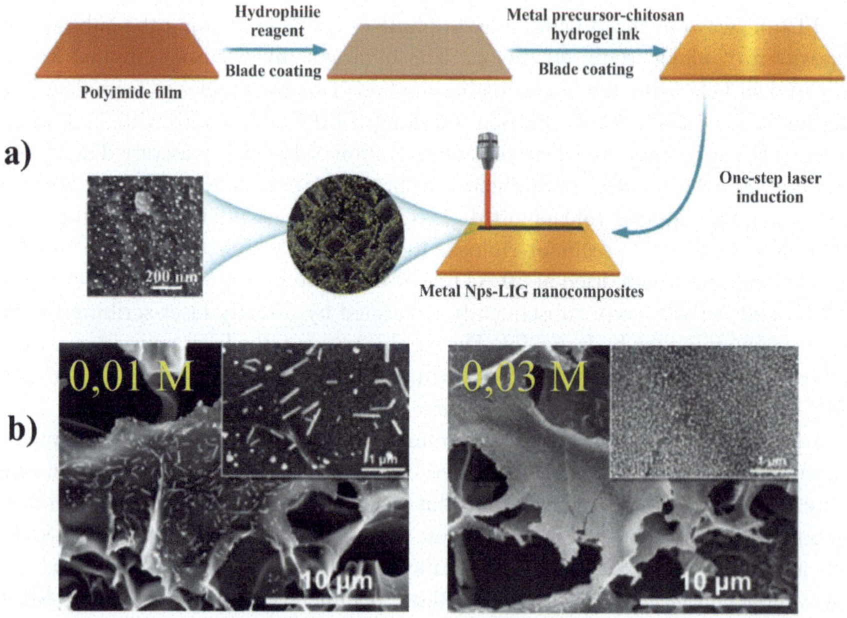

Fig. 6 **a** Schematic representation of the steps proposed by You et al. [13] to produce metal nanoparticle-LIG hybrid nanocomposites using the CO_2 laser; **b** Field emission scanning electron microscopy (FE-SEM) images of the LIG-silver nanoparticle (LIG-AgNP) obtained varying concentrations of silver nitrate solution—imprinted with permission from [51]

nucleate at the defect location on the LIG nanosheets. In the presence of concentrated silver vapor, the clusters aggregate to form nanoparticles. The AgNPs are uniformly anchored on the LIG surface, with their growth limited by the spatial confinement between the LIG layers, preventing excessive size increase. The performance of the LIG-AgNP hybrid electrode was compared to that of the LIG-only electrode. Cyclic voltammetry revealed a more pronounced reduction peak for hydrogen peroxide (H_2O_2), a crucial byproduct of various enzyme-catalyzed reactions in the human body, on the LIG-AgNP electrode, indicating the superior performance of the hybrid material and its high capacity for biosensing.

So far, we have explored the formation of nanoparticles on PI film through laser scribing, a process that, while effective, requires pre-treatment to improve the film's wettability for the deposition of metal precursors. A more versatile alternative is offered by using hydrophilic substrates, such as paper, which provide inherent benefits for material processing. Arantes et al. [52] demonstrated an innovative approach to fabricating a paper-based electrochemical device (ePAD) utilizing laser-scribed electrodes modified with AuNPs on kraft paper. Their process unfolds in two key stages. First, a CO_2 laser initiates pyrolysis, shaping the electrode geometry on the paper substrate. Following this, 30 μL of a 10 mmol L^{-1} $HAuCl_4$ solution is applied to the working electrode in three 10 μL increments, with each layer dried under an infrared lamp. The AuNPs are then synthesized by irradiating the $HAuCl_4$-coated surface with the CO_2 laser. They observed that the laser power and scan rate directly influence the electrode's electrochemical behavior, which the authors assessed using a $[Fe(CN)_6]^{3-/4-}$ redox probe. Through optimization, they identified the ideal conditions—a laser power of 8% and a scan rate of 6 mm s^{-1}. Unlike the behavior observed with PI film, where lower concentrations of metal precursors yielded better results, here, increasing the $HAuCl_4$ concentration enhanced the electrode's performance. This improvement stems from the formation of well-defined, nanometer-sized gold nanoparticles (AuNPs), which contribute to the electrode's superior electrochemical properties. The resulting Au-modified ePAD (LSAu-ePAD) exhibited a notable reduction in charge transfer resistance, a more reversible electrochemical response, and a remarkable 13-fold increase in cathodic and anodic peak currents, compared to the unmodified LS-ePAD. This fact highlights the influence of hybrid materials on electrode performance.

In a subsequent study, Paixão's research group [53] explored the incorporation of various metal cations onto the surface of kraft paper to promote nanoparticle formation. Their approach closely mirrored that of Arantes et al. (2023). The process began with the engraving of a three-electrode system on the paper surface. Following this, 10 μL of solutions containing different metal cations—$ZnSO_4$, $NiCl_2$, $CoCl_2$, $HAuCl_4$, $AgNO_3$, or $CuSO_4$ (each at a concentration of 100 μmol L^{-1})—were carefully deposited onto the WE. Once the solutions were applied, the electrodes were dried in an oven at 60 °C for 10 min to ensure a uniform coating. The process was then refined by a second laser scribing step in raster mode, using a power of approximately 14.1 W and a lateral speed of 2.64 mm s^{-1}. As expected, during this step, the metal cations were reduced, and the nanoparticles were formed, along with graphene-like material obtained by the carbonation of the paper. The SEM image demonstrated the versatility of the laser scribing method, enabling the formation of different nanoparticles that were well distributed throughout the surface of the electrodes. Veloso et al. [54] proposed a method combining 3D printing with laser scribing to improve the conductivity of electrodes in a time-efficient manner. One of the main challenges with 3D-printed electrodes is their poor electrical conductivity due to the insulating thermoplastic (PLA) mixed with conductive materials like carbon black (CB) in low ratios (10–20 wt%). While filaments with higher conductive content exist, their fabrication is complex and time-consuming. To overcome this, the authors used a CO_2 laser to create a nanostructured gold film directly on the 3D-printed CB/PLA electrode. After printing, the working electrode (WE) surface was pre-treated with the laser to expose the conductive material and improve wettability. Then, 10 μL of a 30 mmol L^{-1} $HAuCl_4$ solution was drop-cast onto the WE and dried under infrared light. Laser scanning was employed to promote the formation of gold nanoparticles (AuNPs). The study carefully evaluated the effects of laser power, focal distance, and beam scan rate on the electrode's electrochemical performance. The findings revealed that while laser power and scan rate had minimal impact on performance, powers exceeding 1.0 W caused material ablation. Focal distance, studied between 7 and 13 mm, proved to be a crucial parameter, with the highest current observed at a distance of 9 mm. This integrated method demonstrates how post-processing treatments, particularly laser scribing, can significantly enhance the conductivity of 3D-printed electrodes, paving the way for more efficient and functional electrochemical devices.

The fabrication of hybrid materials combining laser-induced LIG with metal nanoparticles through CO_2 laser scribing has emerged as a significant advancement in material science by introducing nanoparticle precursors into substrates such as polyimide. By directly scribing PLA paper with a laser and then scribing their surfaces, it is possible to create nanocomposites that integrate graphene with metal nanoparticles. This process not only enhances the structural and functional properties of the materials but also introduces novel functionalities. The resulting metal nanoparticle-LIG nanocomposites exhibit several key advantages, including large surface areas, high conductivity, and good biocompatibility. Additionally, these nanocomposites can be patterned into various shapes, offering versatility in design and application. The inclusion of metal nanoparticles significantly enhances the electrochemical catalytic activity of the materials due to their small size and large surface area [55]. Our understanding of laser-induced graphene (LIG) for nanoparticle synthesis is still developing, with only a few examples in the literature involving the deposition of nanoparticles on its surface. For instance, Torati et al. [56] developed an electrochemical biosensor to detect miRNA-141, a crucial biomarker for prostate cancer (PCa). This sensor utilized gold nanoparticles (AuNPs) deposited onto laser-induced graphene (LIG) electrodes. The LIG was patterned on polyimide (PI) film affixed to a polyethylene terephthalate (PET) sheet, and AuNPs were electrochemically deposited to increase the effective surface area and enhance electron transfer kinetics. Similarly, Lahcen et al. [57] introduced a novel biosensing platform using laser-scribed graphene (LSG) electrodes modified with nanostructured gold and molecularly imprinted polymers (MIP). This biomimetic sensing platform was designed to detect human epidermal growth factor receptor 2 (Her-2), a critical breast cancer biomarker. The combination of nanostructured gold and MIP in this system significantly improved the sensor's sensitivity and selectivity for Her-2 detection. In summary, the use of CO_2 laser scribing to create hybrid materials from LIG and nanoparticles has opened new avenues for developing advanced sustainable materials with enhanced electrochemical and catalytic properties. These innovations hold great potential for various applications, particularly in cancer detection, offering improved sensitivity, selectivity, and environmental compatibility.

3 Revolutionizing Cancer Imaging with Nanotechnology

Cancer remains a significant concern. While there have been considerable advances in treatment and diagnosis, we cannot afford to reduce our efforts in pursuing a deeper understanding and improving the quality of life for patients. Moreover, being more ambitious, we aim to find a cure. As mentioned above, the incorporation of nanotechnology into medicine, known as nanomedicine, has led to significant improvements in recent times, including lower detection limits for biomarkers, increased treatment efficacy, and reduced side effects, among other benefits [58]. In this context, Anselmo and Mitragotri [59] provided an excellent description of nanoparticles with medical applications that have already been approved by the U.S. Food and Drug Administration (FDA). Some examples of nanoparticles used in controlled drug release, imaging probes, and contrast agents include Patisiran/ONPATTRO® (the first FDA-approved RNAi therapeutic/nanoparticle developed and marketed by Alnylam), VYXEOSTM (two chemotherapy drugs/nanoparticle, developed and marketed by Jazz Pharmaceuticals), and NBTXR3/Hensify (a radio-enhancing nanoparticle that works synergistically with standard radiation therapy for oncology, developed and marketed by Nanobiotix) [59, 60].

Additionally, Ferumoxytol is currently the only FDA-approved superparamagnetic iron oxide nanoparticle (SPION) for treating iron-deficiency anemia in patients with chronic kidney disease. Beyond its approved use, Ferumoxytol is also used off-label as an MRI angiography agent, particularly for patients with renal failure who are unable to receive gadolinium-based contrast agents. It is essential to note that while iron oxide nanoparticles were the first nanomaterials reported for imaging applications, several formulations approved by the FDA or in Europe between 1993 and 2001 were unfortunately withdrawn due to concerns about toxicity [61].

Imaging techniques are the most employed methods for identifying a wide range of cancers based on various phenotypic characteristics associated with tumor tissues. These techniques are frequently used for screening, staging, and monitoring tumor progression due to their accessibility and noninvasive nature. The effectiveness of an imaging modality is primarily influenced by the growth rate of the solid tumor, which can be described using several mathematical models. Mathematical modeling of proliferation enables prediction of the spatiotemporal growth of tumors and treatment response for cancer [62]. Several models reported in the literature, including the Gompertz, exponential, logistic, surface, linear, Mendelsohn, and Bertalanffy models, are commonly used for this purpose [63, 64]. Traditional imaging techniques, as illustrated in Fig. 7, are employed for cancer detection, including computed tomography (CT), X-ray, magnetic resonance imaging (MRI), diffusion-weighted MRI [65, 66], MR elastography [67], MR spectroscopy [68], MR cholangiography [69], ultrasound (US), and endoscopy [70]. These methods produce detailed images of anatomical structures and are extensively used in clinical settings for tumor imaging [61].

Additionally, clinical molecular imaging (MI) techniques, such as positron emission tomography (PET) and single-photon emission computed tomography (SPECT), have made significant advancements in tumor detection and characterization. A growing number of FDA-approved small molecule and antibody-based radiotracers are now available, further enhancing these methods. The PET technique operates by detecting the radioactivity emitted by a tracer, with radiation levels comparable to those of CT scans. PET imaging is frequently used to measure glucose consumption in different areas of the body, utilizing the radiolabeled glucose analog 18-fluorodeoxyglucose (FDG) [71]. The image created by the PET scan is based on the Warburg effect, which states that glucose is metabolized more rapidly in malignant

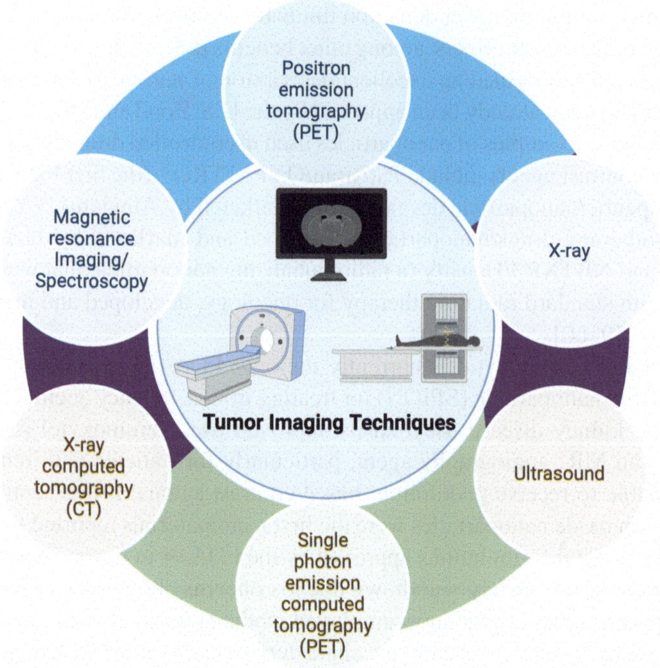

Fig. 7 Schematic representation of the standard imaging methodologies used for cancer detection

tumors compared to benign ones. This method is widely employed for whole-body scans to stage cancer. However, in cases such as pancreatic cancer, the sensitivity and specificity of 18F-FDG PET/CT are limited in detecting small, resectable, and curable lesions (around 2–3 mm). Additionally, the physiological uptake of glucose by normal and inflamed tissues can result in false-positive findings, which may lead to a similar appearance in conditions such as pancreatitis and pancreatic ductal adenocarcinoma [72].

Molecular imaging enables the noninvasive monitoring and measurement of biological processes, similar to a biopsy, but in real-time, offering the potential for sequential, longitudinal monitoring. However, despite these advancements, each of these techniques still has limitations in tumor imaging, including insufficient sensitivity in detection [61, 64]. Additionally, the high cost of diagnostic methods often restricts their accessibility to specific patient groups. Traditional diagnostic imaging techniques are particularly challenging for low-income patients and those in low- and middle-income countries, mainly due to their cost and inadequate medical infrastructure [64]. Many people worldwide live in resource-poor regions and, therefore, have limited access to conventional diagnostic tools. Delays in diagnosis are a key factor contributing to lower survival rates. To address this, some countries have introduced clinical guidelines aimed at standardizing the diagnostic process, clearly outlining the timeframe from diagnosis to the start of treatment, and enhancing clinical outcomes in cancer care [73].

Some of the key limitations include:

- PET (Positron Emission Tomography) is limited by its low spatial resolution, challenges in detecting small cancerous lesions, and the requirement for radioactive compound administration, which increases radiation exposure [64].
- CT (Computed Tomography): Constrained by radiation exposure and the lack of functional information [64].
- MRI (Magnetic Resonance Imaging): This procedure is contraindicated for patients with specific internal or external devices. It involves lengthy steps and may cause discomfort for patients with claustrophobia [64].
- MRS (Magnetic Resonance Spectroscopy) faces challenges such as lengthy procedures, high costs, and a lack of anatomical detail [64].

Accurate cancer diagnosis leads to more effective treatments, improves quality of life by minimizing treatment side effects, and advances the potential for a cure. Therefore, it is essential to enhance sensitivity and address the limitations of current diagnostic methodologies.

When selecting agents for diagnostic imaging techniques, they should possess key characteristics, including high specificity, strong binding affinity, low toxicity, stability (e.g., resistance to plasma enzymes), rapid clearance from non-target tissues, affordability, and approval for clinical use. The radionuclide must also have an appropriate half-life, depending on the intended application. Additionally, properties such as molecular size, charge, specific activity, lipophilicity, stability, and metabolism of the radiolabeled compounds are crucial in determining the specificity of each biological target [71]. Compared to traditional contrast agents, the extended blood circulation time of nanoparticle-based contrast agents is a key factor in enhancing their contrast signal. Modifying nanoparticles to increase circulation time is, therefore, essential for optimizing imaging performance. The most common approach involves encapsulating hydrophobic nanomaterials in a polyethylene glycol (PEG) shell, which significantly improves their solubility and extends their time in the bloodstream [74]. Nanoparticles exhibit unique characteristics—including their composition, structure, shape, and functionality—that make them highly suitable for specialized applications, such as cancer imaging probes, diagnostic tools, and therapeutic agents. Their large surface area enables the effective encapsulation of a significant number of molecules. Furthermore, nanoparticles can be functionalized with various biological substrates, such as RNA, DNA, aptamers, and antibodies, thereby enhancing their potential for theranostic (therapeutic and diagnostic) applications [19].

4 Nanoparticles: A Promising Tool in Biomedical Imaging

To overcome the primary limitation of current imaging techniques, such as X-rays, MRI, and CT scans—namely, their inability to detect small clusters of malignant cells in primary or metastatic sites—quantum dots have been explored as biofunctionalized labels for cancer imaging [75]. Quantum dots (QDs), often referred to as "artificial atoms," have become a key focus in cancer nanomedicine. These semiconductor nanocrystals exhibit unique quantum size effects in their optical properties due to the confinement of charge carriers in all three spatial dimensions [75]. Composed of a relatively small number of atoms (ranging from 100 to 10,000) from semiconductor materials belonging to groups II–VI, III–V, and IV–VI of the periodic table, their diminutive size gives rise to the term "dots," while "quantum" refers to their behavior being governed by quantum mechanics. QDs are considered zero-dimensional nanomaterials because the charge carriers are tightly confined in all three directions [75, 76]. With ultrasmall sizes ranging from 2 to 10 nm, these fluorescent nanomaterials can efficiently accumulate at target sites via the enhanced permeability and retention (EPR) effect while also being cleared from the body through renal excretion [77, 78]. In addition to these features, QDs exhibit superior signal brightness, the ability to simultaneously excite multiple fluorescent colors, and size-dependent emission properties [75]. Moreover, the surface modification of quantum dots (QDs) further enhances their potential in cancer imaging by minimizing toxicity [79]. Some of the well-known quantum dots include carbon quantum dots, selenium quantum dots, silver quantum dots, silicon quantum dots, and gold quantum dots [80]. Fluorescent graphene quantum dots (GQDs) are readily internalized by tumor cells, enabling effective bioimaging [81]. GQDs have been reported for use in both in vitro (tumor cell imaging, targeted cell imaging, and general cell imaging) and in vivo imaging applications [80]. CQDs modified with $-NH_2$, –COOH, –CHO, and $-N(CH_3)_2$ groups have been demonstrated to exhibit low toxicity at concentrations up to 200 $\mu g\ mL^{-1}$; however, CQDs modified with –OH groups have exhibited some cytotoxicity at concentrations around 100 $\mu g\ mL^{-1}$ [82].

Badrigilan et al. [83] developed a theranostic probe that combines superparamagnetic iron oxide (SPIO) and bismuth oxide (Bi_2O_3), both of which are coated with graphene quantum dots (GQDs). The proposed nanoparticles exhibit strong near-infrared (NIR) absorbance, providing high-contrast dual-modality imaging that enhances both CT and MR dual-modal imaging-guided photothermal therapy (PTT) for early cancer treatment. Additionally, a dual-mode bioimaging and chemo-photothermal therapy nanoprobe was proposed by Xuan et al. [84]. The probe, consisting of 50 nm gold nanosphere clusters coated with graphene quantum dots (Au@GQD) and functionalized with folic acid, was loaded with doxorubicin. The authors reported promising results for this combined therapeutic approach.

Silica nanoparticles have been utilized in cancer imaging diagnostics. Among them, mesoporous silica nanoparticles (MSNs) have attracted growing interest due to their well-defined pore size and structure, high surface area, excellent biocompatibility and biodegradability, ease of surface functionalization, and stable dispersions in aqueous solutions [85]. Vandghanooni et al. [86] reviewed advancements in aptamer-conjugated MSNs for enhanced cancer therapy and imaging, highlighting significant achievements despite several key challenges. Although aptamers are considered promising candidate ligands for targeted cancer therapy, they face limitations such as vulnerability to nuclease enzymes, which can compromise their overall efficacy. Building on the potential of silica nanoparticles, Chung and coworkers [70] developed targeted near-infrared (NIR) fluorescent silica nanoparticles (FSNs) and optimized their size to 50–200 nm, functionalizing their surface with an antibody against carcinoembryonic antigen (CEA) to create CEA-FSNs. These nanoparticles demonstrated stronger binding to HT29 cells, with smaller particles exhibiting more efficient internalization than larger ones. As shown in Fig. 8, the authors illustrated that polyps were detected using white-light endoscopy, with NIR fluorescence signals observed in excised tissues after topical application of CEA-FSNs. Immunofluorescence analysis of tissue sections suggested that these particles could serve as key molecular imaging agents for the early detection of colorectal cancer.

Additionally, the sentinel lymph node (SLN)—the first lymph node that a primary tumor may metastasize through lymphatic vessels—was successfully detected using silicon nanoparticles. Traditionally, tracers such as indocyanine green (ICG), methylene blue (MB), and ^{99m}Tc-radiocolloid are administered to label SLNs and guide lymph node resection. However, these tracers have notable drawbacks, including long-term color residue, radiation damage to tissues, and poor accumulation in SLNs. More critically, none of the current SLN tracers can distinguish metastatic SLNs from normal ones. The authors introduced SiNPs-based exosome (SiNPs@EXO) probes for detecting metastatic SLNs by loading silicon nanoparticles into cancer

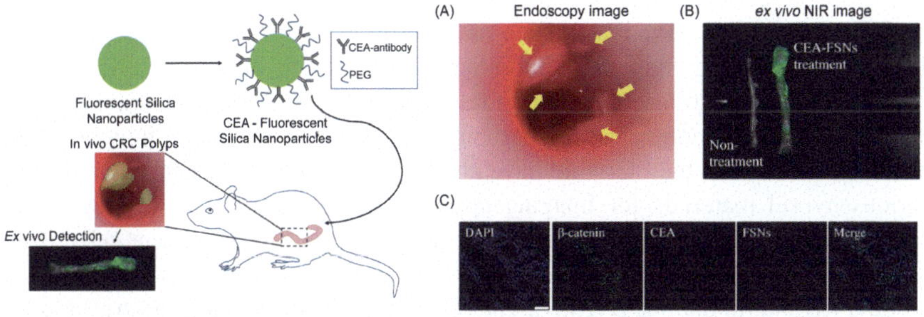

Fig. 8 The specific accumulation of CEA-FSNs in PIRC rats through macroscopic and microscopic imaging of excised tissues. **a** Ab-FSNs were applied topically to polyps (indicated by yellow arrows) observed via white-light endoscopy in the rat colon. **b** Following two PBS washes, a NIRF signal was detected in the excised intestinal tissue. **c** The targeted binding of CEA-FSNs to the polyps was further confirmed in tissue sections using immunofluorescence imaging. The NIRF signals from CEA-FSNs (purple) aligned closely with those from a CRC marker (β-catenin, green) and CEA (red) (scale bar = 100 μm). Reprinted with permission from Ref. [70]

cell-derived exosomes. They reported that the prepared SiNPs@EXO could differentiate between normal and metastatic lymph nodes via fluorescence imaging. Furthermore, the SiNPs@EXO tracers have been proven safe in both in vitro and in vivo studies [87].

Moreover, metallic nanoparticles, particularly gold nanoparticles, play a significant role in diagnostic imaging. Gold is biologically inert and chemically stable; therefore, it is widely explored for its biocompatibility [88]. Gold nanoparticles (GNs), like many precious metals, exhibit a distinctive optical characteristic known as surface plasmon resonance (SPR), enabling their use in near-infrared (NIR)-resonant biomedical imaging techniques such as magnetic resonance imaging (MRI), photoacoustic imaging (PAI), fluorescence imaging, and X-ray scatter imaging. Additionally, GNs produce heat when exposed to NIR laser light, making them ideal for photothermal therapy in cancer treatment. They are inherently low in toxicity and non-immunogenic [89]. Gold nanoparticles exhibiting quenching properties are crucial for fluorescence imaging, particularly when the need arises to toggle fluorescence between "on" and "off" states. Gold nanoparticles, ranging in size from a few nanometers to several hundred, demonstrate outstanding fluorescence quenching efficiency, making them highly valuable for this purpose.

Additionally, AuNPs offer numerous other advantages, including photostability, biocompatibility, flexible synthesis, adjustable sizes, strong light-scattering capabilities in dark-field microscopy, and ease of surface modification [19, 88]. GNPs can scatter light across their plasmon wavelength ranges, producing radiation intensities 10^5–10^6 fold higher than those of traditional fluorochromes. Recently, various GNP contrast agents featuring diverse shapes, such as nanocubes, nanospheres, nanoshells, nanorods, and nanocages, have garnered significant attention for their optical imaging capabilities in pancreatic cells and tissues. The optical properties of GNPs arise from phenomena such as absorption, scattering, or luminescence. Their light-scattering characteristics enhance the imaging of cancer cells through various optical technologies, including reflectance confocal microscopy, dark-field microscopy, optical coherence tomography, and photoacoustic (PA) imaging [90]. Numerous studies in the literature have employed modified gold nanoparticles and various hybrid materials for imaging applications. Hejazi and colleagues developed a nanomaterial-based bioimaging method to detect MCF-7 cancer cells using hyaluronic acid (HA)-functionalized gold nanoparticles stabilized with polyetherimide (PEI) and rhodamine B (RhB). The HA used in the functionalization allows for specific binding to CD44-overexpressing cancer cells through the HA/CD44 interaction, which has a high formation constant. The AuNPs/PEI/RhB-HA materials exhibited promising properties, demonstrating high physical stability and significant quantum yield, making these nanoparticles excellent candidates for in vivo cancer diagnostic applications [91]. Maximenko et al. encapsulated Fe_3O_4 cores within continuous SiO_2 shells, preserving their structural integrity and strong magnetic properties while allowing the shell surface to be coated with small gold nanoparticles (GNPs) and cisplatin (cPt). Cisplatin is a platinum-based coordination compound with a square planar geometry, known for its potent anticancer efficacy against various tumors, and is widely used in chemotherapy. The engineered nanoparticles exhibit remarkable T_2 contrast properties for magnetic resonance imaging (MRI). The functionalized $Fe_3O_4@SiO_2@GNPs$, measuring 40 nm in diameter, were evaluated for their potential in photothermal cancer therapy. The authors observed an enhancement in the cytotoxic effects of $Fe_3O_4@SiO_2@GNPs$ due to the synergistic interactions among cPt, GNPs, and laser irradiation, positioning these nanoparticles as promising agents for MRI-guided chemo-photothermal cancer treatment [92]. Although various studies in the literature reveal that numerous gold-based theranostic platforms have been developed, primarily focusing on animal models, only a limited number show considerable promise for clinical cancer therapy [93].

Magnetic nanoparticles (MNPs) mark a significant breakthrough in biomedicine thanks to their nanoscale dimensions and intrinsic ability to respond to external magnetic fields. These nanoparticles, synthesized from pure metals such as iron (Fe), cobalt (Co), and nickel (Ni), as well as certain rare earth metals or composed of a combination of metals and polymers, are utilized across a diverse array of applications. These include magnetic hyperthermia for cancer treatment, MRI-guided drug delivery, cell tracking, tissue engineering, bioseparation, and contrast enhancement in magnetic resonance imaging (MRI). Their versatility and precision render them indispensable tools in contemporary medical diagnostics and therapies [94, 95].

Magnetic resonance imaging (MRI) continues to face challenges due to its inherent low sensitivity, which limits its effectiveness in diagnosing early-stage diseases and detecting small lesions. In MRI, the relaxation time of water protons is measured, and contrast agents can modify local magnetic fields, accelerating proton relaxation processes to enhance MRI sensitivity and diagnostic accuracy. The diagnostic resolution of MRI can be significantly enhanced using contrast agents, which improve image contrast by shortening either the longitudinal (T1) or transverse (T2) relaxation times of water protons. T1 contrast agents enhance the T1 signal in T1-weighted imaging, producing a positive or brighter contrast, whereas T2 contrast agents decrease the T2 signal in T2-weighted imaging, resulting in a negative or darker contrast [96]. T1 contrast agents, typically composed of paramagnetic ions, preserve magnetic homogeneity and anatomical detail, allowing for precise detection in imaging. T1 agents—particularly gadolinium (Gd) complexes—are more frequently used in clinical practice. Free Gd ions are cytotoxic and tend to accumulate in the liver, spleen, and bones. To mitigate this toxicity, gadolinium (Gd) is subjected to a chelation process, where large organic molecules form a stable complex around the ion. This chelation significantly reduces the risk of Gd-induced toxicity. The resulting stable complex is primarily excreted through the kidneys [97]. In this context, iron oxide nanoparticles (IONPs) are particularly noteworthy for their magnetic properties and biodegradability. IONPs have been extensively studied as T2 contrast agents in MRI due to their ability to shorten transverse relaxation times efficiently. Notably, IONPs offer long blood circulation times, low toxicity, and versatile surface chemistry, making them highly suitable for various biomedical applications. Furthermore, IONPs are naturally cleared from the body via a degradation pathway, ultimately being metabolized into hemoglobin [96, 98]. To address the limitations of MRI, Lu et al. [98] developed a novel contrast agent, $SPIO@SiO_2@MnO_2$ (SSM), illustrated in Fig. 9, which exhibits reduced T1 and T2 contrast intensity under normal physiological conditions. In the acidic environment of tumors or inflamed tissue, the manganese dioxide (MnO_2) layer decomposes into magnetically active Mn^{2+}, enhancing T1-weighted signals and sequentially recovering both T1 and T2 signals. Additionally, the quenching and activation of T1 and T2 contrast can be precisely controlled by the silicon dioxide (SiO_2) intermediate layer situated between the SPIO core and the MnO_2 layer, providing enhanced regulation of imaging accuracy. The authors reported that the proposed SSM agent was successfully utilized to trace small liver metastases, approximately 0.5 mm in diameter, and effectively monitor tissue inflammation [98].

As discussed in the previous paragraphs, current imaging modalities for cancer diagnosis often lack the necessary accuracy, highlighting the urgent need for more

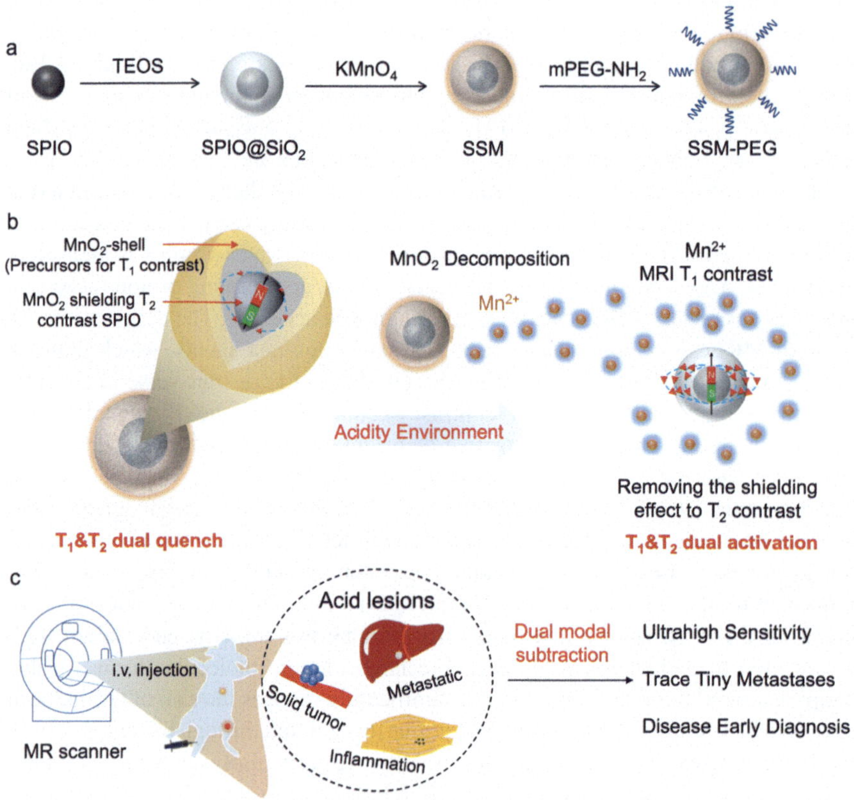

Fig. 9 **a** The author developed a contrast agent, $SPIO@SiO_2@MnO_2$ (SSM), where SPIO stands for superparamagnetic iron oxide and TEOS for tetraethyl orthosilicate. **b** In normal physiological conditions, $SPIO@SiO_2@MnO_2$ exhibits weak T1 and T2 contrast, as the MnO_2 layer suppresses the T2 signal of SPIO. However, in the acidic environment of tumors or inflamed tissues, the MnO_2 layer decomposes into magnetically active Mn^{2+} (T1-weighted), restoring both the T1 and T2 signals. **c** The use of dual-contrast-enhanced subtraction imaging (DESI) further improves imaging quality for disease diagnosis—imprinted permission from Ref. [98]

advanced imaging tools. Nanoparticles (NPs) present remarkable opportunities for enhancing imaging, screening, and early diagnosis in cancer patients. A diverse array of organic and inorganic nanoparticles, including liposomes and metallic nanoparticles, have significantly advanced imaging methodologies for cancer detection. These nanoparticles can be tailored with various biomolecules, such as DNA, aptamers, and antibodies, which improve their targeting accuracy and biocompatibility. Despite these advancements, challenges remain in mitigating immune responses and addressing potential adverse side effects and toxicity. Therefore, it is essential to intensify research efforts focused on optimizing these nanoparticle systems. Literature has highlighted promising results that demonstrate the potential of these techniques to improve detection limits. While some nanoparticles are already commercially available and widely used as contrast agents, ongoing research is crucial to enhance their capabilities further and ensure their practical application in clinical settings. Continued exploration in this field holds great promise for revolutionizing cancer detection and treatment strategies [99, 100].

5 Biogenic Nanoparticles: A Promising Tool for Biocompatible, Sustainable, and Eco-Friendly Synthesis

Metal nanoparticles have been utilized for centuries due to their unique and versatile properties, making them invaluable across a wide range of applications [1]. As previously mentioned, these nanoparticles are employed in various technological fields, including chemistry, healthcare, and environmental sciences. Nanoparticle synthesis techniques are generally classified based on the origin of the materials being converted to the nanoscale [2]. The methods that reduce bulk materials into smaller particles through physical or chemical means are classified as "Top-Down" approaches. Examples of these techniques include electrospinning, laser ablation, sputtering, lithography, and mechanical milling [2].

In contrast, "Bottom-Up" methods involve assembling elemental or molecular precursors to form nanoparticles. Common examples of these techniques include hydrothermal synthesis and chemical vapor deposition (CVD) [2, 3]. Figure 10 illustrates the key differences between the Bottom-Up and Top-Down methods discussed. Beyond these classifications, the literature also categorizes synthesis techniques as physical, chemical, and biogenic approaches (often referred to as green or biological), as depicted in the figure. Physical methods offer advantages such as producing large quantities of particles with high purity; however, they tend to be costly and energy-intensive. Chemical synthesis methods are generally simpler and can yield high outputs, but they often involve toxic reagents, thereby increasing safety and environmental risks [2, 101]. Scientists and engineers have long drawn inspiration from nature, a practice that persists as they strive to replicate and even enhance the remarkable properties exhibited by biological systems [102]. In this context, green

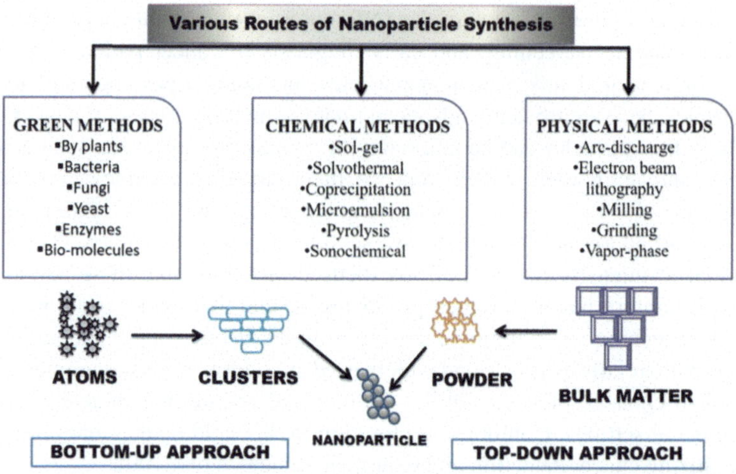

Fig. 10 Examples of "bottom-up" and "up-down" synthesis methods. The figure illustrates these two approaches for nanoparticle synthesis, along with various techniques for physical, chemical, and green synthesis methods. Reprinted with permission from Copyright @ Singh et al. [103], 2021 (BY-NC license)

synthesis methods are straightforward, safe, and non-toxic, making them highly advantageous across various fields. Green synthesized nanoparticles play a significant role in biomedical and environmental applications due to their remarkable stability, high yield, enhanced biocompatibility, and reduced toxicity.

Additionally, as the focus on sustainable development grows, there is increasing interest in biological synthesis because it minimizes the use of harmful chemical agents and conserves raw materials [101–103]. The "biogenic" refers to the utilization of biological elements—such as plants, bacteria, or fungi—as reducing and stabilizing agents during nanoparticle synthesis. Figure 10 illustrates the key stages of the bioreduction process [103]. Ahmad Bhat et al. [102] offer a comprehensive overview of the synthesis of biogenic nanoparticles and their application in breast cancer treatment, highlighting the promising potential of these methods for future therapeutic interventions.

In some cases, two components are used in the synthesis: the metal cation salt and the biological material, which can come from plants, algae, bacteria, or fungi (Fig. 11). The combination of these precursors promotes the reduction of the metal salt, leading to the nucleation process, in which the reduced atoms group together, followed by the growth process, in which there is an increase in concentrated atoms due to the nucleation effect. During these processes, the action of various bioactive components, such as alkaloids, phenolic compounds, ketones, vitamins, and carbohydrates (generally obtained from extracts), promotes the stabilization of the nanostructure [104, 105]. It should be noted that these processes are governed by parameters such as reaction time and temperature, pH, and concentration of the precursors mentioned [104].

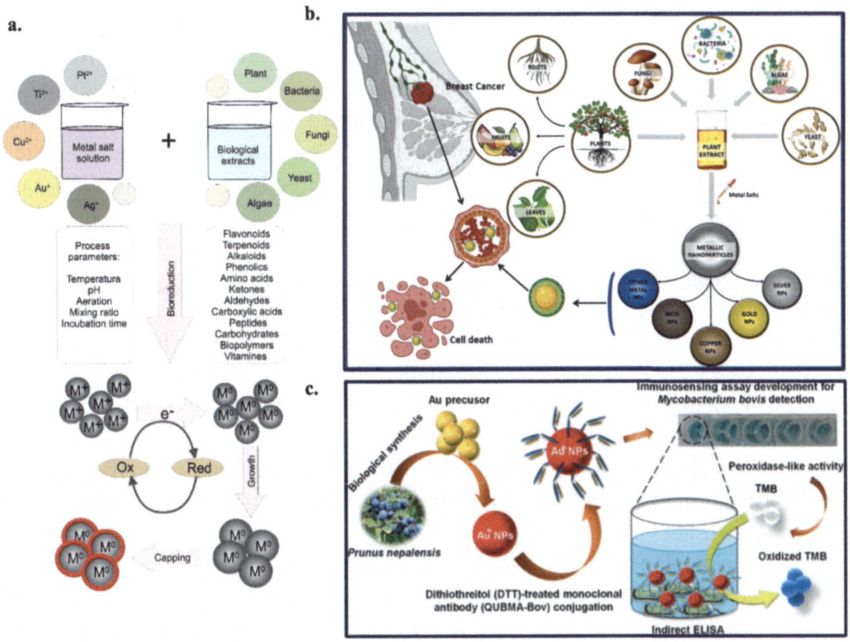

Fig. 11 **a** This schematic diagram illustrates the mechanisms involved in the synthesis of biogenic nanoparticles (BNPs), highlighting the presence of various reagents in the reactions. Specifically, it depicts the interactions between metal salt solutions and biological extracts. Reprinted with permission from Copyright @ Nikolova et al. (2023) (CC BY license). **b** The green synthesis of various metallic nanoparticles using different biological agents highlights their potential activity against breast cancer—imprinted permission from Ref. [102], **c** Additionally, biogenic nanoparticles have been utilized in biosensors for immunosensing assays aimed at detecting *Mycobacterium bovis*. Reprinted with permission from Ref. [105]

In addition, by using bioactive compounds as key elements in reducing and stabilizing the NPs, these syntheses have an eco-friendly approach. It could also be a field of study regarding the reuse of food waste on different scales as a source of bioactive compounds for synthesis [105]. Characteristically, biogenetically synthesized NPs possess a high surface area, catalytic activity, good dispersibility, conductivity, biocompatibility, and optical properties (Fig. 11b, c). Furthermore, biological precursors used in the method add biocompatibility, which reduces toxicity and preserves the effects promoted by the NPs [106, 107]. The versatility of NMs is already well established, and there is a wide range of applications for BMNPs, including the Internet of Things, agriculture, antimicrobial agents, environmental applications, bioimaging, and biomedical fields [103]. About cancer treatment, studies on the anticancer action of BMNPs are recurrent. For example, the use of *Carissa carandás* and *Euphrasia officinalis* leaves as sources of extracts for synthesizing AgNPs has been studied, and, in general, they exhibit a spherical morphology and have been shown to act on different types of cancer. In the first case, the use of AgNPs demonstrated therapeutic action against renal cancer pre-induced by diethylnitrosamine (DEN) in an in vivo model (albino Wistar rats) [108].

In addition, the studies conducted by Singh and collaborators in 2017 demonstrated a rapid and straightforward method for synthesizing Ag and AuNPs using aqueous extracts of *E. officinalis* L. The effects studied are antimicrobial (good performance) and, as the focus of this chapter, anticancer. For AuNPs, this effect is pronounced for human cervical cancer cell lines (HeLa). For AgNPs, anticancer effects have been observed in HeLa and human lung cancer cell lines (A549) [109].

In addition to leaves, other plant organs can be used to extract bioactive compounds. For example, the fruits of *Annona squamosa* L. have been studied for the synthesis of AgNPs, which, when applied to prostate adenocarcinoma (PC-3) and ovarian adenocarcinoma (SKOV3) cell lines, showed a cytotoxic effect on these cancer cells [110]. For gastric cancer cell lines (AGN), the anticancer effect is also observed, and altering the homeostasis of cancer cells has been observed, leading to apoptosis upon exposure to AgNPs obtained from Jamzad (*Satureja rechingeri*) aqueous extract [111].

In addition to plant sources, bacteria are used to produce MNPBs. For example, Zaki and collaborators produced AgNPs using filamentous cyanobacteria (*Plectonema*) with $AgNO_3$ as the source of silver. The results of the AgNPs are available for the A549 cell line, indicating cytotoxicity for this cell line while not affecting peripheral blood mononuclear cells (PBMC), a sign of safety for use [14]. Another successful approach is the use of the aqueous extract of algae (*Sargassum wightii*) for the biosynthesis of MgO nanoparticles (MgONPs). These NM improve apoptosis in A540 cells [15]. Considering the variety of metals and oxide metals used to obtain such NMs, studies have already been reported with CuONPs, ZnONPs, TiO2NPs, FeNPs, and others [5]. Approaches for CuONPs demonstrate efficiency in studies using carcinoma cells, highlighting the effect of inhibited growth as the generation of reactive oxygen species (ROS) occurs in the presence of the NPs [16]. In this work, nanoparticles (NPs) are obtained from bean extract and $CuSO_4$ using the hydrothermal method. Similarly, in Kabir and collaborators' work, oxidative stress was also a key parameter involved in the anticancer effect observed in the human breast cancer cell line (MCF-7), which was associated with the activation of the FAS pathway leading to apoptosis [17]. The NMs studied by these authors were silver/silver chloride NPs (Ag/AgCl-NPs) obtained using *Zizyphus mauritiana* fruit extract (prepared with Tris–HCl) and $AgNO_3$ salt from synthesis precursors.

As noted in the examples above, although biogenic nanoparticles have primarily been focused on anticancer applications [112], they also hold significant potential in electrochemistry. For instance, biogenic nanoparticles have been utilized to modify electrodes for studying electron transfer reactions and for capacitor construction [107]. They have been synthesized using propolis extract to create gold nanoparticles for the detection of As(III) [113], *Phyllanthus emblica* leaf extract for producing Mn_3O_4 nanoparticles for urea detection [114], and Pt–Ag bimetallic nanoparticles derived from the *C. longa* plant using a green synthesis method for dopamine determination [115]. Despite their promising properties, the application of these compounds in electrochemical biosensors for detecting cancer biomarkers remains largely unexplored. In the context of imaging diagnostics, Haque et al. [116] synthesized biogenic silver nanoparticles using the leaf extract of *Zinnia elegans* (ZE), chosen for its ability to emit fluorescence in the near-infrared (NIR) region. The authors reported that AgZE demonstrated biocompatibility and selective anticancer activity against various cancer cell lines, established through multiple assays. The fluorescence properties of AgZE were observed in the NIR region (excitation: 710 nm, emission: 820 nm) in both normal and tumor-bearing C57BL6/J mice.

Additionally, the biodistribution of silver in AgZE was analyzed across various vital organs using inductively coupled plasma optical emission spectrometry (ICP-OES). These results suggest that AgZE has the potential to be an effective cancer therapeutic and a NIR-based noninvasive imaging agent in the future. Biogenic magnetic nanoparticles also show promise in cancer treatment and imaging diagnostics [117]. These nanoparticles offer biodegradable properties and tunable surface modifications, making them particularly useful for biomedical applications such as in vivo magnetic resonance imaging (MRI), magnetic hyperthermia for cancer treatment, and tissue-specific delivery of therapeutic agents [117].

Schuerie et al. [118] reported the first use of magnetic nanoparticles displaying genetically encoded pH-low insertion peptide (pHLIP), a peptide designed to enhance MRI contrast by targeting the extracellular acidity associated with tumors. The authors demonstrated selectivity in vivo, with functionalized magnetosomes accumulating at tumor sites and producing measurable reductions in T2 relaxation times. The pHLIP may serve as a key component in customized peptide combinations on magnetosomes, acting as tumor-selective T2 contrast agents. Biogenic nanoparticles hold tremendous potential as noninvasive tools for treating and monitoring various diseases, including cancer, marking a growing trend in this field. This progress is tied to advancements in synthesis methods and an increasing focus on pharmacokinetic and pharmacodynamic studies to understand better the biocompatibility of these nanostructured platforms [109, 119]. Furthermore, innovations in targeted drug delivery systems using biogenic nanoparticles are promising, as they enhance the distribution of therapeutic agents, aiming to increase the effectiveness against cancer cells while minimizing harm to healthy tissue [120].

6 The Promising but Challenging Path of Nanoparticles in Cancer Imaging and Treatment Monitoring.

Nanoparticles have undeniably revolutionized medicine, leading to significant advancements in both anticancer treatments and early detection strategies. In cancer therapy, they play vital roles in drug delivery, encapsulation, targeted radiopharmaceuticals, contrast agents, and enhancing the sensitivity of electrochemical biosensors. However, the path toward effective cancer detection through imaging remains a significant challenge. The long-term effects of contrast agents need careful management to improve signal clarity in imaging systems while minimizing adverse impacts on therapeutic outcomes. Optimizing the properties of nanoparticles is essential for enhancing image sensitivity and resolution. Biocompatibility and purification are critical issues that must be addressed despite the encouraging results from innovative biosynthesis methods utilizing natural sources such as plants, fungi, microorganisms, and fruits. However, these biogenic nanoparticles are sustainable and offer potential for large-scale commercialization, providing cost-effective and readily available alternatives to traditional synthesis methods [121]. However, they may harbor impurities or contaminants, necessitating rigorous purification for specific applications.

Additionally, the costs associated with large-scale production and economic feasibility remain significant obstacles to clinical application. The properties of nanoparticles can vary widely based on factors such as source material, extraction methods, and environmental conditions, posing challenges for standardization and quality control. Enhancing the stability and production of biogenic nanoparticles can be achieved by employing genetically engineered microbes, which offer better control over the synthesis of enzymes, proteins, and macromolecules. However, this approach may also increase production costs. Continued efforts to develop new synthesis methodologies and deepen our understanding of the biocompatibility and immunological response mechanisms associated with these nanomaterials are essential. While research into biogenic nanoparticles is promising, other types, such as virus particles, protein nanoparticles, and synthetic exosome-like nanoparticles, are also being explored, contributing to the diverse landscape of nanoparticle applications in cancer detection and treatment. Nonetheless, these alternatives also can trigger immunological responses and often entail higher production costs, making them less economically viable.

On the other hand, nanoparticles produced on substrates such as polyamide and sustainable materials like paper present cost-effective and eco-friendly options that streamline the manufacturing process. This area of research is ongoing, necessitating the exploration of new biomaterials and optimizations related to laser processing to achieve precise control over shape, size, reproducibility, and concentration. In conclusion, while the path ahead is long, the potential of nanoparticles in the fight against cancer is promising.

References

1. Osman, A.I., Zhang, Y., Farghali, M., Rashwan, A.K., Eltaweil, A.S., Abd El-Monaem, E.M., Mohamed, I.M.A., Badr, M.M., Ihara, I., Rooney, D.W., Yap, P.S.: Synthesis of Green Nanoparticles for Energy, Biomedical, Environmental, Agricultural, and Food Applications: A Review. Springer International Publishing (2024). https://doi.org/10.1007/s10311-023-01682-3
2. Altammar, K.A.: A review on nanoparticles: characteristics, synthesis, applications, and challenges. Front. Microbiol. **14**, 1–20 (2023). https://doi.org/10.3389/fmicb.2023.1155622

3. Kumar, J.A., Krithiga, T., Manigandan, S., Sathish, S., Renita, A.A., Prakash, P., Prasad, B.S.N., Kumar, T.R.P., Rajasimman, M., Hosseini-Bandegharaei, A., Prabu, D., Crispin, S.: A focus to green synthesis of metal/metal based oxide nanoparticles: various mechanisms and applications towards ecological approach. J. Clean. Prod. **324**, 129198 (2021). https://doi.org/10.1016/j.jclepro.2021.129198
4. Wang, B., Hu, S., Teng, Y., Chen, J., Wang, H., Xu, Y., Wang, K., Xu, J., Cheng, Y., Gao, X.: Current advance of nanotechnology in diagnosis and treatment for malignant tumors. Signal Transduct. Target. Ther. **9** (2024). https://doi.org/10.1038/s41392-024-01889-y
5. Dessale, M., Mengistu, G., Mengist, H.M.: Nanotechnology: a promising approach for cancer diagnosis, therapeutics and theragnosis. Int. J. Nanomed. **17**, 3735–3749 (2022). https://doi.org/10.2147/IJN.S378074
6. Giri, P.M., Banerjee, A., Layek, B.: A recent review on cancer nanomedicine. Cancers (Basel) **15**, 2256 (2023). https://doi.org/10.3390/cancers15082256
7. Materón, E.M., Miyazaki, C.M., Carr, O., Joshi, N., Picciani, P.H.S., Dalmaschio, C.J., Davis, F., Shimizu, F.M.: Magnetic nanoparticles in biomedical applications: a review. Appl. Surf. Sci. Adv. **6**, 100163 (2021). https://doi.org/10.1016/j.apsadv.2021.100163
8. Bajpai, S., Tiwary, S.K., Sonker, M., Joshi, A., Gupta, V., Kumar, Y., Shreyash, N., Biswas, S.: Recent advances in nanoparticle-based cancer treatment: a review. ACS Appl. Nano Mater. **4**, 6441–6470 (2021). https://doi.org/10.1021/acsanm.1c00779
9. Bhardwaj, H., Archana, N., Noumani, A., Himanshu, J.K., Chakravorty, S., Solanki, P.R.: Recent advancement in the detection of potential cancer biomarkers using the nanomaterial integrated electrochemical sensing technique: a detailed review. Mater. Adv. **5**, 475–503 (2024). https://doi.org/10.1039/d3ma00621b
10. Zhou, Y., Tao, L., Qiu, J., Xu, J., Yang, X., Zhang, Y., Tian, X., Guan, X., Cen, X., Zhao, Y.: Tumor biomarkers for diagnosis, prognosis and targeted therapy. Signal Transduct. Target. Ther. **9** (2024). https://doi.org/10.1038/s41392-024-01823-2
11. Ranjan, P., Parihar, A., Jain, S., Kumar, N., Dhand, C., Murali, S., Mishra, D., Sanghi, S.K., Chaurasia, J.P., Srivastava, A.K., Khan, R.: Biosensor-based diagnostic approaches for various cellular biomarkers of breast cancer: a comprehensive review. Anal. Biochem. **610**, 113996 (2020). https://doi.org/10.1016/j.ab.2020.113996
12. Naresh, V., Lee, N.: A review on biosensors and recent development of nanostructured materials-enabled biosensors. Sensors **21**, 1109 (2021). https://doi.org/10.3390/s21041109
13. Wang, X., Zhou, J., Wang, H.: Bioreceptors as the key components for electrochemical biosensing in medicine. Cell Rep. Phys. Sci. **5**, 101801 (2024). https://doi.org/10.1016/j.xcrp.2024.101801
14. Huang, X., Liu, Y., Yung, B., Xiong, Y., Chen, X.: Nanotechnology-enhanced No-wash biosensors for in vitro diagnostics of cancer. ACS Nano **11**, 5238–5292 (2017). https://doi.org/10.1021/acsnano.7b02618
15. Ushiba, S., Nakano, T., Shinagawa, A., Miyakawa, N., Kato, T., Yofu, K., Ono, T., Kanai, Y., Tani, S., Kimura, M., Matsumoto, K.: Biosensing with surface-charge-modulated graphene field-effect transistors beyond nonlinear electrolytic screening. ACS Omega **8**, 49270–49277 (2023). https://doi.org/10.1021/acsomega.3c07650
16. Wu, J., Liu, H., Chen, W., Ma, B., Ju, H.: Device integration of electrochemical biosensors. Nat. Rev. Bioeng. **1**, 346–360 (2023). https://doi.org/10.1038/s44222-023-00032-w
17. An, J., Park, H., Ju, M., Woo, Y., Seo, Y., Min, J., Lee, T.: An updated review on the development of a nanomaterial-based field-effect transistor-type biosensors to detect exosomes for cancer diagnosis. Talanta **279**, 126604 (2024). https://doi.org/10.1016/j.talanta.2024.126604
18. Boukherroub, R., Szunerits, S.: The future of nanotechnology-driven electrochemical and electrical point-of-care devices and diagnostic tests. Annu. Rev. Anal. Chem. (Palo Alto. Calif). **17**, 173–195 (2024). https://doi.org/10.1146/annurev-anchem-061622-012029
19. Lan, H., Jamil, M., Ke, G., Dong, N.: The role of nanoparticles and nanomaterials in cancer diagnosis and treatment: a comprehensive review. Am. J. Cancer Res. **13**, 5751–5784 (2023). http://www.ncbi.nlm.nih.gov/pubmed/38187049%0A; http://www.pubmedcentral.nih.gov/articlerender.fcgi?artid=PMC10767363

20. Mittal, S., Kaur, H., Gautam, N., Mantha, A.K.: Biosensors for breast cancer diagnosis: a review of bioreceptors, biotransducers and signal amplification strategies. Biosens. Bioelectron. **88**, 217–231 (2017). https://doi.org/10.1016/j.bios.2016.08.028
21. Sharifianjazi, F., Jafari Rad, A., Bakhtiari, A., Niazvand, F., Esmaeilkhanian, A., Bazli, L., Abniki, M., Irani, M., Moghanian, A.: Biosensors and nanotechnology for cancer diagnosis (lung and bronchus, breast, prostate, and colon): a systematic review. Biomed. Mater. **17**, 12002 (2022). https://doi.org/10.1088/1748-605X/ac41fd
22. Subba, S.H., Park, S.Y.: In situ cancer-cell-triggered visible changes in mechanical properties, electroconductivity, and adhesiveness of a MnO_2@PD-based mineralized hydrogel. ACS Appl. Mater. Interfaces **15**, 38357–38366 (2023). https://doi.org/10.1021/acsami.3c08501
23. Huang, X., Ni, Z., Su, H., Shang, Y., Liu, H., He, Y., Meng, H., Dong, Y.: Cellulose nanocrystalline and sodium benzenesulfonate-doped polypyrrole nano-hydrogel/Au composites for ultrasensitive detection of carcinoembryonic antigen. New J. Chem. **45**, 5551–5560 (2021). https://doi.org/10.1039/D1NJ00360G
24. Xu, X., Zhang, Z., Yu, H., Shen, T., Pan, H., Chang, D.: Dual-modal aptasensor for sensitive detection of non-small cell lung cancer exosomes utilizing two-dimensional nanopaper Co@g-C3N4@PB. ACS Omega **9**, 34493–34506 (2024). https://doi.org/10.1021/acsomega.4c02346
25. Díaz-Ayala, R., López-Nieves, M., Colón Berlingeri, E.S., Cabrera, C.R., Cunci, L., González, C.I., Escobar, P.F.: Test strip platform spin-off for telomerase activity detection: development of an electrochemical biosensor. ACS Omega **7**, 9964–9972 (2022). https://doi.org/10.1021/acsomega.2c00713
26. Chao, L., Liang, Y., Hu, X., Shi, H., Xia, T., Zhang, H., Xia, H.: Recent advances in field effect transistor biosensor technology for cancer detection: a mini review. J. Phys. D Appl. Phys. **55**, 153001 (2022). https://doi.org/10.1088/1361-6463/ac3f5a
27. Novodchuk, I., Bajcsy, M., Yavuz, M.: Graphene-based field effect transistor biosensors for breast cancer detection: a review on biosensing strategies. Carbon N. Y. **172**, 431–453 (2021). https://doi.org/10.1016/j.carbon.2020.10.048
28. Rubtsova, M., Presnova, G., Presnov, D., Krupenin, V., Grigorenko, V., Egorov, A.: Biosensor based on a nanowire field-effect transistor for the determination of prostate specific antigen. Procedia Technol. **27**, 234–235 (2017). https://doi.org/10.1016/j.protcy.2017.04.099
29. Wu, D., Yu, Y., Jin, D., Xiao, M.M., Zhang, Z.Y., Zhang, G.J.: Dual-aptamer modified graphene field-effect transistor nanosensor for label-free and specific detection of hepatocellular carcinoma-derived microvesicles. Anal. Chem. **92**, 4006–4015 (2020). https://doi.org/10.1021/acs.analchem.9b05531
30. Cai, B., Huang, L., Zhang, H., Sun, Z., Zhang, Z., Zhang, G.-J.: Gold nanoparticles-decorated graphene field-effect transistor biosensor for femtomolar MicroRNA detection. Biosens. Bioelectron. **74**, 329–334 (2015). https://doi.org/10.1016/j.bios.2015.06.068
31. Hetsch, F., Zhao, N., Kershaw, S.V., Rogach, A.L.: Quantum dot field effect transistors. Mater. Today **16**, 312–325 (2013). https://doi.org/10.1016/j.mattod.2013.08.011
32. Li, K., Tu, J., Zhang, Y., Jin, D., Li, T., Li, J., Ni, W., Xiao, M.-M., Zhang, Z.-Y., Zhang, G.-J.: Ultrasensitive detection of exosomal miRNA with PMO-graphene quantum dots-functionalized field-effect transistor biosensor. IScience **25** (2022). https://doi.org/10.1016/j.isci.2022.104522
33. Chen, H.-C., Qiu, J.-T., Yang, F.-L., Liu, Y.-C., Chen, M.-C., Tsai, R.-Y., Yang, H.-W., Lin, C.-Y., Lin, C.-C., Wu, T.-S., Tu, Y.-M., Xiao, M.-C., Ho, C.-H., Huang, C.-C., Lai, C.-S., Hua, M.-Y.: Magnetic-composite-modified polycrystalline silicon nanowire field-effect transistor for vascular endothelial growth factor detection and cancer diagnosis. Anal. Chem. **86**, 9443–9450 (2014). https://doi.org/10.1021/ac5001898
34. Rajesh, Gao, Z., Vishnubhotla, R., Ducos, P., Serrano, M.D., Ping, J., Robinson, M.K., Johnson, A.T.C.: Genetically engineered antibody functionalized platinum nanoparticles modified CVD-graphene nanohybrid transistor for the detection of breast cancer biomarker, HER3. Adv. Mater. Interfaces. **3**, 1600124 (2016). https://doi.org/10.1002/admi.201600124
35. Sadighbayan, D., Hasanzadeh, M., Ghafar-Zadeh, E.: Biosensing based on field-effect transistors (FET): recent progress and challenges. TrAC Trends Anal. Chem. **133**, 116067 (2020). https://doi.org/10.1016/j.trac.2020.116067

36. Yu, Y., Li, Y.-T., Jin, D., Yang, F., Wu, D., Xiao, M.-M., Zhang, H., Zhang, Z.-Y., Zhang, G.-J.: Electrical and label-free quantification of exosomes with a reduced graphene oxide field effect transistor biosensor. Anal. Chem. **91**, 10679–10686 (2019). https://doi.org/10.1021/acs.analchem.9b01950
37. Liang, Y., Xiao, M., Wu, D., Lin, Y., Liu, L., He, J., Zhang, G., Peng, L.-M., Zhang, Z.: Wafer-scale uniform carbon nanotube transistors for ultrasensitive and label-free detection of disease biomarkers. ACS Nano **14**, 8866–8874 (2020). https://doi.org/10.1021/acsnano.0c03523
38. Schumann, M., Sauerbrey, R., Smayling, M.C.: Permanent increase of the electrical conductivity of polymers induced by ultraviolet laser radiation. Appl. Phys. Lett. **58**, 428–430 (1991). https://doi.org/10.1063/1.104624
39. Lin, J., Peng, Z., Liu, Y., Ruiz-Zepeda, F., Ye, R., Samuel, E.L.G., Yacaman, M.J., Yakobson, B.I., Tour, J.M.: Laser-induced porous graphene films from commercial polymers. Nat. Commun. **5**, 5714 (2014). https://doi.org/10.1038/ncomms6714
40. Parmeggiani, M., Zaccagnini, P., Stassi, S., Fontana, M., Bianco, S., Nicosia, C., Pirri, C.F., Lamberti, A.: PDMS/polyimide composite as an elastomeric substrate for multifunctional laser-induced graphene electrodes. ACS Appl. Mater. Interfaces **11**, 33221–33230 (2019). https://doi.org/10.1021/acsami.9b10408
41. Duy, L.X., Peng, Z., Li, Y., Zhang, J., Ji, Y., Tour, J.M.: Laser-induced graphene fibers. Carbon N. Y. **126**, 472–479 (2018). https://doi.org/10.1016/j.carbon.2017.10.036
42. Zhang, J., Ren, M., Wang, L., Li, Y., Yakobson, B.I., Tour, J.M.: Oxidized laser-induced graphene for efficient oxygen electrocatalysis. Adv. Mater. **30**, 1707319 (2018). https://doi.org/10.1002/adma.201707319
43. Barber, R., Cameron, S., Devine, A., McCombe, A., Kirsty Pourshahidi, L., Cundell, J., Roy, S., Mathur, A., Casimero, C., Papakonstantinou, P., Davis, J.: Laser induced graphene sensors for assessing pH: application to wound management. Electrochem. Commun. **123**, 106914 (2021). https://doi.org/10.1016/j.elecom.2020.106914
44. Luo, S., Hoang, P.T., Liu, T.: Direct laser writing for creating porous graphitic structures and their use for flexible and highly sensitive sensor and sensor arrays. Carbon N. Y. **96**, 522–531 (2016). https://doi.org/10.1016/j.carbon.2015.09.076
45. Xu, G., Jarjes, Z.A., Wang, H.-W., Phillips, A.R.J., Kilmartin, P.A., Travas-Sejdic, J.: Detection of neurotransmitters by three-dimensional laser-scribed graphene grass electrodes. ACS Appl. Mater. Interfaces **10**, 42136–42145 (2018). https://doi.org/10.1021/acsami.8b16692
46. Wan, Z., Umer, M., Lobino, M., Thiel, D., Nguyen, N.-T., Trinchi, A., Shiddiky, M.J.A., Gao, Y., Li, Q.: Laser induced self-N-doped porous graphene as an electrochemical biosensor for femtomolar miRNA detection. Carbon N. Y. **163**, 385–394 (2020). https://doi.org/10.1016/j.carbon.2020.03.043
47. Zahed, M.A., Barman, S.C., Das, P.S., Sharifuzzaman, M., Yoon, H.S., Yoon, S.H., Park, J.Y.: Highly flexible and conductive poly (3, 4-ethylene dioxythiophene)-poly (styrene sulfonate) anchored 3-dimensional porous graphene network-based electrochemical biosensor for glucose and pH detection in human perspiration. Biosens. Bioelectron. **160**, 112220 (2020). https://doi.org/10.1016/j.bios.2020.112220
48. Ye, R., Peng, Z., Wang, T., Xu, Y., Zhang, J., Li, Y., Nilewski, L.G., Lin, J., Tour, J.M.: In situ formation of metal oxide nanocrystals embedded in laser-induced graphene. ACS Nano **9**, 9244–9251 (2015). https://doi.org/10.1021/acsnano.5b04138
49. Clerici, F., Fontana, M., Bianco, S., Serrapede, M., Perrucci, F., Ferrero, S., Tresso, E., Lamberti, A.: In situ MoS2 decoration of laser-induced graphene as flexible supercapacitor electrodes. ACS Appl. Mater. Interfaces **8**, 10459–10465 (2016). https://doi.org/10.1021/acsami.6b00808
50. You, Z., Qiu, Q., Chen, H., Feng, Y., Wang, X., Wang, Y., Ying, Y.: Laser-induced noble metal nanoparticle-graphene composites enabled flexible biosensor for pathogen detection. Biosens. Bioelectron. **150**, 111896 (2020). https://doi.org/10.1016/j.bios.2019.111896
51. Zhao, G., Wang, F., Zhang, Y., Sui, Y., Liu, P., Zhang, Z., Xu, C., Yang, C.: High-performance hydrogen peroxide micro-sensors based on laser-induced fabrication of graphene@Ag electrodes. Appl. Surf. Sci. **565**, 150565 (2021). https://doi.org/10.1016/j.apsusc.2021.150565

52. Arantes, I.V.S., Ataide, V.N., Ameku, W.A., Gongoni, J.L.M., Selva, J.S.G., Nogueira, H.P., Bertotti, M., Paixão, T.R.L.C.: Laser-induced fabrication of gold nanoparticles onto paper substrates and their application on paper-based electroanalytical devices. Sens. Diagn. **2**, 111–121 (2023). https://doi.org/10.1039/D2SD00176D
53. Gongoni, J.L.M., Chumanov, G., Paixão, T.R.L.C., Garcia, C.D.: Au-modified carbon electrodes produced by laser scribing for electrochemical analysis of probiotic activity. Anal. Sens. **4**, e202300056 (2024). https://doi.org/10.1002/anse.202300056
54. Veloso, W.B., Meloni, G.N., Arantes, I.V.S., Pradela-Filho, L.A., Muñoz, R.A.A., Paixão, T.R.L.C.: Gold film deposition by infrared laser photothermal treatment on 3D-printed electrodes: electrochemical performance enhancement and application. Analyst **149**, 3900–3909 (2024). https://doi.org/10.1039/D4AN00669K
55. Hui, X., Xuan, X., Kim, J., Park, J.Y.: A highly flexible and selective dopamine sensor based on Pt-Au nanoparticle-modified laser-induced graphene. Electrochim. Acta **328**, 135066 (2019). https://doi.org/10.1016/j.electacta.2019.135066
56. Torati, S.R., Hanson, B., Shinde, M., Slaughter, G.: Gold-deposited laser-induced graphene electrode for detection of miRNA-141. IEEE Sens. J. **24**, 2154–2161 (2024). https://doi.org/10.1109/JSEN.2023.3336165
57. Lahcen, A.A., Rauf, S., Aljedaibi, A., de Oliveira Filho, J.I., Beduk, T., Mani, V., Alshareef, H.N., Salama, K.N.: Laser-scribed graphene sensor based on gold nanostructures and molecularly imprinted polymers: application for Her-2 cancer biomarker detection. Sens. Actuat. B Chem. **347**, 130556 (2021). https://doi.org/10.1016/j.snb.2021.130556
58. Chehelgerdi, M., Chehelgerdi, M., Allela, O.Q.B., Pecho, R.D.C., Jayasankar, N., Rao, D.P., Thamaraikani, T., Vasanthan, M., Viktor, P., Lakshmaiya, N., Saadh, M.J., Amajd, A., Abo-Zaid, M.A., Castillo-Acobo, R.Y., Ismail, A.H., Amin, A.H., Akhavan-Sigari, R.: Progressing nanotechnology to improve targeted cancer treatment: overcoming hurdles in its clinical implementation. Mol. Cancer. **22** (2023). https://doi.org/10.1186/s12943-023-01865-0
59. Anselmo, A.C., Mitragotri, S.: Nanoparticles in the clinic: an update. Bioeng. Transl. Med. **4**, 1–16 (2019). https://doi.org/10.1002/btm2.10143
60. Augustine, R., Al Mamun, A., Hasan, A., Salam, S.A., Chandrasekaran, R., Ahmed, R., Thakor, A.S.: Imaging cancer cells with nanostructures: prospects of nanotechnology driven noninvasive cancer diagnosis. Adv. Colloid Interface Sci. **294**, 102457 (2021). https://doi.org/10.1016/j.cis.2021.102457
61. Liu, C.H., Grodzinski, P.: Nanotechnology for cancer imaging: advances, challenges, and clinical opportunities. Radiol. Imaging Cancer **3** (2021). https://doi.org/10.1148/rycan.2021200052
62. Yankeelov, T.E., Atuegwu, N., Hormuth, D., Weis, J.A., Barnes, S.L., Miga, M.I., Rericha, E.C., Quaranta, V.: Clinically relevant modeling of tumor growth and treatment response. Sci. Transl. Med. **5**, 1–5 (2013). https://doi.org/10.1126/scitranslmed.3005686
63. Jarrett, A.M., Lima, E.A.B.F., Hormuth, D.A., McKenna, M.T., Feng, X., Ekrut, D.A., Resende, A.C.M., Brock, A., Yankeelov, T.E.: Mathematical models of tumor cell proliferation: a review of the literature. (2018). https://doi.org/10.1080/14737140.2018.1527689
64. Pulumati, A., Pulumati, A., Dwarakanath, B.S., Verma, A., Papineni, R.V.L.: Technological advancements in cancer diagnostics: improvements and limitations. Cancer Rep. **6**, 1–17 (2023). https://doi.org/10.1002/cnr2.1764
65. Petersson-Sjögren, M., Chan, H.F., Collier, G.J., Norquay, G., Olsson, L.E., Wollmer, P., Löndahl, J., Wild, J.M.: Airspace Dimension Assessment (AiDA) by inhaled nanoparticles: benchmarking with hyperpolarised 129Xe diffusion-weighted lung MRI. Sci. Rep. **11**, 1–9 (2021). https://doi.org/10.1038/s41598-021-83975-7
66. Simón, M., Jørgensen, J.T., Norregaard, K., Kjaer, A.: 18F-FDG positron emission tomography and diffusion-weighted magnetic resonance imaging for response evaluation of nanoparticle-mediated photothermal therapy. Sci. Rep. **10**, 1–9 (2020). https://doi.org/10.1038/s41598-020-64617-w
67. Akkaya, H.E., Erden, A., Öz, D.K., Ünal, S., Erden, İ: Magnetic resonance elastography: basic principles, technique, and clinical applications in the liver. Diagn. Interv. Radiol. **24**, 328–335 (2018). https://doi.org/10.5152/dir.2018.18186

68. Weinberg, B.D., Kuruva, M., Shim, H., Mullins, M.E.: Clinical applications of magnetic resonance spectroscopy in brain tumors: from diagnosis to treatment. Radiol. Clin. North Am. **59**, 349–362 (2021). https://doi.org/10.1016/j.rcl.2021.01.004
69. Kierans, A.S., Costello, J., Qayyum, A., Taouli, B., Venkatesh, S.K., Yoon, J.H., Bali, M.A., Bolan, C.W., Lee, J.M., Marks, R.M., El Homsi, M., Miller, F.H.: Imaging cholangiocarcinoma: CT and MRI techniques. Abdom. Radiol. (2024). https://doi.org/10.1007/s00261-024-04216-9
70. Chung, S.-J., Hadrick, K., Nafiujjaman, M., Apu, E.H., Hill, M.L., Nurunnabi, M., Contag, C.H., Kim, T.: Targeted biodegradable near-infrared fluorescent nanoparticles for colorectal cancer imaging. ACS Appl. Bio Mater. (2024). https://doi.org/10.1021/acsabm.4c00072
71. Crişan, G., Moldovean-cioroianu, N.S., Timaru, D.G., Andrieş, G., Căinap, C., Chiş, V.: Radiopharmaceuticals for PET and SPECT Imaging: a literature review over the last decade. Int J Mol Sci **23** (2022). https://doi.org/10.3390/ijms23095023
72. Huang, W., Huang, G., An, S., Liu, J.: Molecular imaging of pancreatic ductal adenocarcinoma. View **4**, 1–18 (2023). https://doi.org/10.1002/VIW.20230029
73. Barata, F., Fidalgo, P., Figueiredo, S., Tonin, F.S., Duarte-Ramos, F.: Limitations and perceived delays for diagnosis and staging of lung cancer in Portugal: a nationwide survey analysis. PLoS One **16**, e0252529 (2021). https://doi.org/10.1371/journal.pone.0252529
74. Han, X., Xu, K., Taratula, O., Farsad, K.: Applications of nanoparticles in biomedical imaging. Nanoscale **11**, 799–819 (2019). https://doi.org/10.1039/c8nr07769j
75. Bharathala, S., Sharma, P.: Biomedical applications of nanoparticles. Elsevier (2019). https://doi.org/10.1016/B978-0-12-818823-1.00008-9
76. Hamidu, A., Pitt, W.G., Husseini, G.A.: Recent breakthroughs in using quantum dots for cancer imaging and drug delivery purposes. Nanomaterials **13**, 2566 (2023). https://doi.org/10.3390/nano13182566
77. Afshari, M.J., Li, C., Zeng, J., Cui, J., Wu, S., Gao, M.: Self-illuminating NIR-II bioluminescence imaging probe based on silver sulfide quantum dots. ACS Nano **16**, 16824–16832 (2022). https://doi.org/10.1021/acsnano.2c06667
78. Roy, S., Bobde, Y., Ghosh, B., Chakraborty, C.: Targeted bioimaging of cancer cells using free folic acid-sensitive molybdenum disulfide quantum dots through fluorescence "turn-off,." ACS Appl. Bio Mater. **4**, 2839–2849 (2021). https://doi.org/10.1021/acsabm.1c00090
79. Al-Hetty, H.R.A.K., Jalil, A.T., Al-Tamimi, J.H.Z., Shakier, H.G., Kandeel, M., Saleh, M.M., Naderifar, M.: Engineering and surface modification of carbon quantum dots for cancer bioimaging. Inorg. Chem. Commun. **149**, 110433 (2023). https://doi.org/10.1016/j.inoche.2023.110433
80. Henna, T.K., Pramod, K.: Graphene quantum dots redefine nanobiomedicine. Mater. Sci. Eng. C. **110** (2020). https://doi.org/10.1016/j.msec.2020.110651
81. Fan, Z., Li, S., Yuan, F., Fan, L.: Fluorescent graphene quantum dots for biosensing and bioimaging. RSC Adv. **5**, 19773–19789 (2015). https://doi.org/10.1039/c4ra17131d
82. Wang, H., Yang, S., Chen, L., Li, Y., He, P., Wang, G., Dong, H., Ma, P., Ding, G.: Tumor diagnosis using carbon-based quantum dots: Detection based on the hallmarks of cancer. Bioact. Mater. **33**, 174–222 (2024). https://doi.org/10.1016/j.bioactmat.2023.10.004
83. Badrigilan, S., Shaabani, B., Gharehaghaji, N., Mesbahi, A.: Iron oxide/bismuth oxide nanocomposites coated by graphene quantum dots: "three-in-one" theranostic agents for simultaneous CT/MR imaging-guided in vitro photothermal therapy. Photodiagn. Photodyn. Ther. **25**, 504–514 (2019). https://doi.org/10.1016/j.pdpdt.2018.10.021
84. Xuan, Y., Zhang, R.Y., Zhao, D.H., Zhang, X.S., An, J., Cheng, K., Hou, X.L., Song, X.L., Di Zhao, Y., Yang, X.Q.: Ultrafast synthesis of gold nanosphere cluster coated by graphene quantum dot for active targeting PA/CT imaging and near-infrared laser/pH-triggered chemo-photothermal synergistic tumor therapy. Chem. Eng. J. **369**, 87–99 (2019). https://doi.org/10.1016/j.cej.2019.03.035
85. Ahmadi, F., Sodagar-Taleghani, A., Ebrahimnejad, P., Pouya Hadipour Moghaddam, S., Ebrahimnejad, F., Asare-Addo, K., Nokhodchi, A.: A review on the latest developments of mesoporous silica nanoparticles as a promising platform for diagnosis and treatment of cancer. Int. J. Pharm. **625**, 122099 (2022). https://doi.org/10.1016/j.ijpharm.2022.122099

86. Vandghanooni, S., Barar, J., Eskandani, M., Omidi, Y.: Aptamer-conjugated mesoporous silica nanoparticles for simultaneous imaging and therapy of cancer. TrAC—Trends Anal. Chem. **123**, 115759 (2020). https://doi.org/10.1016/j.trac.2019.115759
87. Han, J., Zhang, L., Cui, M., Su, Y., He, Y.: Rapid and accurate detection of lymph node metastases enabled through fluorescent silicon nanoparticles-based exosome probes. Anal. Chem. **93**, 10122–10131 (2021). https://doi.org/10.1021/acs.analchem.1c01010
88. Luo, D., Wang, X., Burda, C., Basilion, J.P.: Recent development of gold nanoparticles as contrast agents for cancer diagnosis. Cancers (Basel) **13**, 1825 (2021). https://doi.org/10.3390/cancers13081825
89. Fan, M., Han, Y., Gao, S., Yan, H., Cao, L., Li, Z., Liang, X.J., Zhang, J.: Ultrasmall gold nanoparticles in cancer diagnosis and therapy. Theranostics **10**, 494–4957 (2020). https://doi.org/10.7150/thno.42471
90. Banstola, A., Emami, F., Jeong, J.H., Yook, S.: Current applications of gold nanoparticles for medical imaging and as treatment agents for managing pancreatic cancer. Macromol. Res. **26**, 955–964 (2018). https://doi.org/10.1007/s13233-018-6139-4
91. Hejazi, M., Arshadi, S., Amini, M., Baradaran, B., Shahbazi-Derakhshi, P., Sameti, P., Soleymani, J., Mokhtarzadeh, A., Tavangar, S.M.: Hyaluronic acid-functionalized gold nanoparticles as a cancer diagnostic probe for targeted bioimaging applications. Microchem. J. **193**, 108953 (2023). https://doi.org/10.1016/j.microc.2023.108953
92. Maximenko, A., Depciuch, J., Łopuszyńska, N., Stec, M., Światkowska-Warkocka, Ż., Bayev, V., Zieliński, P.M., Baran, J., Fedotova, J., Węglarz, W.P., Parlinska-Wojtan, M.: Fe3O4@SiO2@Au nanoparticles for MRI-guided chemo/NIR photothermal therapy of cancer cells. RSC Adv. **10**, 26508–26520 (2020). https://doi.org/10.1039/d0ra03699d
93. Li, X., Zhang, Y., Liu, G.K., Luo, Z., Zhou, L., Xue, Y., Liu, M.: Recent progress in the applications of gold-based nanoparticles towards tumor-targeted imaging and therapy. RSC Adv. **12**, 7635–7651 (2022). https://doi.org/10.1039/d2ra00566b
94. Farzin, A., Etesami, S.A., Quint, J., Memic, A., Tamayol, A.: Magnetic nanoparticles in cancer therapy and diagnosis. Adv. Healthc. Mater. **9**, 139–148 (2020). https://doi.org/10.1002/adhm.201901058
95. Li, Z., Bai, R., Yi, J., Zhou, H., Xian, J., Chen, C.: Designing smart iron oxide nanoparticles for MR imaging of tumors. Chem. Biomed. Imaging **1**, 315–339 (2023). https://doi.org/10.1021/cbmi.3c00026
96. Jeon, M., Halbert, M.V., Stephen, Z.R., Zhang, M.: Iron oxide nanoparticles as t 1 contrast agents for magnetic resonance imaging: fundamentals, challenges, applications, and prospectives. Adv. Mater. **33**, 1–41 (2021). https://doi.org/10.1002/adma.201906539
97. Estelrich, J., Sánchez-Martín, M.J., Busquets, M.A.: Nanoparticles in magnetic resonance imaging: from simple to dual contrast agents. Int. J. Nanomed. **10**, 1727–1741 (2015). https://doi.org/10.2147/IJN.S76501
98. Lu, H., Chen, A., Zhang, X., Wei, Z., Cao, R., Zhu, Y., Lu, J., Wang, Z., Tian, L.: A pH-responsive T1-T2 dual-modal MRI contrast agent for cancer imaging. Nat. Commun. **13** (2022). https://doi.org/10.1038/s41467-022-35655-x
99. Pavitra, E., Dariya, B., Srivani, G., Kang, S.M., Alam, A., Sudhir, P.R., Kamal, M.A., Raju, G.S.R., Han, Y.K., Lakkakula, B.V.K.S., Nagaraju, G.P., Huh, Y.S.: Engineered nanoparticles for imaging and drug delivery in colorectal cancer. Semin. Cancer Biol. **69**, 293–306 (2021). https://doi.org/10.1016/j.semcancer.2019.06.017
100. Yang, K., Zhang, S., He, J., Nie, Z.: Polymers and inorganic nanoparticles: a winning combination towards assembled nanostructures for cancer imaging and therapy. Nano Today **36**, 101046 (2021). https://doi.org/10.1016/j.nantod.2020.101046
101. Sharma, N.K., Vishwakarma, J., Rai, S., Alomar, T.S., Almasoud, N., Bhattarai, A.: Green route synthesis and characterization techniques of silver nanoparticles and their biological adeptness. ACS Omega **7**, 27004–27020 (2022). https://doi.org/10.1021/acsomega.2c01400
102. Bhat, S.A., Kumar, V., Dhanjal, D.S., Gandhi, Y., Mishra, S.K., Singh, S., Webster, T.J., Ramamurthy, P.C.: Biogenic Nanoparticles: Pioneering a New Era in Breast Cancer Therapeutics—A Comprehensive Review. Springer US (2024). https://doi.org/10.1186/s11671-024-04072-y

103. Singh, K.R., Nayak, V., Singh, J., Singh, A.K., Singh, R.P.: Potentialities of bioinspired metal and metal oxide nanoparticles in biomedical sciences. RSC Adv. **11**, 24722–24746 (2021). https://doi.org/10.1039/d1ra04273d
104. Nikolova, M.P., Joshi, P.B., Chavali, M.S.: Updates on biogenic metallic and metal oxide nanoparticles: therapy, drug delivery and cytotoxicity. Pharmaceutics **15** (2023). https://doi.org/10.3390/pharmaceutics15061650
105. Si, A., Pal, K., Kralj, S., El-Sayyad, G.S., de Souza, F.G., Narayanan, T.: Sustainable preparation of gold nanoparticles via green chemistry approach for biogenic applications. Mater. Today Chem. **17**, 100327 (2020). https://doi.org/10.1016/j.mtchem.2020.100327
106. Nyandoro, V.O., Masioge, H.K., Malago, Z.L.: Biogenic synthesis of metal nanoparticles: promoting green nanotechnology and sustainable development goals. Clean Technol. Environ. Policy (2024). https://doi.org/10.1007/s10098-024-02966-0
107. Heinemann, M.G., Rosa, C.H., Rosa, G.R., Dias, D.: Biogenic synthesis of gold and silver nanoparticles used in environmental applications: a review. Trends Environ. Anal. Chem. **30** (2021). https://doi.org/10.1016/j.teac.2021.e00129
108. Singh, D., Chaudhary, D., Kumar, V., Verma, A.: Amelioration of diethylnitrosamine (DEN) induced renal oxidative stress and inflammation by Carissa carandas embedded silver nanoparticles in rodents. Toxicol. Rep. **8**, 636–645 (2021). https://doi.org/10.1016/j.toxrep.2021.03.014
109. Singh, H., Du, J., Singh, P., Yi, T.H.: Ecofriendly synthesis of silver and gold nanoparticles by Euphrasia officinalis leaf extract and its biomedical applications. Artif. Cells Nanomed. Biotechnol. **46**, 1163–1170 (2018). https://doi.org/10.1080/21691401.2017.1362417
110. Mokhtar, F.A., Selim, N.M., Elhawary, S.S., Abd El Hadi, S.R., Hetta, M.H., Albalawi, M.A., Shati, A.A., Alfaifi, M.Y., Elbehairi, S.E.I., Fahmy, L.I., Ibrahim, R.M.: Green biosynthesis of silver nanoparticles using Annona glabra and Annona squamosa extracts with antimicrobial, anticancer, apoptosis potentials, assisted by in silico modeling, and metabolic profiling. Pharmaceuticals **15**, 1–21 (2022). https://doi.org/10.3390/ph15111354
111. Shiripoure Ganjineh Ketab, R., Tafvizi, F., Khodarahmi, P.: Biosynthesis and chemical characterization of silver nanoparticles using Satureja Rechingeri Jamzad and their apoptotic effects on AGS gastric cancer cells. J. Clust. Sci. **32**, 1389–1399 (2021). https://doi.org/10.1007/s10876-020-01903-7
112. Chaudhary, V., Sonu, Chowdhury, R., Thukral, P., Pathania, D., Saklani, S., Lucky, Rustagi, S., Gautam, A., Mishra, Y.K., Singh, P., Kaushik, A.: Biogenic green metal nano systems as efficient anticancer agents. Environ. Res. **229**, 115933 (2023). https://doi.org/10.1016/j.envres.2023.115933
113. Wu, Z., Shen, X., Liu, Y., Shi, M., Li, J., Zhang, Z.: Biogenic fabrication of gold nanoparticles for treatment of gastric cancer and electrochemical sensing of As(III). Alexandria Eng. J. **90**, 216–221 (2024). https://doi.org/10.1016/j.aej.2024.01.057
114. Prajapati, J.P., Singh, P., Singh, K.R.B., Singh, J., Mallick, S., Singh, R.P.: Biogenic synthesis of Mn3O4 NPs using Phyllanthus emblica leaf extract for electrochemical sensing of urea. J. Mol. Struct. **1307**, 137918 (2024). https://doi.org/10.1016/j.molstruc.2024.137918
115. Bekmezci, M., Ozturk, H., Akin, M., Bayat, R., Sen, F., Darabi, R., Karimi-Maleh, H.: Bimetallic biogenic Pt-Ag nanoparticle and their application for electrochemical dopamine sensor. Biosensors **13** (2023). https://doi.org/10.3390/bios13050531
116. Haque, S., Norbert, C.C., Acharyya, R., Mukherjee, S., Kathirvel, M., Patra, C.R.: Biosynthesized silver nanoparticles for cancer therapy and in vivo bioimaging. Cancers (Basel) **13** (2021). https://doi.org/10.3390/cancers13236114
117. Nayak, V., Singh, K.R.B., Verma, R., Pandey, M.D., Singh, J., Pratap Singh, R.: Recent advancements of biogenic iron nanoparticles in cancer theranostics. Mater. Lett. **313**, 131769 (2022). https://doi.org/10.1016/j.matlet.2022.131769
118. Schuerle, S., Furubayashi, M., Soleimany, A.P., Gwisai, T., Huang, W., Voigt, C., Bhatia, S.N.: Genetic encoding of targeted magnetic resonance imaging contrast agents for tumor imaging. ACS Synth. Biol. **9**, 392–401 (2020). https://doi.org/10.1021/acssynbio.9b00416

119. Kah, G., Chandran, R., Abrahamse, H.: Biogenic silver nanoparticles for targeted cancer therapy and enhancing photodynamic therapy. Cells **12**, 2012 (2023). https://doi.org/10.3390/cells12152012
120. Qian, Z., He, K., Feng, R., Chen, J., Li, B., Zhang, Y., Yu, S., Tang, K., Gan, N., Wu, Y.-X.: Intelligent biogenic missile for two-photon fluorescence imaging-guided combined photodynamic therapy and chemotherapy in tumors. Anal. Chem. **96**, 6674–6682 (2024). https://doi.org/10.1021/acs.analchem.4c00074
121. Pathania, D., Sharma, M., Sonu, Kumar, S., Thakur, P., Torino, E., Janas, D., Thakur, S.: Essential oil derived biosynthesis of metallic nano-particles: implementations above essence. Sustain. Mater. Technol. **30**, e00352 (2021). https://doi.org/10.1016/j.susmat.2021.e00352

Clinical Translation and Regulatory Landscape

Shubham Mehta, Gautam Patel, Pavan Karkare, and Jignesh Patel

Abstract The structure of the drug transport that uses Nano has become a device to change in cancer. But a successful combination in clinical practice, how to improve in the test tubes and in the standard body efforts for identifying clinical experiments and designs related to important achievements for translation success, including biological compatibility, profile, launching drugs and target setting techniques. The comprehensive inspection of landscapes, rules, and regulations of important agencies, including FDA and EMA, the work frame of the class and indicators for the approval of nano doctors also summarize the challenges of dealing with risk only particles. A case study of nano particles that have been accepted to present valuable in-depth information and important obstacles and events of their scientific travel, focusing on classes for the development of fate, ethics and society, including personal safety that is affected, transparent in clinical experiments and equally accessible. The end of bankruptcy emphasizes the procurement of fate, progress in science, supervision, mutual efforts between stakeholders and innovation to improve the translation method. This comprehensive analysis provides a guide for researchers, physicians, and regulators in navigating the complex area of nanoparticle-based cancer care.

Keywords Clinical translation · Nanoparticle-based drug delivery · Regulatory frameworks (FDA/EMA) · Ethical and societal implications

S. Mehta · G. Patel (✉) · P. Karkare
Chemistry Department, Parul Institute of Applied Science, Parul University, Vadodara, Gujarat, India
e-mail: gautampatel1573@yahoo.co.in

J. Patel
Clinical Research Department, Parul Sevashram Hospital, Parul University, Vadodara, Gujarat, India

N. J. Joshi and G. Patel (eds.), *Nanoparticle Delivery in Cancer Care*,
Series in BioEngineering, https://doi.org/10.1007/978-981-95-0457-2_9

1 Introduction

1.1 Brief Overview of Nanoparticle-Based Therapies in Cancer Care

Nearly every nation in the world has identified cancer as one of the major causes of death and a serious danger to life extension [1], and its incidence and mortality rates are rapidly rising globally. There were an estimated 10.3 million cancer deaths and 19.3 million new cancer cases globally in 2020 alone, due to its nature. Because of its nature, cancer starts when proto-oncogenes are damaged or mutated, which leads to uncontrolled cell division and proliferation. Nowadays, the most established and successful methods of treating cancer are still surgery and radiation therapy for the majority of local and non-metastatic malignant tumor diseases [2], with the potential risks for patients being the surgical wound, severe side effects, and anaesthesia [3]. Chemotherapy, which includes hormone and biological treatments, has emerged as a useful adjuvant treatment option for both local and metastatic malignancies [4]. However, the issues that came with it were toxicity, resistance to chemotherapy, and difficulties since the poisonous compounds in medicine not only prevent cancer cells from proliferating quickly but also normal cells like bone marrow and cells in the gastrointestinal tract [5]. Although several novel chemotherapy medications with reduced toxicity have been created as a result of advancements in drug production processes, side effect issues remain unresolved. Some fresh ideas or solutions ought to be put forth in light of this.

The advancement of nanotechnology has led to the production of a number of medications, including those connected to nanoparticles, that were created using this method and have been used in the investigation of cancer treatment [6]. Particles with a single dimension between 1 and 300 nm are known as nanoparticles (NPs), and they are also outfitted with characteristics based on their surface functions and size [7]. The lengthy circulation duration, therapeutic drug carrying capacity, and penetration into deep tumor tissues are among the common characteristics and capabilities of nanoparticles due to their molecular architectures and particle sizes [8, 9]. Furthermore, compared to conventional medications, nanoparticles' modifiability may make it simpler to functionalize them with specific markers. Additionally, taking into account several novel therapeutic approaches that have been put out recently, such as gene therapy and immunotherapy, which have raised the bar for medical proficiency, the Compared to other medications, the functionalization method of nanoparticles appears to be simpler to satisfy the requirement [10–12]. For instance, the creation of lipid- and polymer-based nanoparticles was a key component of RNA-based treatment [13, 14]. Experts have worked hard to create chemotherapeutic drugs, including nanoparticles, for the treatment of cancer because of these benefits.

Lipid-based nanoparticles, polymer-based nanoparticles, carbon-based nanoparticles, and metal nanoparticles were the general categories of nanoparticles [9, 15]. Additionally, the U.S. Food and Drug Administration (FDA) has authorized several medications connected to nanoparticles for clinical use, or they have been developed and assessed in clinical studies [16]. A few of these medications are intended to treat cancer. For instance, a significant amount of research has been done to assess the effectiveness of albumin-bound paclitaxel nanoparticles in treating lung and breast cancers [17–19]. The injection of nanoparticles was also implicated in a few other malignant tumor illnesses. Nevertheless, there hasn't been a comprehensive summary or evaluation of the principles and structure of those nanoparticles or the calibre of those clinical studies. There is no example of how nanoparticle-related medicine has advanced in clinical trials for cancer treatment. In light of these, we decided to conduct this systematic review to evaluate RCTs that used nanoparticles as cancer patient interventions and summarize their results [20], classifications, and side effects to offer recommendations for clinical practice and future research in the areas of nanoparticle cancer clinical applications (Table 1).

Table 1 Clinically approved cancer treatment medications based on nanoparticles [21]

Name	Active pharmaceutical ingredient (s)	Main NPs ingredient (s)	Targeted cancer	Approval (year)
Polymer-based NPs				
Oncaspar	ASNase	PEGylated protein	Acute lymphoblastic leukaemia	FDA (1994)
Genexol-PM	Paclitaxel	Micelle	Metastatic breast cancer; pancreatic cancer	South Korea (2007)
Eligard	Leuprolide acetate	PLGA	Prostate cancer	FDA (2002)
Zinostatin Stimalamer	NCS	SMA	Primary unrespectable hepatocellular carcinoma	Japan (1994)
Lipid-based NPs				
Mepact	Mifamurtide	Liposome	Non-metastatic, resectable osteosarcoma	EMA (2009)
DaunoXome	Daunorubicin	Liposome	HIV-associated Kaposi's sarcoma	FDA (1996)
Vyxeos	Cytarabine:Daunorubicin (5:1 M ratio)	Liposome	High-risk acute myeloid leukemia	FDA (2017) EMA (2018)
Albumin-based NPs				
Abraxane	Paclitaxel	Albumin	Lung cancer; metastatic breast cancer; metastatic pancreatic cancer	FDA (2005) EMA (2008)
Other NPs				
NanoTherm	Not applicable	Iron oxide	Glioblastoma	EMA (2011)

2 Clinical Application of Nanoparticles in the Treatment of Cancer

2.1 Preclinical Development

A Google Scholar search using the terms "nano-formulations, cancer cure" yielded about 3000 results in 0.08 s; more than 18,000 of these articles were based on pharmaceuticals. However, it is disheartening to the researchers that a large portion of this fundamental research was not translated into useful applications. Although the first analogue of the drug Doxil was first brought to the market in 1995, 25 years ago, there are currently very few approved nanomedicines for the treatment of cancer. Although a great deal of research has been done to use nanoparticles for cancer therapeutic applications, very few formulations have made it to the clinical stage in all these years. According to a brief analysis of the NP-based formulations that are presently on the market and undergoing clinical trials, very few nano-formulations have been authorized for use as cancer medications [22]. About 75 nano-formulations are presently undergoing clinical trials, and 16 cancer medications based on nanotechnology have received FDA approval [23]. The five primary cancer nanomedicines currently available on the market are listed in Fig. 1.

2.1.1 Polymeric Micelles, Polymer/Lipids, Metal and Metal Oxide Nanoparticles, and Other Mixtures

The other drugs are metal and metal oxide, polymeric micelles, polymer/lipids, and other conjugates, and they are in different phases of approval or research (Fig. 2). Platinum is one of the metals most frequently used to deliver anticancer medications; cisplatin was the first anticancer medication to be used, and many more were used after it. Platinol was approved by the FDA to treat several cancers in combination with picoplatin, carboplatin, sebriplatin, ormaplatin, oxaliplatin, enloplatin, satraplatin, zeniplatin, carboplatin, and picoplatin [25]. However, the hegemony of platinum and

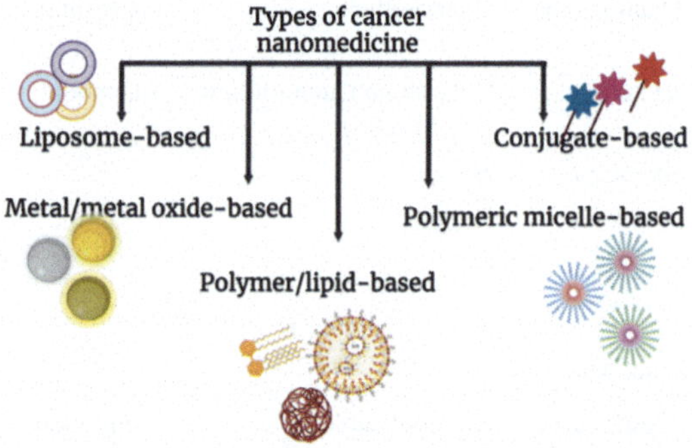

Fig. 1 There are several kinds of nanomedicines that are used to treat cancer [24]

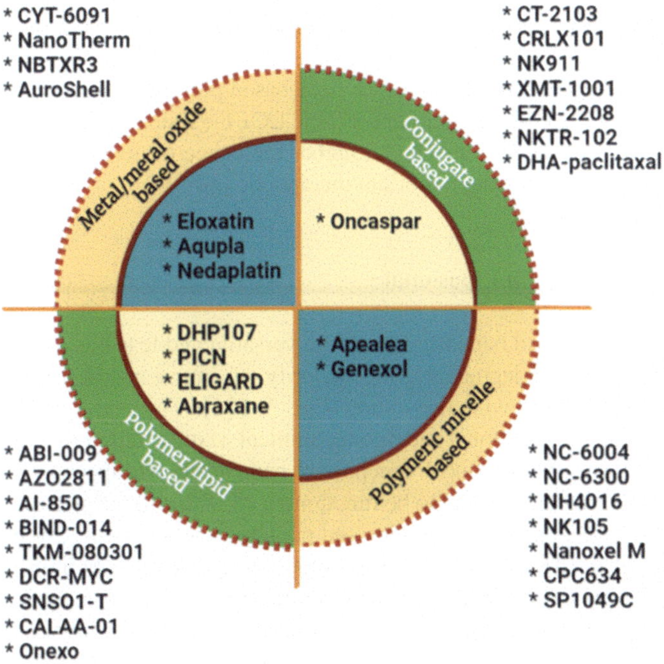

Fig. 2 Medicines used in nano-chemotherapy that are based on polymeric micelles, metal and metal oxides, polymer/lipids, and other mixtures [24]

platinum-containing drugs declined in subsequent years as a number of issues arose from their use in clinical settings.

The medication can be administered from nanoscale vesicles of self-assembled amphiphilic micelles by penetrating the endothelial cell junctions of cancer cells and exploding close to the tumor microenvironment. These polymeric micelles are superior drug administrators compared to liposomes and other lipid nanoparticles because they more efficiently transport the drug carrier within the tumor [26]. When used as drug carriers, core–shell shaped polymeric micelles improve drug solubility. Water-insoluble drugs are encapsulated in the micelle core by chemical or physical conjugation, which increases the drug's accessibility to the site of action [27].

In recent years, a novel nano-drug system was created by taking advantage of advancements in lipid-based polymeric systems. The hybrid structure of the lipid-polymer nanoparticles is made up of a lipid shell around a polymeric core [28]. It is hypothesized that they would combine the advantages of lipid-based nano-carriers, such as their ease of drug release, with other properties including surface modification and functionalization. A receptor-targeting polymer is connected to nanosheets, which serve as a delivery platform for the different components of a nano-conjugate drug system [27–29]. By focusing on the tumor, novel cytotoxic drugs are created to optimize the anticancer impact. One such innovative semisynthetic taxane compound that has no cytotoxic effect until it is converted to paclitaxel is the fatty acid docosahexaenoic acid (DHA–paclitaxel) conjugated with the antitumor medication paclitaxel [30]. In recent years, a novel nano-drug system was created by taking advantage of advancements in lipid-based polymeric systems. In patients with advanced non-small cell lung cancer (NSCLC), the tested dosages of the taxane-conjugated medication demonstrated a higher incidence of myelosuppression than traditional taxane, suggesting important therapeutic utility.

2.1.2 Nanoparticles in Liposomal

Because liposome-based nanodrugs are more versatile and feasible, as well as more biocompatible and biodegradable, the majority of cancer medications based on nanoparticles are made with liposomes (Fig. 3) [31]. The majority of approved medications and medications undergoing clinical testing fall into this category primarily because of liposomes' exceptional capacity to encapsulate hydrophilic and hydrophobic molecules and to be functionalized with a range of molecules such as PEG, antibodies, aptamers, proteins and peptides, carbohydrates, or other small molecules in targeted liposomes [32]. Liposomes' bilayered phospholipid membrane, which is resistant to both chemicals and the immune system, shields drugs inside from proteasomal breakdown and biological inactivation during drug delivery [33]. The primary benefit is that the drug is safely encapsulated within the liposomal core, preventing the drug from affecting the healthy, non-targeted cells that surround the drug target.

2.2 Design Considerations for Translational Success

One area where nanochemodrugs have moved forward in an impressive way is in immunotherapy. Numerous targets for immunotherapy were made available when it was found that many patients with breast cancer and other adenocarcinomas have humoral and cellular immune responses to tumor-associated antigens (TAA) that are overexpressed, like MUC1. Phase III clinical trials are being conducted to test tecemotide, a MUC1-specific cancer immunotherapy, for the treatment of stage IIIA/IIIB non-small cell lung cancer [34]. Similarly, a phase 1 trial for Lipovaxin-MM, a liposomal vaccine that targets dendrically, was initiated in the case of malignant melanoma [35]. Using cyclodextrin polymeric nanoparticle (CDP) technology, CRLX101 is a first-in-class nanopharmaceutical that may transform treatment into beneficial clinical outcomes [36]. The substantial drug-binding effectiveness of the nanoparticle-conjugated albumin (Nab) complex is one of the reasons why the majority of nanochemodrugs fall into this category. Gemcitabine and Nab-paclitaxel's combined therapeutic potential [37], cyclophosphamide [38], and atezolizumab [39] are being tested for early-stage and metastatic breast cancer. Another Nab-paclitaxel combination, ABI-007, has undergone clinical trials for metastatic breast cancer and stage IV NSCLC. Some of the nanoparticles in different

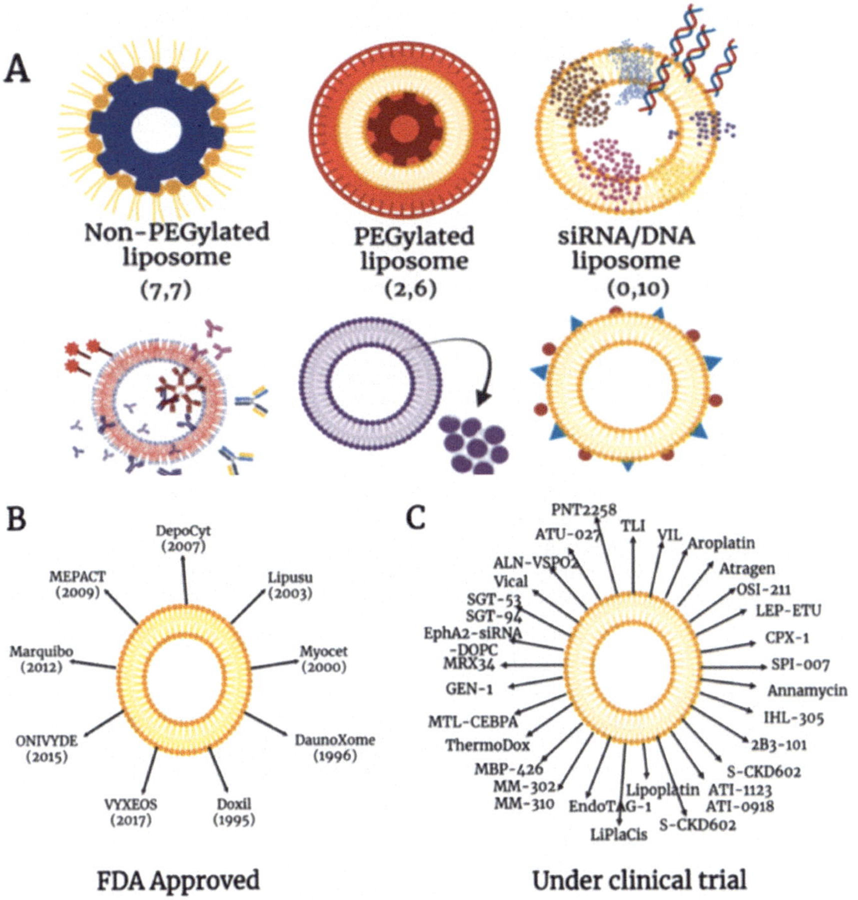

Fig. 3 **a** Different types of nanomedicines based on liposomes. The range of formulations within the category indicates how adaptable the liposomes are for creating nano-drugs. The quantity in parentheses denotes the number of medications that are pending approval and trials, respectively, under every category [24]. **b** Key FDA-approved liposome-based Nano chemotherapeutic medications [24]. **c** A few of the clinically tested liposome-based nano chemotherapeutic medications [24]

shapes that have been clinically studied for various cancer types are included in Table 2.

Aside from that, some gold nanoconstructs have reached the clinical trial phase, although the large body of literature detailing new and strange medicinal formulations involving AuNP has yet to contribute any. This imbalance between preclinical and clinical trial success comes from most research being materials science-oriented with the aim of producing multifunctional NPs with new properties instead of focusing on qualities needed in the clinic [40]. For instance, inorganic and organic NPs that have made it to the clinic are often straightforward and made up of a few components,

Table 2 A few of the nanomedicines that have recently been the subject of clinical trials [24]

Nano drug	Conventional drug	Cancer type	Clinical trials.gov identifier
NC-6004 (NP-cisplatin)	Gemcitabine	Solid tumors	NCT02240238
Docetaxel-PNP	Taxotere	Solid tumors	NCT02274610
Paclitaxel Nab	Alisertib	Adenocarcinoma, pancreatic neoplasms	NCT01677559
	Cyclophosphamide, Trastuzumab, Bevacizumab	Breast cancer	NCT00254592
	Pembrolizumab	Metastatic urothelial carcinoma	NCT03464734
	Carboplatin, Erlotinib hydrochloride	NSCLC	NCT01928160
	Doxorubicin hydrochloride, Cyclophosphamide, Filgrastim, Trastuzumab	Estrogen receptor-positive Breast cancer HER2-positive breast cancer	NCT00407888
	Cetuximab, IMRT (Intensity-modulated radiation therapy)	Head and neck cancer	NCT00736619
	Imiquimod	Male breast cancer, Recurrent breast cancer, Skin metastases, Stage IV breast cancer	NCT00821964
	Gemcitabine, Capecitabine	Pancreatic neoplasms, Pancreatic cancer, Adenocarcinoma	NCT01161186
	Paclical®, Taxol®	Epithelial ovarian cancer, Primary peritoneal cancer, Fallopian tube cancer	NCT00989131
	PIPAC	Peritoneal carcinomatosis, Ovarian cancer, Breast cancer, Stomach cancer, Pancreatic cancer	NCT03304210
ONPATTRO	Patisiran	Transthyretin amyloidosis	NCT03862807
VYXEOS	Cytarabine, daunorubicin	Acute myeloid leukemia	NCT04920500
CRLX101 (cyclodextrin-based polymer)	Camptothecin	NSCLC, Primary peritoneal cancer	NCT01380769

(continued)

Table 2 (continued)

Nano drug	Conventional drug	Cancer type	Clinical trials.gov identifier
AGuIX	Polysiloxane gadolinium-chelates based nanoparticles	Brain cancer	NCT02820454
CPC634 (CriPec®)	Docetaxel	Ovarian cancer	NCT03742713

which facilitates the prediction of them in vivo behaviour and the scaling up of their manufacture. Even though AuNPs might not be hazardous in the short term, long-term adverse effects are a cause for worry. In light of the aforementioned, provide a summary of the therapeutic AuNP clinical environment as of right now. By examining how AuNPs can truly benefit patient therapy by examining the properties of gold nanoconstructs investigated in the clinic (Table 3).

Additionally, one of the most researched plant metabolites, flavonoids, includes a variety of phenolic groups. Experimental investigations of flavonoids as anti-cancer treatments are described in a number of papers. Until recently, there was no direct proof from controlled clinical studies that flavonoids had anti-cancer properties. However, clinical research has started to bridge the gap between laboratory experiments and actual clinical practice in the current state of affairs. Researchers are now attempting to investigate flavonoids therapeutically, with several flavonoids qualifying at various stages of clinical trials, in light of their pre-clinical anti-cancer and cancer-preventive potential (Table 4).

2.3 Challenges in Clinical Translation

The process of getting nano drugs from the clinic to the market still faces significant challenges despite all these advancements. The primary deterring issue is a lack of knowledge regarding the nanoparticle's mode of action with biomolecules. Continuous manufacture of nanoparticles that have long-lasting impacts on cancer cells frequently turns into a problem for pharmaceutical corporations. Pharmaceutical firms have withdrawn certain medications, such as DepoCyt. These businesses attribute the obstacle to technical problems. Before a medicine is marketed as a cancer cure, researchers and pharmaceutical corporations must address several challenges. These study findings aren't decoded and promoted for a few reasons. There are a number of problems that might make the procedure likely to encounter difficulties. Here are some of the topics that are covered.

Table 3 Clinical studies of therapeutic gold nanoparticles [41]

Name	NPs type	Applications	Clinical study
NU-0129	siRNA and thiolated PEG made on the surface of 13-nm AuNP	Therapy for glioblastoma	NCT03020017, Phase 0, Completed (2017–2020)
CNM-Au8	13-nm AuNP in drinking bicarbonate solution	Treatment of various neurodegenerative diseases	NCT02755870, Phase I, Completed (2015–2016) NCT03815916, Phase II, Completed (2019–2021)
CYT-6091	PEGylated 27-nm AuNP functionalized with TNF	Anti-tumor therapy through immune response regulation	NCT00356980, Phase 1, Completed (2006–2009) NCT00436410, Phase 0, Completed (2006–2009)
naNO-COVID	T-cell priming peptide cocktails of coronavirus conjugated on the surface of AuNP	Vaccines against SARS-CoV-2	NCT05113862, Phase I, Active (2022–)
AuroShell	PEGylated 120-nm silica core and 12 to 15-nm gold shell	Photothermal ablation of various types of cancer	NCT00848042, Completed (2008–2014) NCT01679470, Terminated (2012–2014) NCT02680535, Completed (2016–2020) NCT04240639, Recruiting (2020–)
C19-A3 AuNP	Human proinsulin peptide (C19-A3) conjugated with ultrasmall AuNP (<5 nm)	Treatment of the autoimmune disease type 1 diabetes	NCT02837094, Phase I, Active (2016–)
EMX-001 (naNO DENGUE)	T-cell priming peptide cocktails of dengue virus conjugated on the surface of AuNP	Vaccines against dengue fever	NCT04935801, Phase I, Active (2021–)

2.3.1 The Drug's Internalization in Cells

Typically, a medication circulates and builds up at the site of action before being absorbed by the cell. Due mostly to the drug's incapacity to enter tumor cells, this technique continues to be a bottleneck in the development of nano-chemotherapeutic medications [43]. On the other hand, unlike conventional pharmaceuticals, the entire intracellular, paracellular, and transcellular pathways of the nano-drug across any biological membrane are not fully understood, which makes cellular internalization of nano chemotherapeutic medications a barrier to their clinical development. Only once the internalization pathway has been well characterized can the pharmaceutical formulation be released. When an apolipo-protein-based nanomedicines with the capacity to convert zwitter-ionic to cationic charges was

Table 4 Flavonoid nanotechnology and their clinical studies [42]

Flavonoid	Cancer type	Objective	Clinical phase	Clinical study
Watercress	Long-term effects secondary to cancer therapy in adults	Chemopreventive activity of flavonoids in breast cancer and colon cancer	Phase 3	NCT02468882
EGCG	Breast cancer	Clinical study of EGCG for skin prevention in patients with breast cancer undergoing adjuvant radiotherapy	Phase 2	NCT02580279
Catechin	Liver cancer	Effectiveness of catechin in preventing liver cancer in patients with cirrhosis	Phase 1	NCT03278925
EGCG	Lung cancer	Whether ECGC can be applied for esophageal protection in patients with lung cancer undergoing radial radiotherapy	Phase 2	NCT02577393
Sinecatechins	Skin cancer	Topical application of green tea ointment in the management of superficial skin cancer	Phase 2	NCT02029352
Cocoa extract	Cancer	Chemo-preventive activity of flavonoid found in cocoa extract	Phase 4	NCT02422745
Quercetin	Chemotherapy induced oral mucositis	Chemotherapy-induced oral mucositis prevention and treatment in patients with malignancies of the blood	Phase 1	NCT01732393

created, strategic research was conducted to address this conundrum [44]. The tumor microenvironment reacts favourably to these pH-responsive Nano molecules. The effectiveness of the nanomedicines' cellular internalization is enhanced by the surface charge conversion that the nanodrugs produces. Recently, redox-responsive polymeric micelle conjugates have been designed to target proteins, which aids in the release of the loaded medication to the intended region while preventing complicated cellular internalization [45].

2.3.2 Breaking Through Biological Barriers and Increasing Bioavailability

Biological barriers restrict the build-up of nano-therapeutic medications at disease progression sites. Occasionally, the medication molecules just spread and circulate freely throughout the body, leading to harmful side effects. Additionally, this will lessen the impact or reaction that the doctor anticipated the recommended dosages would have [46]. At this stage, the need for an appropriate medication delivery mechanism cannot be overlooked. The distribution of nano-chemo drugs that can pass various biological barriers has been made feasible by nanocarriers, revolutionizing the targeted drug delivery field. According to a recent study, Prunus spinose fruits (PSF) with various phenolic compounds can be used to create biomimetic nanoparticles that might enhance the extract's effectiveness by directing the payload into the lipid bilayer with more build-up in the afflicted tissue. When the nanoparticle was tested on HUVEC cells, it showed improved biocompatibility and reduced toxicity, indicating a novel platform for encapsulating medications with limited stability and bioavailability [47].

2.3.3 The Challenge of Forecasting a Patient's Propensity for Allergic Reactions

Even when patients get the same dosage of medication, it can be exceedingly challenging to anticipate a specific pattern or behaviour. Even while the drug's pertinent mechanisms of action in a diverse patient group throughout the clinical research were remarkably comparable, they can have unanticipated and conflicting adverse effects when they are administered to a large number of individuals globally. Similarly, a straightforward evaluation of adverse effects does not ensure the medication's safety [48]. This is a complicated situation in and of itself since the medication has an endless amount of potential to react with various substances that it encounters in the body, and it is crucial to carefully investigate and evaluate this interaction at the molecular level.

2.3.4 Regulating the Immune System's Reaction to the Novel Drug

When a new chemical is introduced into the body, the immune system's response is unpredictable. This is one of the primary causes of the possibility that current drug therapies could improve survival rates without truly curing the disease. A human's immune system is sophisticated and intricate. It would be a "dream-come-true" scenario if the mechanism could be properly deciphered. Products that regulate the immune system can be developed to treat a variety of diseases. Immunotherapy is a new cancer treatment that modifies the immune system [49]. In this case, a T-cell infusion system was created that will replicate the process that is mediated by the donor's immune system rather than the host's. The process produced the mirror. By

Table 5 There are challenges in applying nanoparticles as clinical cancer treatments [24]

Challenges	References
Spending less time in the body	[51]
Getting regulatory approval to sell the drug is a difficult experience	[52]
Deficient elimination and biodegradation	[53]
Formulation stability problems	[54]
Inability to scale up in vitro results to in vivo tests	[55]
Inability of the drug formulation to selectively adsorb onto the target	[55]
The long process of creating new drugs	[56]

producing a host versus tumor (HVT) effect and a non-toxic host versus graft (HVG) reaction, EffectTM helps treat cancer and other illnesses. Likewise, new concepts that are centered on how to use and exploit the manner in which the human immune system combats cancer and other illnesses may transform drug discovery research in the future. Such curative effects may transform the pharmacology and chemotherapy sectors.

The other problems faced with this operation have been enumerated in Table 5. For a theoretical drug delivery system to be more effective as a therapy, the drug must be delivered to the precise location of the body. The power of a drug to cause apoptosis in cancerous cells is indicative of its activity against the disease. The therapeutic molecule's ability to reach the pharmacological target with the fewest side effects is crucial in this situation. The EPR effect, which results in selective drug accumulation in tumors; the drugs' poor bio-accessibility to tumor tissues; and inadequate cellular absorption, which calls for higher dosages, are some issues that must be resolved for therapy to be effective. Higher dosages raise the risk of multiple drug resistance and increase toxicity to healthy cells. According to recent studies, nanoparticles can particularly help with these issues without endangering healthy cells [50]. They can be used to ensure that cancer cells are eliminated while also effectively preventing unwanted toxicity to healthy cells. Using a variety of delivery methods, nanoparticles are currently being determined to be the most effective option for drug delivery. A graphical depiction of some of these delivery routes can be found in Fig. 4.

3 Regulatory Landscape for Nanoparticles in Cancer Care

3.1 Overview of Regulatory Bodies

Our traditional understanding of how drugs are transported has evolved with the introduction of successful nano-pharmaceuticals, sparking hopes that even more advanced nanosystems will emerge in the future. Furthermore, the competitive pharmaceutical industry is starting to generate generic copies of proprietary nano-pharmaceutical

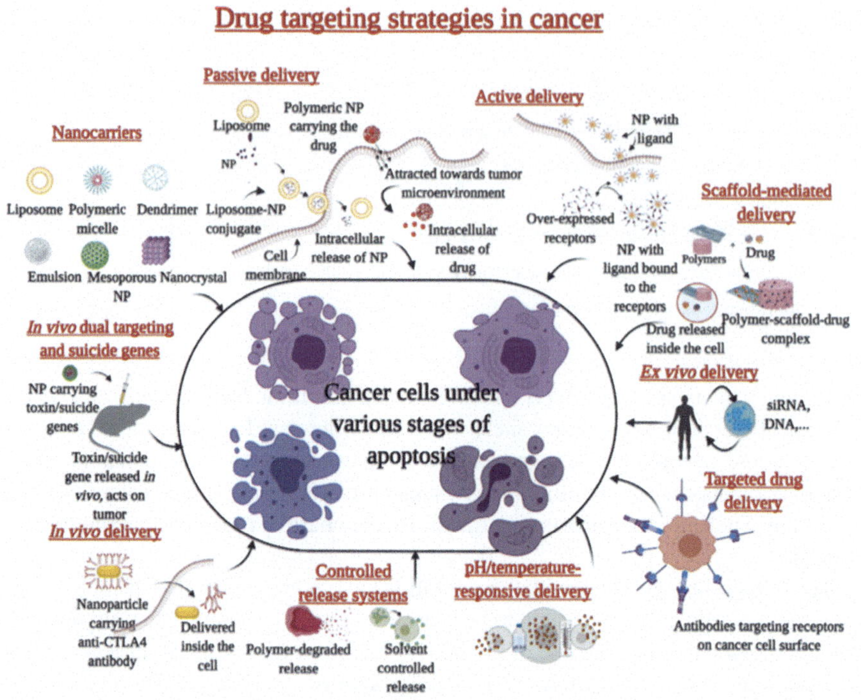

Fig. 4 Diagrammatic illustration of how to use drug targeting to introduce therapeutic compounds into cancer cells. By ensuring that the drug reaches the target cells efficiently, the different drug delivery techniques reduce the possibility of unintentional harm to healthy cells [24]

goods once the patent expires [57]. The pharmaceutical sector must expand in this manner, and regulatory bodies worldwide must establish clear regulatory guidelines. The clinical use of nanotechnology, or nanomedicine, aids in the development of nano-products and the resolution of issues facing the healthcare sector. Personalized medication may be developed by nanomedicine, which offers superior pharmacokinetic and pharmacodynamics effects compared to conventional or micro-sized medicinal products [32, 33]. According to the regulatory clearance of Abraxane® and other comparable drugs generated from nanotechnology, nanomedicine has recently made the transition from the lab to the clinic [58]. Furthermore, the lack of novel pharmaceutical products especially those with nano origins is a difficult issue in the US, EU, and other pharmaceutical markets, which is why generic nanomedicines were approved quickly. Additionally, the nano-formulation patent period is gradually coming to an end, opening the door for the producer to launch generic nanomedicine. Therefore, a thorough grasp of the regulatory data that may be applied to the unique characteristics of nanomedicines is necessary for drug manufacture. Intellectual property, technological concerns, development expenses, ethics, and regulatory approval in the relevant markets are some of the major obstacles to the creation of novel nanomedicine or the launch of generic nanomedicine.

In addition, among the several cancer medications now on the market, doxorubicin HCl liposome injection plays a significant role in cancer treatment. The FDA authorized doxorubicin HCl liposome injection in 1995. Janssen presently markets the product under the names Doxil® in the US and Japan and Caelyx® in other countries. Multiple myeloma, ovarian cancer, and Kaposi's sarcoma associated with AIDS are treated using the formulation. It is also approved in the European Union for breast cancer in patients who are at risk of anthracycline cardiotoxicity [59], in addition to the previously mentioned groups. On BALB/c mice with syngeneic fibrosarcoma, the antitumor effectiveness and safety of Sun Pharmaceutical Industry Ltd.'s doxorubicin HCl liposome injection were assessed. Athymic nude mice that carried MX-1 indicated that it was effective at all tested dosages when compared to the reference medication, Doxil.

Moreover, Paclitaxel human serum albumin protein (HAS)-bound nanoparticles, known as Abraxane, are used for a number of indications, including pancreatic adenocarcinoma, non-small cell lung cancer (NSCLC), and metastatic breast cancer [60]. Instead of doing comprehensive safety testing for the generic form of Abraxane, the FDA solely took into account bioequivalence studies when evaluating ANDA applications [61]. The USFDA granted Panacea Biotec an ANDA in 2018 for submitting a generic version of Abraxane. Subsequently, Abraxis Bioscience and Panacea Biotec, along with its partners Apotex Inc. and Apotex Corporation, agreed to develop and commercialize the generic version in the United States (2017c). A positive judgment was also adopted by the EMA-affiliated Committee for Medicinal Products for Human Use (CHMP), which recommended the marketing authorization of Pazenir (Teva B.V.), a generic form of Abraxane. The EMA mainly took into account quality studies and research on the advantages and disadvantages of the active ingredient in the approved indication.

3.2 Guidelines for Nanomedicine Approvals

3.2.1 India Guidelines

In India, regulation of pharmaceuticals and medical devices is done by the Central Drugs Standard Control Organization (CDSCO). The physicochemical characteristics of the nanomaterial employed in the therapeutic product are taken into account while evaluating nanopharmaceuticals for quality, safety, and efficacy on a case-by-case basis. In 2019, the "Guidelines for Evaluation of Nanopharmaceuticals in India" guideline document was released. The inter-ministerial Expert Committee, which was established by the Department of Biotechnology and consisted of specialists from several Government of India ministries, worked together to create this guideline paper. This guideline's definition of a nanomaterial is comparable to the USFDA and EMA's definitions. The guidelines classify nanopharmaceuticals into the following categories:

(a) Nanomaterial degradability: the material's capacity to break down naturally or not.
(b) Nanomaterial nature: the material may be organic or inorganic, and it may be multi-component, depending on the synthesis.
(c) Nano form of the substance: this group includes either the active pharmaceutical ingredient (API) transformed into a nanoparticulate or nanocrystal form or nanocarriers loaded with the API, such as liposomes, polymeric nanoparticles, etc.

Additionally, included in the guideline are the specifications for information pertaining to preclinical research, clinical trials, drug product stability data, and data needed for nanopharmaceuticals assessment.

3.2.2 USFDA Guidelines

Over the years, the USFDA has not released any recommendations specifically for nanomedicines. The USFDA created guidelines for pharmaceutical products that combine biological products and nanomaterials in 2017. In the past, the USFDA assessed items based on nanotechnology on an individual basis. Materials with a nanoscale range exhibit distinct biological impacts and physicochemical characteristics from those with a larger size. Pharmaceutical formulations' qualities are improved by using this dimension-dependent phenomenon. Due to the drug product's increased potency, decreased toxicity profile, and targeted delivery of the drug molecules at the site of action, several properties of formulations in the nanoscale range have been demonstrated, including increased bioavailability, decreased dose, and decreased dosing frequency. The USFDA has not yet defined the regulatory meanings of the terms "nanotechnology," "nanomaterial," "nanoscale," or any other term pertaining to pharmaceutical goods based on nanoformulations. The USFDA will now question whether a drug product utilizing new technology has an external dimension or surface structure somewhere between the nanoscale range. If the dimensions of the drug product are beyond this range, up to one micrometer, the USFDA will also question how the dimensions influence the physicochemical properties and biological activity of the product.

Along with describing the drug substance, the information should go into detail about the drug product's composition, including lipid, non-lipid, and non-lipid excipients (buffers), physicochemical parameters, manufacturing processes, stability studies, and product-specific guidelines. Clinical pharmacology studies, details on drug release kinetics, and a comparison of the liposome and non-liposome drug products should all be included in the document's human pharmacokinetics section to comprehend the different biopharmaceutical parameters of the novel formulation. The guideline also includes a particular suggestion for the information that should be provided on the label of liposome medicinal products.

3.2.3 European Guidelines

In 2009, the European Medicines Agency (EMA) established the Committee for Medicinal Products for Human Use (CHMP), an ad hoc nanomedicine expert group. The ad hoc group advises on whether to accept or reject nanomedicines by the EU. A product with one or more external dimensions between 1 and 100 nm was referred to as a "nanomaterial" by the European Commission in 2011. The CHMP assists in the creation of technical publications that discuss the present state of EU nanomedicine regulation. The EMA has previously reviewed and approved nanomedicines on an individual basis, much like the USFDA. EMA has released many reflection papers about a particular nanomedicine product, but at this time, the agency has no standards for nanomedicines or nanosimilars. In some areas where scientific information is growing and experience is comparatively limited, the agency also prepares a reflection paper to provide an outline for discussion and clarification. Preparation of block copolymer micelle drug products, intravenous iron-containing nanocolloidal drug products, intravenous liposome drug products, and general surface coating considerations for parenteral administration of coated nanomedicine products are all addressed in reflection papers, which also discuss the data requirements for nanotechnology-based products.

3.3 Challenges in Regulatory Approval

The process of developing a new product with therapeutic utility is a logical and methodical approach that gathers data from studies on the product's physicochemical characteristics and biological activity to "proof of concept" in order to prevent waste of time and resources [62]. When reporting scientific data, there are several critical considerations. Besides regulatory concerns and agency recommendations, upfront research must focus on the stability of manufacture of the nanoformulation as well as its physicochemical characterization [63]. When choosing the appropriate tissue and animal model, the first justifications for the medication should be taken into account, and the biological activity demonstrated in in vitro testing such as cell lines should coincide with the in vivo assays [64]. The many problems that have arisen have been explained by the presentation of the different systems that regulatory agencies have approved [65]. We detail each parameter and the challenges of making NPs for biological applications in the subsequent sections.

3.3.1 Physical–Chemical Description

Unlike the creation of other therapeutic agents, complete physicochemical characterization of NPs is the initial step in assessing their potential for toxicity in biological systems. This early and essential stage of development underpins the principles of efficacy, safety, and quality [66]. Early in the course of product development, appropriate characterization techniques that predict toxicological and therapeutic outcomes and relate effect and biological consequences must be employed to render a product effective in the clinical environment [67]. NPs are spread across all organs and systems and traverse endothelial and epithelial barriers [68]. They are even capable of entering cells through multiple routes and diffusing into cell membranes [69]. Physical and chemical properties of NPs are responsible for their toxicity [69], as is the connection between these properties and undesirable effects such as platelet aggregation and thrombosis. Vascular hemostasis is impacted by and susceptible to thrombosis from carbon nanoparticles used in drug delivery and research [70]. Inflammation and neurological or cardiovascular disorders are additional negative consequences [71]. They can enter cells through the nucleus and mitochondria, changing metabolism, damaging DNA, and resulting in cell death [72].

One of the techniques used to change the physicochemical properties of nanoparticles is functionalization, which entails the attachment of molecules such as folic acid, biotin, oligonucleotides, peptides, monoclonal antibodies, and functional groups onto the particle surface [73]. It allows for the addition or modification of specific properties in the NPs with high precision [74]. The functionalized particles possess good physical properties, anti-agglomeration properties, and non-invasive properties [75]. Non-covalent or covalent bonds are possible for binding. Modifications can be performed in non-covalent conjugation without disturbing the molecular structure. For example, image-guided chemo-photothermal therapy was facilitated by Yue et al.'s (2019) noncovalent functionalization method of cucurbit-7-uril to gold nanoparticles [76]. For covalent bonding, linker molecules such as PEG can be employed to form structures with multiple applications for medicinal and diagnostic use [77]. For example, Chen et al. (2017) designed antigen-binding fragment (Fab)-conjugated antibody micelles using PEG as a spacer to enhance the NPs' surface Fab density and bioactivity [78].

Additionally, their limited approval for human therapies is primarily due to a lack of particle analysis [79]. Drug characteristics should include encapsulation efficiency, evaluating the encapsulated drug, and figuring out the drug's size, shape, content, charge, and surface chemistry. By considering factors like loading, distribution, release kinetics, cell contact, and transport system (Fig. 5) [65]. Specialized techniques to physicochemically and biologically analyse nanomedicines are frequently lacking, and regulations are required to evaluate the safety and quality of these novel products [80]. Every NP may need to be evaluated using new or altered techniques [81].

Selecting the best method for measuring the physicochemical parameters presents challenges even at the start of the characterization process. Dynamic light scattering (DLS) is the most popular technique for determining the diameter, particle size distribution (PSD), and stability of suspended nanoparticles (NPs). On the nanoscale, the procedure conforms with International Standards Guidelines [81], despite the fact that the results cannot be replicated [83]. It can only function if the particles tend to be monodisperse, and its limitations include tracking the lowest diameter

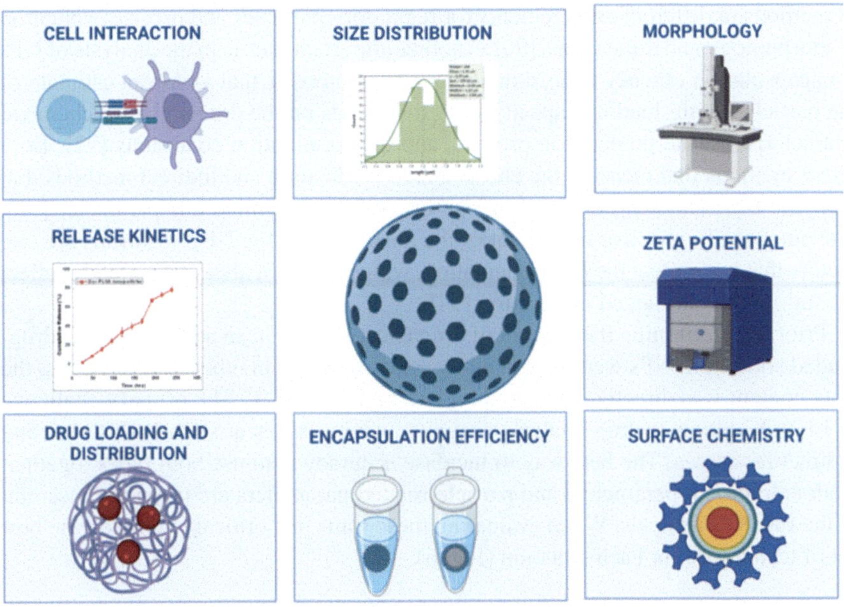

Fig. 5 Key factors to consider when characterizing biological NPs physicochemically [82]

changes [84] and figuring out the size distribution. Using multiple techniques in complementary ways is the best way to estimate size, PSD, and stability [85]. Electron microscopy (EM), which generates images with higher contrast, high atomic number elements, and a resolution of less than 1 nm, is the most efficient technique for figuring out the size and shape of NPs [86]. Numerous modifications have been made to this technique, including transmission electron microscopy (TEM), which allows for the observation of a large population of particles at high lateral resolution [82], scanning electron microscopy (SEM), which detects surface defects and topochemical information [87], and both combined: transmission and scanning electron microscopy (STEM), which allows for the determination of the size distribution, interfacial structure, compositional distribution, and phases of the NPs [88].

However, because these methods rely on a very small, non-representative sample for evaluation, they are unable to detect the presence of agglomerates and accurately determine the organic coating or coverage of individual particles [89]. Particle surface characteristics are altered or destroyed by electron beam degradation caused by SEM analysis [87], but particle thickness cannot be measured by TEM diagnostics [82]. Without the need for an incident electron beam or sample pre-treatment, atomic force microscopy (AFM), a substitute for mass spectrometry (MS), provides a three-dimensional characterization of the NPs by measuring their thickness and monitoring surface heterogeneity [85]. The probe (tip) sets the observation, which overestimates the dimensions and restricts the lateral resolution. The particle conformation is also altered by sample preparation [82]. Regulations suggest the incorporation of a variety of methods in different arrangements for regulatory purposes and precise evaluation of morphology and dimensions [90]. Another important metric in the analysis of NPs is encapsulation efficacy (EE), a qualitative characteristic that yields an estimate of the particles' drug-loading capacity [91]. It depends on the drug's physicochemical characteristics, the preparation process, and the formulation conditions [92]. Both direct methods that measure the encapsulated medication and indirect methods that compute the non-encapsulated drug can be used to determine this parameter [93]. The amount of the active ingredient or the conditions of the NPs' synthesis are two factors that may affect the EE's calculation, which does not always accurately reflect the ratio of the entrapped medication [94].

Prior to determining the amount of drug encapsulated in an aqueous phase, drug-loaded solid lipid NPs were separated by centrifugation; in other investigations, the drug content was directly calculated to determine the EE%. The primary challenge in EE calculation is drug analysis accuracy, where issues arise in both direct and indirect processes. The best way to increase accuracy is to use both types together. Both entrapment parameters and particle-related parameters are taken into account in the full NPs analysis. When evaluating these traits, it's critical to employ the best set of techniques for each situation (Fig. 6).

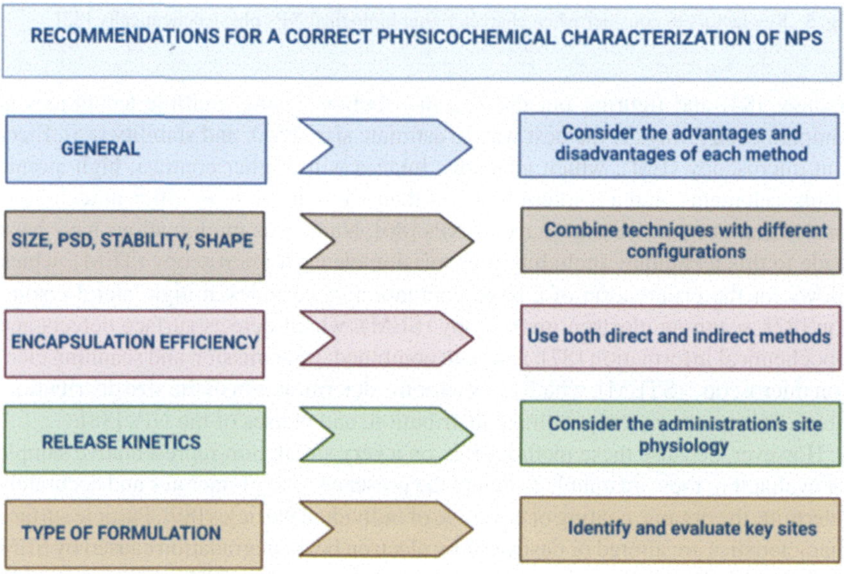

Fig. 6 Suggestions for accurately characterizing nanoformulations physicochemically [82]

4 Strategies to Overcome Barriers in Nanoparticle

4.1 Enhancing Targeting Specificity and Bioavailability

Many medications that go through presystemic metabolism through the lymphatic system can have their bioavailability increased by using nanocarriers such as polymeric nanoparticles, liposomes, dendrimers, nanospheres, nanocapsules, etc. [95]. The mucous layer and the GIT epithelium serve as the two primary entrance points for nanocarriers in the lymphatic system via the oral route. Mucus is a gel-like fluid with epithelial cells lining it. Additionally, epithelial cells serve as barriers for the nanocarriers. Nanocarriers primarily use one of two methods to get through mucosal and/or epithelial barriers: paracellular transport or transcellular transport [96]. Passive diffusion over intercellular gaps is known as paracellular transport. The ideal method for transporting hydrophilic drugs is the paracellular pathway. Junctional complexes either collect more than 1 nm between cells or decrease or prevent macromolecules from passing through. The polymeric nanoparticles are unable to travel by paracellular transport because of this. The drug's paracellular transport can be enhanced by employing permeation enhancers, including calcium chelators, polymers, and cationic and anionic [97]. Transcellular transportation can be either active or absorptive. Nanoparticles can be absorbed by enterocytes or Peyer's patch M cells via a transcellular route. There are four primary processes involved in transcellular transport: (1) phagocytosis, (2) clathrin-mediated endocytosis, (3) micropinocytosis, and (4) caveolin-mediated endocytosis. Enterocytes can assist in moving tiny particles down the transcellular channel, while M-cells can absorb larger particles, such as polymeric nanoparticles [98]. Figure 7 depicts these different drug transportation pathways.

4.1.1 Lipid Nanoparticles Made Absorption Easier

One kind of colloidal medicine carrier is a lipid nanoparticle. Lipoprotein nanoparticles that vary in size from 50 to 1000 nm enter the lymphatic system. A paracellular or intercellular route allows the lipid nanoparticles to enter cells. Size affects how well lipid nanoparticles are absorbed by M-cells [100]. Gastrointestinal barriers keep anticancer drugs such as docetaxel and paclitaxel from being orally accessible; multidrug efflux proteins such as p-glycoproteins are the main reason. Modified lipid nanoparticles containing P-gp inhibitors have been used to address this issue and increase the bioavailability of certain anticancer medications. Lipid nanoparticles are, therefore, more effective for treating many illnesses, including oral cancer treatment [101]. The literature claims that glibenclamide is a solid lipid nanoparticle that was produced by emulsion solvent evaporation. Glibenclamide's systemic availability is increased and its first-pass metabolism is decreased in male Sprague–Dawley rats due to improved lymphatic targeting in their pharmacokinetics [102].

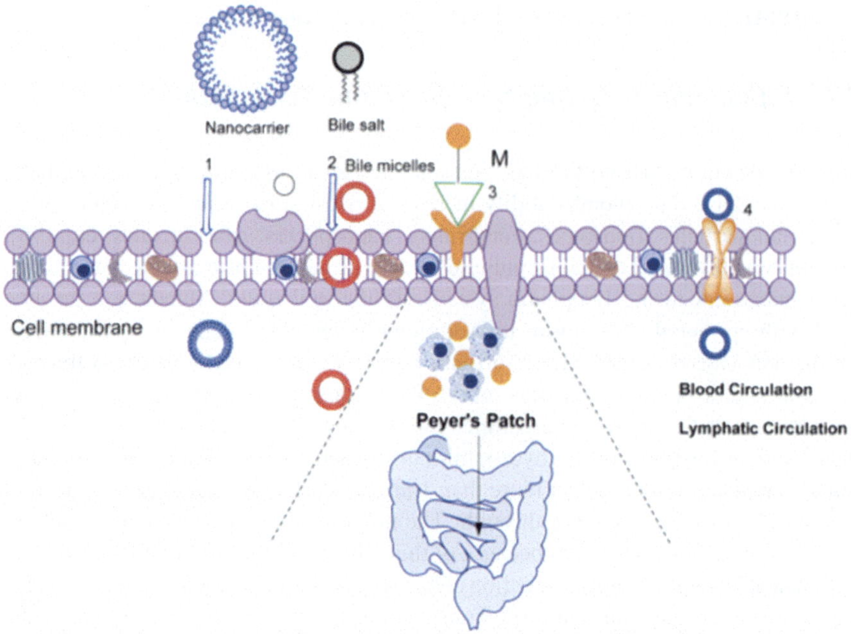

Fig. 7 There are four different routes for medication delivery using nanocarriers: 1. Transcellular 2. M cells mediated 3. Chylomicron-mediated 4. Transporters acted as mediators [99]

4.1.2 Nanoparticles of Polymers Made Absorption Easier

By hydrogen bonding, hydrophobic interactions, or polymer entanglements with mucins, polymeric nanoparticles (NPs) prepared from polymers like poly (sebacic acid), poly (acrylic acid), and poly (lactic acid) facilitate mucoadhesion. Peyer's patches are able to use these polymeric materials more effectively for nanoparticles [103]. Insulin-loaded polymeric nanoparticles prepared from polygalacturonic acid by the water–oil–water solvent evaporation method have been reported. Pharmacokinetic studies on male Sprague–Dawley rats revealed that polymeric nanoparticles improved insulin stability by decreasing first-pass metabolism, improving lymphatic targeting, and increasing bioavailability [104]. Using poly ε-caprolactone as the polymer, the lopinavir-loaded polymeric nanoparticles were successfully prepared using the emulsion solvent evaporation method. The oral bioavailability of lopinavir was nearly quadrupled when it was combined with poly ε-caprolactone nanoparticles, according to pharmacokinetic studies conducted on male Wistar rats that included lymphatic transport [105].

4.2 *Scalable and Reproducible Manufacturing Techniques*

When creating nanoparticles, two methods are frequently employed: top-down and bottom-up approaches. Using the Turkevich method, gas-phase synthesis, block copolymer synthesis, and microbial synthesis, the bottom-up approach nucleates atomic-sized materials into nanoparticles [106]. The top-down method uses milling, laser ablation, and spark ablation to physically break down bulk materials into nano-sized pieces. Because of their unique colloidal structure, LNPs are generally synthesized via a wet chemical bottom-up technique. An outline of the primary techniques for creating LNP is given in this section [81, 82]. These techniques include impingement jet mixing technology, thin film hydration, nanoprecipitation, single/double emulsification, non-solvent emulsification, and microfluidic processes, as shown in Fig. 8. Because it directly affects their physicochemical characteristics, drug loading effectiveness, stability, and behaviour in vivo, the method used for LNP synthesis is essential in defining their therapeutic uses. Every LNP synthesis technique produces nanoparticles with unique properties that have an immediate effect on how well they work in different therapeutic applications [107].

4.2.1 Nanoprecipitation

Since its initial development by Fessi et al. in 1989, the nanoprecipitation technology has mostly been utilized to encapsulate hydrophobic medications (Fig. 8a). This technique encourages the spontaneous development of LNPs through a precipitation process. Under mild magnetic agitation, two miscible solvents—the organic phase and the aqueous phase—are continuously mixed. The aqueous phase contains stabilizers and water, while the organic phase consists of drug molecules, organic solvents, lipophilic surfactants, polymers, and film-forming polymers [84, 85]. LNP is made from polymers that are either biodegradable (such as polylactide (PLA), polylactide-coglycolide (PLGA), or poly-ε-caprolactone (PCL)) or nonbiodegradable. The aqueous/organic phase ratio, lipid/surfactant/drug concentration, and stirring rate are just a few of the variables that can significantly affect the size and efficacy of drug encapsulation in LNPs made by nanoprecipitation. Different LNP sizes can arise from incomplete and inhomogeneous saturated lipid solutions that prevent spontaneous nucleation during the formation of tiny particles [86, 87]. Unevenly small LNP sizes can also result from improper mixing of organic and aqueous solutions before precipitation. Therefore, before any testing or application, it is essential to precisely define the size and surface morphology of the generated LNPs.

One significant issue that affects batch-to-batch variance and lowers the overall quality of the LNPs is incomplete mixing during nanoprecipitation. Careful mixing process optimization is necessary to overcome these problems. Improved homogenization of lipids and medicinal molecules may be attained by regulating important mixing parameters, including speed, duration, and solvent ratio. By ensuring a more even distribution, this fine-tuning produces LNP formulations that are reliable and

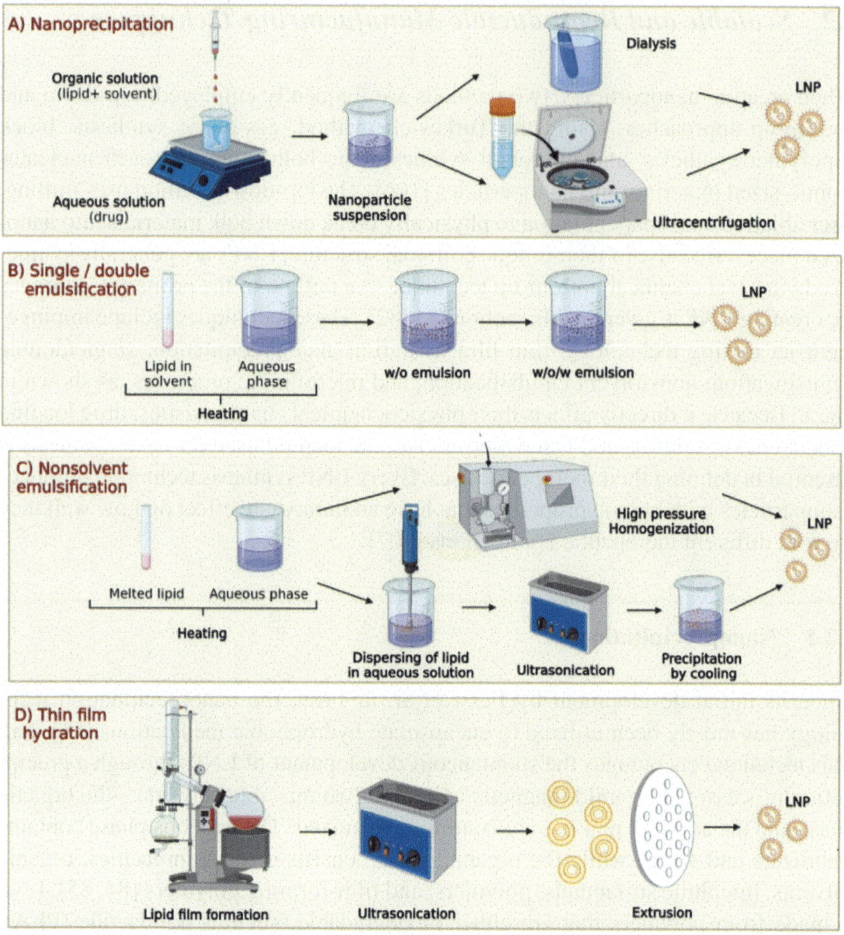

Fig. 8 Procedures for LNP synthesis in both industrial and laboratory settings: **a** Nanoprecipitation [108], **b** single/double emulsification [108], **c** nonsolvent emulsification [108], **d** thin film hydration [108]

effective across batches. To improve the quality and repeatability of LNPs, increase their suitability for large-scale manufacturing, and broaden their potential in a range of therapeutic applications, these nanoprecipitation issues must be resolved.

4.2.2 One or Two Emulsification

A famous technique for creating LNPs containing hydrophobic medicines is the single oil/water solvent emulsification technique, which became developed via Gasko

(Fig. 8b). Water, emulsifiers like polysorbate 20 and polysorbate six hundred, surfactants like Epikuron 2 hundred and Tween 80, and occasional-melting lipids like stearic acid and Compritol 888 ATO are all worried. The solvent combination is warmed, water is introduced, after which the combination is emulsified to generate oil-in-water emulsions. For LNP crystallization, those emulsions are moved to bloodless water while being constantly stirred [109]. The solvent-based emulsification technique, which may be achieved via solvent injection, solvent diffusion, emulsion-solvent evaporation, or solvent displacement, is an extra method for growing LNPs [110]. This approach produces oil-in-water emulsions by way of dissolving lipids and medications which are poorly soluble in water in a natural solvent and then emulsifying the combination with an aqueous solution. After agitating the emulsion to extract the organic solvent, centrifugation is carried out. However, if the lipid solubility in the solvent is low, residual solvent toxicity may be an issue. The emulsification technique has a low dispersion diploma, emulsion instability, lipid insolubility in organic solvents, and the requirement for extra solvent elimination operations, no matter its low mechanical strength enter. To reduce the feasible toxicity of any leftover solvents within the generated LNPs, it's far essential to consist of a solvent elimination section inside the synthesis method, constant with LNPs made utilising nanoprecipitation and microfluidic techniques [111].

4.2.3 Nonsolvent Emulsification

Melted lipids are used as the liquid phase in nonsolvent emulsification procedures, which provide a solvent-free method of producing oil-in-water emulsions (Fig. 8c). Ultrasonication and high-pressure homogenization are often employed techniques for this [112]. Melted lipids, medications, and a heated aqueous phase with surfactants are combined during the homogenization process. For several cycles, the mixture is homogenized at high pressure to produce the appropriate size of nanoparticles. As an alternative, high-speed stirring is used to homogenize the aqueous phase with surfactants and melted lipids [113]. Large particles are then broken down into tiny droplets using Ultrasonication. However, there may be a batch-to-batch fluctuation due to the intricate and multi-step processes involved in the nonsolvent emulsification approach. Differences in homogenization/sonication time, lipid surfactant ratio, drug concentrations, and lipid/surfactant type can all affect the lipid nanoparticles' particle sizes, polydispersion indices, zeta potentials, and drug entrapment efficiency. These changes can also influence the product outcomes [93, 94]. The factors need to be adjusted cautiously to ensure uniform product quality.

4.2.4 Hydration of Thin Films

A popular technique for creating lipid nanoparticles in both commercial and scientific settings is thin film hydration (Fig. 8d). This method creates a thin lipid layer on the flask's sides by dissolving and mixing lipids in an organic solvent until they are homogeneous, after which the solvent is eliminated by rotational evaporation [95, 96]. The lipid solution is repeatedly run through a filter with consistent pore diameters after it has been hydrated, producing lipid nanoparticles of the same size [114]. The mean size and polydispersity of the generated nanoparticles can be influenced by a variety of factors, including temperatures, extrusion force, membrane pore size, applied pressure, and cycle count. Because it uses less energy, has a lesser chance of contamination, and produces a smaller particle size distribution with reliable filters, the extrusion method is preferred. Extruders used in the industry, such as LIPEX and LiposoFast LF-50, have effectively created liposomes carrying a variety of payloads, such as proteins, peptides, small compounds, and bigger molecules [115].

5 Future Prospective

Sudden advances in nanotechnology, material science, and precision medicine promise a lot to the application of nanoparticle-based therapies in oncology. Translational efficiency will be increased a lot as we move forward by employing artificial intelligence (AI) and machine learning (ML) for designing nanoparticles and optimizing clinical trials. The basis of future cancer treatment is likely to be individualized nanomedicine, where therapies are tailored based on a patient's genetic makeup and tumor microenvironment. The boundaries of what can be achieved with nanomedicine will be further expanded by efforts to increase targeted specificity, reduce systemic toxicity, and enable stimuli-responsive drug delivery. Further, green synthesis developments and scalable processes will enhance production sustainability and economy. To have standardized evaluation processes and facilitate faster, safer approval, scientists, industry players, and global regulators need to interact more closely. For clinical translation to proceed better, robust regulatory requirements and foresighted preclinical models need to be established. Ultimately, the integration of nanotechnology, omics technologies, immunotherapy, and real-time diagnostics will provide new tools for all-encompassing and responsive cancer therapy, transforming the clinical setting to encompass more efficient, targeted, and patient-focused treatments.

6 Conclusion

A significant advance in oncology, nanoparticle therapeutics holds out new promise for the effective and targeted treatment of a variety of cancers. They offer a possible substitute for traditional chemotherapy because they can increase drug solubility, extend circulation times, and target tumor tissue selectively by mechanisms like the enhanced permeability and retention (EPR) effect. While promising, however, a number of challenges must be overcome in bringing these nanomedicines from the bench to the clinic. Clinical translation is a complex process that requires careful focus on patient-specific parameters, design optimization, and preclinical development. Advancement is often hindered by barriers such as expensive manufacturing, limited scalability, and heterogeneity of biological responses. In addition, regulatory regimes remain disconnected, with uneven and sometimes incompatible standards for approving nanomedicines in different regions of the world. For this reason, navigating through regulatory hurdles is an essential but challenging aspect of product development. Proactive strategies involving enhancing precision toward targets, establishing scalable and reproducible production technologies, and instilling early liaison with regulatory authorities are necessary for overcoming such barriers. A multidisciplinary approach combining advances in materials science, pharmacology, clinical oncology, and regulatory science will be key to the future success of nanoparticle-based therapy.

Acknowledgements Not applicable.

Authors Contributions SM were involved in the conception and design of this chapter and completed the figures and tables. GP were involved in reading the article and final approval of the submitted chapter. All authors have read and approved the final version of the manuscript.

Funding This work did not receive any funding.

Declaration The authors confirm that they did not receive any financial support or grants during the preparation of this manuscript.

Conflict of Interest The authors of this review assert that they conducted the study without any affiliations or financial relationships that could potentially pose conflicts of interest.

References

1. Sung, H., et al.: Global cancer statistics 2020: GLOBOCAN estimates of incidence and mortality worldwide for 36 cancers in 185 countries. CA Cancer J. Clin. **71**(3), 209–249 (2021)
2. Kalbasi, A., et al.: Tumor-derived CCL2 mediates resistance to radiotherapy in pancreatic ductal adenocarcinoma. Clin. Cancer Res. **23**(1), 137–148 (2017)
3. Qin, S.-Y., Cheng, Y.-J., Lei, Q., Zhang, A.-Q., Zhang, X.-Z.: Combinational strategy for high-performance cancer chemotherapy. Biomaterials **171**, 178–197 (2018)
4. Chabner, B.A., Roberts, T.G.: Chemotherapy and the war on cancer. Nat. Rev. Cancer **5**(1), 65–72 (2005)
5. Pérez-Herrero, E., Fernández-Medarde, A.: Advanced targeted therapies in cancer: drug nanocarriers, the future of chemotherapy. Eur. J. Pharm. Biopharm. **93**, 52–79 (2015)
6. Wang, Y., Li, P., Truong-Dinh Tran, T., Zhang, J., Kong, L.: Manufacturing techniques and surface engineering of polymer based nanoparticles for targeted drug delivery to cancer. Nanomaterials **6**(2), 26 (2016)
7. Gwinn, M.R., Vallyathan, V.: Nanoparticles: health effects—pros and cons. Environ. Health Perspect. **114**(12), 1818–1825 (2006)
8. Liu, J., Zhang, R., Xu, Z.P.: Nanoparticle-based nanomedicines to promote cancer immunotherapy: recent advances and future directions. Small **15**(32) (2019)

9. Mehta, S., Shah, V., Patel, G., Conte-Junior, C.A., Joshi, N.: A holistic review of recent advances in nano-based drug delivery systems for the treatment of triple-negative breast cancer (TNBC). J. Nanoparticle Res. **26**(5), 87 (2024)
10. Amreddy, N., et al.: Recent Advances in Nanoparticle-Based Cancer Drug and Gene Delivery, pp. 115–170. (2018)
11. Mehta, S., Koralkar, N., Patel, G.: Functionalized polymer nanocomposites for biomedical applications. In: Advances in Functionalized Polymer Nanocomposites, pp. 705–746. Elsevier (2024)
12. Shah, V., Bhaliya, J., Patel, G.M., Deshmukh, K.: Advances in polymeric nanocomposites for automotive applications: a review. Polym. Adv. Technol. **33**(10), 3023–3048 (2022)
13. Yang, X., et al.: Enhanced endosomal escape by photothermal activation for improved small interfering RNA delivery and antitumor effect. Int. J. Nanomed. **13**, 4333–4344 (2018)
14. Patel, G.M., Shah, V., Bhaliya, J., Pathan, P., Nikita, K.M.: Polymer-based nanomaterials: an introduction. In: Smart Polymer Nanocomposites, pp. 27–59. Elsevier (2023)
15. Najahi-Missaoui, W., Arnold, R.D., Cummings, B.S.: Safe nanoparticles: are we there yet? Int. J. Mol. Sci. **22**(1), 385 (2020)
16. Bobo, D., Robinson, K.J., Islam, J., Thurecht, K.J., Corrie, S.R.: Nanoparticle-based medicines: a review of FDA-approved materials and clinical trials to date. Pharm. Res. **33**(10), 2373–2387 (2016)
17. Kundranda, M., Niu, J.: Albumin-bound paclitaxel in solid tumors: clinical development and future directions. Drug Des. Dev. Ther. 3767 (2015)
18. Patel, G.M., Bhatt, G.J., Deota, P.T.: Synthesis and characterization of silicon-based hybrid nanoparticles. In: Silicon-Based Hybrid Nanoparticles, pp. 11–43. Elsevier, 2022
19. Bhanderi, K., Joshi, J., Suthar, V., Shah, V., Patel, G.M., Patel, J.: Eco-friendly polymer nanocomposites based on bio-based fillers: preparation, characterizations and potential applications. In: Biodegradable and Biocompatible Polymer Nanocomposites, pp. 173–203. Elsevier (2023)
20. Shah, V., Bhaliya, J., Patel, G.M., Popaliya, M., Mishra, A., Patel, P.R.: Biodegradable polymer nanocomposites as electrode materials for electrochemical double-layer capacitors and hybrid supercapacitor applications. In: Biodegradable and Biocompatible Polymer Nanocomposites, pp. 311–352. Elsevier (2023)
21. Wang, S., et al.: Nanoparticle-based medicines in clinical cancer therapy. Nano Today **45**, 101512 (2022)
22. Ferrari, R., Sponchioni, M., Morbidelli, M., Moscatelli, D.: Polymer nanoparticles for the intravenous delivery of anticancer drugs: the checkpoints on the road from the synthesis to clinical translation. Nanoscale **10**(48), 22701–22719 (2018)
23. He, H., Liu, L., Morin, E.E., Liu, M., Schwendeman, A.: Survey of clinical translation of cancer nanomedicines—lessons learned from successes and failures. Acc. Chem. Res. **52**(9), 2445–2461 (2019)
24. Mundekkad, D., Cho, W.C.: Nanoparticles in clinical translation for cancer therapy. Int. J. Mol. Sci. **23**(3), 1685 (2022)
25. Ndagi, U., Mhlongo, N., Soliman, M.: Metal complexes in cancer therapy—an update from drug design perspective. Drug Des. Dev. Ther. **11**, 599–616 (2017)
26. Ghosh, B., Biswas, S.: Polymeric micelles in cancer therapy: state of the art. J. Control. Rel. **332**, 127–147 (2021)
27. Zhou, Q., Zhang, L., Yang, T., Wu, H.: Stimuli-responsive polymeric micelles for drug delivery and cancer therapy. Int. J. Nanomed. **13**, 2921–2942 (2018)
28. Persano, F., Gigli, G., Leporatti, S.: Lipid-polymer hybrid nanoparticles in cancer therapy: current overview and future directions. Nano Express **2**(1), 012006 (2021)
29. Jana, D., Jia, S., Bindra, A.K., Xing, P., Ding, D., Zhao, Y.: Clearable black phosphorus nanoconjugate for targeted cancer phototheranostics. ACS Appl. Mater. Interfaces **12**(16), 18342–18351 (2020)
30. Payne, M., et al.: DHA-paclitaxel (Taxoprexin) as first-line treatment in patients with stage IIIB or IV non-small cell lung cancer: report of a phase II open-label multicenter trial. J. Thorac. Oncol. **1**(9), 984–990 (2006)

31. Pandey, H., Rani, R., Agarwal, V.: Liposome and their applications in cancer therapy. Brazilian Arch. Biol. Technol. **59** (2016)
32. Olusanya, T., Haj Ahmad, R., Ibegbu, D., Smith, J., Elkordy, A.: Liposomal drug delivery systems and anticancer drugs. Molecules **23**(4), 907 (2018)
33. Rommasi, F., Esfandiari, N.: Liposomal nanomedicine: applications for drug delivery in cancer therapy. Nanoscale Res. Lett. **16**(1), 95 (2021)
34. Wurz, G.T., Kao, C.-J., Wolf, M., DeGregorio, M.W.: Tecemotide: an antigen-specific cancer immunotherapy. Hum. Vaccin. Immunother. **10**(11), 3383–3393 (2014)
35. Gargett, T., et al.: Phase I trial of Lipovaxin-MM, a novel dendritic cell-targeted liposomal vaccine for malignant melanoma. Cancer Immunol. Immunother. **67**(9), 1461–1472 (2018)
36. Young, C., Schluep, T., Hwang, J., Eliasof, S.: CRLX101 (formerly IT-101)—a novel nanopharmaceutical of camptothecin in clinical development. Curr. Bioact. Compd. **7**(1), 8–14 (2011)
37. Yardley, D.A., et al.: Phase II/III weekly nab-paclitaxel plus gemcitabine or carboplatin versus gemcitabine/carboplatin as first-line treatment of patients with metastatic triple-negative breast cancer (the tnAcity study): study protocol for a randomized controlled trial. Trials **16**(1), 575 (2015)
38. Ogino, M., Fujii, T., Koibuchi, Y., Nakazawa, Y., Takata, D., Shirabe, K.: Phase II study of nab-paclitaxel plus cyclophosphamide plus trastuzumab neoadjuvant chemotherapy in early HER-2-positive breast cancer. Anticancer Res. **41**(8), 3899–3904 (2021)
39. Weng, X., et al.: First-line treatment with atezolizumab plus nab-paclitaxel for advanced triple-negative breast cancer. Am. J. Clin. Oncol. **43**(5), 340–348 (2020)
40. Nikita, K., Patel, D., Patel, G.: Additive manufacturing of multifunctional polymer nanocomposites: from 3 <scp>D</scp> to 4 <scp>D</scp>. In: Nanotechnology-Based Additive Manufacturing, pp. 277–313. Wiley (2023)
41. Zhang, R., Kiessling, F., Lammers, T., Pallares, R.M.: Clinical translation of gold nanoparticles. Drug Deliv. Transl. Res. **13**(2), 378–385 (2023)
42. Sharma, T., Singh, D., Mahapatra, A., Mohapatra, P., Sahoo, S., Sahoo, S.K.: Advancements in clinical translation of flavonoid nanoparticles for cancer treatment. OpenNano **8**, 100074 (2022)
43. He, B., Sui, X., Yu, B., Wang, S., Shen, Y., Cong, H.: Recent advances in drug delivery systems for enhancing drug penetration into tumors. Drug Deliv. **27**(1), 1474–1490 (2020)
44. Wang, S., et al.: Zwitterionic-to-cationic charge conversion polyprodrug nanomedicine for enhanced drug delivery. Theranostics **10**(15), 6629–6637 (2020)
45. Du, Y., Wang, S., Zhang, T., He, D., Tu, J., Shen, Y.: Enhanced cytotoxicity of a redox-sensitive hyaluronic acid-based nanomedicine toward different oncocytes via various internalization mechanisms. Drug Deliv. **27**(1), 128–136 (2020)
46. Blanco, E., Shen, H., Ferrari, M.: Principles of nanoparticle design for overcoming biological barriers to drug delivery. Nat. Biotechnol. **33**(9), 941–951 (2015)
47. Tiboni, M., et al.: Prunus spinosa extract loaded in biomimetic nanoparticles evokes in vitro anti-inflammatory and wound healing activities. Nanomaterials **11**(1), 36 (2020)
48. Alshammari, T.M.: Drug safety: the concept, inception and its importance in patients' health. Saudi Pharm. J. **24**(4), 405–412 (2016)
49. Goswitz, V.C., Peter Sawicki, Z.: Cancer therapy based on a mechanism of action for controlling the immune system and the resulting patent portfolio. Recent Pat. Endocr. Metab. Immune Drug Discov. **7**(1), 1–10 (2013)
50. Senapati, S., Mahanta, A.K., Kumar, S., Maiti, P.: Controlled drug delivery vehicles for cancer treatment and their performance. Signal Transduct. Target. Ther. **3**(1), 7 (2018)
51. Rodriguez, P.L., Harada, T., Christian, D.A., Pantano, D.A., Tsai, R.K., Discher, D.E.: Minimal 'self' peptides that inhibit phagocytic clearance and enhance delivery of nanoparticles. Science (80–) **339**(6122), 971–975 (2013)
52. Farjadian, F., Ghasemi, A., Gohari, O., Roointan, A., Karimi, M., Hamblin, M.R.: Nanopharmaceuticals and nanomedicines currently on the market: challenges and opportunities. Nanomedicine **14**(1), 93–126 (2019)

53. Duan, X., Li, Y.: Physicochemical characteristics of nanoparticles affect circulation, biodistribution, cellular internalization, and trafficking. Small **9**(9–10), 1521–1532 (2013)
54. Hafeez, M.N., Celia, C., Petrikaite, V.: Challenges towards targeted drug delivery in cancer nanomedicines. Processes **9**(9), 1527 (2021)
55. Agrahari, V., Agrahari, V.: Facilitating the translation of nanomedicines to a clinical product: challenges and opportunities. Drug Discov. Today **23**(5), 974–991 (2018)
56. DiMasi, J.A., Feldman, L., Seckler, A., Wilson, A.: Trends in risks associated with new drug development: success rates for investigational drugs. Clin. Pharmacol. Ther. **87**(3), 272–277 (2010)
57. Paliwal, R., Babu, R.J., Palakurthi, S.: Nanomedicine scale-up technologies: feasibilities and challenges. AAPS PharmSciTech **15**(6), 1527–1534 (2014)
58. Ho, D., Wang, P., Kee, T.: Artificial intelligence in nanomedicine. Nanoscale Horizons **4**(2), 365–377 (2019)
59. Burade, V., Bhowmick, S., Maiti, K., Zalawadia, R., Ruan, H., Thennati, R.: Lipodox® (generic doxorubicin hydrochloride liposome injection): in vivo efficacy and bioequivalence versus Caelyx® (doxorubicin hydrochloride liposome injection) in human mammary carcinoma (MX-1) xenograft and syngeneic fibrosarcoma (WEHI 164) mouse mode. BMC Cancer **17**(1), 405 (2017)
60. Liu, Z., Chen, X.: Simple bioconjugate chemistry serves great clinical advances: albumin as a versatile platform for diagnosis and precision therapy. Chem. Soc. Rev. **45**(5), 1432–1456 (2016)
61. Fromer, M.J.: ODAC decides all thumbs way down for Abraxane because sponsor requests to omit randomized trial. Oncol. Times **28**(18), 24 (2006)
62. Kashoki, M., et al.: A comparison of <scp>EMA</scp> and <scp>FDA</scp> decisions for new drug marketing applications 2014–2016: concordance, discordance, and why. Clin. Pharmacol. Ther. **107**(1), 195–202 (2020)
63. Clogston, J.D., Hackley, V.A., Prina-Mello, A., Puri, S., Sonzini, S., Soo, P.L.: Sizing up the next generation of nanomedicines. Pharm. Res. **37**(1), 6 (2020)
64. Cicha, I., et al.: From design to the clinic: practical guidelines for translating cardiovascular nanomedicine. Cardiovasc. Res. **114**(13), 1714–1727 (2018)
65. Sainz, V., et al.: Regulatory aspects on nanomedicines. Biochem. Biophys. Res. Commun. **468**(3), 504–510 (2015)
66. de Vlieger, J.S.B., et al.: Report of the AAPS Guidance Forum on the FDA Draft Guidance for Industry: 'drug products, including biological products, that contain nanomaterials.' AAPS J. **21**(4), 56 (2019)
67. Cho, E.J., Holback, H., Liu, K.C., Abouelmagd, S.A., Park, J., Yeo, Y.: Nanoparticle characterization: state of the art, challenges, and emerging technologies. Mol. Pharm. **10**(6), 2093–2110 (2013)
68. Oberdörster, G., et al.: Principles for characterizing the potential human health effects from exposure to nanomaterials: elements of a screening strategy. Part. Fibre Toxicol. **2**(1), 8 (2005)
69. Sukhanova, A., Bozrova, S., Sokolov, P., Berestovoy, M., Karaulov, A., Nabiev, I.: Dependence of nanoparticle toxicity on their physical and chemical properties. Nanoscale Res. Lett. **13**(1), 44 (2018)
70. Radomski, A., et al.: Nanoparticle-induced platelet aggregation and vascular thrombosis. Br. J. Pharmacol. **146**(6), 882–893 (2005)
71. Madl, A.K., Plummer, L.E., Carosino, C., Pinkerton, K.E.: Nanoparticles, lung injury, and the role of oxidant stress. Annu. Rev. Physiol. **76**(1), 447–465 (2014)
72. Barua, S., Mitragotri, S.: Challenges associated with penetration of nanoparticles across cell and tissue barriers: a review of current status and future prospects. Nano Today **9**(2), 223–243 (2014)
73. Aravind, A., et al.: AS1411 aptamer tagged PLGA-lecithin-PEG nanoparticles for tumor cell targeting and drug delivery. Biotechnol. Bioeng. **109**(11), 2920–2931 (2012)
74. Sanità, G., Carrese, B., Lamberti, A.: Nanoparticle surface functionalization: how to improve biocompatibility and cellular internalization. Front. Mol. Biosci. **7** (2020)

75. Yue, L., Sun, C., Cheng, Q., Ding, Y., Wei, J., Wang, R.: Gold nanorods with a noncovalently tailorable surface for multi-modality image-guided chemo-photothermal cancer therapy. Chem. Commun. **55**(90), 13506–13509 (2019)
76. Jung, E., et al.: Molecularly engineered theranostic nanoparticles for thrombosed vessels: H_2O_2-activatable contrast-enhanced photoacoustic imaging and antithrombotic therapy. ACS Nano **12**(1), 392–401 (2018)
77. Chen, S., et al.: Controlled Fab installation onto polymeric micelle nanoparticles for tuned bioactivity. Sci. Technol. Adv. Mater. **18**(1), 666–680 (2017)
78. Fornaguera, C., Solans, C.: Characterization of polymeric nanoparticle dispersions for biomedical applications: size, surface charge and stability. Pharm. Nanotechnol. **6**(3), 147–164 (2018)
79. Okay, S.: Single-molecule characterization of drug delivery systems. Assay Drug Dev. Technol. **18**(1), 56–63 (2020)
80. Musazzi, U.M., Marini, V., Casiraghi, A., Minghetti, P.: Is the European regulatory framework sufficient to assure the safety of citizens using health products containing nanomaterials? Drug Discov. Today **22**(6), 870–882 (2017)
81. Gioria, S., et al.: Are existing standard methods suitable for the evaluation of nanomedicines: some case studies. Nanomedicine **13**(5), 539–554 (2018)
82. Ramos, T.I., et al.: The Hitchhiker's guide to human therapeutic nanoparticle development. Pharmaceutics **14**(2), 247 (2022)
83. Caputo, F., Clogston, J., Calzolai, L., Rösslein, M., Prina-Mello, A.: Measuring particle size distribution of nanoparticle enabled medicinal products, the joint view of EUNCL and NCI-NCL. A step by step approach combining orthogonal measurements with increasing complexity. J. Control. Rel. **299**, 31–43 (2019)
84. Watanabe, A., Takagi, M., Murata, S., Kato, M.: Stability and drug release studies of an antimycotic nanomedicine using HPLC, dynamic light scattering and atomic force microscopy. J. Pharm. Biomed. Anal. **148**, 149–155 (2018)
85. Mourdikoudis, S., Pallares, R.M., Thanh, N.T.K.: Characterization techniques for nanoparticles: comparison and complementarity upon studying nanoparticle properties. Nanoscale **10**(27), 12871–12934 (2018)
86. Teulon, J.-M., et al.: On the operational aspects of measuring nanoparticle sizes. Nanomaterials **9**(1), 18 (2018)
87. Rydz, J., Šišková, A., Andicsová Eckstein, A.: Scanning electron microscopy and atomic force microscopy: topographic and dynamical surface studies of blends, composites, and hybrid functional materials for sustainable future. Adv. Mater. Sci. Eng. **2019**, 1–16 (2019)
88. Su, D.: Advanced electron microscopy characterization of nanomaterials for catalysis. Green Energy Environ. **2**(2), 70–83 (2017)
89. Robson, A.-L., et al.: Advantages and limitations of current imaging techniques for characterizing liposome morphology. Front. Pharmacol. **9** (2018)
90. Pujals, S., Albertazzi, L.: Super-resolution microscopy for nanomedicine research. ACS Nano **13**(9), 9707–9712 (2019)
91. El-Say, K.M.: Maximizing the encapsulation efficiency and the bioavailability of controlled-release cetirizine microspheres using Draper–Lin small composite design. Drug Des. Devel. Ther. 825 (2016)
92. Dey, S., Pramanik, S., Malgope, A.: Formulation and optimization of sustained release Stavudine microspheres using response surface methodology. ISRN Pharm. **2011**, 1–7 (2011)
93. Gaikwad, V.L., Choudhari, P.B., Bhatia, N.M., Bhatia, M.S.: Characterization of pharmaceutical nanocarriers: in vitro and in vivo studies. In: Nanomaterials for Drug Delivery and Therapy, pp. 33–58. Elsevier (2019)
94. Daneshmand, S., et al.: Encapsulation challenges, the substantial issue in solid lipid nanoparticles characterization. J. Cell. Biochem. **119**(6), 4251–4264 (2018)
95. Harde, H., Das, M., Jain, S.: Solid lipid nanoparticles: an oral bioavailability enhancer vehicle. Expert Opin. Drug Deliv. **8**(11), 1407–1424 (2011)

96. Mahmood Hussain, M.: A proposed model for the assembly of chylomicrons. Atherosclerosis **148**(1), 1–15 (2000)
97. Hynes, R.O.: Integrins. Cell **110**(6), 673–687 (2002)
98. Ikomi, F., Kawai, Y., Ohhashi, T.: Recent advance in lymph dynamic analysis in lymphatics and lymph nodes. Ann. Vasc. Dis. **5**(3), 258–268 (2012)
99. Pandya, P., Giram, P., Bhole, R.P., Chang, H.-I., Raut, S.Y.: Nanocarriers based oral lymphatic drug targeting: strategic bioavailability enhancement approaches. J. Drug Deliv. Sci. Technol. **64**, 102585 (2021)
100. Jain, S., Rathi, V.V., Jain, A.K., Das, M., Godugu, C.: Folate-decorated PLGA nanoparticles as a rationally designed vehicle for the oral delivery of insulin. Nanomedicine **7**(9), 1311–1337 (2012)
101. Jaiswal, M., Dudhe, R., Sharma, P.K.: Nanoemulsion: an advanced mode of drug delivery system. 3 Biotech **5**(2), 123–127 (2015)
102. Javed, I., et al.: Lecithin-gold hybrid nanocarriers as efficient and pH selective vehicles for oral delivery of diacerein—in-vitro and in-vivo study. Colloids Surf. B Biointerfaces **141**, 1–9 (2016)
103. Jepson, M.A., Clark, M.A., Hirst, B.H.: M cell targeting by lectins: a strategy for mucosal vaccination and drug delivery. Adv. Drug Deliv. Rev. **56**(4), 511–525 (2004)
104. Joshi, G., Kumar, A., Sawant, K.: Bioavailability enhancement, Caco-2 cells uptake and intestinal transport of orally administered lopinavir-loaded PLGA nanoparticles. Drug Deliv. **23**(9), 3492–3504 (2016)
105. Jung, C., Hugot, J.-P., Barreau, F.: Peyer's patches: the immune sensors of the intestine. Int. J. Inflam. **2010**, 1–12 (2010)
106. Shafique, M., et al.: Formulation development of lipid polymer hybrid nanoparticles of doxorubicin and its in-vitro, in-vivo and computational evaluation. Front. Pharmacol. **14** (2023)
107. Martínez Rivas, C.J., et al.: Nanoprecipitation process: from encapsulation to drug delivery. Int. J. Pharm. **532**(1), 66–81 (2017)
108. Mehta, M., Bui, T.A., Yang, X., Aksoy, Y., Goldys, E.M., Deng, W.: Lipid-based nanoparticles for drug/gene delivery: an overview of the production techniques and difficulties encountered in their industrial development. ACS Mater. Au **3**(6), 600–619 (2023)
109. Katara, R., Majumdar, D.K.: Eudragit RL 100-based nanoparticulate system of aceclofenac for ocular delivery. Colloids Surf. B Biointerfaces **103**, 455–462 (2013)
110. Dong, Y., Ng, W.K., Shen, S., Kim, S., Tan, R.B.H.: Solid lipid nanoparticles: continuous and potential large-scale nanoprecipitation production in static mixers. Colloids Surf. B Biointerfaces **94**, 68–72 (2012)
111. Gupta, A., Eral, H.B., Hatton, T.A., Doyle, P.S.: Nanoemulsions: formation, properties and applications. Soft Matter **12**(11), 2826–2841 (2016)
112. Ganesan, P., Narayanasamy, D.: Lipid nanoparticles: different preparation techniques, characterization, hurdles, and strategies for the production of solid lipid nanoparticles and nanostructured lipid carriers for oral drug delivery. Sustain. Chem. Pharm. **6**, 37–56 (2017)
113. Battaglia, L., Trotta, M., Gallarate, M., Carlotti, M.E., Zara, G.P., Bargoni, A.: Solid lipid nanoparticles formed by solvent-in-water emulsion–diffusion technique: development and influence on insulin stability. J. Microencapsul. **24**(7), 672–684 (2007)
114. Tenchov, R., Bird, R., Curtze, A.E., Zhou, Q.: Lipid nanoparticles—from liposomes to mRNA vaccine delivery, a landscape of research diversity and advancement. ACS Nano **15**(11), 16982–17015 (2021)
115. Berger, N., Sachse, A., Bender, J., Schubert, R., Brandl, M.: Filter extrusion of liposomes using different devices: comparison of liposome size, encapsulation efficiency, and process characteristics. Int. J. Pharm. **223**(1–2), 55–68 (2001)

Biocompatibility and Toxicity of Nanoparticles in Cancer Care

Nagina Rehman, Muhammad Asad, Tehseen Fatima, and Muhammad Abdul Basit

Abstract Nanoparticles have emerged as a transformative tool in cancer care due to their novel approach for drug delivery and diagnostics. Materials having at least one dimension in the range of 1–100 nm are termed as nanomaterials (NMs). The increased surface area and a more targeted drug delivery system make nanoparticles (NPs) ideal for biological applications, but at the same time there is also a potential for them to induce toxicity within the cells. More research is required in the field of biocompatibility before they can be safely used for clinical applications. Understanding biocompatibility and toxicity of NMs is important for advancing from experimental stage to safe and effective cancer treatments.

Keywords Nanomaterials · Nanoparticles · Drug · Cancer · Biocompatibility · Toxicity

1 Introduction

Nanotechnology involves the dealing of matter at nanoscale, ranging from 1 to 100 nm. The term nanotechnology was first used in 1974 [1], and now incorporates a diverse number of applications including electronics, medicine, and materials science [2]. This technology has revolutionized the industry and scientific world due to its interdisciplinary growth [3]. Furthermore, NMs possess different behaviors as compared to their bulk counterparts, this is due to their extremely small size which provides a greater surface area to volume ratio. These materials have at least one

N. Rehman
Department of Zoology, University of Mianwali, Mianwali, Pakistan

M. Asad · T. Fatima
Department of Zoology, Division of Science and Technology, University of Education, Lahore, Pakistan

M. A. Basit (✉)
Department of Materials Science and Engineering, Institute of Space Technology, Islamabad, Pakistan
e-mail: ab_saim@hotmail.com; a.basit@ist.edu.pk

N. J. Joshi and G. Patel (eds.), *Nanoparticle Delivery in Cancer Care*,
Series in BioEngineering, https://doi.org/10.1007/978-981-95-0457-2_10

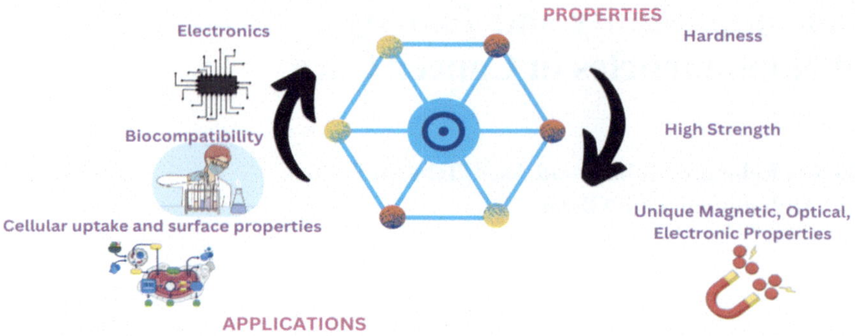

Fig. 1 Utilization of NMs in various advanced applications

dimension in the nanometer range [4], and can possess unique properties, including hardness, high strength, diffusivity, making them promising candidates for biomedical applications [5]. Their scope also includes energy storage, energy conversion, and photovoltaics [6], which are depicted in Fig. 1.

In the realm of medical materials, NMs have the tendency to introduce unique magnetic, optical and electronic properties [7]. The recent trends of nanotechnology dominate molecular mechanisms, health risks, toxicological effects and their safer applications [8].

2 Nanoparticles Used in Cancer Care

Carbon-based NPs, such as graphene nanosheets, carbon nanotubes, and nanodiamonds have had successful trials and proved themselves useful in the field of advanced biotechnological applications. Some of their major applications are in analytical chemistry, biomedical research, antibacterial properties and environmental cleaning [9–11], they are also beginning to show potential for cancer therapy by targeting micro-environments for tumor control [12, 13]. Carbon NMs researched for cell compatibility are shown in Fig. 2.

Silica-based NPs are used in various fields such as drug delivery, environmental remediation and biomedical applications. Currently, silica-based NPs are trending among medicinal applications [15]. Carbon and silica based NMs are expected to contribute to the innovative applications in medical fields. Furthermore, the evolution of silica gel into NMs indicates their broad-spectrum advancements [16]. Their targeted drug delivery makes them of significant attention in cancer research. The porous hybrid silica NPs can be effectively used to target cancer cells and imaging agents [17]. In oncology, more advanced luminescent silica NPs have been used to diagnose cancer [18]. Furthermore, mesoporous silica NPs are also used for cancer treatment [19].

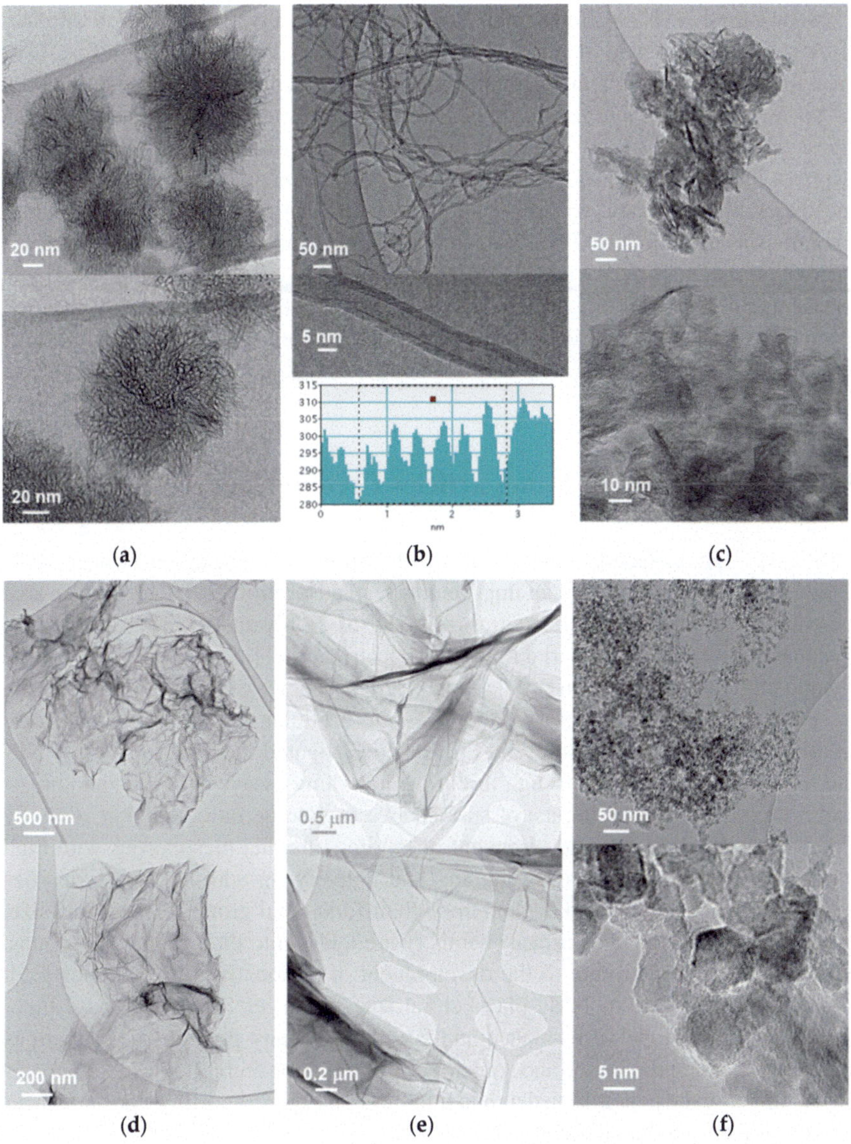

Fig. 2 TEM images of carbon nanomaterials tested for cell compatibility: **a** CNH, **b** CNT, **c** CNP, **d** RGO, **e** GO and **f** ND. Inset in (**b**) shows image analysis of a MWCNT with six concentric nanotubes. Reprinted with permission from [14] Copyrights MDPI 2020

Metal-based NMs also have unique properties and a diverse number of applications, particularly in fields such as fuel cells, metal–air batteries, water splitting, CO_2 reduction, as well as in chemical, pharmaceutical, petroleum, energy, and automobile industries [20]. Their surface properties are very important in coordinating surface interactions [21]. They are used for the production of clean energy due to their progress in electrocatalysis [22]. Metallic NPs contribute to diagnosing, imaging and therapizing for various forms of cancer. In cancer treatment, nano medicines are a non-invasive option [23]. These NPs work for cancer immunotherapy and are currently ongoing clinical trials [24]. They are also used for targeted drug delivery in cancer treatment, exhibiting their multifunctional capabilities [25]. The utilization of metallic NPs is a significant leap in improving cancer treatment [26]. Various metallic NPs used in cancer treatment are shown in Fig. 3.

3 Characteristics of Nanoparticles Used in Cancer Care

NMs have high surface to volume ratio and exhibit quantum effect at nano scale [28]. Their surface area plays an important role in adsorption [29]. The surface area increases as the size of NMs decreases [30]. Self-assembly and specificity of surface area make them a useful material for battery electrodes [31]. Size and surface charge of nano particles also helps in cellular uptake. Mesospheres of silica and polystyrene are used in ovarian cancer for endocytosis, intracellular trafficking and biocompatibility [32]. A previous study by [33] has investigated the effect of Fe_3O_4 NPs and their biocompatibility through in vitro cytotoxicity assays with glia, and breast cancer cells showing the impact of surface area to volume ratio. Another study by [34] has shed light on the development of tailored zirconia mesoporous with high surface area optimized for drug loading and delivery. In addition, the properties and applications of NPs also depend upon their functional groups. The shape, size, chemical composition and surface chemistry are relevant to their functional groups. All these aspects contribute to the development of various key technologies [35]. Surface modifications of NMs play central role in cancer therapy because these modifications improve performance and specificity. Many previous investigations have demonstrated the importance of surface modification technologies for nanocarriers to increase the penetration depth in tumors [36]. Figure 4 portrays the anticancer activity of plant mediated metallic NPs.

The application of nanotechnology in regenerative medicine has led to improvements in the ability to image stem cells, as well as constructing materials that mirror the stem cell milieu and promo the prospect of tissue remodeling [37, 38]. Advanced NMs bring more light in cancer therapy since they can provide better drug vectors, diagnostics, and therapeutic applications than the traditional therapy [39].

Nano magnetic materials have also been applied in order to act as an intermediary for Electromagnetic Stimulation on stem cells and cancer cells, which further reinforces the use of nanotechnology on regenerative medical studies. A few benefits of NPs in cancer therapy can be seen in Fig. 5.

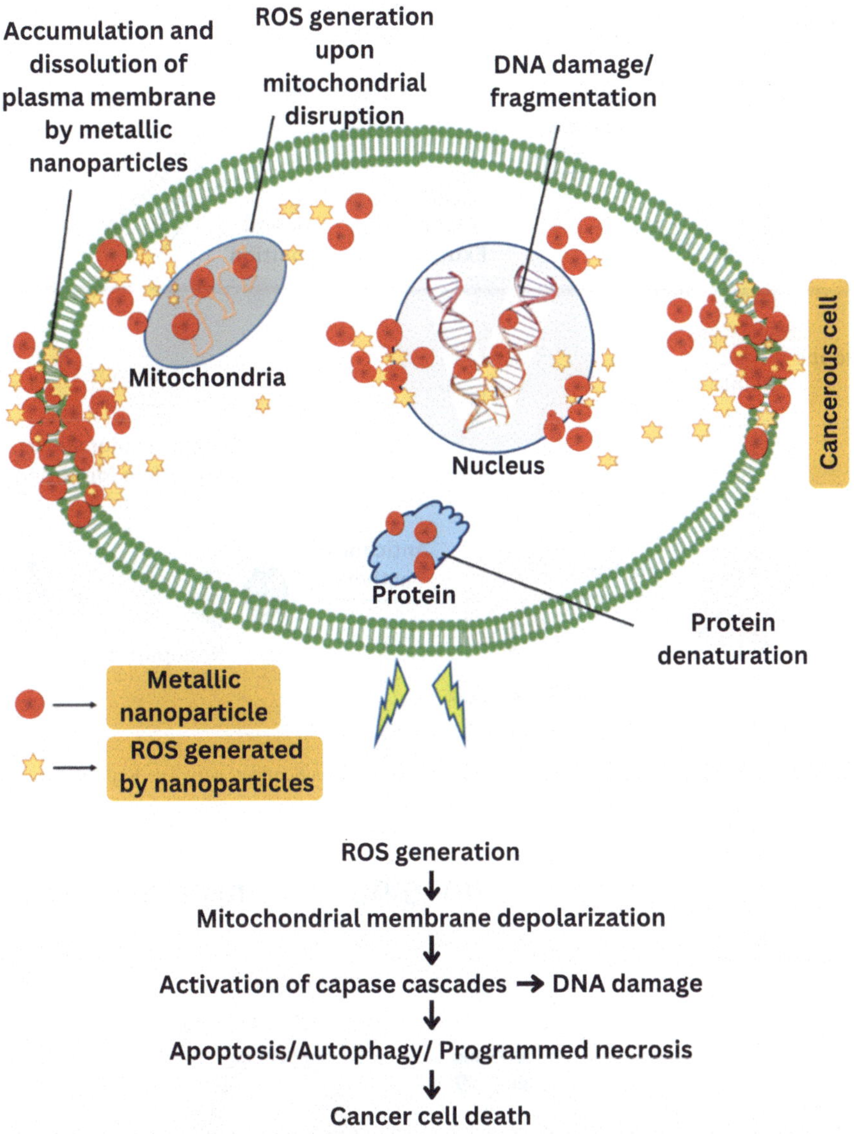

Fig. 3 Proposed mechanism of green synthesized metallic NPs mediated cytotoxicity in cancerous cells. Reprinted with permission from [27]. Copyrights MDPI 2021

4 Blood–Brain Barrier (BBB) and Brain Drug Delivery

The blood–brain barrier provides a formidable challenge to the transport of molecules into brain which restricts the efficacy of treatments of brain tumors. However, biophysical barrier restricts the efficiency of treatments to brain cancers by not

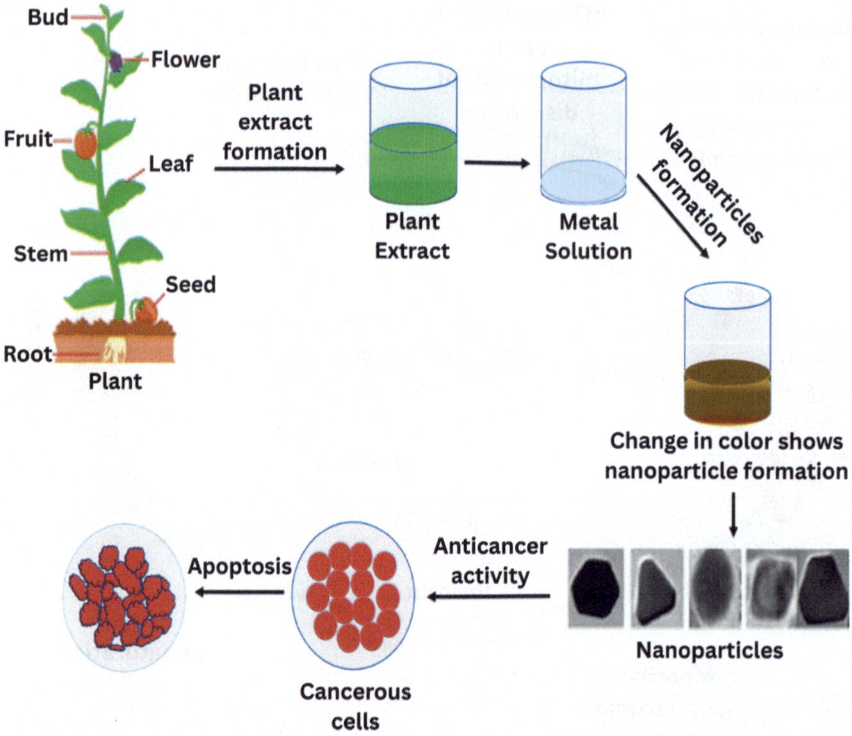

Fig. 4 Plant mediated metallic NPs synthesis and their anticancer activity. Reprinted with permission from [27]. Copyrights MDPI 2021

Fig. 5 Applications of iron oxide nanoparticles in cancer therapies. Reprinted with permission from [40]. Copyrights MDPI 2023

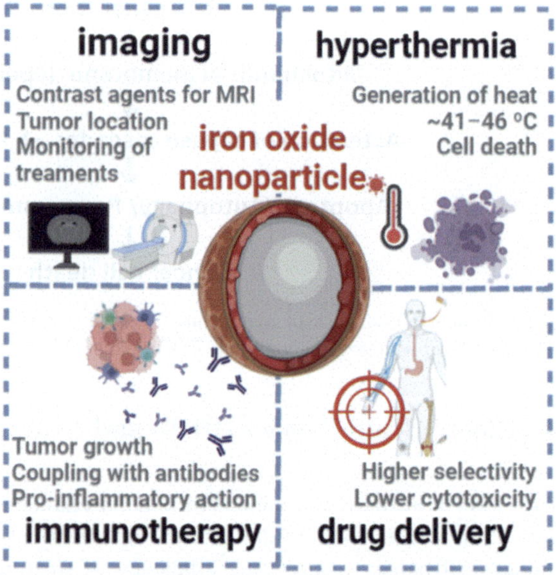

allowing drugs to penetrate beyond the blood–brain barrier. In recent years, nanotechnology has provided approaches to drug delivery system across the blood–brain barrier [41]. In the same respect, multitargeted nanoparticles have been realized to deliver combinations of drugs to the brain through the blood–brain barrier for brain metastases of triple-negative breast cancer cells which underlines the application of nanotechnology in handling challenges associated with the brain cancer [42].

For example, the biological efficiency and toxicities of functionalized silica-coated iron oxide nanoparticles within human astrocytes and the application of glutathione incorporated docetaxel linked nanoparticles to target the brain for cancer therapy have been established [43]. Also, synthesized metal nanoparticles exert cytotoxic effects on cancer cells but do not have toxic effects on astrocytes, meaning that they could be applied in cancer treatment [44].

Pericytes act as key regulators of the tumor microenvironment, contributing to regulation of microvascular stability and nanoparticle penetration in cancer treatment. Investigations have established that isolating pericyte coverage may increase the uptake of nanoparticles in tumor xenografts making it possible to treat aggressive cancers such as pancreatic cancer [42]. Treatment involving selective destruction of tumor vascular pericytes through nanoparticles conjugated with peptides have indicated potential in the anti-angiogenic therapy of metastatic melanoma. It is also important to learn the interactions between pericytes and cancer cells so as to enhance the nanocarriers penetration to the tumor microenvironment with the aim of improving the outcome of treatment [45].

Microglial active nanoparticles have become a potential tool to target the microglial cells in cancer treatment. Microglial cells of the central nervous system are involved in the immune response to stimuli [46], gold nanoparticles can engage with microglia which indicates that nanoparticles have the capability to control the activity of the immune reactions in brains [47]. Also, nanoparticles from brain-cortex microglia-derived exosomes have been successfully used to restrain the invasive character of cancer cells, which indicates their possible application as therapies for glioma [48]. Polymeric nanoparticle-based targeting of anti-inflammatory agents to microglial cells has also been looked at as a viable approach at addressing neuroinflammatory diseases [49].

Nanoparticles produce useful advancements in cancer research with regard to the modulation of tight junctions in cancer cells. Gold nanoparticles have been demonstrated to potentiate the action of cetuximab in colon carcinoma cells through alteration of tight junction [50]. Hydroxycamptothecin conjugated Fe_3O_4 nanoparticles also cause apoptosis in lung cancer cells with the activation of caspase-8, and the disuse of tight junctions [42]. These results open the possibility of making use of nanoparticles for targeted delivery toward cancer cell tight junctions to enhance therapeutic effects [51]. Metal nanoparticles could alter the tight junction associated proteins in the human epidermal keratinocytes indicating how nanoparticles disrupt tight junction regulation [52]. The characteristics of nanoparticles also influence the interaction of nanoparticles with blood vessels and affect tight junctions, as shown in laboratory experiments analyzing nanoparticles and endothelial tight junctions [53].

Lipophilic molecules have been well-incorporated into nanoparticles, which serve as good candidates for cancer drug delivery. Chitosan based nanoparticles for encapsulation of lipophilic anticancer agents have been understood to exhibit efficacy in targeting cancerous cells [54]. Lipid-polymer hybrid nanoparticles have been explored for the targeted delivery of water soluble and water insoluble forms of doxorubicin in to breast cancer, showing that these nanoparticle carriers offer a great promise in cancer chemotherapy [55]. In addition, there is the application of the solid lipid nanoparticles containing lipophilic derivatives of doxorubicin that have shown improved drug targeting in the drug resistant cancer cells implying.

Nanoparticles act as potential carriers for cancer targeted drug delivery and they are shown to improve the cellular uptake and change the endocytic process. Gao et al. have demonstrated ligand-functionalized nanoparticles to enhance cell uptake, alter the dominant pathway of particle internalization from macropinocytosis to clathrin-mediated endocytosis, and improve distribution and internalization in glioma cells [56]. In a similar fashion, Lu et al. have indicated that energy dependent endocytosis mechanism of cancer cells allows the uptake of mesoporous silica nanoparticles thus advocating for their use in delivery of anticancer drugs [57]. These studies emphasized the need for knowledge regarding the uptake nanoparticle by the cancer cells, so that an optimal therapeutic effectiveness can be achieved [56, 57].

Nanoparticles are considered as potential therapeutic agents in cancer treatment due to their ability of targeted drug delivery. Clathrin-mediated endocytosis (CME) is considered to be among the most important processes for the uptake of nanoparticles by cancer cells and operations of their therapeutic properties. Other researches have indicated that CME has a central role in the internalization of various nanoparticles into cancer cells with particles such as gold nanoparticles, hydroxyapatite nanoparticles and magnetic nanoparticles. For instance, the study has showed that size and shape characteristics of hydroxyapatite nanoparticles affect the process of CME in melanoma cells, thereby promising anti-tumor effects.

Nanoparticles have come into the forefront of targeted drug delivery in cancer treatment and the pattern of internalization of nanoparticles has been found to be very significant. With regards to the mechanism through which nanoparticles are internalized into cancer cells, it has been established that both clathrin dependent and caveolin dependent pathways are viable. Such mechano-signaling pathways have been reported to influence the nanoparticle uptake within the cancerous cells. Moreover, the transport of therapeutic nanoparticles across blood-endothelial barrier has been shown to occur through caveolae-mediated process, and this shows the probability of using this pathway for drug delivery [58]. In addition, it has described that caveolae-mediated endocytosis could be applied to effective siRNA delivery of polymeric nanoparticles to macrophages and there are good potentials on the targeted delivery on cancer treatment [59]. The complex transport mechanism of the BBB can be seen in Fig. 6.

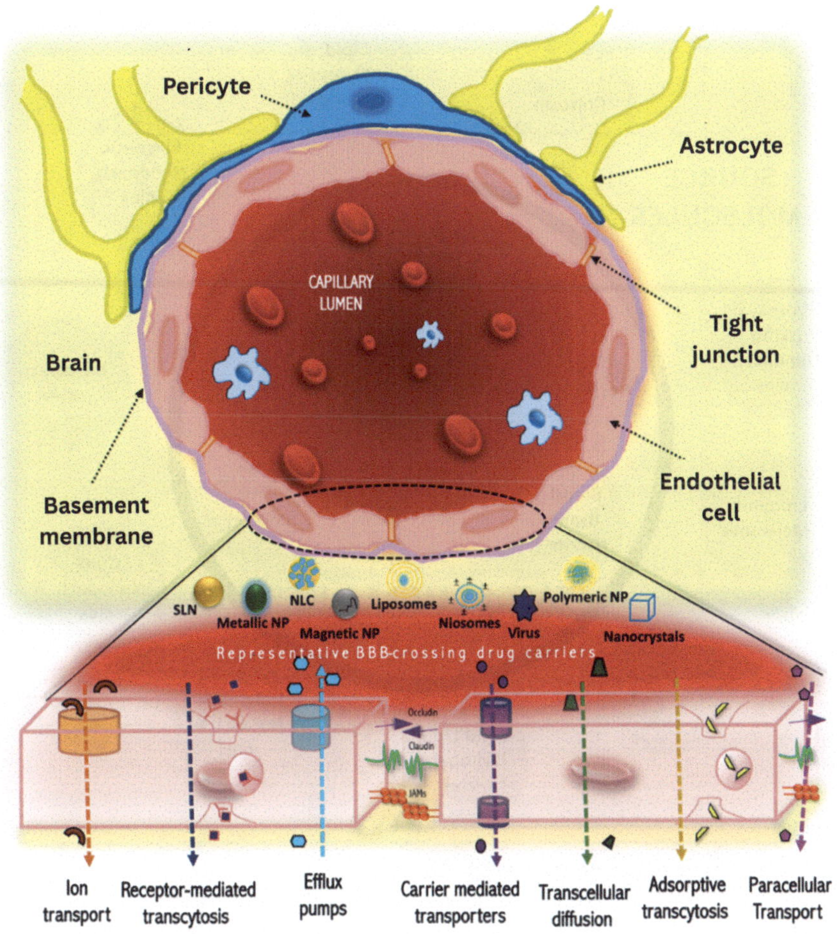

Fig. 6 Structure and transport mechanisms of the blood–brain barrier (BBB). Reprinted with permission from [60]. Copyrights MDPI 2020

5 Toxicity and Biocompatibility

See Fig. 7.

5.1 Biocompatibility

Despite a wide range of advantages of nanotechnology in the field of medicine, it is worth noting that reduced particle size makes them more reactive with living cells, raising the issue of potential toxicity. This results in higher cytotoxic changes

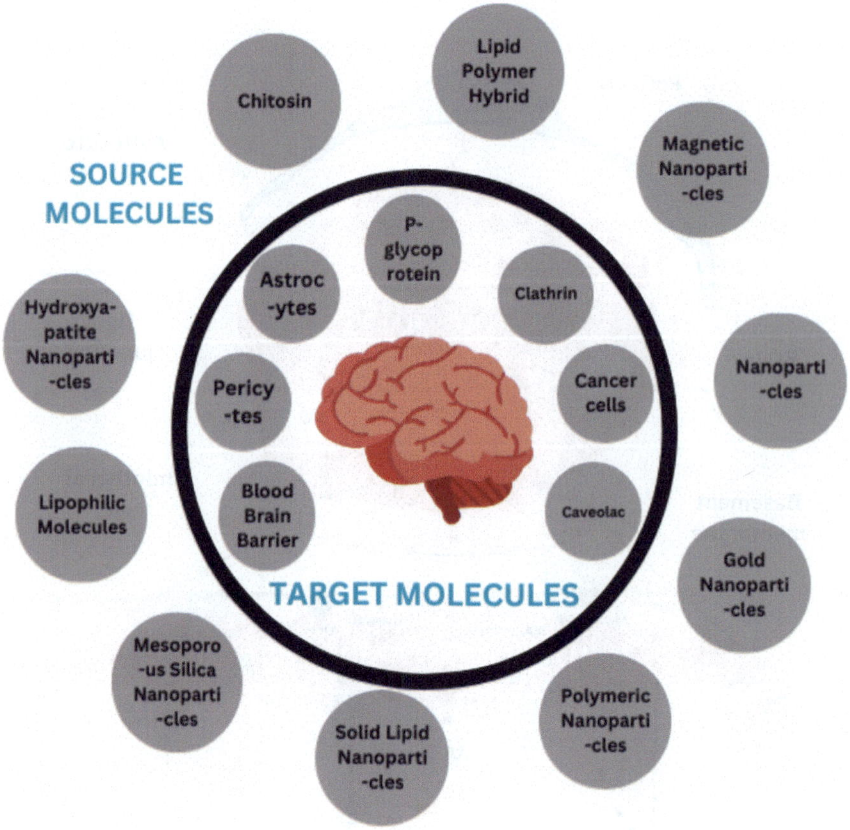

Fig. 7 Blood–brain barrier (BBB) and brain drug delivery

in normal cells due to swift internalization as compared to particles with larger diameters. The size and shape of NPs are crucial factors affecting biocompatibility. NPs have received substantial attention in cancer related research because of their promising uses in theragnosis, drug delivery and imaging. Gold NPs have been widely explored due to their biocompatibility and readily available optical characteristics suitable for cancer theragnosis [61]. In the same respect, green synthesis of silver nanoparticles has revealed certain potency in respect to biocompatibility, stability and anticancer properties [62]. Concerning Meso-Silica NPs, research studies relating to biocompatibility, biodistribution, as well as performance of drug delivery in cancer treatment have been carried out and yielded positive results for the effectiveness of NPs in animal models [63] (Fig. 8).

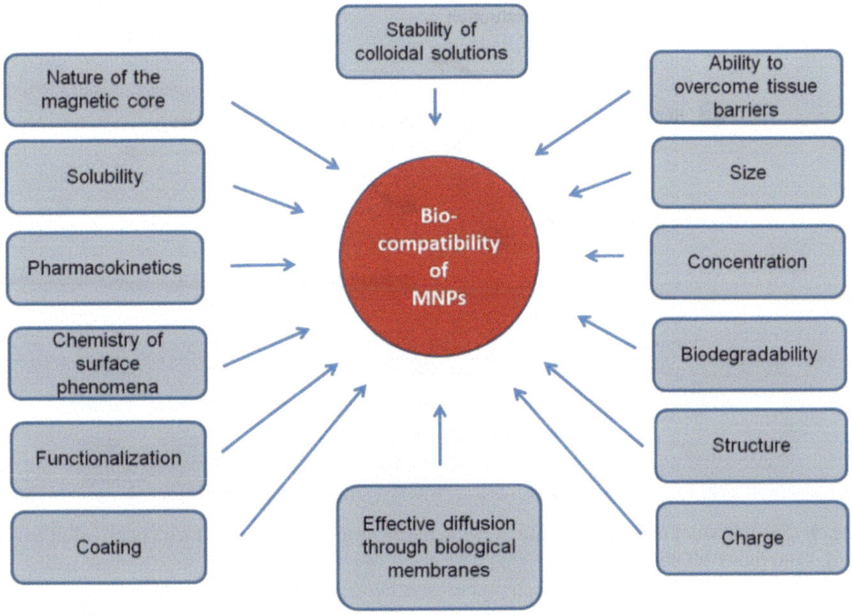

Fig. 8 Physical and chemical factors of magnetic nanoparticles, determining their biocompatibility. Reprinted with permission from [64]. Copyrights MDPI 2020

5.2 *Nanotoxicology*

Nanotoxicology is a relatively new branch of toxicology which deals with toxicologic aspects of nanomaterials on living beings and ecosystems. The toxicity of nanomaterials has been overlooked in the past, but now there is a growing concern about toxicity of nanomaterials. Due to the very small size, their surface area to volume ratio is exceptionally high providing greater chemical and biological reactivity which in turn poses a greater risk of toxicity than their corresponding larger bulk particles. Due to growing developments in the subject of nanotechnology, it is extremely critical to analyze safety issues concerning nanoparticles.

Recent research in nanotoxicology have contributed significantly to the knowledge of cytotoxicity of nanomaterials and the need to establish the risks associated with nanoparticles in numerous applications in biomedicine [65]. This is an emerging field that is critical for developing a better understanding of NP-biological interfaces for cancer treatment and other biomedical applications (Fig. 9).

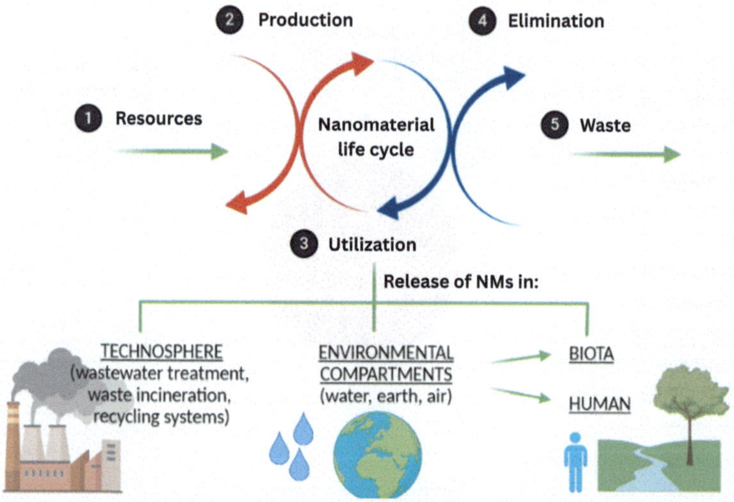

Fig. 9 The impact of nanomaterial life cycle on the environment. Reprinted with permission from [66]. Copyrights MDPI 2020

5.3 *Nanoparticles Safety*

Nanoparticles have attracted significant interest in cancer study because of their diagnostic and therapeutic potentials. Successful trials have proved nanocarriers to be capable of delivering drugs to tumor tissues [67]. Recent research regarding nanoparticles has proved to be very useful because of their ability to transport drugs to target materials. However, the legalities on the usage of nanoparticles in cancer treatment are still hazy and subject to change. Nanomaterials can have adverse effects on biological life and environment, it is important to consider their toxicity at all stages, i.e. production, application, and disposal. Besides human life, they also pose a hazard to agriculture and marine life [68]. Tests to determine toxicity can be performed on both, cell cultures (in vitro) and in living organisms (in vivo) such as fish and mice [69, 70]. However, there is no internationally recognized standard for verifying nanoparticles' toxicity, which can cause confusion and loopholes in comparison of the results. The most commonly used nanotoxicity model are small rodents like mice and rats due to their greater resemblance to humans, and ease of handling [71].

6 In Vivo and In Vitro Testing

Nanoparticles have become popular in cancer research because of their versatility in both, culture and live models. Nanoparticles can be useful for in vitro cytotoxicity and apoptosis in cancer cells [72]. Based on In vivo results, nanoparticles introduce cytotoxicity, and are potentially useful for cancer therapies [73]. Comparing the result of toxicity of nanoparticles plays a significant role in understanding the behavior of the nanoparticles in cancer as well as the development of potential treatment methods [74].

The binding of plasma proteins to curcumin nanoparticles has great significance in the efficacy of cancer treatment due to its contribution to tumor targeting and improving the general pharmacokinetics of the drug molecules [75]. Biodistribution studies shed light on nanoparticles' in vivo characteristics and their distribution inside an organism [63, 76]. Other aspects of the clearance and biodistribution of polymeric nanoparticles have also been investigated by others in order to increase their likelihood of being used therapeutically [77]. As active targeting and controlled pharmacokinetics and biodistribution of engineered nanoparticles, there is a potential for enhancement of cancer therapy outcomes [78].

NPs have been proved to have a great application in cancer research and therapy in virtue of their renal clearance and tumor targeting imaging with high tumor-to-background ratio. Renal clearable nanoparticles have helped to extend the frontiers of optimizing biomedical imaging and cancer therapy by increasing the use of nanomaterials in cancer treatment and biomedical imaging. Research has shown in various studies that renal clearance of imaging nanoparticles is effective and rapid tumor targeting is possible. Furthermore, the synthesis of nanomedicine with improved renal clearance is being investigated to overcome possible toxicity in the case of cancer therapy. The renal clearance of nanoparticles provides a route for their excretion from the body, making them valuable tools for cancer therapy.

Several nanoparticles have been considered in cancer research mainly due to their characteristics and their ability to destruct tumor cells. It is thus imperative to comprehend the hepatic clearance of nanoparticles with a view of designing appropriate cancer therapy processes.

Nanoparticles are a relatively new concept in cancer research. Knowing about the distribution of nanoparticles within a cell is particularly important for research in cancer treatment [79] (Fig. 10).

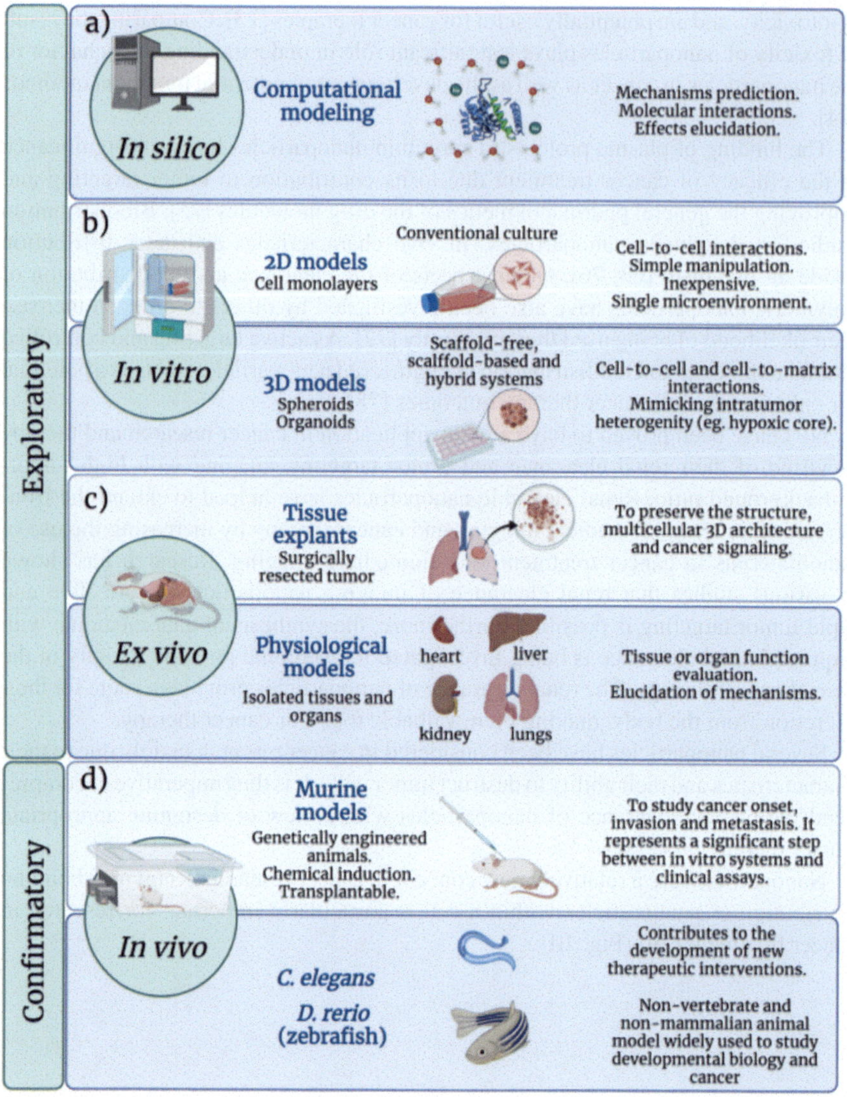

Fig. 10 Features of **a** in silico **b** in vitro, **c** ex vivo, and **d** in vivo MNP evaluation models. Reprinted with permission from [80] Copyrights MDPI 2021

References

1. Ramsden, J.J.: What is nanotechnology? In: Nanotechnology, pp. 1–18. Elsevier (2016). https://doi.org/10.1016/B978-0-323-39311-9.00007-8
2. Bhushan, B. (ed.): Springer Handbook of Nanotechnology. Springer Berlin Heidelberg, Berlin, Heidelberg (2017). https://doi.org/10.1007/978-3-662-54357-3

3. Emerich, D.F., Thanos, C.G.: Nanotechnology and medicine. Expert Opin. Biol. Ther. **3**(4), 655–663 (2003). https://doi.org/10.1517/14712598.3.4.655
4. Koch, C.C.: Nanostructured materials: an overview. In: Zhu, Y.T. (ed.) Bulk Nanostructured Materials, 1st ed., pp. 1–20. Wiley (2009). https://doi.org/10.1002/9783527626892.ch1
5. Suryanarayana, C.: Recent developments in nanostructured materials. Adv. Eng. Mater. **7**(11), 983–992 (2005). https://doi.org/10.1002/adem.200500135
6. Logothetidis, S. (ed.): Nanostructured materials and their applications. In: NanoScience and Technology. Springer Berlin Heidelberg, Berlin, Heidelberg (2012). https://doi.org/10.1007/978-3-642-22227-6
7. Singh, S.K., Kulkarni, P.P., Dash, D.: Biomedical applications of nanomaterials: an overview. In: Bagchi, D., Bagchi, M., Moriyama, H., Shahidi, F. (eds.) Bio-Nanotechnology, 1st ed., pp. 1–32. Wiley (2013). https://doi.org/10.1002/9781118451915.ch1
8. Lanone, S., Boczkowski, J.: Biomedical applications and potential health risks of nanomaterials: molecular mechanisms. Curr. Mol. Med. **6**(6), 651–663 (2006). https://doi.org/10.2174/156652406778195026
9. Mauter, M.S., Elimelech, M.: Environmental applications of carbon-based nanomaterials. Environ. Sci. Technol. **42**(16), 5843–5859 (2008). https://doi.org/10.1021/es8006904
10. Maiti, D., Tong, X., Mou, X., Yang, K.: Carbon-based nanomaterials for biomedical applications: a recent study. Front. Pharmacol. **9**, 1401 (2019). https://doi.org/10.3389/fphar.2018.01401
11. Scida, K., Stege, P.W., Haby, G., Messina, G.A., García, C.D.: Recent applications of carbon-based nanomaterials in analytical chemistry: critical review. Anal. Chim. Acta **691**(1–2), 6–17 (2011). https://doi.org/10.1016/j.aca.2011.02.025
12. Saleem, J., Wang, L., Chen, C.: Carbon-based nanomaterials for cancer therapy via targeting tumor microenvironment. Adv. Healthc. Mater. **7**(20), 1800525 (2018). https://doi.org/10.1002/adhm.201800525
13. Bianco, A., Kostarelos, K., Prato, M.: Opportunities and challenges of carbon-based nanomaterials for cancer therapy. Expert Opin. Drug Deliv. **5**(3), 331–342 (2008). https://doi.org/10.1517/17425247.5.3.331
14. Garriga, R.: Toxicity of Carbon Nanomaterials and Their Potential Application as Drug Delivery Systems: In Vitro Studies in Caco-2 and MCF-7 Cell Lines. MDPI (2020). https://doi.org/10.3390/nano10081617
15. Nayl, A.A., Abd-Elhamid, A.I., Aly, A.A., Bräse, S.: Recent progress in the applications of silica-based nanoparticles. RSC Adv. **12**(22), 13706–13726 (2022). https://doi.org/10.1039/D2RA01587K
16. Li, H., Chen, L., Li, X., Sun, D., Zhang, H.: Recent progress on asymmetric carbon- and silica-based nanomaterials: from synthetic strategies to their applications. Nano-Micro Lett. **14**(1), 45 (2022). https://doi.org/10.1007/s40820-021-00789-y
17. Rosenholm, J.M., et al.: Targeting of porous hybrid silica nanoparticles to cancer cells. ACS Nano **3**(1), 197–206 (2009). https://doi.org/10.1021/nn800781r
18. Arap, W., et al.: Luminescent silica nanoparticles for cancer diagnosis. Curr. Med. Chem. **20**(17), 2195–2211 (2013). https://doi.org/10.2174/0929867311320170005
19. Koohi Moftakhari Esfahani, M., Alavi, S.E., Cabot, P.J., Islam, N., Izake, E.L.: Application of mesoporous silica nanoparticles in cancer therapy and delivery of repurposed anthelmintics for cancer therapy. Pharmaceutics **14**(8), 1579 (2022). https://doi.org/10.3390/pharmaceutics14081579
20. Chen, Y., et al.: Two-dimensional metal nanomaterials: synthesis, properties, and applications. Chem. Rev. **118**(13), 6409–6455 (2018). https://doi.org/10.1021/acs.chemrev.7b00727
21. Liu, P., Qin, R., Fu, G., Zheng, N.: Surface coordination chemistry of metal nanomaterials. J. Am. Chem. Soc. **139**(6), 2122–2131 (2017). https://doi.org/10.1021/jacs.6b10978
22. Xu, H., Shang, H., Wang, C., Du, Y.: Low-dimensional metallic nanomaterials for advanced electrocatalysis. Adv. Funct. Mater. **30**(50), 2006317 (2020). https://doi.org/10.1002/adfm.202006317

23. Khursheed, R., et al.: Biomedical applications of metallic nanoparticles in cancer: current status and future perspectives. Biomed. Pharmacother. **150**, 112951 (2022). https://doi.org/10.1016/j.biopha.2022.112951
24. Evans, E.R., Bugga, P., Asthana, V., Drezek, R.: Metallic nanoparticles for cancer immunotherapy. Mater. Today **21**(6), 673–685 (2018). https://doi.org/10.1016/j.mattod.2017.11.022
25. Desai, N., Momin, M., Khan, T., Gharat, S., Ningthoujam, R.S., Omri, A.: Metallic nanoparticles as drug delivery system for the treatment of cancer. Expert Opin. Drug Deliv. **18**(9), 1261–1290 (2021). https://doi.org/10.1080/17425247.2021.1912008
26. Ahmad, M.Z., et al.: Metallic nanoparticles: technology overview and drug delivery applications in oncology. Expert Opin. Drug Deliv. **7**(8), 927–942 (2010). https://doi.org/10.1517/17425247.2010.498473
27. Andleeb, A., et al.: A systematic review of biosynthesized metallic nanoparticles as a promising anti-cancer-strategy. Cancers **13**(11), 2818 (2021). https://doi.org/10.3390/cancers13112818
28. Asha, A.B., Narain, R.: Nanomaterials properties. In: Polymer Science and Nanotechnology, pp. 343–359. Elsevier (2020). https://doi.org/10.1016/B978-0-12-816806-6.00015-7
29. Shaji, A., Zachariah, A.K.: Surface area analysis of nanomaterials. In: Thermal and Rheological Measurement Techniques for Nanomaterials Characterization, pp. 197–231. Elsevier (2017). https://doi.org/10.1016/B978-0-323-46139-9.00009-8
30. Palchoudhury, S., Baalousha, M., Lead, J.R.: Methods for measuring concentration (mass, surface area and number) of nanomaterials. In: Frontiers of Nanoscience, vol. 8, pp. 153–181. Elsevier (2015). https://doi.org/10.1016/B978-0-08-099948-7.00005-1
31. Royston, E., Ghosh, A., Kofinas, P., Harris, M.T., Culver, J.N.: Self-assembly of virus-structured high surface area nanomaterials and their application as battery electrodes. Langmuir **24**(3), 906–912 (2008). https://doi.org/10.1021/la7016424
32. Isidoro, C., Maneerat, E., Giovia, A., Carlo, F., Caputo, G.: Biocompatibility, endocytosis, and intracellular trafficking of mesoporous silica and polystyrene nanoparticles in ovarian cancer cells: effects of size and surface charge groups. Int. J. Nanomed. 4147 (2012). https://doi.org/10.2147/IJN.S33803
33. Ankamwar, B., et al.: Biocompatibility of Fe_3O_4 nanoparticles evaluated by in vitro cytotoxicity assays using normal, glia and breast cancer cells. Nanotechnology **21**(7), 075102 (2010). https://doi.org/10.1088/0957-4484/21/7/075102
34. Sponchia, G., et al.: Biocompatible tailored zirconia mesoporous nanoparticles with high surface area for theranostic applications. J. Mater. Chem. B **3**(36), 7300–7306 (2015). https://doi.org/10.1039/C5TB01424G
35. Geißler, D., Nirmalananthan-Budau, N., Scholtz, L., Tavernaro, I., Resch-Genger, U.: Analyzing the surface of functional nanomaterials—how to quantify the total and derivatizable number of functional groups and ligands. Microchim. Acta **188**(10), 321 (2021). https://doi.org/10.1007/s00604-021-04960-5
36. Li, Z., et al.: Applications of surface modification technologies in nanomedicine for deep tumor penetration. Adv. Sci. **8**(1), 2002589 (2021). https://doi.org/10.1002/advs.202002589
37. Solanki, A., Kim, J.D., Lee, K.-B.: Nanotechnology for regenerative medicine: nanomaterials for stem cell imaging. Nanomedicine **3**(4), 567–578 (2008). https://doi.org/10.2217/17435889.3.4.567
38. Engel, E., Michiardi, A., Navarro, M., Lacroix, D., Planell, J.A.: Nanotechnology in regenerative medicine: the materials side. Trends Biotechnol. **26**(1), 39–47 (2008). https://doi.org/10.1016/j.tibtech.2007.10.005
39. Medina-Reyes, E.I., Garcia-Viacobo, D., Carrero-Martinez, F.A., Chirino, Y.I.: Applications and risks of nanomaterials used in regenerative medicine, delivery systems, theranostics, and therapy. Crit. Rev. Ther. Drug Carr. Syst. **34**(1), 35–61. (2017). https://doi.org/10.1615/CritRevTherDrugCarrierSyst.2017016983
40. Sell, M., et al.: Application of nanoparticles in cancer treatment: a concise review. Nanomaterials **13**(21), 2887 (2023). https://doi.org/10.3390/nano13212887

41. Tang, W., Fan, W., Lau, J., Deng, L., Shen, Z., Chen, X.: Emerging blood–brain-barrier-crossing nanotechnology for brain cancer theranostics. Chem. Soc. Rev. **48**(11), 2967–3014 (2019). https://doi.org/10.1039/C8CS00805A
42. Zhang, T., et al.: Multitargeted nanoparticles deliver synergistic drugs across the blood-brain barrier to brain metastases of triple negative breast cancer cells and tumor-associated macrophages. Adv. Healthc. Mater. **8**(18), 1900543 (2019). https://doi.org/10.1002/adhm.201900543
43. Grover, A., Hirani, A., Pathak, Y., Sutariya, V.: Brain-targeted delivery of docetaxel by glutathione-coated nanoparticles for brain cancer. AAPS PharmSciTech **15**(6), 1562–1568 (2014). https://doi.org/10.1208/s12249-014-0165-0
44. Vodyanoy, V.J., Daniels, Y., Pustovyy, O., MacCrehan, W., Muramoto, S., Stan, G.S.: Engineered metal nanoparticles in the sub-nanomolar levels kill cancer cells. Int. J. Nanomed. 1567 (2016). https://doi.org/10.2147/IJN.S101463
45. Jiang, Z., et al.: Pericytes in the tumor microenvironment. Cancer Lett. **556**, 216074 (2023). https://doi.org/10.1016/j.canlet.2023.216074
46. Albanese, A., Sykes, E.A., Chan, W.C.W.: Rough around the edges: the inflammatory response of microglial cells to spiky nanoparticles. ACS Nano **4**(5), 2490–2493 (2010). https://doi.org/10.1021/nn100776z
47. Hutter, E., et al.: Microglial response to gold nanoparticles. ACS Nano **4**(5), 2595–2606 (2010). https://doi.org/10.1021/nn901869f
48. Murgoci, A., et al.: Brain-cortex microglia-derived exosomes: nanoparticles for glioma therapy. ChemPhysChem **19**(10), 1205–1214 (2018). https://doi.org/10.1002/cphc.201701198
49. Cahalane, C., et al.: Targeted delivery of anti-inflammatory and imaging agents to microglial cells with polymeric nanoparticles. Mol. Pharm. **17**(6), 1816–1826 (2020). https://doi.org/10.1021/acs.molpharmaceut.9b00489
50. Leve, F., Bonfim, D.P., Fontes, G., Morgado-Díaz, J.A.: Gold nanoparticles regulate tight junctions and improve cetuximab effect in colon cancer cells. Nanomedicine **14**(12), 1565–1578 (2019). https://doi.org/10.2217/nnm-2019-0023
51. Shipunova, V.O., et al.: Targeting cancer cell tight junctions enhances PLGA-based photothermal sensitizers' performance in vitro and in vivo. Pharmaceutics **14**(1), 43 (2021). https://doi.org/10.3390/pharmaceutics14010043
52. Yuan, J., Zhang, Y., Zhang, Y., Mo, Y., Zhang, Q.: Effects of metal nanoparticles on tight junction-associated proteins via HIF-1α/miR-29b/MMPs pathway in human epidermal keratinocytes. Part. Fibre Toxicol. **18**(1), 13 (2021). https://doi.org/10.1186/s12989-021-00405-2
53. Zhang, Y., Yang, W.-X.: Tight junction between endothelial cells: the interaction between nanoparticles and blood vessels. Beilstein J. Nanotechnol. **7**, 675–684 (2016). https://doi.org/10.3762/bjnano.7.60
54. Abruzzo, A., et al.: Chitosan nanoparticles for lipophilic anticancer drug delivery: development, characterization and in vitro studies on HT29 cancer cells. Colloids Surf. B Biointerfaces **145**, 362–372 (2016). https://doi.org/10.1016/j.colsurfb.2016.05.023
55. Tahir, N., et al.: Lipid-polymer hybrid nanoparticles for controlled delivery of hydrophilic and lipophilic doxorubicin for breast cancer therapy. Int. J. Nanomed. **14**, 4961–4974 (2019). https://doi.org/10.2147/IJN.S209325
56. Gao, H., et al.: Ligand modified nanoparticles increases cell uptake, alters endocytosis and elevates glioma distribution and internalization. Sci. Rep. **3**(1), 2534 (2013). https://doi.org/10.1038/srep02534
57. Lu, J., et al.: Mesoporous silica nanoparticles for cancer therapy: energy-dependent cellular uptake and delivery of paclitaxel to cancer cells. NanoBiotechnology **3**(2), 89–95 (2007). https://doi.org/10.1007/s12030-008-9003-3
58. Wang, Z.: Caveolae-mediated delivery of therapeutic nanoparticles across blood-endothelial barrier. Austin J. Anal. Pharm. Chem. **1**(4), 1018 (2014)
59. Song, Y., Wu, Y., Xu, L., Jiang, T., Tang, C., Yin, C.: Caveolae-mediated endocytosis drives robust siRNA delivery of polymeric nanoparticles to macrophages. ACS Nano **15**(5), 8267–8282 (2021). https://doi.org/10.1021/acsnano.0c08596

60. Cortés, H., et al.: A reevaluation of chitosan-decorated nanoparticles to cross the blood-brain barrier. Membranes **10**(9), 212 (2020). https://doi.org/10.3390/membranes10090212
61. Kang, M.S., Lee, S.Y., Kim, K.S., Han, D.-W.: State of the art biocompatible gold nanoparticles for cancer theragnosis. Pharmaceutics **12**(8), 701 (2020). https://doi.org/10.3390/pharmaceutics12080701
62. Jadhav, K., et al.: Phytosynthesis of silver nanoparticles: characterization, biocompatibility studies, and anticancer activity. ACS Biomater. Sci. Eng. **4**(3), 892–899 (2018). https://doi.org/10.1021/acsbiomaterials.7b00707
63. Lu, J., Liong, M., Li, Z., Zink, J.I., Tamanoi, F.: Biocompatibility, biodistribution, and drug-delivery efficiency of mesoporous silica nanoparticles for cancer therapy in animals. Small **6**(16), 1794–1805 (2010). https://doi.org/10.1002/smll.201000538
64. Zamay, G.S., Zamay, T.N., Lukyanenko, K.A., Kichkailo, A.S.: Aptamers increase biocompatibility and reduce the toxicity of magnetic nanoparticles used in biomedicine. Biomedicines **8**(3), 59 (2020). https://doi.org/10.3390/biomedicines8030059
65. Saifi, M.A., Khan, W., Godugu, C.: Cytotoxicity of nanomaterials: using nanotoxicology to address the safety concerns of nanoparticles. Pharm. Nanotechnol. **6**(1), 3–16 (2018). https://doi.org/10.2174/2211738505666171023152928
66. Zielińska, A., et al.: Nanotoxicology and nanosafety: safety-by-design and testing at a glance. Int. J. Environ. Res. Public Health **17**(13), 4657 (2020). https://doi.org/10.3390/ijerph17134657
67. Wolfram, J., et al.: Safety of nanoparticles in medicine. Curr. Drug Targets **16**(14), 1671–1681 (2015). https://doi.org/10.2174/1389450115666140804124808
68. Inbathamizh, L., Harsha Varthan, M.K., Kumar, R.S.R., Rohinth, M., Tawfeeq Ahmed, Z.H.: Nanotechnology: ethical impacts, health issues, and safety issues. In: Malik, J.A., Sadiq Mohamed, M.J. (eds.) Modern Nanotechnology, pp. 455–477. Springer Nature Switzerland, Cham (2023). https://doi.org/10.1007/978-3-031-31104-8_20
69. Bian, W.-P., Pei, D.-S.: Zebrafish model for safety and toxicity testing of nutraceuticals. In: Nutraceuticals, pp. 333–339. Elsevier (2016). https://doi.org/10.1016/B978-0-12-802147-7.00025-5
70. Hadrup, N., et al.: Subacute oral toxicity investigation of selenium nanoparticles and selenite in rats. Drug. Chem. Toxicol. **42**(1), 76–83 (2019). https://doi.org/10.1080/01480545.2018.1491589
71. Richmond, J.: Refinement, reduction, and replacement of animal use for regulatory testing: future improvements and implementation within the regulatory framework. ILAR J. **43**(Suppl. 1), S63–S68 (2002). https://doi.org/10.1093/ilar.43.Suppl_1.S63
72. Kumar, V., Sharma, N., Maitra, S.S.: In vitro and in vivo toxicity assessment of nanoparticles. Int. Nano Lett. **7**(4), 243–256 (2017). https://doi.org/10.1007/s40089-017-0221-3
73. Cha, K.E., Myung, H.: Cytotoxic effects of nanoparticles assessed in vitro and in vivo. J. Microbiol. Biotechnol. **17**(9), 1573–1578 (2007)
74. Mahmoudi, M., Hofmann, H., Rothen-Rutishauser, B., Petri-Fink, A.: Assessing the in vitro and in vivo toxicity of superparamagnetic iron oxide nanoparticles. Chem. Rev. **112**(4), 2323–2338 (2012). https://doi.org/10.1021/cr2002596
75. Yallapu, M.M., Ebeling, M.C., Jaggi, M., Chauhan, S.C.: Plasma proteins interaction with curcumin nanoparticles: implications in cancer therapeutics. Curr. Drug Metab. **14**(4), 504–515 (2013). https://doi.org/10.2174/1389200211314040012
76. Sa, L.T.M., et al.: Biodistribution of nanoparticles: initial considerations. J. Pharm. Biomed. Anal. **70**, 602–604 (2012). https://doi.org/10.1016/j.jpba.2012.06.008
77. Alexis, F., Pridgen, E., Molnar, L.K., Farokhzad, O.C.: Factors affecting the clearance and biodistribution of polymeric nanoparticles. Mol. Pharm. **5**(4), 505–515 (2008). https://doi.org/10.1021/mp800051m
78. Arvizo, R.R., et al.: Modulating pharmacokinetics, tumor uptake and biodistribution by engineered nanoparticles. PLoS ONE **6**(9), e24374 (2011). https://doi.org/10.1371/journal.pone.0024374
79. Chou, L.Y.T., Ming, K., Chan, W.C.W.: Strategies for the intracellular delivery of nanoparticles. Chem. Soc. Rev. **40**(1), 233–245 (2011). https://doi.org/10.1039/C0CS00003E

80. Tinajero-Diaz, E., Salado-Leza, D., Gonzalez, C., Martinez Velazquez, M., Lopez, Z., Bravo-Madrigal, J., Knauth, P., Flores-Hernandez, F.Y., Herrera-Rodriguez, S.E., Navarro, R.E. and Cabrera-Wrooman, A., Krötzsch Zaira, E., Carvajal, Y. G., Hernández-Gutiérrez, R.: Green metallic nanoparticles for cancer therapy: evaluation models and cancer applications. Pharmaceutics **13**(10), 1719 (2021) https://doi.org/10.3390/pharmaceutics13101719

Personalized Cancer Treatment with Nanotechnology

Keyur Bhatt and Nihal Patel

Abstract Nanotechnology holds the potential to transform cancer treatment by embracing personalized medicine. It delves into the exceptional characteristics of nanoparticles, specifically highlighting their precision in targeting and delivering therapeutic agents directly to cancer cells. This targeted strategy offers significant advantages over conventional therapies by reducing damage to healthy tissues while potentially improving treatment efficacy. Various applications of nanotechnology in personalized cancer treatment are elucidated. Firstly, nanotechnology facilitates the development of drug delivery systems, where nanoparticles can be customized to transport specific drugs and release them exclusively at cancerous sites. This precise delivery mechanism not only enhances the potency of drugs but also reduces systemic toxicity. Secondly, the integration of therapeutic and diagnostic functions in theragnostic nanoparticles allows for real-time feedback on treatment effectiveness alongside drug administration. Additionally, nanoparticles show promise in boosting the body's immune response against cancer by activating the immune system to recognize and eliminate cancerous cells. Despite these advancements, the chapter acknowledges the challenges in translating nanotechnology-based therapies from experimental settings to clinical practice. It discusses ongoing research efforts aimed at overcoming these obstacles, laying the groundwork for a future where personalized cancer treatment using nanotechnology becomes a tangible reality.

Keywords Nanotechnology · Cancer treatment · Personalized medicine · Nanoparticles · Drug delivery systems · Immunotherapy

K. Bhatt (✉) · N. Patel
Department of Chemistry, Faculty of Science, Ganpat University, Mehsana, Gujarat, India
e-mail: kdb01@ganpatuniversity.ac.in; drkdbhatt@outlook.com

N. J. Joshi and G. Patel (eds.), *Nanoparticle Delivery in Cancer Care*,
Series in BioEngineering, https://doi.org/10.1007/978-981-95-0457-2_11

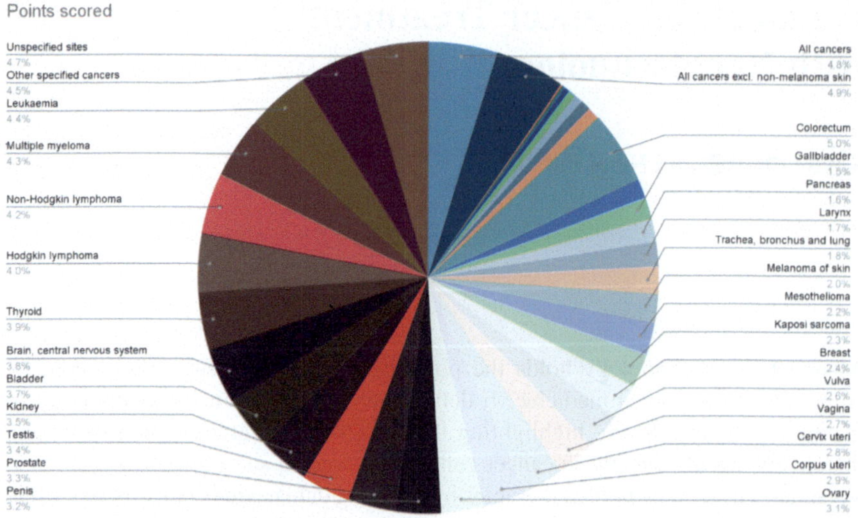

Fig. 1 Key cancer data and key figures on IARC: 2020–2021 [WHO]

1 Introduction

1.1 The Worldwide Cancer Burden

According to the World Health Organization (WHO), cancer is one of the main causes of morbidity and mortality globally, taking the lives of around 10 million people each year [1, 2]. The illness is caused by aberrant cells proliferating out of control, which can spread to other organs and invade nearby tissues. Lung, breast, colorectal, prostate, and liver cancers are common cancers that vary in frequency among various populations and geographical areas [3].

Cancer's heterogeneity is the foundation of its complexity. Despite having the same main site of origin, no two patients appear with the same form of cancer [4, 5]. This variability is caused by a variety of factors, including tumor microenvironment, genetic abnormalities, epigenetic modifications, and patient-specific traits like age, lifestyle, and comorbidities. Traditional cancer treatments, which frequently take a one-size-fits-all approach, face serious difficulties as a result of this intricacy [6–8] (Fig. 1).

1.2 Difficulties with Traditional Cancer Treatment

Chemotherapy, radiation therapy, and surgery are the mainstays of traditional cancer therapies [9]. Even though these treatments have saved many lives, they have some significant drawbacks: Chemotherapy medications frequently target both healthy and malignant cells, exhibiting a lack of selectivity. Significant adverse effects include nausea, exhaustion, hair loss, and immunosuppression follow from this [10, 11]. Over time, tumors may become resistant to chemotherapy, decreasing its effectiveness and increasing the risk of disease recurrence [12, 13]. A treatment that works for one patient might not work for another because of the genetic and genomic variability of cancer cells within a single tumor [14, 15]. The doses needed to eradicate cancer cells frequently approach causing harm to healthy tissues because radiation treatment and chemotherapy have limited therapeutic windows [16, 17]. Innovative strategies that can offer more accurate, efficient, and individualized cancer treatment are desperately needed in light of these difficulties.

1.3 A Revolution in Personalized Medicine

Precision medicine, sometimes referred to as personalized medicine, is a departure from the conventional "one-size-fits-all" approach to a more individualized approach that takes into account each person's unique genetic, environmental, and behavioral characteristics. The goal of personalized medicine is to pinpoint the precise biomarkers and genetic mutations causing a patient's cancer in order to develop treatments that specifically address these anomalies [18, 19]. Next-generation sequencing (NGS) and other advanced technologies provide a thorough genomic examination of a patient's tumor, detecting copy number variations, gene fusions, and mutations that can be targeted with certain treatments [20, 21].

The choice of targeted treatments or immunotherapies is influenced by biomarkers such EGFR mutations, HER2 amplification, and PD-L1 expression. Patients receive medicines tailored to address the detected changes based on their genetic and molecular profiles, increasing effectiveness and minimizing needless side effects [22]. The need for new predictive biomarkers, the scarcity of focused treatments, and the high expense of genomic testing are some of the obstacles that still need to be overcome despite the promise of personalized medicine. This is where the development of tailored cancer treatment is greatly aided by nanotechnology.

1.4 Nanotechnology's Function in the Treatment of Cancer

The manipulation of materials at the nanoscale (1–100 nm), where they display distinct physical, chemical, and biological properties, is known as nanotechnology. Because of these characteristics, nanoparticles (NPs) are excellent options for treating cancer and have a number of benefits over traditional treatments [23].

1.4.1 Accurate Medication Administration

The capacity of nanotechnology to precisely deliver therapeutic drugs to cancer cells is one of its most important accomplishments. By limiting the amount of hazardous medications that reach healthy cells, this targeted delivery lowers adverse effects and improves therapeutic efficacy. Because tumors have poor lymphatic drainage and leaky vasculature, nanoparticles can gather in the tumor microenvironment in a specific way. Drug delivery to malignant tissues is improved by this passive targeting technique. Ligands, antibodies, or peptides that identify and attach to particular receptors overexpressed on cancer cells—such as HER2 in breast cancer or folate receptors in ovarian cancer—can be used to functionalize nanoparticles.

1.4.2 Theragnostics: Integrating Diagnosis and Therapy

Real-time tracking of medication distribution, tumor response, and treatment effectiveness is made possible by theragnostic nanoparticles, which combine therapeutic and diagnostic capabilities into a single platform. This dual feature enhances results by enabling tailored therapy modifications based on real-time input.

Superparamagnetic iron oxide nanoparticles (SPIONs) enable simultaneous imaging and treatment by acting as MRI contrast agents and chemotherapeutic medication carriers.

1.4.3 Changing Immune System Functions

By influencing the immune system to identify and destroy cancer cells, nanoparticles can help improve cancer immunotherapy. To strengthen the immune response against cancers, they can administer cytokines, cancer vaccines, or immune checkpoint inhibitors.

Liposomal nanoparticles that deliver anti-PD-1 or anti-CTLA-4 antibodies have demonstrated potential in improving T-cell activation and combating cancer cells' ability to evade the immune system. By delivering tumor antigens and adjuvants to dendritic cells, nanoparticles can elicit a strong immune response against the tumor.

1.5 Nanotechnology's Benefits for Personalized Medicine

Nanotechnology is the perfect platform for individualized cancer treatment since it provides a number of clear benefits [24]. Hydrophobic medications can be encapsulated by nanoparticles, which increases their solubility and stability in biological settings. It is possible to design nanoparticles so that they release medications in response to particular stimuli, such as temperature, pH, or enzymes found in the tumor microenvironment. Nanoparticles lessen systemic exposure and related toxicities by precisely targeting medications to cancer cells. Nanoparticles can perform a variety of tasks on a single platform, including immunological regulation, imaging, and drug delivery.

1.6 Opportunities and Difficulties

Although nanotechnology has a lot of potential, a few obstacles need to be overcome before its full potential in tailored cancer treatment can be realized.

1.6.1 Safety and Biocompatibility

It is crucial to guarantee the long-term safety and biocompatibility of nanoparticles. Certain nanoparticles have the potential to cause immunological responses, build up in tissues that are not their intended target, or show surprising toxicity. To assess their safety characteristics, extensive preclinical and clinical research is required.

1.6.2 Obstacles in Regulation

Compared to traditional medications, nanomedicines have more complicated regulatory pathways that call for thorough characterisation, safety evaluation, and clinical validation. To expedite the approval process, cooperation between industry players, researchers, and regulatory bodies is crucial.

1.6.3 Manufacturing and Scalability

Another major problem is creating cost-effective and scalable manufacturing methods for nanoparticles. Clinical translation depends on standardizing production procedures and guaranteeing batch-to-batch uniformity.

1.7 Prospects for the Future

With continuous research aimed at resolving present issues and investigating novel uses, the future of nanotechnology in customized cancer treatment appears bright. Integrating immunotherapy, imaging, and medication delivery onto a single platform for all-encompassing cancer treatment. Using artificial intelligence (AI) to create and refine nanoparticle compositions according to patient-specific information can increase accuracy and effectiveness. Using nanoparticles to deliver gene-editing agents like CRISPR-Cas9 precisely, creating new therapeutic options for cancer.

1.8 Final Thoughts

Because nanotechnology offers safe, effective, and focused therapeutic options, it is redefining individualized cancer treatment. Nanotechnology has the potential to revolutionize cancer treatment and enhance patient outcomes by overcoming the drawbacks of traditional treatments and utilizing the special qualities of nanoparticles. Nanotechnology-based therapies are anticipated to become a key component of personalized medicine as research and development proceed, opening the door to a more accurate, efficient, and patient-centered cancer treatment system in the future.

2 Nanotechnology: Fundamentals and Applications

In nanotechnology, matter is manipulated at the nanoscale, usually at a size of 1–100 nm (nm). Materials at this scale have distinct physical, chemical, and biological characteristics that set them apart from their counterparts in bulk. These characteristics make nanotechnology ideal for use in medicine, especially oncology, where minimum toxicity, precision, and targeted delivery are critical. This section explores the basic properties of nanoparticles and their various uses in the treatment of cancer [25] (Table 1 and Fig. 2).

Table 1 Summary of nanoparticles in cancer treatment

Nanoparticle type	Key features	Example/benefit
Liposomes	Spherical vesicles with phospholipid bilayers; encapsulate hydrophilic and hydrophobic drugs	• **Doxil® (liposomal doxorubicin)**: reduced cardiotoxicity and improved tumor targeting
Polymeric nanoparticles	Biodegradable polymers (PLGA, PEG); offer controlled and sustained drug release	• **Paclitaxel-loaded PLGA NPs**: enhanced absorption, reduced toxicity, customizable drug release
Metallic nanoparticles	Gold, silver, platinum nanoparticles with optical and thermal properties for therapy and diagnosis	• **Gold NPs**: used in photothermal therapy and imaging for localized tumor ablation
Magnetic nanoparticles	Iron oxide-based; used for imaging and hyperthermia cancer treatment	• **SPIONs**: enhance MRI imaging and provide non-invasive hyperthermia therapy

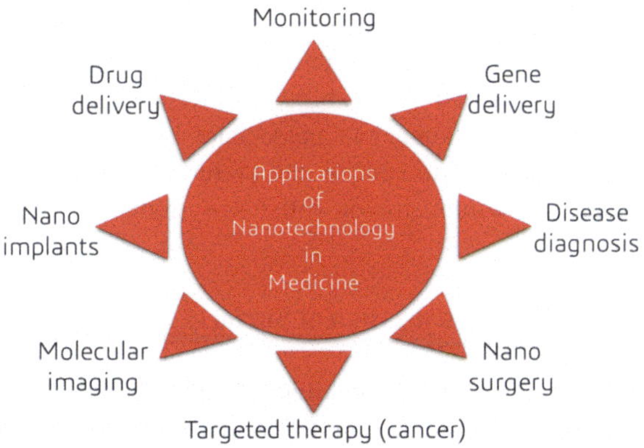

Fig. 2 Applications pf nanotechnology in medicine

2.1 Particular Properties of Nanoparticles

Nanoparticles (NPs) have unique properties that make them useful in medicine, particularly in the treatment of cancer. Compared to traditional treatment procedures, their special qualities allow for improved drug distribution, penetration of biological barriers, and contact with cellular components.

2.1.1 High Volume to Surface Area Ratio

Nanoparticles' high surface area-to-volume ratio is one of its most crucial properties. A material's surface atom proportion rises dramatically in relation to its internal atom proportion as its size shrinks to the nanoscale. The high surface area of nanoparticles enables them to carry a greater payload of therapeutic agents, such as imaging agents, chemotherapeutic medicines, and biomolecules like DNA and RNA. This results in enhanced drug loading.

The solubility and bioavailability of poorly water-soluble medications can be enhanced by nanoparticles because of their small size and surface area. Hydrophobic medications such as paclitaxel, for instance, become more soluble when encapsulated in nanoparticles, which enables efficient distribution in watery biological environments.

2.1.2 Enhanced Permeability

Larger molecules usually cannot pass through biological barriers, whereas nanoparticles can. This feature is very helpful for getting medications to places that are hard to get to, such tumors or the brain. The majority of medications cannot pass across the blood–brain barrier (BBB), which is a selective barrier. Treatment for brain cancers and neurodegenerative illnesses is made possible by nanoparticles' ability to traverse the blood–brain barrier, particularly those coated with polyethylene glycol (PEG).

The Enhanced Permeability and Retention (EPR) effect refers to the presence of leaky vasculature and inadequate lymphatic drainage in tumors. The ability of nanoparticles to concentrate preferentially in tumor tissues allows for targeted medication administration with less systemic harm.

2.1.3 Functionalization of Surfaces

A range of chemicals, including as ligands, antibodies, peptides, and polymers, can be added to nanoparticles to improve their therapeutic efficacy and targeted specificity. Active targeting is made possible through functionalization with ligands that attach to particular receptors that are overexpressed on cancer cells.

Medications are delivered to HER2-positive cells only by HER2-targeted nanoparticles for breast cancer, preserving healthy tissues. When functionalized with enzyme-responsive or pH-sensitive polymers, nanoparticles can release their payload in response to certain stimuli in the tumor microenvironment. The pharmacokinetics and biodistribution of nanoparticles are also improved by surface functionalization, which increases their concentration in target tissues and lengthens their circulation duration.

2.2 Types of Nanoparticles Used to Treat Cancer

For the treatment of cancer, several kinds of nanoparticles have been created; each has special qualities and uses. These consist of metallic, polymeric, magnetic, and liposome nanoparticles.

2.2.1 The Liposome

Liposomes are spherical vesicles that contain medicinal substances and are made up of one or more phospholipid bilayers. They are among the most well-known drug delivery systems for nanoparticles.

Water-soluble (hydrophilic) and fat-soluble (hydrophobic) medications can both be encapsulated by liposomes, offering a versatile approach to drug administration. By encasing chemotherapeutic drugs, liposomes minimize adverse effects by limiting their exposure to healthy tissues. Doxorubicin liposomal, or Doxil®: The FDA-approved medicine Doxil®, which wraps doxorubicin in a PEGylated liposome to improve drug delivery to tumors and decrease cardiotoxicity, is used to treat Kaposi's sarcoma and ovarian cancer.

2.2.2 Polymeric Nanoparticles

Poly(lactic-co-glycolic acid) (PLGA), polylactic acid (PLA), and polyethylene glycol (PEG) are examples of biodegradable and biocompatible polymers that make up polymeric nanoparticles. They provide sustained and regulated medication release, which makes them appropriate for long-term cancer treatment. Over time, a controlled and sustained release of pharmaceuticals is made possible by the breakdown of polymers in response to physiological circumstances. Polymeric nanoparticles' size, surface charge, and drug release profile can be altered to suit certain medical uses. Paclitaxel-encapsulated PLGA nanoparticles are widely employed in cancer treatment; they have demonstrated better absorption and lower toxicity when compared to traditional formulations.

2.2.3 Nanoparticles Made of Metal

Gold, silver, and platinum nanoparticles are examples of metallic nanoparticles with special optical and electrical characteristics that make them appropriate for cancer treatment and diagnosis. Through localized hyperthermia caused by gold nanoparticles' (AuNPs') ability to absorb and transform near-infrared (NIR) light into heat, photothermal therapy (PTT) can kill cancer cells specifically. Diagnostics and imaging: Metallic nanoparticles are perfect for imaging methods like surface-enhanced Raman spectroscopy (SERS) and optical coherence tomography (OCT) because of their potent scattering and absorption capabilities.

Gold nanoparticles are used as contrast agents in imaging and in photothermal therapy to locally ablate malignancies. Very biocompatible and stable. multifunctionality for diagnostics and treatment.

2.2.4 Magnetic Substances

Applications of magnetic nanoparticles, especially those made of iron oxide, in imaging and cancer treatment have drawn a lot of interest. When used as contrast agents in magnetic resonance imaging (MRI), iron oxide nanoparticles improve the visibility of malignancies. In order to treat hyperthermia, magnetic nanoparticles that are exposed to an alternating magnetic field can produce heat, which kills cancer cells while leaving healthy tissues unharmed. Hyperthermia therapy is used to treat a variety of malignancies, while superparamagnetic iron oxide nanoparticles (SPIONs) are used as contrast agents in MRIs. Capability for non-invasive imaging and therapy. Safe and biodegradable for human consumption.

Nanoparticles present a flexible and potent cancer treatment platform that may be able to get around many of the drawbacks of traditional treatments. Their special qualities, including as their high surface area-to-volume ratio, improved permeability, and surface functionalization, allow for multifunctionality, controlled release, and accurate drug administration. Each of the following nanoparticle types—liposomes, polymeric, metallic, and magnetic—brings unique benefits that make them appropriate for a range of medical and diagnostic uses. The creation of innovative nanoparticles with improved targeting, safety, and efficacy resulting from ongoing nanotechnology research will further transform cancer treatment and move us closer to more individualized and potent treatments.

3 Precision Targeting with Nanoparticles

One of the main benefits of using nanotechnology to treat cancer is precision targeting, which makes it possible to deliver therapeutic drugs to cancer cells only while preserving healthy tissues. Nanoparticles can be designed to collect precisely in tumor sites, improving therapy efficacy and lowering systemic toxicity, in contrast to traditional therapies that disperse medications throughout the body in an unspecific manner.

A better therapeutic index Increased drug concentration at the tumor location reduces negative effects while improving therapeutic results. Negative effects are lessened since healthy tissues are not exposed to as many cytotoxic chemicals. Traditional therapy resistance can be addressed by focusing on particular biochemical pathways or tumor microenvironments.

The Enhanced Permeability and Retention (EPR) effect, one of the distinct anatomical and physiological features of tumors, is exploited by passive targeting. Nanoparticles can extravasate and gather in the tumor interstitium due to the tumor vasculature's leaking nature.

Nanoparticles are kept in the tumor microenvironment for extended periods of time due to tumors' ineffective lymphatic drainage. Delivery of chemotherapy drugs, including doxorubicin, that are contained in polymeric nanoparticles or liposomes. In contrast to free doxorubicin, Doxil® (liposomal doxorubicin) selectively accumulates in solid tumors by taking advantage of the EPR effect, which lowers cardiotoxicity (Fig. 3).

Drug administration is inconsistent due to the wide variation in the EPR effect between patients and tumor types. Because not all cancers have high levels of vascular permeability, passive targeting is less successful. Functionalizing nanoparticles with ligands, antibodies, or peptides that selectively bind to receptors that are overexpressed on cancer cells is known as active targeting. Targeting moieties attached to nanoparticles allow them to selectively bind to and be internalized by cancer cells by recognizing particular receptors. By means of receptor-mediated endocytosis, active targeting enhances the cellular uptake of nanoparticles.

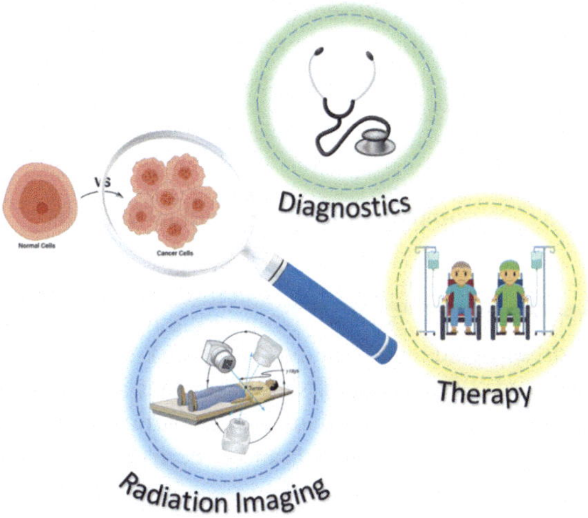

Fig. 3 From diagnosis to therapy and radiation imaging, advancing cancer treatment with precision and care

3.1 Nanoparticles Targeted to HER2

In some breast tumors, HER2 is overexpressed. Trastuzumab and other anti-HER2 antibody-conjugated nanoparticles specifically target HER2-positive cancer cells. Administration of doxorubicin or paclitaxel for breast cancer that is HER2-positive. Breast, lung, and ovarian malignancies all overexpress folate receptors. Drugs are delivered to folate receptor-positive cancer cells preferentially by folate-conjugated nanoparticles.

Lung, colorectal, and head and neck cancers frequently overexpress the epidermal growth factor receptor (EGFR). Drug delivery to malignancies that express EGFR is enhanced by nanoparticles functionalized with anti-EGFR ligands or antibodies.

3.1.1 Functionalization of the Surface

To increase their selectivity for cancer cells, nanoparticles can be functionalized with a variety of targeting moieties: Monoclonal antibodies (mAbs) that target EGFR or HER2 include cetuximab and trastuzumab, respectively. Tiny peptides that attach to integrins that are overexpressed on cancer cells, such as RGD (arginine-glycine-aspartic acid). Single-stranded nucleic acids known as aptamers attach themselves selectively to target proteins on cancer cells.

3.1.2 Dimensions and Form

The biodistribution, circulation duration, and cellular absorption of nanoparticles are influenced by their size and shape. To maximize tumor accumulation through the EPR effect and prevent quick clearance by the reticuloendothelial system (RES), nanoparticles with a diameter of 10–100 nm are ideal. The targeting efficacy of spherical, rod-, and disc-shaped nanoparticles is influenced by their distinct interactions with cells and tissues.

3.1.3 Nanoparticles that Respond to Stimuli

Certain triggers in the tumor microenvironment can cause nanoparticles to release their therapeutic payload. Drugs are released by pH-sensitive nanoparticles in the acidic environment of tumors (pH ~ 6.5), whereas normal tissues (pH ~ 7.4) are left unaltered. When tumor-specific enzymes such matrix metalloproteinases (MMPs) are present, enzyme-responsive nanoparticles release medications. Frequently used in the treatment of localized cancer, these particles release medications when exposed to extreme heat.

1. *The use of chemotherapy*

By delivering chemotherapeutic medicines precisely to cancer cells, precision-targeted nanoparticles increase their effectiveness while lowering systemic toxicity. When compared to non-targeted formulations, docetaxel-loaded PLGA nanoparticles functionalized with anti-HER2 antibodies have shown enhanced cytotoxicity in HER2-positive breast cancer cells.

2. *The use of genes*

Targeted gene therapy is made possible by the very selective delivery of gene-editing tools, such CRISPR-Cas9, to cancer cells using nanoparticles. CRISPR components are delivered by lipid nanoparticles (LNPs) to eliminate oncogenes in liver cancer.

3. *Immunotherapy*

By delivering cancer vaccines or immune checkpoint inhibitors to immune cells or tumor locations, nanoparticles improve cancer immunotherapy. Checkpoint inhibitors are delivered to the tumor microenvironment precisely by nanoparticles functionalized with anti-PD-L1 antibodies, which improves the immune response.

Targeted nanoparticle effectiveness may be diminished by differences in receptor expression between distinct tumor cells. Some nanoparticles may accumulate in healthy tissues even after functionalization, which could result in off-target consequences. The intricate structure and functionality of nanoparticles provide difficulties for both regulatory approval and large-scale production. Combination therapy involves the use of precisely targeted nanoparticles in conjunction with other therapeutic approaches, including radiotherapy or immunotherapy, to achieve complementary outcomes. Applying AI to forecast patient-specific targeting tactics and optimize nanoparticle design. Creating patient-specific nanoparticles using proteomic and genetic information to provide genuinely customized cancer treatment is known as personalized nanomedicine.

Table 2 Overview of targeting mechanisms in nanotechnology-based cancer therapy

Targeting type	Mechanism	Example	Benefit
Passive targeting	Exploits the enhanced permeability and retention (EPR) effect in tumors	**Doxil® (liposomal doxorubicin)**	Reduced cardiotoxicity
Active targeting	Functionalized with ligands, antibodies, or peptides to bind specific receptors	**HER2-targeted nanoparticles**	Selective delivery to HER2-positive breast cancer cells

An important development in cancer treatment is precision targeting with nanoparticles, which has the potential to increase effectiveness, lower toxicity, and improve patient outcomes. Precision-targeted nanoparticles have the potential to become a key component of individualized cancer treatment by utilizing passive and active targeting mechanisms, refining nanoparticle design, and resolving present issues (Table 2).

4 Enhancing Drug Delivery Systems

One of the biggest challenges in cancer treatment is efficient drug delivery. The therapeutic efficiency of anticancer medications is limited by problems with traditional delivery methods, such as oral or intravenous administration, such as poor solubility, low bioavailability, systemic toxicity, and non-specific distribution. Drug delivery systems based on nanotechnology provide novel solutions by enhancing the therapeutic agents' pharmacokinetics, targeting, and controlled release. This section examines the ways in which nanoparticles (NPs) improve drug delivery methods and the results of cancer treatment (Fig. 4).

4.1 Drawbacks of Traditional Medication Administration

Systemic administration, in which medications move throughout the body and interact with both healthy and malignant areas, is the foundation of conventional chemotherapy. This non-specific distribution has a number of drawbacks. The dose needed for efficacy is near the dose that results in toxicity since many chemotherapeutic drugs have narrow therapeutic windows. Serious adverse effects include nausea, hair loss, and immunological suppression result with this. In aqueous biological conditions, many anticancer medications, including docetaxel and paclitaxel, have poor bioavailability and low solubility due to their hydrophobic nature.

The half-life and therapeutic efficacy of conventional medications are typically diminished by their quick removal from the bloodstream. Because medications are

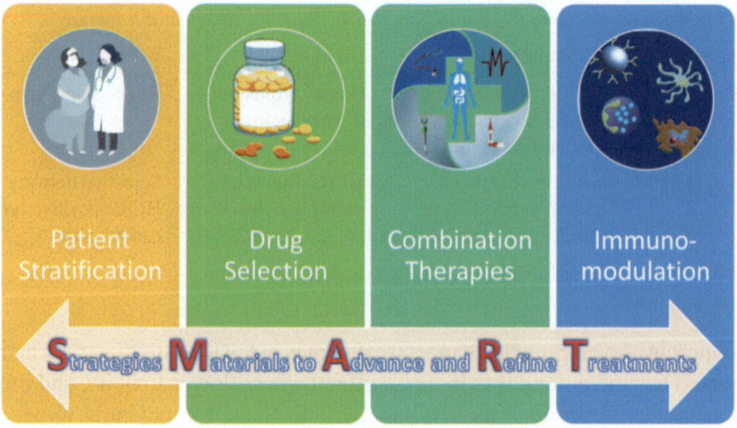

Fig. 4 Strategies materials to advance and refine treatments

dispersed evenly throughout the body, there is little buildup at the tumor site and a significant risk of off-target effects.

4.2 Drug Delivery Using Nanotechnology

A flexible framework for getting around the drawbacks of conventional medication delivery methods is provided by nanoparticles. Through the manipulation of nanoparticles' size, shape, composition, and surface characteristics, scientists can create delivery systems that enhance drug solubility, stability, circulation time, and targeting effectiveness.

4.3 Essential Elements of Drug Delivery Systems Based on Nanoparticles

4.3.1 Better Stability and Solubility

Hydrophobic medications can be encapsulated in nanoparticles to increase their stability and solubility in biological settings. Hydrophobic medications are frequently encapsulated in liposomes and solid lipid nanoparticles (SLNs) to improve their solubility and bioavailability.

Hydrophobic medications can be encapsulated in biodegradable polymers like poly(lactic-co-glycolic acid) (PLGA) and polyethylene glycol (PEG), which prolongs their circulation and prevents breakdown. When compared to free paclitaxel, paclitaxel-loaded PLGA nanoparticles have demonstrated improved solubility, extended circulation, and greater tumor accumulation.

4.3.2 Managed and Extended Discharge

It is possible to design nanoparticles to release their therapeutic payload in a regulated and prolonged way, guaranteeing a constant medication concentration at the intended location. The ability of nanoparticles to release medications in reaction to the tumor's acidic environment (pH ~ 6.5) while protecting healthy tissues (pH ~ 7.4) is known as pH-sensitive release. Drugs can be released from enzyme-sensitive nanoparticles in response to tumor-specific enzymes, such as matrix metalloproteinases (MMPs). Often utilized in the treatment of cancer, thermoresponsive nanoparticles release medications when exposed to high temperatures, or hyperthermia. These liposomes lessen systemic toxicity by releasing doxorubicin only in the acidic tumor microenvironment.

4.3.3 Improved Targeting of Tumors

Tumor targeting is enhanced by nanoparticles via two main mechanisms

Passive Targeting (EPR Effect): Because tumor tissues have poor lymphatic drainage and leaky vasculature, the Enhanced Permeability and Retention (EPR) effect enables nanoparticles to aggregate preferentially in these tissues. The drug's local concentration at the tumor site is raised by this passive targeting.

Reduces cardiotoxicity by delivering doxorubicin just to tumors by taking use of the EPR effect. The process of functionalizing nanoparticles with ligands, antibodies, or peptides that identify and bind to particular receptors that are overexpressed on cancer cells is known as "active targeting." This improves therapeutic effectiveness and cellular absorption.

Active targeting examples include

Drug delivery to HER2-positive breast cancer cells is possible with HER2-targeted nanoparticles. Folate-targeted nanoparticles: These particles target the overexpressed folate receptors found in lung and ovarian malignancies.

4.3.4 Diminished Systemic Hazard

Nanoparticles limit systemic exposure by delivering medications precisely to cancer cells, reducing side effects and enhancing patient quality of life.

Adding polyethylene glycol (PEG) to the surface of nanoparticles decreases the immune system's ability to recognize and eliminate them, extending their half-life and promoting the growth of tumors. PEGylated Liposomal Doxorubicin (Doxil®): Shows better therapeutic results and less cardiotoxicity than free doxorubicin.

4.4 Uses of Drug Delivery Systems Based on Nanoparticles

4.4.1 The Use of Chemotherapy

Nanoparticles improve the solubility, targeting, and controlled release of chemotherapeutic drugs, increasing their effectiveness and safety. Enhance the solubility and therapeutic index of docetaxel, a frequently used chemotherapeutic drug for prostate and breast cancer, by using docetaxel-loaded nanoparticles [26].

4.4.2 Combination Treatment

To produce synergistic effects, nanoparticles can co-deliver several therapeutic compounds, including immunomodulators and chemotherapeutic medications. In preclinical models, lipid nanoparticles that deliver both doxorubicin and siRNA targeting oncogenes have demonstrated increased anti-tumor activity.

4.4.3 Gene Therapy

Targeted gene therapy is made possible by the delivery of gene-editing tools, including CRISPR-Cas9, to cancer cells via nanoparticles. CRISPR components were successfully given to preclinical liver cancer models in order to knock out oncogenes [26].

4.5 Issues and Prospects for the Future

Notwithstanding its benefits, a number of issues need to be resolved in order to fully utilize the potential of drug delivery systems based on nanoparticles: Clinical translation depends on the long-term safety and biocompatibility of nanoparticles.

Scalability and Manufacturing: Large-scale production of nanoparticles requires the development of scalable and economical manufacturing techniques. The intricate structure and functionality of nanoparticles make regulatory approval difficult. Using proteomic and genetic information to create patient-specific nanoparticles for tailored cancer treatment is known as personalized nanomedicine. Using artificial intelligence (AI) to forecast patient-specific drug delivery tactics and optimize nanoparticle design.

With their increased solubility, controlled release, superior targeting, and less toxicity, nanoparticles provide a revolutionary approach to drug delivery in cancer treatment. Nanotechnology-based platforms have the potential to transform cancer treatment and enhance patient outcomes by resolving the issues with traditional drug delivery methods. Drug delivery methods based on nanoparticles are anticipated to be crucial to the development of customized cancer treatment in the future as research and development proceed.

5 Theragnostic Nanoparticles: Integrating Therapy and Diagnosis

By integrating therapeutic and diagnostic capabilities onto a single platform, nanotechnology holds the potential to completely transform the way cancer is treated. By combining the delivery of therapeutic drugs with real-time monitoring capabilities, this method known as theragnostics offers a more individualized and effective cancer treatment plan. Theragnostic nanoparticles allow clinicians to assess therapy efficacy in real time, allowing for prompt modifications to treatment plans by offering both diagnostic and treatment input. The idea of theragnostic nanoparticles, their modes of action, uses in cancer treatment, and the difficulties and potential of this novel strategy will all be covered in this section.

Therapy (treatment) and diagnostics are two essential components of cancer treatment that are combined in theragnostics. The objective is to develop nanoparticles that can both transport medications straight to cancer cells and provide tumor diagnostic data. With the help of these dual-purpose nanoparticles, doctors can keep an eye on the course of treatment and modify their plans in real time. Personalized therapy, which adapts therapies to the unique features of a patient's tumor, may improve results and reduce needless side effects. The use of theragnostic nanoparticles makes this possible [27].

5.1 Theragnostic Nanoparticle Mechanisms

Theragnostic nanoparticles combine diagnostic and therapeutic properties in a number of ways. With the addition of imaging capabilities to view tumor sites and track treatment outcomes, these nanoparticles can transport chemotherapeutic medications, gene treatments, or other therapeutic substances [28, 29].

5.1.1 Imaging and Targeted Drug Delivery

Targeted therapy and imaging are combined in a single system using theragnostic nanoparticles. Therapeutic medicines can be delivered directly to the tumor by targeting receptors that are overexpressed on cancer cells with particular ligands or antibodies attached to the surface of nanoparticles. Simultaneously, the system incorporates imaging agents like gold nanoparticles, magnetic nanoparticles, or fluorescent dyes, which enable real-time tracking of tumor behavior and drug delivery [28, 30, 31].

These could be biologic treatments like antibodies, gene-editing technologies like CRISPR-Cas9, or chemotherapeutic medications. Imaging agents, such as quantum dots for fluorescence imaging or superparamagnetic iron oxide nanoparticles (SPIONs) for magnetic resonance imaging (MRI), are used to monitor the distribution and position of nanoparticles throughout the body [32].

5.1.2 Tracking the Reaction to Treatment

Real-time therapy monitoring is made possible by theragnostic nanoparticles. This feedback system aids in determining whether the medicine is being released correctly, whether the tumor is responding to treatment, and whether the nanoparticles are reaching the tumor site. Doctors can more successfully modify the treatment plan by keeping an eye on the tumor microenvironment.

For instance, based on imaging feedback, the therapeutic dose or medication type can be quickly changed if the tumor is not responding to treatment. Non-invasive imaging the dispersion of nanoparticles in the body can be seen using methods including magnetic resonance imaging (MRI), PET (Positron Emission Tomography), and fluorescence imaging. This allows for ongoing insights about the size, progression, and response to treatment of the tumor.

5.2 Theragnostic Nanoparticle Applications in Cancer Therapy

By increasing both therapeutic efficacy and diagnostic accuracy, theragnostic nanoparticles may advance cancer treatment [28].

5.2.1 Drug Delivery Guided by Imaging

Drug delivery and precise tumor imaging are two possible uses for nanoparticles. These nanoparticles guarantee that chemotherapeutic medications are not only delivered precisely to the cancer cells but are also monitored during the course of treatment by combining both imaging agents and chemotherapeutic drugs.

For instance, magnetic nanoparticles are employed to administer chemotherapeutic medications and do MRI-based imaging. An external magnetic field can be used to direct the nanoparticles to the tumor site, and the magnetic nanoparticles' MRI signal aids in determining the drug's location and concentration.

5.2.2 Imaging and Therapy in Multiple Modes

Theragnostic nanoparticles can provide a multi-modal approach to cancer treatment by combining several therapeutic agents (such as chemotherapy, gene therapy, and immunotherapy) with diverse imaging modalities (such as MRI, CT scans, and fluorescence). This connection makes it easier to monitor tumor response more precisely and modify treatments as necessary.

For instance, gold nanoparticles can be employed as contrast agents in CT or photoacoustic imaging to show the size and location of the tumor while also being utilized in photothermal therapy, which uses light to heat up tumor cells and kill them.

5.2.3 Immunotherapy for Cancer

Apart from administering drugs directly, theragnostic nanoparticles can also be utilized to deliver immuno-modulating substances like cytokines, cancer vaccines, or checkpoint inhibitors to strengthen the immune response to cancer. Immune checkpoint inhibitors can be blocked and the immune system can be reactivated against the tumor using nanoparticles functionalized with anti-PD-L1 antibodies. By adding imaging agents, physicians may track how well the nanoparticles are influencing the immune system and modify the treatment strategy as necessary.

5.2.4 Gene Therapy

With the added advantage of real-time gene therapy success monitoring, nanoparticles present a viable delivery system for gene-editing tools (such as CRISPR-Cas9) to cancer cells. CRISPR components are delivered using lipid nanoparticles to target particular oncogenes in tumor cells. To see the delivery and determine whether the gene-editing process was successful, imaging tools are employed.

5.3 *Difficulties in the Development of Theragnostic Nanoparticles*

Theragnostic nanoparticles show promise, however there are still a number of obstacles preventing their broad clinical use [28].

5.3.1 Toxicity and Biocompatibility

For nanoparticles to prevent immunological reactions or harmful effects on healthy tissues, they must be biocompatible. To make sure that nanoparticles don't build up in non-target tissues or cause long-term damage, extensive preclinical and clinical research is necessary.

5.3.2 Design Complexity

Compared to conventional drug administration methods, the design and synthesis of theragnostic nanoparticles are intrinsically more complicated. To guarantee that both therapeutic compounds and imaging modalities work well together and that the nanoparticle stays stable in vivo, careful design is needed when combining them into a single nanoparticle.

5.3.3 Manufacturing and Regulatory Obstacles

Compared to traditional medications or imaging agents, the regulatory mechanisms for theragnostic nanoparticles are more intricate. Strict safety and effectiveness requirements must be met by nanoparticles, and increasing production for clinical usage poses problems with reproducibility, cost, and consistency.

5.4 *Theragnostic Nanoparticles' Prospects for the Future*

Theragnostic nanoparticles appear to have a bright future as research on enhancing their scalability, safety, and effectiveness continues. Among the upcoming developments are.

5.4.1 Customized Healthcare

Theragnostic nanoparticles will play a crucial role in individualized cancer treatment as research advances. Better patient outcomes can result from doctors delivering more targeted medicines while tracking the tumor's response by customizing nanoparticles to the genetic profile of the tumor.

5.4.2 Advanced Technologies for Imaging

More precise and thorough tumor monitoring will be possible because to developments in imaging technologies, such as multimodal imaging, which combines MRI, PET, and fluorescence. This will facilitate more efficient treatment planning [33].

5.4.3 Integrating Various Therapeutic Approaches

In order to improve treatment efficacy and minimize the number of injections needed, theragnostic nanoparticles may be further tuned to deliver several therapies (such as immunotherapy, gene therapy, and chemotherapy) in a single nanoparticle system.

By combining diagnosis and treatment into one system, therapeutic nanoparticles provide a revolutionary approach to cancer treatment. By facilitating real-time treatment progress monitoring and enabling prompt therapeutic modifications, these nanoparticles hold promise for more individualized, effective, and efficient cancer therapy. The development of theragnostic nanoparticles has enormous potential for the future of cancer treatment, bringing us one step closer to genuinely individualized medicine, even though there are still issues with design, safety, and regulatory approval [34].

6 Immunotherapy Approaches Using Nanotechnology

One of the most promising oncology treatment approaches is immunotherapy, which uses the body's immune system to combat cancer. However, issues like insufficient immune response, off-target effects, and poor drug delivery frequently restrict its clinical use. By improving tumor targeting, delivering immune-stimulating drugs more precisely and precisely, and increasing the immune system's capacity to identify and eliminate cancer cells, nanotechnology presents novel ways to increase the efficacy of immunotherapy. This section examines the different ways that nanotechnology is transforming cancer immunotherapy [35, 36] (Table 3).

6.1 Nanoparticles in Immunotherapy for Cancer

It is possible to design nanoparticles to transport immune-modulating substances including cytokines, cancer vaccines, and checkpoint inhibitors straight to the tumor site. Nanoparticles can improve the immune system's capacity to combat cancer

Table 3 Overview of nanotechnology-based immunotherapy approaches

Approach	Mechanism	Example	Benefit
1. Delivery of checkpoint inhibitors	Nanoparticles deliver anti-PD-1 or anti-CTLA-4 antibodies to block immune checkpoint pathways	**PD-L1-targeted nanoparticles**	Enhanced tumor targeting and improved immune activation
2. Cancer vaccines	Encapsulation of tumor antigens and adjuvants for targeted delivery to immune cells, especially dendritic cells	**Liposomes carrying tumor-associated antigens (e.g., HER2)**	Enhanced immune response and activation of cytotoxic T cells
3. Immunomodulation	Nanoparticles present tumor-associated antigens or deliver immunomodulatory agents like cytokines	**Nanoparticles displaying carcinoembryonic antigen (CEA)** or delivering **IL-2**	Improved immune system activation and increased tumor infiltration by immune cells
4. Lipid nanoparticles (LNPs)	Delivery of RNA-based therapies, including mRNA vaccines and immune checkpoint inhibitors	**mRNA vaccines encoding tumor antigens (HER2, HPV)**	Activation of both innate and adaptive immune responses
5. Dendritic cell-targeting nanoparticles	Direct delivery of antigens and adjuvants to dendritic cells for enhanced T-cell activation	**Nanoparticles functionalized with mannose or DEC-205 antibodies**	Stronger and more specific T-cell response against tumors
Challenges			
Safety and biocompatibility	Ensuring nanoparticles are non-toxic and do not cause unwanted immune responses	Engineering biocompatible nanoparticles	Reduced off-target effects and improved safety profile
Effective targeting and delivery	Developing strategies to selectively target tumor and immune cells	Advanced surface functionalization	Increased efficiency in immune cell targeting and tumor infiltration
Regulatory barriers	Extensive preclinical and clinical testing for safety and efficacy	Compliance with regulatory standards	Ensures therapeutic safety and effectiveness, but may increase development time and costs

while reducing systemic toxicity by better targeting and releasing these chemicals [37].

6.1.1 Delivery of Checkpoint Inhibitors

Anti-PD-1 and anti-CTLA-4 antibodies are examples of immune checkpoint inhibitors that have demonstrated great potential in the treatment of cancer. However, the ability of the immune system to react and their ability to reach the tumor site frequently restrict their clinical efficacy. These inhibitors can be better delivered by nanoparticles, which will increase the immune cells' absorption of them in the tumor microenvironment.

Immune checkpoint pathways within the tumor microenvironment can be blocked by using nanoparticles functionalized with antibodies that target PD-L1 or CTLA-4. This makes it easier for the immune system to identify and combat cancer cells.

6.1.2 Vaccines for Cancer

The goal of cancer vaccines is to activate the immune system to seek out and eliminate cancerous cells. However, two major obstacles are inadequate immune activation and poor tumor antigen delivery. By encapsulating tumor antigens and adjuvants, nanoparticles can effectively transfer these substances to immune cells—especially dendritic cells, which are essential for triggering a strong immunological response [38]. To improve the immune response to cancer, liposomes and polymeric nanoparticles, for instance, can carry immune-boosting substances and tumor-associated antigens straight to dendritic cells.

6.2 Using Nanoparticles for Immunomodulation

In addition to delivering vaccinations and immune checkpoint inhibitors, nanoparticles can be engineered to actively alter the immune system, enhancing its capacity to identify and combat tumor cells [39].

6.2.1 Presentation of Tumor-Associated Antigens

To improve immune cell antigen presentation, nanoparticles can be designed to display tumor-associated antigens on their surface. Both CD8+ T cells (cytotoxic T cells) and CD4+ T helper cells, which are crucial for the immune response to cancer, are stimulated by this method. Some T cell responses against colorectal and breast cancer can be triggered by nanoparticles displaying the carcinoembryonic antigen (CEA) or HER2.

6.2.2 Delivery of Immunomodulatory Agents

Additionally, cytokines like GM-CSF or interleukins (IL-2, IL-12) that stimulate immune cell activation and proliferation can be delivered via nanoparticles, strengthening the immune response against tumors. However, because these cytokines broadly activate the immune system, systemic delivery of these cytokines may result in serious side effects. By delivering cytokines precisely to the tumor, nanoparticles help improve therapy results by lowering off-target effects. IL-2 or GM-CSF can be encapsulated by nanoparticles and released locally within the tumor microenvironment, increasing the activity of TILs and encouraging the death of cancer cells.

6.2.3 Immunotherapy Delivery Systems Based on Nanoparticles

Because of their adaptability, nanoparticles can be used to create a variety of delivery systems that integrate many therapeutic approaches, increasing the overall effectiveness of cancer immunotherapy.

LNPs, or lipid nanoparticles

Lipid nanoparticles are frequently employed to deliver small compounds and RNA-based treatments, including mRNA vaccinations. They can be made more functional to enhance immune cell absorption and targeting. When used in conjunction with other immune-modulating drugs, LNPs are very helpful for delivering immune checkpoint inhibitors and mRNA-based vaccines. LNPs, for instance, have the ability to deliver mRNA vaccines that encode tumor antigens (such as HER2 or HPV vaccines) and have immune-boosting properties that stimulate both innate and adaptive immune responses.

6.3 Nanoparticles of Dendritic Cells

Delivery systems based on nanoparticles can directly target dendritic cells, which are essential for triggering the immune response. Antigens and adjuvants can be delivered by dendritic cell-targeting nanoparticles, eliciting a more powerful and targeted immune response. Dendritic cells can be targeted by nanoparticles functionalized with mannose or DEC-205 antibodies, which also improve the presentation of tumor antigens and elicit a stronger T-cell response [40].

6.4 Difficulties and Prospects

Although immunotherapy based on nanotechnology has showed a lot of promise, there are a few issues that need to be resolved.

6.4.1 Safety and Biocompatibility

For nanoparticles to be successful in therapeutic settings, they must be biocompatible, non-toxic, and immune-system-inhibiting. In order to prevent adverse consequences, nanoparticles should be engineered to prevent off-target immune responses.

6.4.2 Effective Targeting and Delivery

It is still very difficult to target tumor cells and immune cells that contribute to cancer immunity specifically. Two important areas of continuing study are increasing the effectiveness of immune activation and creating more complex targeted tactics.

6.4.3 Obstacles in Regulation

Before being authorized for clinical use, immunotherapy nanoparticles must pass strict safety and effectiveness requirements. This calls for a great deal of preclinical and clinical testing, which can be expensive and time-consuming.

Immunotherapies based on nanotechnology offer a promising new avenue for the treatment of cancer. Nanoparticles can enhance the immune system's capacity to identify and combat cancer cells by improving the administration of immunomodulatory drugs, cancer vaccines, and immune checkpoint inhibitors. In the near future, more efficient and individualized cancer immunotherapy may be possible thanks to continued research into nanoparticle design and delivery methods, even though targeting, safety, and regulatory approval issues still exist.

6.5 Difficulties in Clinical Translation of Cancer Treatments Based on Nanotechnology

Although nanotechnology has a lot of potential to transform the way cancer is treated, there are a lot of obstacles that need to be overcome before it can be widely used in clinical settings. These difficulties include budgetary constraints, technological constraints, safety issues, and regulatory obstacles. Here, we examine some of the most important problems that need to be solved if nanomedicine is to fulfill its full potential in the treatment of cancer.

6.5.1 Toxicity and Biocompatibility

Making sure nanoparticle-based treatments are biocompatible and reducing toxicity are two of the main issues. Although they are good at targeting and delivering drugs, the elements that make up nanoparticles can have unexpected effects on the body. Nanoparticles' distinct qualities, like their size and surface features, can trigger inflammation, activate the immune system, and cause unintended buildup in tissues that are not their intended target.

Interaction with the immunological System: Nanoparticles, particularly those derived from inorganic materials, have the ability to trigger immunological responses that activate the complement system, potentially resulting in allergic reactions or inflammation. If nanoparticles are not biodegradable or effectively bioeliminated, their gradual buildup in organs such as the liver, spleen, or lungs may result in toxicity or other long-term health effects.

Extensive preclinical testing is necessary to assess the long-term safety and toxicity of nanoparticles in order to allay these worries. Additionally, biocompatible materials that are biodegradable and harmlessly removed from the body must be developed.

6.5.2 Issues with Standardization and Manufacturing

One of the biggest challenges still facing the industry is the regulated, repeatable, and scalable production of nanoparticles. To guarantee consistent therapeutic effects, nanoparticle formulations must adhere to stringent requirements for size, homogeneity, surface charge, and drug loading capacity. To satisfy the requirements of clinical applications, the manufacturing process also needs to be compatible with large-scale production.

Even minor adjustments to a nanoparticle's size or surface characteristics can have a big impact on how it behaves in the body. For example, a nanoparticle's size may affect its capacity to accumulate in tumors via the increased permeability and retention (EPR) effect; even little changes can alter the effectiveness of treatment.

A lot of formulations of nanoparticles that perform well in preclinical models don't scale up to large production. Cost-effectiveness, product consistency, and adherence to Good Manufacturing Practice (GMP) guidelines are all requirements for large-scale production.

7 Challenges in Clinical Translation

One of the biggest challenges in clinical translation is creating consistent, repeatable processes for producing nanoparticles at a scale that guarantees their clinical relevance.

7.1 *Obstacles in Regulation*

Since nanotechnology-based treatments do not fit into the conventional medication categories set by regulatory bodies such as the FDA or EMA, they present particular regulatory issues. The uniqueness and multifunctionality of nanoparticles, which combine therapeutic and diagnostic qualities, make the approval procedure for nanomedicines more complicated.

Before approving a product, regulatory bodies need proof of its safety and effectiveness. However, because of the special characteristics of nanoparticles, namely their capacity to penetrate biological barriers and build up in tumors, additional safety regulations are required. Standardized procedures for assessing the immunogenicity, biodistribution, and pharmacokinetics of nanoparticles are currently lacking.

The approval process for medications based on nanoparticles can be more time-consuming than for conventional treatments due to the intricacy of their evaluation. This may cause patients to miss out on potentially life-saving treatments. One of the most important obstacles to clinical translation is developing a regulatory framework that can accurately assess the advantages and disadvantages of nanomedicines while maintaining safety.

7.1.1 Financial and Cost Obstacles

It is costly to develop, produce, and commercialize cancer treatments based on nanotechnology. Clinical trials, nanoparticle production scaling up, and research and development (R&D) can be prohibitively expensive.

Nanoparticle design calls for advanced materials and technology, which can raise development expenses. Additionally, the expense of clinical trials is raised by the intricate safety and regulatory testing needed for nanoparticles. The high cost of manufacture may restrict access to nanoparticle-based medicines, even if they are shown to be successful. Patients in low-resource environments or those without full health insurance may not be able to afford the high production costs of many nanomedicines. Making nanomedicines accessible and inexpensive for all patients requires figuring out how to lower production costs without sacrificing the products' quality or safety.

8 Future Perspectives and Directions

8.1 Future Prospects and Pathways in Cancer Treatments Based on Nanotechnology

With a number of new avenues positioned to improve the efficacy, accuracy, and accessibility of therapies based on nanoparticles, the use of nanotechnology in cancer treatment has enormous promise for the future. The next generation of cancer medicines will be driven by ongoing nanomedicine research and innovation, revolutionizing the way we detect, treat, and track cancer. The following are some of the main areas where cancer treatments based on nanotechnology are anticipated to develop.

8.2 Customized Nanomedicine

The creation of customized cancer treatments will become more crucial as cancer research progresses. Treatment regimens can be greatly customized by engineering nanoparticles to target particular molecular markers specific to each patient's malignancy.

Based on a patient's immune system traits, tumor biomarkers, and genetic profile, customized nanoparticles can be created. By precisely delivering medications to the cancer cells, these nanoparticles might reduce side effects and enhance therapeutic results.

By combining therapeutic nanoparticles (theragnostics) with diagnostic instruments, physicians may track the development of tumors in real time and modify treatment plans as necessary. AI algorithms could be used to further refine this individualized technique by analyzing patient data and customizing nanoparticle design for each instance. Because medicines will be carefully tailored to the genetic and molecular makeup of each patient's cancer, personalized nanomedicine promises more effective treatments with fewer adverse effects.

8.3 Approaches to Multimodal Therapy

To increase their efficacy, future nanoparticle-based treatments are anticipated to integrate several therapeutic approaches onto a single platform. These multimodal approaches might use a single nanoparticle system to administer radiation therapy, immunotherapy, gene therapy, and chemotherapy all at once.

Nanoparticles may have a synergistic effect that strengthens the immune system while targeting cancer cells by combining treatments like chemotherapy with immunotherapy or gene therapy. This may be especially useful for treating tumors that are resistant to traditional therapies. By combining therapeutic nanoparticles with imaging capabilities, real-time tumor response monitoring is made possible, allowing physicians to modify the medication type or dosage in response to the tumor's progression.

By targeting cancer cells from several different directions at once, multimodal therapy platforms have the potential to provide faster therapeutic responses and more effective treatment regimens.

8.4 Progress in the Design of Nanoparticles

Developments in nanoparticle engineering and design will be crucial to the use of nanotechnology in cancer treatment in the future. Scientists are trying to create nanoparticles with better qualities that will increase their safety and medicinal effectiveness.

The upcoming generation of nanoparticles will be more sensitive to the tumor microenvironment and will only release their therapeutic payloads in response to particular environmental cues, such as pH, temperature, or particular enzymes. Nanoparticles may develop into nano-robots that can carry out intricate operations including drug delivery, tissue mending, and malignant cell detection. At the cellular level, these cutting-edge devices would provide exact control over cancer treatment. Enhancements in the design of nanoparticles will improve the specificity and functionality of nanomedicines, increasing their efficacy and patient safety.

8.5 Overcoming Cost and Manufacturing Obstacles

For nanomedicines to be widely used in clinical settings, it will be essential to lower the cost and increase the scalability of nanoparticle production. Novel approaches to nanomanufacturing, such 3D printing and microfluidics, present encouraging ways to produce nanoparticles in an economical and repeatable way.

By producing nanoparticles at scale more quickly and effectively, advanced manufacturing techniques will lower prices and increase availability. By working together, scientists, pharmaceutical firms, and government organizations will try to lower the cost of nanomedicines and increase their accessibility for a wider range of patients.

The advantages of cancer medicines based on nanotechnology will become more widely available by removing the manufacturing and budgetary obstacles, making them an essential part of cancer treatment plans in the future.

9 Conclusion

Although nanotechnology has the potential to completely transform the treatment of cancer, there are a number of important obstacles that must be addressed before it can be widely used in clinical settings. Assuring the biocompatibility and long-term safety of nanoparticles, resolving possible toxicity concerns brought on by their distinct size and surface characteristics, and creating scalable, repeatable manufacturing procedures that can satisfy clinical requirements are some of the main challenges. Furthermore, because nanoparticles combine both therapeutic and diagnostic properties, the regulatory environment for nanomedicines is still complicated and calls for new safety and efficacy requirements. Notwithstanding these difficulties, customized nanomedicine has enormous promise for improving the accuracy of cancer treatments by customizing nanoparticles to target certain genetic abnormalities and biomarkers in individual patients. Multimodal therapies, which combine chemotherapy, immunotherapy, and gene therapy in a single nanoparticle platform, are also suggested by the future of nanotechnology in cancer treatment. This approach improves treatment efficacy by targeting cancer from several perspectives. The development of intelligent nanoparticles that can react to changes in the tumor microenvironment, such as pH or temperature, is another crucial advancement in nanoparticle design. This will provide accurate and regulated medication delivery. Furthermore, advancements in nanomanufacturing—such as automated procedures and scalable production techniques—are crucial to lowering the exorbitant cost and expanding the patient base for nanoparticle-based treatments. The future of cancer treatments based on nanotechnology appears to be quite bright as long as these obstacles are removed. Nanomedicine has the potential to provide more individualized, effective treatments with fewer side effects by improving targeted delivery methods, refining nanoparticle formulations, and combining diagnostics and therapeutic capabilities (theragnostics). This could completely change the way cancer is treated. The application of nanotechnology in clinical oncology may prove to be a crucial component in the battle against cancer as these issues are resolved, providing patients with hope for better results, increased survival rates, and an enhanced quality of life.

References

1. Sung, H., et al.: Global cancer statistics 2020: GLOBOCAN estimates of incidence and mortality worldwide for 36 cancers in 185 countries. CA: Cancer J. Clin. **71**(3), 209–249 (2021)
2. Jemal, A., et al.: Global cancer statistics. CA: Cancer J. Clin. **61**(2), 69–90 (2011)
3. Kamangar, F., Dores, G.M., Anderson, W.F.: Patterns of cancer incidence, mortality, and prevalence across five continents: defining priorities to reduce cancer disparities in different geographic regions of the world. J. Clin. Oncol. **24**(14), 2137–2150 (2006)
4. Chen, C., et al.: Identifying primary tumor site of origin for liver metastases via a combination of handcrafted and deep learning features. J. Pathol.: Clin. Res. **10**(1), e344 (2024)

5. Sánchez-Guixé, M., et al.: Origins of second malignancies in children and mutational footprint of chemotherapy in normal tissues. Cancer Discov. **14**(6), 953–964 (2024)
6. Batulan, Z., et al.: Multiple chronic conditions. In: Advancing Research on Chronic Conditions in Women. National Academies Press (US) (2024)
7. Zhang, H., et al.: Big data in breast cancer: towards precision treatment. Digital health **10**, 20552076241293696 (2024)
8. Huuhtanen, J.: Immunomonitoring in Cancer and Autoimmune Disorders at the Single-Cell level (2024)
9. Dabighi, M.M.: Current perspectives in cancer treatment: a comprehensive. The Journal **2**(01), 10–13 (2023)
10. Adugna, A., et al.: Medical nanoscale materials for virus-induced cervical cancer therapeutic modalities: for targeting delivery. OpenNano 100221 (2024)
11. David, B.G.: Nanotechnology for Targeted Cancer Therapy
12. Duan, C., et al.: Overcoming cancer multi-drug resistance (MDR): reasons, mechanisms, nanotherapeutic solutions, and challenges. Biomed. Pharmacother. **162**, 114643 (2023)
13. Li, J., Zhu, L., Kwok, H.F.: Nanotechnology-based approaches overcome lung cancer drug resistance through diagnosis and treatment. Drug Resist. Updates **66**, 100904 (2023)
14. Guo, L., et al.: Breast cancer heterogeneity and its implication in personalized precision therapy. Exp. Hematol. Oncol. **12**(1), 3 (2023)
15. Veneziani, A.C., et al.: Heterogeneity and treatment landscape of ovarian carcinoma. Nat. Rev. Clin. Oncol. **20**(12), 820–842 (2023)
16. Liu, B., et al.: Exploring treatment options in cancer: tumor treatment strategies. Signal Transduct. Target. Ther. **9**(1), 175 (2024)
17. Kaur, R., Bhardwaj, A., Gupta, S.: Cancer treatment therapies: traditional to modern approaches to combat cancers. Mol. Biol. Rep. **50**(11), 9663–9676 (2023)
18. Ciechanover, A.: The revolution of personalized medicine: are we going to cure all diseases and at what price? In: 2023-Sustainable Industrial Processing Summit. Flogen Star Outreach (2023)
19. Bao, Y., et al.: Beyond blood: advancing the frontiers of liquid biopsy in oncology and personalized medicine. Cancer Sci. **115**(4), 1060–1072 (2024)
20. Sasirekhamani, M., Jasmin, S.J.: 34. Bioinformatics and personalized medicine: utilizing genomic data and computational approaches to tailor medical treatments to individual patients. In: Mosaic of Ideas: Multidisciplinary Reflections, p. 343 (2024)
21. Purba, S.Y.: Precision medicine: utilizing genomic insights for customized therapeutic strategies. J. Acad. Sci. **1**(5), 497–504 (2024)
22. Addissouky, T., et al.: Revolutionary innovations in diabetes research: from biomarkers to genomic medicine. Iran. J. Diabetes Obes. (2023)
23. Alrushaid, N., et al.: Nanotechnology in cancer diagnosis and treatment. Pharmaceutics **15**(3), 1025 (2023)
24. Shrestha, B., Tang, L., Hood, R.L.: Nanotechnology for personalized medicine. In: Nanomedicine, pp. 555–603. Springer (2023)
25. Kumar, R., Kumar, M., Luthra, G.: Fundamental approaches and applications of nanotechnology: a mini review. Mater. Today: Proc. (2023)
26. Haleem, A., et al.: Applications of nanotechnology in medical field: a brief review. Glob. Health J. **7**(2), 70–77 (2023)
27. Dristant, U., et al.: An overview of polymeric nanoparticles-based drug delivery system in cancer treatment. Technol. Cancer Res. Treat. **22**, 15330338231152084 (2023)
28. Al-Thani, A.N., et al.: Nanoparticles in cancer theragnostic and drug delivery: a comprehensive review. Life Sci. 122899 (2024)
29. Hosseini, S., et al.: Theranostic polymeric nanoparticles as a new approach in cancer therapy and diagnosis: a review. Mater. Today Chem. **29**, 101400 (2023)
30. Altinbasak, I., et al.: Theranostic nanogels: multifunctional agents for simultaneous therapeutic delivery and diagnostic imaging. Nanoscale (2024)

31. Singh, D.: Exploiting nuclear-mitochondrial cross-talk in theranostics: enhancing drug delivery and diagnostic precision. Mitochondrion **75**, 101839 (2024)
32. Yang, J., et al.: Exceedingly small magnetic iron oxide nanoparticles for T1-weighted magnetic resonance imaging and imaging-guided therapy of tumors. Small **19**(49), 2302856 (2023)
33. Truong, T.T., et al.: Precision-engineered metal and metal-oxide nanoparticles for biomedical imaging and healthcare applications. Adv. Colloid Interface Sci. 103263 (2024)
34. Goldfried, M.R.: On the history of therapeutic integration. Behav. Ther. **13**(5), 572–593 (1982)
35. Croitoru, G.-A., et al.: Nanostructured drug delivery systems in immunotherapy: an updated overview of nanotechnology-based therapeutic innovations. Appl. Sci. **14**(19), 8948 (2024)
36. Shao, J., Jin, Y., Jin, C.: A new approach to overcoming resistance to immunotherapy: nanotechnology. Front. Oncol. **13**, 1210245 (2023)
37. Lu, Q., et al.: Nanoparticles in tumor microenvironment remodeling and cancer immunotherapy. J. Hematol. Oncol. **17**(1), 16 (2024)
38. Sahin, U., Türeci, Ö.: Personalized vaccines for cancer immunotherapy. Science **359**(6382), 1355–1360 (2018)
39. Panigaj, M., et al.: Therapeutic immunomodulation by rationally designed nucleic acids and nucleic acid nanoparticles. Front. Immunol. **14**, 1053550 (2023)
40. Cruz, L.J., et al.: Targeting nanoparticles to dendritic cells for immunotherapy. Methods Enzymol. **509**, 143–163 (2012)

Emerging Nanotechnologies Transforming Cancer Therapeutic

Ankita Doshi, Lalit Singh, and Brijesh Shah

Abstract The advancement of novel cancer therapies is critical to addressing the limitations of traditional treatments. Nanotechnology has emerged as a transformative approach, offering precision and efficiency in cancer treatment and diagnostics. This work explores the foundational principles of nanotechnology, emphasizing key nanomaterials and mechanisms for drug delivery. It delves into the application of nanotechnologies in cancer therapy, including targeted drug delivery, enhanced chemotherapy, immunotherapy, RNA-based treatments, and hyperthermia. Additionally, the study highlights advancements in cancer diagnostics, focusing on early detection, nanosensors, and imaging technologies. Despite its potential, the translation of nanotechnology into clinical practice faces significant challenges, including biocompatibility, safety concerns, and regulatory hurdles. Looking forward, personalized nanomedicine, combination therapies, and global accessibility present promising future directions. This discussion underscores nanotechnology's pivotal role in revolutionizing cancer treatment while addressing critical barriers to its widespread adoption.

Keywords Nanotechnology in cancer · Targeted drug delivery · Cancer nanomedicine · Nanocarriers · Gene therapy · Photothermal therapy

---

A. Doshi
Department of Biomedical and Life Sciences, School of Science, Navrachana University, Vadodara, GJ, India

L. Singh · B. Shah (✉)
Department of Chemistry, School of Sciences, Navrachana University, Vadodara, GJ, India
e-mail: brijeshshah27@gmail.com

N. J. Joshi and G. Patel (eds.), *Nanoparticle Delivery in Cancer Care*,
Series in BioEngineering, https://doi.org/10.1007/978-981-95-0457-2_12

1 Introduction

1.1 The Need for Novel Cancer Therapies

Cancer rates has grown by unprecedented rates, costing 10 million lives every year. Currently one of the most common methods for cancer treatment, chemotherapy lacks in multiple aspects of treatment ranging from accurate drug delivery for tumour site to tumour targeting. This leads to reduced efficiency rates and unsure results. Also, the micro-scale environment around tumour may vary individual to individual [1] Pain and discomfort during cancer treatments and advanced treatments is very prevalent with 55% patients experiencing it in early stages and 66% in advanced stages of cancer [2]. Also, collateral damage to non-cancerous while targeting the cancerous continuously deteriorate the life expectancy of the patient [3, 4]. Emerging and more efficient solutions can be found in nanotechnology which facilitates enhanced localized drug efficacy, improved imaging with refined radiation therapy. Also, limiting systemic toxicity and improved targeting capability gives them the edge over the existing treatments and remedies [3, 5, 6]. In nanoparticles category, smart nanoparticles have also emerged as more viable options. Unlike nanoparticles, they can be triggered by target—specific sites with accurate drug delivery and specific stimuli [7]. These nanoparticles collectively aggregate at the target site and release their loads which is both efficient and smarter [8]. Recent developments in cancer therapy of theragnostic and smart nanoparticles have been unlocked due to co-delivering therapeutics and reagents capabilities [1, 9].

1.2 Nanotechnology's Role in Cancer Treatment

Nanotechnology has contributed significantly in cancer therapeutics by offering unique advantages over the conventional drug delivery systems. The ability to target tumour tissues, enhanced drug solubility rates and therapeutic potency has revolutionized the cancer treatment process, providing an effective and efficient treatment option. Features driving nanoparticles in the cancer therapeutics are their larger effective surface area, small size with standout physicochemical characteristics, enabling efficient drug loading and systematic controlled release. Overall, ensuring the accuracy with minimised side effect on healthy tissues. Tremendous potential in penetration levels by active and passive pathways in tumour tissues, and reaching intercellular targets, reflecting potential to overcome biological barriers [10]. Also, cancer imaging techniques have been improved by unique set of optical and magnetic properties, providing high resolution and sensitivity control. Furthermore, nanotech has been implemented as semiconductors quantum dots, depicting size-dependent fluorescent emissions with high image stability, making them highly potent for tumour visualization, also enabling accurate staging and real time tumour monitoring and treatment response (Fig. 1) [11, 12].

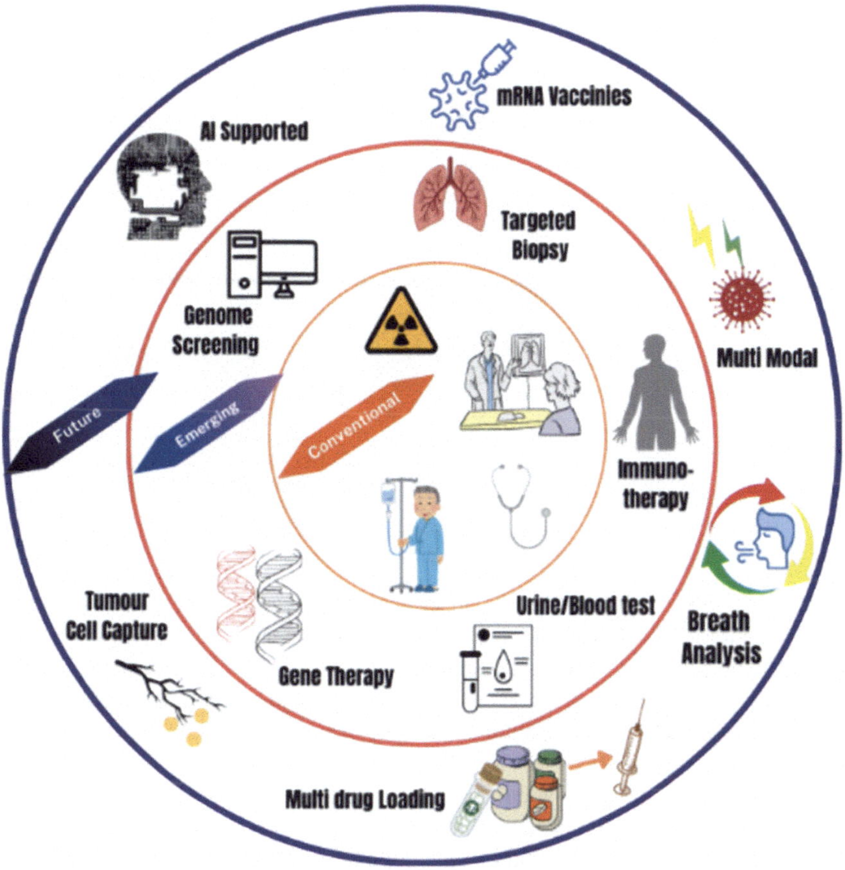

Fig. 1 Few conventional, emerging and future cancer therapies and diagnostics

2 Nanotechnology Basics

As advancements in cancer therapeutics, nanotechnology has become more apparent choice as the research work has been increased rapidly [13]. Enhanced permeation and retention effects, where the nanoparticles effectively accumulate within the tumour vasculature is passive targeting for tumours. Multiple tumours have open blood vessels with pores appropriate for nanoparticles to pass through them and settles within the tumour tissues [14]. Moreover, effective permeation and retentions is not a wholesome solution as protection against the drug effect for healthy tissues is not sufficient. There are physiological obstacles that prevent nanoparticles from reaching their target, even without a diseased state, and can become even more complex to navigate for cancer patients [15]. Protein and lipid adsorption, blood flow rate, coronas, and phagocytic cells can reduce stability and delivery capability. Interstitial pressure and extracellular matrices can also limit access to a

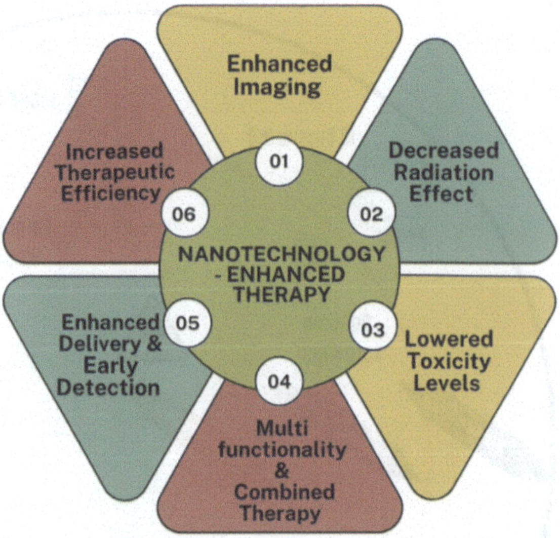

Fig. 2 Compared to traditional anti-cancer medications, radiation therapy, diagnostics, and imaging, nanotechnology offers numerous benefits. Nanomedicines can increase treatment efficacy at the target site while reducing systemic toxicity through focused administration

tumour [3, 16]. Currently approved cancer nano-medicines are predominantly liposomal formulations and drug conjugates (protein, polymer, and/or antibody) focused on improving pharmacokinetics and pharmacodynamics (PK/PD) of the free drug and utilizing passive targeting. There are many clinical studies currently investigating nanomaterials for therapeutic and diagnostic applications, including imaging modalities (Fig. 2) [3, 17].

There are multiple physiological obstacles preventing nanoparticles from reaching target, without the diseased sate, making the process complicated for cancer treatment. The internal pressure within extracellular matrices, phagocytic cells and protein—lipid adsorption can reduce stability and delivery capability and accessibility to the tumour [18]. The differences in cancer types also complicates the targeting ability that poses a need to altercate the formulation accordingly. Improvement in pharmacokinetics profiling, solubility and stability, bio stability and availability have been greatly improved by first generation nanomedicines [19]. New aspects of nanoparticles and related technology can be explored by incorporating specialized and unique designs and functions ensuring next generational transformation in treatments by combining nanotechnology with triggered drug release, combination therapies, gene therapy, novel immune therapy. With rapidly developing scientific insights over cancer survival mechanisms, nanomedicine will be playing a pivotal role in enhancing the bioimaging and diagnostic capabilities to halt metastasis. Medicines and drugs are categorised in vast ranges of concentrations and dose distribution, posing a difficulty to optimize dose concentration for the combination therapies techniques, and increased complexities by cancer types and patients' individuality. Nanoparticles which are biomacromolecules, lipid based, inorganic, carbon based and hydrogel possess ability to formulate and combine with vast range of therapeutics with broader chemical potency and characteristics [3, 20]. Nanocarriers are implied for specific targeting by utilizing drug conjugated tethered with a molecule having high cancerous cell affinity and lower for the healthy surrounding cells. Hence, lowering the probability of systemic toxicity inside [21]. Nano-carrier material must be equipped with drug protection properties from degradation while obtaining elongated circulations rate and avoiding the reticuloendothelial system uptake. Confirming an efficient route to deliver target cells [22].

Hence, formulation of nanotechnology requires effective and careful selection of choice of ligand, the carrier material and the potent concentration of ligand over the nanocarrier. Specific mechanisms of actions of nano medicines is required due to presence of multipole routes, like intracellular and outer cellular membrane diffusion delivery to optimize the nano formulation. Furthermore, adding to specific targeting, implementation of stimuli-responsive activation can significantly enhance the therapeutic specificity. Designing nano carriers to release under specific parameters like glucose levels, pH range, oxidative/reductive conditions, ion concentration and specific enzymes or by external stimulations (magnetic and electric and radiations) can complement the drug release specificity to precise chemical or biological conditions environment in and around the tumour [23, 24]. For example: pH responsive peptide based nano particles formulated to morph into fibrils around the tumour microenvironment, exhibiting extensive fluorescent signals enhancing photodynamic therapy [3, 25].

2.1 Nanotechnology Approaches in Medicines

When compared to normal materials, nanoparticles exhibit particular and distinctive features. The large surface area to volume ratio of the nanoparticle's aids in the high drug loading [26]. Hundreds of medicinal molecules are contained within a single vehicle by polymeric nanoparticles. The size-dependent characteristics of magnetic materials and quantum confinement in semiconductors allow for the application of nanomaterials in the medical arena. The extensive usage of nanoparticles in fluorescence biology has led to their incorporation into imaging methods [27, 28].

Nanomachinery: This seeks to describe the quantitative components at the molecular scale. A thorough understanding of biological mechanisms, early diagnosis, and treatment are crucial components of nanomachinery. The creation of nanomachines and nanorobots is one of the most exciting areas of advanced nanotechnology. The researchers were particularly interested in the unique features, capacities, and ability to perform a variety of jobs. Human immunodeficiency virus (HIV), fever, ovulation, and fast pregnancy tests all use them [27, 29].

Nano imaging: Quantum dots (Q dots) are a well-known example of a semiconducting nanocrystal. The electrons in the Q dots become excited and undergo a transition when exposed to ultraviolet light. An electron releases energy in the form of fluorescence light when it transitions from an excited state to its ground state. The size of the Q-dots is the only factor influencing the energy difference between two states. Numerous cells and biological processes were located and identified using a broad spectrum of distinct colours. The detection power of quantum dots is at least ten times that of organic fluorescence molecules, making them extremely powerful detectors [29]. They are currently only utilized for in vitro research diagnostics due to toxicity concerns [27].

Drug delivery: There is no need for inorganic materials because a variety of compounds can be transported with the aid of solid fat or polymer nanoparticles. There is less room to use other particles, particularly inorganic ones. Polymer nanoparticles were extensively employed in the 1970s as a delivery system for proteins, peptides, and anti-tumour drugs. The emergence of fat, dendrimer, felluren, and nanocrystal nanoparticles occurred later in the mid-1990s [27, 30]. Despite the promising future of drug delivery systems, several challenges included the creation of suitable techniques to improve the specificity of delivery systems for target cells and to more effectively transport active ingredients to the cell's destination. Inadequate control over the bioavailability of active ingredients in the target tissue is another barrier. Because of its large surface area and adsorbing ability, carbon nanotubes have been successfully used in medicine. They turned out to be a great way to deliver medications straight into cells without requiring the body to metabolize them [31]. A novel substance that releases Nitric Oxide (NO) has surfaced and is thought to possess antibacterial and wound-healing properties. Because of its regulatory, protective, and harmful properties, researchers have expressed interest in the design and synthesis of NO-donating medications and materials [27].

Nanoparticle as antibacterial agent: The use of antibiotics to treat the majority of bacterial illnesses transformed the medical field. Unfortunately, a lot of bacteria have evolved in the modern era to avoid the antibiotics that are currently in use. Acinetobacter baumannii, Pseudomonas aeruginosa, Enterobacter species, Staphylococcus aureus, Klebsiella pneumoniae, and Enterococcus faecium are the most prevalent antibiotic-resistant bacteria. Because of its unique characteristics, gold nanoparticles are used as an antibacterial agent. The synthesis of gold nanoparticles and their potential uses in medication delivery and cancer treatment piqued the researchers' curiosity [27].

2.2 Mechanisms of Nanocarrier-Based Drug Delivery

Stimuli: A stimulus-responsive nanoparticle functions much like a toolkit. Smart nanoparticles can change their shape, structure, solubility, and surface. Also, the charge association, or dissociation behaviour in response to internal or external stimuli enhances endosomal escape, cell uptake, and payload release [32]. Two approaches involve using responsive nanomaterials, such as light and heat-sensitive gold nanoparticles, and then modifying the nanoparticles through polymerization or functional group linkage [1].

Endogenous triggers: pH. The nanoparticles that target tumour extracellular and intracellular pH triggers have been the most extensively studied to date. Practically all solid tumours can accelerate glycolysis because the large amount of lactic acid produced provides sufficient energy for tumour cells. It is also known as the Warburg effect, which converts glucose directly into lactic acid [1, 33].

Enzyme: Enzymes are abundant in tissues and organs, allowing the human body to function normally. The tumour microenvironment has abnormal expression of enzymes such as matrix metalloproteinases, cathepsins, phospholipases, and oxidoreductases because tumour cells grow faster than normal organs and require more enzymes for function [1].

Redox: The redox-sensitive system is another potential strategy for endogenous stimuli-dependent smart delivery of therapeutic drugs. Tumour cells usually have relatively higher oxidative stress than normal cells because they contain reactive species. The byproduct of aerobic metabolism is reactive oxygen, including superoxide dismutase, glutathione (GSH), and hydrogen peroxide [34].

Light: Light therapy is a significant breakthrough in disease treatment. Phototherapy uses sunlight's visible and invisible light to prevent and treat diseases. Photoresponsive smart nanoparticles are gaining popularity as controlled drug delivery systems for anticancer therapy due to their ability to release encapsulated medicines on demand through light irradiation [1].

Exogenous stimuli Temperature: Since diseased/tumour tissues have the physiological characteristics of higher temperature (40–42 °C) than normal tissues (37 °C) [35].

Ultrasound: Ultrasound technology is clinically recognized for diagnosing and treating diseases. Ultrasound is useful for pinpointing specific locations. Benefits of this method include improved spatiotemporal control, safety, non-invasiveness, tuneable frequency, and deep tissue penetration. It has become a popular external trigger method for cancer treatment [1].

Electric field: Electric field is a non-invasive stimulating factor, that allows for the improvement of pharmacological treatment efficacy. Clinical applications of electric fields have been used to treat cancer. For instance, the US FDA has approved the use of electric fields to treat glioblastoma, opening the door to the use of electric fields in the treatment of cancer [1, 36].

Target: Currently, the EPR effect brought on by the extravasation of the nanoscale therapeutic compounds at tumour locations is primarily responsible for the vascular tumour. In order to improve therapeutic efficacy and reduce systemic toxicity, hundreds of passive targeting techniques based on the EPR effect have been studied. However, the EPR effect might not be advantageous for all solid tumours because the tumour blood flow and vascular permeability differ significantly from the normal value [37]. Furthermore, compared to rodents, humans do not experience the EPR effect as well. 'Missile-like' nanoparticles for active tumour-targeting are more appealing than passive targeting. Tumour cells have a number of distinct receptors on their surface, including peptides, antibodies, transferrin, folic acid, etc. [1].

Modification of antibodies: Tumour surface-specific antibody-based targeting has consistently garnered interest, and therapeutically effective cancer-targeted antibodies have been demonstrated. Epidermal growth factor receptor (EGFR) antibody. EGFR is a cell surface receptor expressed in many different types of cells, especially overexpressed in cancer cells. Researchers have shown that the overexpression of EGFR is closely related to the differentiation and migration of cancer cells. Cetuximab is a chimeric monoclonal antibody that selectively binds to EGFR, which can effectively block the activation of EGFR and its downstream signalling pathways, inhibiting therefore EGFR-related development and cancer progression [1]. P-glycoprotein (P-gp) antibody. Tumour cell multidrug resistance (MDR) is one of the primary causes resulting in chemotherapy failure. It is now established that the principal mechanism through which MDR functions is the efflux of p-glycoprotein (P-gp). P-gp is an energy-dependent transporter that can efflux many compounds of varied structure out of the cell in the reverse direction [1]. P-gp is overexpressed in tumour drug-resistant cells, and the P-gp antibody-modified nano-drug carrier system offers a targeted approach to the efficient capture of such cells. Ma et al. designed a P-gp antibody-modified porous hydrogel particle for drug-resistant tumour cell capture and found that it can adsorb more drug-resistant tumour cells. This strategy endows the hydrogel material with the ability to efficiently capture and detect drug-resistant tumour cells [1, 38] (Table 1).

Table 1 Typical peptide alterations of smart nanoparticles for cancer treatment

Cancer/tumour type	Peptide type	Ligand used	Nanocarriers	Target receptors	Refs.
Non-Hodgkin lymphoma	Therapeutic	Tat-POSH-C	Micelles	–	[1, 39]
Melanoma	Cell penetrating	TAT	Liposomes	CPP's (cell penetrating peptides)	[1, 40]
Colorectal carcinoma	Cell penetrating	Octa arginine	Lipoplex	CPP's (cell penetrating peptides)	[1, 41]
Brest cancers	Targeting peptide	Her2	Hyaluronic acid–micelles	Epidermal growth factor receptor-2	[1, 42]
Prostate tumour	Targeting peptide	DUP1	Polymeric micelles	PSMA (prostate specific membrane antigen)	[1, 43]
Glioblastoma	Targeting peptide	Angi-pep-2	RBC membranes (red blood cells)	Low density LPS's (lipoprotein receptors)	[1, 44]

Payload

The ultimate objective of developing smart nanoparticles is to improve patient care by increasing therapeutic effect, lowering toxicity, and extending survival. To accomplish, to achieve these aims, scientists and medical professionals have developed a number of strategies to get around the inherent flaws in the medications that are now on the market. As contemporary biomedicine advances, smart nanoparticles can carry a wider range of payloads, including small molecule medications, nucleic acids, peptides, proteins, and even living cells [1].

Peptides and proteins: Despite the fact that small molecule drug delivery research has a strong basis, small molecule medications' target constitutes only 2–5% of the human genome, necessitating the creation of alternate treatments [1]. Proteins with over 50 amino acids and peptides with 2–50 amino acids are more selective for human-specific targets. Peptides and protein medicines have higher biological activity and lower toxicity than small compounds because of their bigger molecular size and varied quaternary structure, which improves their interaction with particular protein pockets [1, 45].

The nucleic acids: the number of pharmacological targets has significantly increased thanks to peptide and protein medicines. But because drugs containing nucleic acids can be used to silence or repair damaged genes and encourage the development of beneficial therapeutic genes since they have a greater capacity to control gene expression. Nucleic acids that are naked have a limited half-life because they are extremely susceptible to nuclease destruction. At the same time, the human immune system is removing and adjusting to external DNA and RNA [1].

3 Nanotechnologies in Cancer Therapy

Although certain targeted treatments can show tumour specificity, their practical applicability is limited by PK/PD proper biodistribution or connections. The potency and specificity of tumour necrosis factor-related apoptosis-inducing ligand (TRAIL) toward malignant cells while sparing healthy cells make it the perfect anti-cancer treatment [46]. However, due to the short half-life and quick renal clearance of the off-targeted TRAIL, it finds it difficult to progress past preclinical. Also, any therapeutic's PK/PD parameters could be improved by nano formulation, which would allow for drug repurposing [3, 46]. Stimuli-responsive carriers are made to release their payload in response to particular triggers, such as pH or temperature changes or the overexpression of particular TME-specific enzymes, elevated intracellular glutathione levels, and external stimuli like radiation, ultrasonic, magnetic fields, etc. [47].

In this regard, medication release within the TME or other targeted locations can be used to accomplish precise delivery. One of the most often altered or removed genes in breast cancer is TP53; in up to 44% of TNBC, the mutation is found, whereas in 15% of ER-positive breast tumours. Presently, there are two main approaches used in cancer immunotherapy: modifying effector immune cells with monoclonal antibodies (mAbs) and encouraging the co-engagement of interaction between T cells and tumour cells by the use of bispecific T cell-engaging antibodies or chimeric antigen-receptor-T cells. The future of cancer immunotherapy may lie in combining the two approaches into a single system. This was recently shown in a versatile antibody immobilization nanoplatform that was created by attaching anti-IgG (Fc specific) antibody (αFc) to the surface of a nanoparticle (αFc-NP), enabling the immunization of two different types of monoclonal antibodies [48, 49]. In a number of murine tumour models, immunomodulating nano-adaptors (imNAs) performed better than a combination of mAbs in T cell and natural killer cell responses as well as macrophage-driven immune responses [3]. It has been demonstrated that nanomaterials are very helpful for co-delivering several chemotherapy drugs. The biological characteristics of drugs might vary greatly. Co-delivery inside a single carrier can normalize dispersion and delivery since it is distinctly distinct from its synergistic complement. In the clinic today, anti-PD-1/PD-L1 antibodies are used to disrupt the immunological checkpoint, which reverses T cell fatigue and malfunction and has been shown to be effective in treating cancer [3].

3.1 Nanomedicine in Immunotherapy

The adaptive immune system's detection of tumour-associated antigens (TAAs) or tumour-specific antigens (TSAs) serves as the foundation for immunotherapy. Immuno-checkpoint inhibitors, CAR-T cell therapy, and cancer vaccines are examples of groundbreaking advancements in the field of immunotherapies for cancer [50]. While TSAs are only identified in tumour cells, TAAs are present in all cell types and are usually overexpressed in tumour cells. Antigen-presenting cells (APCs) absorb the tumour related protein and break it down into tiny protein fragments to produce tumour-directed immune responses [51]. The T cell receptors (TCR) recognize the HLA-peptide complex after it binds to patient-specific human leukocyte antigen (HLA) molecules, and the T cell then causes tumour cell death [50].

Numerous clinical investigations have employed the NY-ESO-1 antigen, which has been shown to be effective in cancer treatment by eliciting a better immune response and producing favourable results in some trials. A subpopulation of immune cells known as invariant natural killer T (iNKT) cells are able to identify glycolipid antigens that are presented by the non-polymorphic MHC class I-like molecule CD1d. When activated, they effectively generate cytokines that increase cytotoxic T cell responses and stimulate other immune cells. Even at low dosages, iNKT agonists exhibit strong adjuvant effects when given concurrently [3].

Currently, there are two main approaches used in cancer immunotherapy: modifying effector immune cells with monoclonal antibodies (mAbs) and promoting T cell and tumour cell co-engagement with chimeric antigen-T cells or bispecific T cell-engaging antibodies. The future of cancer immunotherapy may lie in combining the two approaches into a single system. This was recently shown in a versatile antibody immobilization nanoplatform that was created by attaching anti-IgG (Fc specific) antibody (αFc) to the surface of a nanoparticle (αFc-NP), enabling the immunization of two different types of monoclonal antibodies [48, 49]. In a number of murine tumour models, immunomodulating nano-adaptors (imNAs) performed better than a combination of mAbs in T cell and natural killer cell responses as well as macrophage-driven immune responses [3].

3.2 Gene Delivery and RNA-Based Therapies

Delivering nucleic acids to express pro-apoptotic proteins, replace defective genes, down-regulate or quiet oncogenic pathways, create anti-cancer cytokines, and/or stimulate the immune system against cancer are all important aspects of gene therapy's role in the fight against cancer [52]. A significant obstacle in gene delivery is getting nucleic acids to the intended location without degrading them. The first siRNA-delivering liposome approved by the FDA in 2019 was Patisiran (ONPATTRO®), which delivers siRNA against the gene that causes the production of the transthyretin protein, which can result in hereditary transthyretin amyloidosis. Clinical translation of gene therapy still faces difficulties in finding safe and effective delivery systems. In terms of gene delivery, recombinant viral vectors are better than nonviral vectors, however, they have drawbacks as well, including immune response, large-scale production, gene size restriction, restricted cell tropisms, and inability to modify the surface without sacrificing vector integrity [53]. In comparison to viral vectors, non-viral vectors can have a lower transfection capacity despite being synthetically dynamic, having low immunogenicity, and being easier to produce on a large scale.

In the US, Rexin-G was the first targeted gene therapy vector to be granted orphan drug priority and fast track classification for a number of cancer indications. Targeting aberrant Signature (SIG) proteins in malignancies, Rexin-G is a replication-incompetent retroviral vector that uses a mysterious collagen-binding motif on its envelope (ClinicalTrials.gov Identifier: NCT00504998) [54]. NG-641 is an oncolytic adenoviral vector that contains four genes: a bi-specific FAP-targeted T-cell activator that activates T-cells to destroy fibroblasts, as well as three additional genes for recruiting and activating T-cells (CXCL9, CXCL10, and interferon alpha) [3, 55].

The lipid bilayer membrane of exosomes, which range in diameter from 30 to 100 nm, contains proteins, mRNA, lncRNA, miRNA, DNA, and other cellular constituents [55]. Through membrane fusion, exosomes can enter recipient cells and alter translation and transcription [55]. They have enormous potential for cancer treatment since they are very stable and biocompatible, show tumour homing, and are modifiable. To deliver siRNA or shRNA specific to oncogenic KRASG12D (iExosomes), a frequent mutation in pancreatic cancer, exosomes produced from normal fibroblast-like mesenchymal cells were engineered (ClinicalTrials.gov Identifier: NCT03608631). Targeting oncogenic Kras with greater efficiency than liposomes, iExosomes rely on CD47 and are aided by micropinocytosis [56]. mRNA cancer vaccines, which target TSAs, are a new tool in the fight against cancer. These antigens can be swiftly identified using bioinformatics methods and next-generation sequencing, and they can be incorporated into mRNA vaccines, which have gained popularity recently due to the success of mRNA SARS-CoV-2 vaccines. Compared to DNA vaccines, mRNA vaccines offer a number of benefits, such as increased protein expression levels, quick and transient protein expression, a simpler production technique, and no genomic integration [3].

Nano formulation is necessary for effective delivery since nucleic acid-based treatments are prone to rapid degradation and inadequate cellular absorption. The BNT111/Melanoma FixVac mRNA vaccine, a liposomal formulation of mRNA encoded against four different malignant melanoma-associated antigens—NY-ESO-1, melanoma-associated antigen A3, tyrosinase, and transmembrane phosphatase with tensin homology—is being tested in humans for the first time in the Lipo-MERIT trial (ClinicalTrials.gov Identifier: NCT02410733). Also, patients with ovarian cancer will receive an intravenous vaccination before and during (neo)adjuvant treatment as part of a research to examine a liposomal mRNA vaccine (W_ova1 vaccine) that delivers three ovarian cancer TSA RNAs (ClinicalTrials.gov Identifier: NCT04163094) [3].

All things considered; mRNA vaccines developed in nanocarriers have demonstrated early clinical promise through targeted delivery to APCs. In addition to being stand-alone Immunotherapeutics that stimulate the immune system against particular antigens, these nano vaccines have also been used in conjunction with checkpoint antibodies in a number of recent trials, which should improve therapeutic outcomes [3, 57].

3.3 Hyperthermia and Photothermal Therapy

A promising technique for removing malignant tissue is hyperthermia, which uses targeted heat to destroy cells. It has been demonstrated that some nanomaterials can intensify the effects of hyperthermia. Nanomaterials having a high-absorption cross-section that can transform an external energy source into heat can produce uniform and selective hyperthermia. Multi-modal therapies, including photothermal, photodynamic, magnetic, and radiation, have the potential to eradicate cancer through physical activation. In particular, it has been widely practiced to employ carbon and gold nanoparticles to produce hyperthermal effects when exposed to near-infrared light (NIR). With medicines attached to the surface or encapsulated inside, they can also be used for triggered drug release, in which the payload releases in response to temperature or radiation changes. In order to localize treatment and/or physically ablate malignant cells, this method can combine a combinatorial approach by employing therapeutic drugs in combination with an external trigger. Recent research has shown that activatable polymeric pro-nanoagonist (APNA), which is activated by second near-infrared (NIR-II) light that permits deep-tissue penetration and serves as a technique for future radiotherapy, causes immunogenic cell death and tumour ablation through the photothermal effect [3].

Photothermal regulation of heat-sensitive TRPV1 or TRPV2 ion channels is a unique method of using nanomaterials with radiation to modulate cell stemness [58]. In both in vitro and in vivo models, this recent study shows that NIR photoactive nanocarbon complexes can induce TRPV2 overexpression in cancer cells, disrupt intracellular Ca_2^+ regulation, and suppress Wnt/β-catenin signalling, leading to the death of cancer cells and inhibition of stemness. Although photothermal agents (PTAs) have demonstrated encouraging outcomes in clinical trials, their quick degradation restricts the photothermal stability needed for effective therapy However, because they deteriorate more slowly, those with excellent photothermal stability raise more safety concerns [58]. Few PTAs have strong photothermal stability and quick degradation at the moment. It has recently been demonstrated that black phosphorus's (BP) natural capacity to capture Cu_2^+ can both speed up BP degradation and improve photothermal stability [1].

4 Advancements in Cancer Diagnostics

Nanomaterials have numerous benefits that can enhance cancer diagnostics and treatments, and they are incredibly flexible and adaptable. However, the potential advantages must be evaluated against aspects like production cost, scalability, safety, and complexity of nano formulations. Costs, production requirements, and testing criteria all rise in tandem with design and material complexity [59]. The environmental impact of the nanomaterials themselves, as well as the costs of energy and product manufacturing, is becoming a more significant consideration for commercial production.

Even without the problem of ambiguous regulatory requirements, there are numerous problems when comparing the costs and benefits of nanomedicine. It is not as easy as comparing apples to apples, though, as the production costs of nanomedicine can be significantly higher than those of traditional medications, depending on formulation and complexity [59]. Quality of life is typically only considered during clinical trials, according to a recent study, it is frequently not a co-primary endpoint and was evaluated until death in only 5 out of 149 studies (3.4%), with only 1 out of 74 studies focusing on metastatic or incurable cancers [3, 60]. The study and its conclusions were criticized by many in the field of science, but there is a lot of promise in using artificial intelligence to direct fundamental research. In addition to cross-referencing with current and new research to optimize for strategy and imitation, a wide range of factors can be utilized to assess the possibility of clinical translation for nanotechnologies. Machine learning applications can significantly benefit nanomedicine by analysing patient tumour profiles, choosing the best material based on drug target, mechanistic characteristics, and personalized prognoses, and analysing medication response to nanoparticle design. In the near future, nanotechnology for cancer diagnosis, chemotherapy, and radiotherapy stands to advance significantly, making the cancer landscape much more bearable for patients and physicians. Despite cancer's dynamic character, innovation is still advancing and the convergence of several technologies holds hope for success [3].

4.1 Early Detection with Nanotechnology

Developing cancer medications is an expensive undertaking, to put it mildly. The majority of drug candidates fail clinical trials, and the expense of creating a successful medication can approach the billion-dollar mark [61]. Only 50 new small molecule anti-cancer medications were approved by the FDA between 2015 and 2020, despite the fact that the entire oncology pipeline comprises hundreds of compounds in late-stage research annually [3]. Metastases account for 90% of cancer-related fatalities. Even the average patient cost is much higher for treating late-stage cancer than for treating early-stage cancer [62]. There are countless advantages to early detection and routine screening, especially since certain cancers only show symptoms in their advanced stages and screening techniques can be used to assess and tailor treatment for individual patients [62].

Conventional diagnostic techniques generally use traditional imaging techniques, such as X-rays, CT scans, MRIs, and ultrasounds, along with the Papanicolaou test to identify cervical cancer, occult blood detection for colon cancer, and detection of prostate-specific antigen levels in blood samples. However, rather than comprehensive screening of various cancer types, currently available cancer screenings are usually limited to a subset of individuals based on age and/or risk level, and are specific to the aforementioned cancers [63]. Despite the fact that certain metastatic cancers may be easy to identify, standard imaging makes it practically hard to tell based on where the onset of metastases is taking place. The following step is usually surgical resection, which depends heavily on the tumour margin and cannot ensure total removal of all cancerous cells, particularly while trying to preserve as much healthy tissue as possible [3].

4.2 Nano Sensors and Imaging Techniques

Nanomaterials can significantly enhance tumour detection by tumour targeting and targeted intrusion detection, while traditional imaging techniques are unable to diagnose tumours early physical–chemical characteristics that can improve the signal. New imaging platforms and methods with greater sensitivity and no potential negative effects can be made possible by certain nanomaterials [3]. The prognosis worsens when the most aggressive CTCs eventually spread to other tissues and develop into metastatic cancers. Early and precise cancer diagnosis can be greatly aided by the use of CTCs, and in-depth investigations can pinpoint certain biomarkers that can be used to infer patient prognosis and treatment response [64]. However, for CTC detection to be a valid clinical diagnostic/prognostic tool for cancer, a number of issues must be resolved. The percentage of CTCs in circulation during the early stages of cancer is quite small, and their heterogeneity makes it challenging to separate and examine them. For accurate, reliable, and consistent CTC detection, nanotechnology offers a number of benefits. For CTC detection, a variety of nanostructures have been employed, including metal nanoparticles, carbon-based, magnetic, polymeric, and quantum dots [65].

The serum prostate state-specific antigen test is used for prostate cancer screening, which has a high probability of false positives (80%), which results in needless biopsies and abuse [66]. There is a significant unmet demand for precision screening for prostate cancer given the test's frequency. For example, a Phase 1 clinical trial employing machine learning for a urine multi-marker microfluidic biosensor is presently underway (ClinicalTrials.gov Identifier: NCT04825002) [3]. The non-invasiveness and patient compliance of diagnostic screening technologies are critical to their performance. Particularly in contrast to biopsies or blood samples, testing urine, saliva, or breath is non-invasive. Even in a healthy state, exhaled breath contains trace amounts of volatile organic compounds (VOCs); nevertheless, in a sick state, the content and quantities of these molecules can identify the type and stage of cancer [67]. Also, to capture volatile organic compounds (VOCs), the gold nanoparticles can also be coupled with organic molecules. Gas chromatography can then be used for analysis by mass spectrometry and motograph. For extensive clinical use, the Na-nose offers the following benefits: low cost, simplicity of use, high reproducibility, and real-time detection (ClinicalTrials.gov Identifier: NCT03967652) [3].

Traditional imaging devices have made progress in preoperative diagnosis and staging, but there is still opportunity for improvement in terms of sensitivity, resolution, intraoperative techniques, and activity. Improvements to image-guided procedures are still being made by incorporating specific targeting, optically-active materials, and nano-sized probes for alternative modes of imaging [3, 68].

5 Challenges in Translating Nanotechnology to the Clinic

The development in the field of nanotechnology is enormously growing at a global scale due to its promising and advanced approach in almost all the areas. With the flourishing knowledge, advancement in frontier cancer treatment is remarkable which suggests its effectiveness in the field of therapeutics including In Vivo and In Vitro. Despite this, there are very few nano-formulations that could enter clinical trials [69]. Translating biomedicine nanotechnology from bench to bedside faces multiple challenges such as regulatory hurdles in the process for bringing molecule to clinical trials, Economic factors, manufacturing and Scalability and certain biological challenges [70].

The biological challenges are primarily due to factors like heterogeneity in diseases, patient genetic variability, preclinical models and barriers to delivery. For instance, variations in the type of tumour, tumour microenvironments, vascular permeability, immune responses, nanoparticle polydispersity, nanoparticle distribution and uptake can lead to unpredictable therapeutic outcomes and so impact efficacy of the nanomedicine [71–74]. One of such challenging condition includes failure of CALAA-01, the first experimental therapeutic targeted nanoparticle with siRNA for treating the cancer. Despite promising preclinical results, it failed in clinical trials due to poor efficacy driven by variable patient responses [75]. Besides this, animal models used for the In Vivo studies are also raises issue at preclinical level due to not efficient enough to replicate the complexity of human diseases accurately. For example, the vasculature and the immune response of the mice tumour may not resonate completely with the human and may mislead the conclusions about a nanomedicine's effectiveness. Also, the long-term treatment, biocompatibility and toxicity caused by nano-formulations need to be thoroughly evaluated leading to lengthy preclinical studies and ultimately delay the development process of nanomedicine [76]. On contrary, Abraxane® (albumin-bound paclitaxel), exceptionally succeeded the preclinical trials and translated into FDA approval by overcoming solubility issues, demonstrating the potential for targeted design. Finally, the targeted delivery to the desired tissue may reduce due to the presence of barriers such as the blood–brain barrier (BBB), liver sequestration, and rapid clearance by the mononuclear phagocyte system (MPS) that may create significant obstacles for nanoparticles path decreasing the therapeutic impact of nanomedicines. To address this issue, most commonly modifications in the surface chemistry is done. For example, Barenholz demonstrated the PEGylation of Doxil® that enhanced the circulation time, but faced challenges in active tumour targeting, underscoring the complexity of achieving precise delivery [77]. So overall, addressing these biological challenges is crucial for advancing the clinical success of nanomedicine.

Another major clinical issue is the manufacturing and scalability of nanomedicines at commercial level. The property of polydispersity due to varied nucleation size and time, poses primary concern in the consistency of nanoparticle production. Achieving mono-dispersiveness in the nano-formulation is so significant that even a minor deviation may vary the therapeutic efficacy and increase cytotoxicity. To address this issue, continuous flow synthesis and microfluidics techniques are employed to increase consistency over various batches even though being economically demanding [78]. Moreover, scalability also demands high-purity materials and advanced instrumentation that collectively makes nanotechnology a costlier affair. For example, the liposome for Doxil® posed technical challenges with the uniformity of various batches while scalability. TO overcome the challenges, high quality protocols and quality checks were carried out which lead to very high production cost [77]. Additionally, the critical check point on the nanomedicine production is kept by regulatory bodies which mandate the detailed physicochemical properties and biological performances to ensure efficiency, safety, reducing cost and facilitate the widespread adoption of nanomedicine therapies.

The challenge related to regulatory affairs include certain ambiguities in guidelines related to involvement of multidiscipline like chemistry, biology, material science and many more that makes regulators task tough enough to decide whether they should be considering it as drugs or devices. Zelnate® is one of such examples, a DNA vaccine adjuvant nanoparticle [79]. Furthermore, Safety Evaluation poses challenges including traditional toxicity testing methods as the results of the short-term study may not resonate with the long-term study and may result in causing chronic toxicity. The failure of AuroLase™ gold nanoparticle therapy is well known example of the same [80]. Intellectual Property Issues further complicate the landscape of nanomedicine as the invention often involves many collaborations across academia and industry that may lead to complexity and disputes over patent rights. Such disputes remarkably hamper the process of translation of invention from bench-to-bed side. The known example is the mRNA technology where multiple collaborators claim the rights to fundamental findings leading to complex legality and hindering further progress [81].

It is indeed a fact that economy plays an important role in the growth and uptake of nanomedicines as well. In fact, it poses a great investment cost challenge because quite a majority of setups for the development of nanomedicine require specialized facilities and advanced synthesis techniques. This indeed makes nanomedicine development costly as compared to conventional drugs. Another challenge is related to the market feasibility of nanomedicines because most of them are targeting diseases that are either rare or complex with fairly limited patients in volume, and therefore, the production as well as a commercial market of profit is low. However, not to be ruled out from such examples are some successes like NanoTherm® for glioblastoma which promises a future in niche markets, although limited reach [82]. Competing with conventional therapies is another hurdle: nanomedicines have to demonstrate clear advantages in terms of efficacy, safety, and cost-effectiveness for acceptance. Thus, to further use nanomedicine technologies as sustainable advancements, there has to be much overcoming the economic enabling features barrier.

Innovative and collaborative approaches would still be needed to address challenges in nanomedicine. Disease-driven approaches are focused on designing nanoparticles that operated by specific mechanisms of diseases stimulus-driven system wherein release occurs in tissues undergoing specific pH or temperature changes. Advances in preclinical testing using humanized models and organ-on-chip systems have closed the gap between what is done in the laboratory and expected clinical outcomes. They afford more realistic predictions of nanoparticle behaviour [83, 84]. Alliances between academia, industry, and the regulatory world such as NCL-funded communities have guaranteed the efficient compliance in and innovation through pooling resources and expertise for research to clinical use.

5.1 Future Directions in Nanomedicine: Personalized Approaches, Combination Therapies, and Global Cancer Treatment

Nanomedicine has completely transformed modern medicine. It has emerged with cutting-edge solutions addressing the most important challenges in treating diseases. As cancer is one of the most widespread causes of mortality worldwide, the demand for innovative therapeutic procedures, which further enhance patient outcomes while increasing worldwide accessibility to care, is critical. The future of nanomedicine will hinge on three main engines: personalized therapies; new combination therapy innovations; and greatness in global reach of cancer delivery.

5.2 Personalized Nanomedicine

Personalized nanomedicine is an unparalleled feature of modern cancer therapies in which treatments are designed specifically for a given patient's profile. Whose genetic, molecular and environmental data can be applied for the nanomedicine construction with greater efficacy and reduced side effects. More recent advances in nanotechnology have enabled this targeted delivery of therapeutic nanoparticles to cancer cells, with off-target effects occurring at a minimum. Functionalized nanoparticles decorated with specific cancer biomarker moieties show enormous promise in preclinical research [5, 85].

For instance, for breast cancer, HER2-targeted nanoparticles demonstrated delivery of the drug to the target site and reduce systemic toxicity [86]. Besides, the novel approaches of proteomics and single-cell genomics have largely contributed to the improvement of personalized nanomedicines through the detailed understanding of tumour heterogeneity, leading to targeted and effective treatments. AI is also taking a pioneering step in revolutionizing the field of nanomedicine through its ability to analyze huge datasets and the outcome which can be applied further for effective design of nanoparticle specific to the patients. Machine learning algorithms can be helpful in predicting models enabling the tailoring the drug delivery systems by fine-tuning the necessity of the nanoparticles such as size, surface chemistry, charge and ensure their compatibility with the unique biological environment of each patient by identifying significant patterns in patient specific dataset [87]. However, challenges in scaling patient-specific nano formulation production and establishing robust regulatory frameworks persist. Collaboration across disciplines and emerging technologies enable tailored nanomedicine.

5.3 Combination Therapies

Nanotechnology is transforming combinatorial therapy approach in the field of cancer treatment. Combination therapies aid by targeting multiple pathways simultaneously and co-deliver multiple therapeutic agents effectively which collectively enhance the efficacy of the formulation [88, 89]. To minimize drug resistance, chemotherapeutics, immunomodulators, or genetic materials can be used in combinations to enable synergistic effects For example, liposomal nanoparticles delivering both doxorubicin and siRNA have demonstrated improved outcomes in preclinical models of multidrug-resistant cancers [90]. In immunotherapy, nanoparticles elevate the delivery and stability of immune-modulating agents. Nanoparticles can transport immune modulators like cytokines directly to tumours, boosting their effectiveness [91]. Nano vaccines deliver tumour antigens to dendritic cells that have the potential of inducing very robust immune anti-tumour responses [92]. Furthermore, this novel strategy is also associated with the combination of photothermal and photodynamic therapies.—Using Gold nanoparticles, which can generate heat when laser irradiated, one can specifically destroy cancer cells and healthy tissues alike [93, 94].

Chemotherapeutics or-as additional combinations with immune therapy will even strengthen the treatment results. Thus, multidrug resistance remains one of the most significant challenges the biomedical field is facing today across the world. In modern times, the nanotechnology approaches to cancers have been focused on multi-targeting mechanisms of resistance. For example, polymeric nanoparticles dually loaded with inhibitor of P-glycoprotein and chemotherapeutic agents demonstrated a reversal of resistance, enhancing therapeutical efficacy [95]. These advances point towards the increased success rate of nanotechnology in the creation of more effective and versatile combination cancer therapies.

5.4 Global Cancer Treatment

The increasing prevalence of cancer in populations around the world reinforces the urgent need for treatment methods that can be effective, but also affordable to the entire populace. Nanomedicine offers the best approach to watery lower to middle income countries, where technology-based advanced cancer therapies may be inaccessible [96]. Recent developments in nanoparticle synthesis methods here include microfluidic platforms and green nanotechnology, and they have triggered the facilitation of cheaper manufacturing. Green synthesis is in part economic due to the extent to which it reduces usage of costly reagents and, in the main, it is much eco-friendly [97]. Such strategies will render nanomedicine affordable and sustainable even in resource-poor contexts. The advances in nanoparticle-based biosensors have indeed thrown a paradigm shift in the techniques for cancer diagnosis because they provide an effective and specific detection of cancer biomarkers in the blood or saliva samples to achieve rapid and accurate results.

This approach alters the chances of early detection of cancer and helps in planning and tailoring the type of diagnostics to be used for efficacious treatment [98, 99]. Attaining equitable access to nanomedicine requires addressing economic or logistical barriers; thus, a public–private approach to partnership and global health programs could contribute significantly to progress in the development and delivery of nano-formulations. It can also stimulate incorporation of nanomedicine into national health policy regarding cancer care, thereby hastening acceptance in low and middle-income countries and opening up more inclusive and accessible avenues for global cancer treatment.

6 Conclusion

Nano-formulations have very specialized properties making them highly potent for delivery of drugs in target therapy in cancer treatment. The future of nanomedicine is therein personalizing treatment and optimizing combinations of therapies while promoting equal access to superlative advanced cancer care globally. Personalized nanomedicine should drive individualized combinations of therapeutic treatment, maximizing therapeutics and minimizing adverse effects with an eye toward a more patient-cantered approach to treatment. Outcomes of combination therapies with nanotechnology integrated in them will present new horizons toward dealing with the serious problems of drug resistance and poor outcomes with chemotherapy treatment courses. Advances in cost-effective manufacturing and scalable platforms may finally enable nanomedicine to reach populations worldwide, closing the breach between innovation and equity. Interdisciplinary collaboration and sustained investment will be pivotal for the total realization of what nanomedicine can do, namely transforming global cancer treatment. This chapter highlights the transformative and translational prospects of nanomedicine regarding the advantages and disadvantages along with its application in nanotechnology for the treatment of cancer, challenges, and the continuous need for research and innovation to overcome these existing barriers. This will help in improving clinical outcomes at a global level.

References

1. Sun, L., Liu, H., Ye, Y., Lei, Y., Islam, R., Tan, S., Tong, R., Miao, Y.B., Cai, L.: Smart nanoparticles for cancer therapy. In: Signal Transduction and Targeted Therapy, vol. 8, issue 1. Springer Nature (2023). https://doi.org/10.1038/s41392-023-01642-x
2. Iwase, S., Kawaguchi, T., Tokoro, A., Yamada, K., Kanai, Y., Matsuda, Y., Kashiwaya, Y., Okuma, K., Inada, S., Ariyoshi, K., Miyaji, T., Azuma, K., Ishiki, H., Unezaki, S., Yamaguchi, T.: Assessment of cancer-related fatigue, pain, and quality of life in cancer patients at palliative care team referral: a multicenter observational study (JORTC PAL-09). PLoS ONE **10**(8), e0134022 (2015). https://doi.org/10.1371/journal.pone.0134022

3. Kemp, J.A., Kwon, Y.J.: Cancer nanotechnology: current status and perspectives. In: Nano Convergence, vol. 8, issue 1. Korea Nano Technology Research Society (2021). https://doi.org/10.1186/s40580-021-00282-7
4. Lorusso, D., Bria, E., Costantini, A., Di Maio, M., Rosti, G., Mancuso, A.: Patients' perception of chemotherapy side effects: expectations, doctor-patient communication and impact on quality of life—an Italian survey. Eur. J. Cancer Care **26**(2), e12618 (2016). https://doi.org/10.1111/ecc.12618
5. Shi, J., Kantoff, P.W., Wooster, R., Farokhzad, O.C.: Cancer nanomedicine: progress, challenges, and opportunities. Nat. Rev. Cancer **17**(1), 20–37 (2017)
6. Falagan-Lotsch, P., Grzincic, E.M., Murphy, C.J.: New advances in nanotechnology-based diagnosis and therapeutics for breast cancer: an assessment of active-targeting inorganic nanoplatforms. Bioconjug. Chem. **28**(1), 135–152 (2016). https://doi.org/10.1021/acs.bioconjchem.6b00591
7. Gupta, P.K.: Drug targeting in cancer chemotherapy: a clinical perspective. J. Pharm. Sci. **79**(11), 949–962 (1990). https://doi.org/10.1002/jps.2600791102
8. Senapati, S., Mahanta, A.K., Kumar, S., Maiti, P.: Controlled drug delivery vehicles for cancer treatment and their performance. Signal Transduct. Target. Ther. **3**(1) (2018). https://doi.org/10.1038/s41392-017-0004-3
9. Dahmana, N., Kowalczuk, L., Gabriel, D., Behar-Cohen, F., Gurny, R., Kalia, Y.N.: Ocular biodistribution of spironolactone after a single intravitreal injection of a biodegradable sustained-release polymer in rats. Mol. Pharm. **17**(1), 59–69 (2019). https://doi.org/10.1021/acs.molpharmaceut.9b00707
10. Miao, X., Yang, W., Feng, T., Lin, J., Huang, P.: Drug nanocrystals for cancer therapy. Wiley Interdiscip. Rev. Nanomedicine Nanobiotechnology **10**(3) (2017). https://doi.org/10.1002/wnan.1499
11. Patel, A., Patel, K., Patel, V., Rajput, M.S., Patel, R., Rajput, A.: Nanocrystals: an emerging paradigm for cancer therapeutics. Futur. J. Pharm. Sci. **10**(1) (2024). https://doi.org/10.1186/s43094-024-00579-4
12. Joseph, E., Singhvi, G.: Multifunctional nanocrystals for cancer therapy: a potential nanocarrier. In: Elsevier eBooks, pp. 91–116 (2019). https://doi.org/10.1016/b978-0-12-816505-8.00007-2
13. Wicki, A., Witzigmann, D., Balasubramanian, V., Huwyler, J.: Nanomedicine in cancer therapy: challenges, opportunities, and clinical applications. J. Control. Release **200**, 138–157 (2014). https://doi.org/10.1016/j.jconrel.2014.12.030
14. Izci, M., Maksoudian, C., Manshian, B.B., Soenen, S.J.: The use of alternative strategies for enhanced nanoparticle delivery to solid tumors. Chem. Rev. **121**(3), 1746–1803 (2021). https://doi.org/10.1021/acs.chemrev.0c00779
15. Kang, H., Rho, S., Stiles, W.R., Hu, S., Baek, Y., Hwang, D.W., Kashiwagi, S., Kim, M.S., Choi, H.S.: Size-dependent EPR effect of polymeric nanoparticles on tumor targeting. Adv. Healthc. Mater. **9**(1) (2019). https://doi.org/10.1002/adhm.201901223
16. Gustafson, H.H., Holt-Casper, D., Grainger, D.W., Ghandehari, H.: Nanoparticle uptake: the phagocyte problem. Nano Today **10**(4), 487–510 (2015). https://doi.org/10.1016/j.nantod.2015.06.006
17. Beltrán-Gracia, E., López-Camacho, A., Higuera-Ciapara, I., Velázquez-Fernández, J.B., Vallejo-Cardona, A.A.: Nanomedicine review: clinical developments in liposomal applications. Cancer Nanotechnol. **10**(1) (2019). https://doi.org/10.1186/s12645-019-0055-y
18. Walkey, C.D., Olsen, J.B., Guo, H., Emili, A., Chan, W.C.W.: Nanoparticle size and surface chemistry determine serum protein adsorption and macrophage uptake. J. Am. Chem. Soc. **134**(4), 2139–2147 (2011). https://doi.org/10.1021/ja2084338
19. Ouyang, B., Poon, W., Zhang, Y., Lin, Z.P., Kingston, B.R., Tavares, A.J., Zhang, Y., Chen, J., Valic, M.S., Syed, A.M., MacMillan, P., Couture-Senécal, J., Zheng, G., Chan, W.C.W.: The dose threshold for nanoparticle tumour delivery. Nat. Mater. **19**(12), 1362–1371 (2020). https://doi.org/10.1038/s41563-020-0755-z
20. Wang, S., Zhu, B., Liu, M., Zhang, L., Yu, J., Zhou, M.: Direct Z-scheme ZnO/CdS hierarchical photocatalyst for enhanced photocatalytic H_2-production activity. Appl. Catal. B Environ. Energy **243**, 19–26 (2018). https://doi.org/10.1016/j.apcatb.2018.10.019

21. Yoo, J.D., Bae, S.M., Seo, J., Jeon, I.S., Vadevoo, S.M.P., Kim, S., Kim, I., Lee, B., Kim, S.: Designed ferritin nanocages displaying trimeric TRAIL and tumor-targeting peptides confer superior anti-tumor efficacy. Sci. Rep. **10**(1) (2020). https://doi.org/10.1038/s41598-020-77095-x
22. Rao, L., Bu, L., Xu, J., Cai, B., Yu, G., Yu, X., He, Z., Huang, Q., Li, A., Guo, S., Zhang, W., Liu, W., Sun, Z., Wang, H., Wang, T., Zhao, X.: Red blood cell membrane as a biomimetic nanocoating for prolonged circulation time and reduced accelerated blood clearance. Small **11**(46), 6225–6236 (2015). https://doi.org/10.1002/smll.201502388
23. Kolosnjaj-Tabi, J., Gibot, L., Fourquaux, I., Golzio, M., Rols, M.: Electric field-responsive nanoparticles and electric fields: physical, chemical, biological mechanisms and therapeutic prospects. Adv. Drug Deliv. Rev. **138**, 56–67 (2018). https://doi.org/10.1016/j.addr.2018.10.017
24. Kamaly, N., Yameen, B., Wu, J., Farokhzad, O.C.: Degradable controlled-release polymers and polymeric nanoparticles: mechanisms of controlling drug release. Chem. Rev. **116**(4), 2602–2663 (2016). https://doi.org/10.1021/acs.chemrev.5b00346
25. Sun, B., Chang, R., Cao, S., Yuan, C., Zhao, L., Yang, H., Li, J., Yan, X., Van Hest, J.C.M.: Acid-activatable transmorphic peptide-based nanomaterials for photodynamic therapy. Angew. Chem. Int. Ed. **59**(46), 20582–20588 (2020). https://doi.org/10.1002/anie.202008708
26. Kataoka, K., Harada, A., Nagasaki, Y.: Block copolymer micelles for drug delivery: design, characterization and biological significance. Adv. Drug Deliv. Rev. **64**, 37–48 (2012). https://doi.org/10.1016/j.addr.2012.09.013
27. Damodharan, J.: Nanomaterials in medicine—an overview. Mater. Today: Proc. **37**(Part 2), 383–385 (2020). https://doi.org/10.1016/j.matpr.2020.05.380
28. Alivisatos, P.: The use of nanocrystals in biological detection. Nat. Biotechnol. **22**(1), 47–52 (2003). https://doi.org/10.1038/nbt927
29. Ghuman, J., Zunszain, P.A., Petitpas, I., Bhattacharya, A.A., Otagiri, M., Curry, S.: Structural basis of the drug-binding specificity of human serum albumin. J. Mol. Biol. **353**(1), 38–52 (2005). https://doi.org/10.1016/j.jmb.2005.07.075
30. Goldberg, M., Langer, R., Jia, X.: Nanostructured materials for applications in drug delivery and tissue engineering. J. Biomater. Sci. Polym. Ed. **18**(3), 241–268 (2007). https://doi.org/10.1163/156856207779996931
31. Shi, J., Votruba, A.R., Farokhzad, O.C., Langer, R.: Nanotechnology in drug delivery and tissue engineering: from discovery to applications. Nano Lett. **10**(9), 3223–3230 (2010). https://doi.org/10.1021/nl102184c
32. Kaminskas, L.M., Kelly, B.D., McLeod, V.M., Sberna, G., Owen, D.J., Boyd, B.J., Porter, C.J.: Characterisation and tumour targeting of PEGylated polylysine dendrimers bearing doxorubicin via a pH labile linker. J. Control. Release **152**(2), 241–248 (2011). https://doi.org/10.1016/j.jconrel.2011.02.005
33. Hossen, S., Hossain, M.K., Basher, M., Mia, M., Rahman, M., Uddin, M.J.: Smart nanocarrier-based drug delivery systems for cancer therapy and toxicity studies: a review. J. Adv. Res. **15**, 1–18 (2018). https://doi.org/10.1016/j.jare.2018.06.005
34. Mo, R., Gu, Z.: Tumor microenvironment and intracellular signal-activated nanomaterials for anticancer drug delivery. Mater. Today **19**(5), 274–283 (2015). https://doi.org/10.1016/j.mattod.2015.11.025
35. Liu, D., Yang, F., Xiong, F., Gu, N.: The smart drug delivery system and its clinical potential. Theranostics **6**(9), 1306–1323 (2016). https://doi.org/10.7150/thno.14858
36. Aguilar, A.A., Ho, M.C., Chang, E., Carlson, K.W., Natarajan, A., Marciano, T., Bomzon, Z., Patel, C.B.: Permeabilizing cell membranes with electric fields. Cancers **13**(9), 2283 (2021). https://doi.org/10.3390/cancers13092283
37. Maeda, H.: The 35th anniversary of the discovery of EPR effect: a new wave of nanomedicines for tumor-targeted drug delivery—personal remarks and future prospects. J. Pers. Med. **11**(3), 229 (2021). https://doi.org/10.3390/jpm11030229
38. Ma, X., Zhao, Z., Wang, H., Liu, Y., Xu, Y., Zhang, J., Chen, B., Li, L., Zhao, Y.: P-Glycoprotein antibody decorated porous hydrogel particles for capture and release of drug-resistant tumor cells. Adv. Healthc. Mater. **8**(13) (2019). https://doi.org/10.1002/adhm.201900136

39. Smith, J.D., Cardwell, L.N., Porciani, D., Nolla, A., Cornelison, B.T., Schulte, M.C., Gallazzi, F., Burke, D.H., Daniels, M.A., Ulery, B.D.: Therapeutic peptide delivery via aptamer-displaying, disulfide-linked peptide amphiphile micelles. Mol. Syst. Des. Eng. **5**(1), 269–283 (2019). https://doi.org/10.1039/c9me00092e
40. Zou, L., Ding, W., Zhang, Y., Cheng, S., Li, F., Ruan, R., Wei, P., Qiu, B.: Peptide-modified vemurafenib-loaded liposomes for targeted inhibition of melanoma via the skin. Biomaterials **182**, 1–12 (2018). https://doi.org/10.1016/j.biomaterials.2018.08.013
41. Singh, T., Kang, D.H., Kim, T.W., Kong, H.J., Ryu, J.S., Jeon, S., Ahn, T.S., Jeong, D., Baek, M.J., Im, J.: Intracellular delivery of oxaliplatin conjugate via cell penetrating peptide for the treatment of colorectal carcinoma in vitro and in vivo. Int. J. Pharm. **606**, 120904 (2021). https://doi.org/10.1016/j.ijpharm.2021.120904
42. Chen, Q., Long, M., Qiu, L., Zhu, M., Li, Z., Qiao, M., Hu, H., Zhao, X., Chen, D.: Decoration of pH-sensitive copolymer micelles with tumor-specific peptide for enhanced cellular uptake of doxorubicin. Int. J. Nanomed. **11**, 5415–5427 (2016). https://doi.org/10.2147/ijn.s111950
43. Chen, H., Wu, F., Li, J., Jiang, X., Cai, L., Li, X.: DUP1 peptide modified micelle efficiently targeted delivery paclitaxel and enhance mitochondrial apoptosis on PSMA-negative prostate cancer cells. SpringerPlus **5**(1) (2016). https://doi.org/10.1186/s40064-016-1992-0
44. Liu, Y., Zou, Y., Feng, C., Lee, A., Yin, J., Chung, R., Park, J.B., Rizos, H., Tao, W., Zheng, M., Farokhzad, O.C., Shi, B.: Charge conversional biomimetic nanocomplexes as a multifunctional platform for boosting orthotopic glioblastoma RNAI therapy. Nano Lett. **20**(3), 1637–1646 (2020). https://doi.org/10.1021/acs.nanolett.9b04683
45. Lau, J.L., Dunn, M.K.: Therapeutic peptides: historical perspectives, current development trends, and future directions. Bioorg. Med. Chem. **26**(10), 2700–2707 (2017). https://doi.org/10.1016/j.bmc.2017.06.052
46. Tuthill, M.H., Montinaro, A., Zinngrebe, J., Prieske, K., Draber, P., Prieske, S., Newsom-Davis, T., Von Karstedt, S., Graves, J., Walczak, H.: TRAIL-R2-specific antibodies and recombinant TRAIL can synergise to kill cancer cells. Oncogene **34**(16), 2138–2144 (2014). https://doi.org/10.1038/onc.2014.156
47. Van Der Meel, R., Sulheim, E., Shi, Y., Kiessling, F., Mulder, W.J.M., Lammers, T.: Smart cancer nanomedicine. Nat. Nanotechnol. **14**(11), 1007–1017 (2019). https://doi.org/10.1038/s41565-019-0567-y
48. Wang, W., Zhao, J., Hao, C., Hu, S., Chen, C., Cao, Y., Xu, Z., Guo, J., Xu, L., Sun, M., Xu, C., Kuang, H.: The development of chiral nanoparticles to target NK cells and CD8+ T cells for cancer immunotherapy. Adv. Mater. **34**(16) (2022). https://doi.org/10.1002/adma.202109354
49. Wang, W., Zhao, J., Hao, C., Hu, S., Chen, C., Cao, Y., Xu, Z., Guo, J., Xu, L., Sun, M., Xu, C., Kuang, H.: The development of chiral nanoparticles to target NK cells and CD8+ T cells for cancer immunotherapy. Adv. Mater. **34**(16) (2022). https://doi.org/10.1002/adma.202109354
50. Stauss, H.J., Morris, E.C., Abken, H.: Cancer gene therapy with T cell receptors and chimeric antigen receptors. Curr. Opin. Pharmacol. **24**, 113–118 (2015). https://doi.org/10.1016/j.coph.2015.08.006
51. Liu, W., Zou, M., Liu, T., Zeng, J., Li, X., Yu, W., Li, C., Ye, J., Song, W., Feng, J., Zhang, X.: Cytomembrane nanovaccines show therapeutic effects by mimicking tumor cells and antigen presenting cells. Nat. Commun. **10**(1) (2019). https://doi.org/10.1038/s41467-019-11157-1
52. Zhu, C., Fu, S., Shi, Q., Du, D., Lin, Y.: Single-atom electrocatalysts. Angew. Chem. Int. Ed. **56**(45), 13944–13960 (2017). https://doi.org/10.1002/anie.201703864
53. Lugin, M.L., Lee, R.T., Kwon, Y.J.: Synthetically engineered adeno-associated virus for efficient, safe, and versatile gene therapy applications. ACS Nano **14**(11), 14262–14283 (2020). https://doi.org/10.1021/acsnano.0c03850
54. Chawla, S.P., Bruckner, H., Morse, M.A., Assudani, N., Hall, F.L., Gordon, E.M.: A phase I–II study using Rexin-G tumor-targeted retrovector encoding a dominant-negative cyclin G1 inhibitor for advanced pancreatic cancer. Mol. Ther.—Oncolytics, **12**, 56–67 (2018). https://doi.org/10.1016/j.omto.2018.12.005
55. Champion, B.R., Besneux, M., Patsalidou, M., Silva, A., Zonca, M., Marino, N., Di Genova, G., Illingworth, S., Fedele, S., Slater, L., Lilley, F., Plumb, D., West, K., Cockle, P., Brown, A.:

Abstract 5013: NG-641: an oncolytic T-SIGn virus targeting cancer-associated fibroblasts in the stromal microenvironment of human carcinomas. Cancer Res. **79**(13_Supplement), 5013 (2019). https://doi.org/10.1158/1538-7445.am2019-5013
56. Kamerkar, S., LeBleu, V.S., Sugimoto, H., Yang, S., Ruivo, C.F., Melo, S.A., Lee, J.J., Kalluri, R.: Exosomes facilitate therapeutic targeting of oncogenic KRAS in pancreatic cancer. Nature **546**(7659), 498–503 (2017). https://doi.org/10.1038/nature22341
57. Mi, Y., Hagan, C.T., Vincent, B.G., Wang, A.Z.: Emerging nano-/microapproaches for cancer immunotherapy. Adv. Sci. **6**(6) (2019). https://doi.org/10.1002/advs.201801847
58. Yu, Y., Yang, X., Reghu, S., Kaul, S.C., Wadhwa, R., Miyako, E.: Photothermogenetic inhibition of cancer stemness by near-infrared-light-activatable nanocomplexes. Nat. Commun. **11**(1) (2020). https://doi.org/10.1038/s41467-020-17768-3
59. Stavis, S.M., Fagan, J.A., Stopa, M., Liddle, J.A.: Nanoparticle manufacturing—heterogeneity through processes to products. ACS Appl. Nano Mater. **1**(9), 4358–4385 (2018). https://doi.org/10.1021/acsanm.8b01239
60. Wilson, M.K., Friedlander, M.L., Joly, F., Oza, A.M.: A systematic review of health-related quality of life reporting in ovarian cancer phase III clinical trials: room to improve. Oncologist **23**(2), 203–213 (2017). https://doi.org/10.1634/theoncologist.2017-0297
61. Mullard, A.: New drugs cost US $2.6 billion to develop. Nat. Rev. Drug Discov. **13**(12), 877 (2014). https://doi.org/10.1038/nrd4507
62. Kakushadze, Z., Raghubanshi, R., Yu, W.: Estimating cost savings from early cancer diagnosis. Data **2**(3), 30 (2017). https://doi.org/10.3390/data2030030
63. Bevers, T.B., Anderson, B.O., Bonaccio, E., Borgen, P.I., Buys, S., Daly, M.B., Dempsey, P.J., Farrar, W.B., Fleming, I., Garber, J.E., Harris, R.E., Helvie, M., Hoover, S., Krontiras, H., Shaw, S., Singletary, E., Skinner, C.S., Smith, M.L., Tsangaris, T.N., et al.: Breast cancer screening and diagnosis clinical practice guidelines. J. Natl. Compr. Canc. Netw. **4**(5), 480 (2006). https://doi.org/10.6004/jnccn.2006.0040
64. Pereira-Veiga, T., Martínez-Fernández, M., Abuin, C., Piñeiro, R., Cebey, V., Cueva, J., Palacios, P., Blanco, C., Muinelo-Romay, L., Abalo, A., Costa, C., López-López, R.: CTCs expression profiling for advanced breast cancer monitoring. Cancers **11**(12), 1941 (2019). https://doi.org/10.3390/cancers11121941
65. Xu, D., Guo, D., Zhang, J., Tan, X., Deng, Z., Hou, X., Wang, S.: Innovative tumor interstitial fluid-triggered carbon dot-docetaxel nanoassemblies for targeted drug delivery and imaging of HER2-positive breast cancer. Int. J. Pharm. **657**, 124145 (2024). https://doi.org/10.1016/j.ijpharm.2024.124145
66. Hoffman, R.M., Gilliland, F.D., Adams-Cameron, M., Hunt, W.C., Key, C.R.: Prostate-specific antigen testing accuracy in community practice. BMC Fam. Pract. **3**(1) (2002). https://doi.org/10.1186/1471-2296-3-19
67. Singh, B., Maharjan, S., Pan, D.C., Zhao, Z., Gao, Y., Zhang, Y.S., Mitragotri, S.: Imiquimod-gemcitabine nanoparticles harness immune cells to suppress breast cancer. Biomaterials **280**, 121302 (2021). https://doi.org/10.1016/j.biomaterials.2021.121302
68. Pisal, D.S., Kosloski, M.P., Balu-Iyer, S.V.: Delivery of therapeutic proteins. J. Pharm. Sci. **99**(6), 2557–2575 (2010). https://doi.org/10.1002/jps.22054
69. Gaspar, R., Duncan, R.: Polymeric carriers: preclinical safety and the regulatory implications for design and development of polymer therapeutics. Adv. Drug Deliv. Rev. **61**, 1220–1231 (2009). https://doi.org/10.1016/j.addr.2009.06.003
70. Satrialdi, Takano, Y., Hirata, E.E., Ushijima, N., Harashima, H., Yamada, Y.: An effective in vivo mitochondria targeting nanocarrier combined with a π-extended porphyrin type photosensitizer. Nanoscale Adv. **3**, 5919–5927 (2021)
71. Singh, H.P., Kaur, A., Kaur, I., Buttar, H.S., Bhullar, S.K.: Gold nanoparticles: a promising therapeutic approach. Biomed. Rev. 23–36 (2015)
72. Ma, Y., Li, W., Zhou, Z., Qin, X., Wang, D., Gao, Y., et al.: Peptide aptamer coassembly nanocarrier for cancer therapy. Bioconjug. Chem. **30**, 536–540 (2019)
73. Zeeshan, F., Madheswaran, T., Panneerselvam, J., Taliyan, R., Kesharwani, P.: Human serum albumin as multifunctional nanocarrier for cancer therapy. J. Pharm. Sci. **110**(9), 3111–3117 (2021)

74. Bhattacharya, S., Saindane, D., Prajapati, B.G.: Liposomal drug delivery and its potential impact on cancer research. Anticancer Agents Med. Chem. **22**, 2671–2683 (2022)
75. Zuckerman, J.E., Gritli, I., Tolcher, A., Heidel, J.D., Lim, D., Morgan, R., et al.: Correlating animal and human phase Ia/Ib clinical data with CALAA-01, a targeted, polymer-based nanoparticle containing siRNA. Proc. Natl. Acad. Sci. **111**(31), 11449–11454 (2014)
76. Zhao, C., Lv, H., Tao, S., Zhang, T., Xu, N., Zhu, L.: Exosomes: promising nano carrier for cancer therapy. Nano Sel. **3**, 919–929 (2022)
77. Barenholz, Y.C.: Doxil®—the first FDA-approved nano-drug: lessons learned. J. Control. Release **160**(2), 117–134 (2012)
78. Gimondi, S., Ferreira, H., Reis, R.L., Neves, N.M.: Microfluidic devices: a tool for nanoparticle synthesis and performance evaluation. ACS Nano **17**(15), 14205–14228 (2023)
79. Ilg, T.: Investigations on the molecular mode of action of the novel immunostimulator ZelNate: activation of the cGAS-STING pathway in mammalian cells. Mol. Immunol. **90**, 182–189 (2017)
80. Hirsch, L.R., Stafford, R.J., Bankson, J.A., Sershen, S.R., Rivera, B., Price, R.E., et al.: Nanoshell-mediated near-infrared thermal therapy of tumors under magnetic resonance guidance. Proc. Natl. Acad. Sci. **100**(23), 13549–13554 (2003)
81. Aquino-Jarquin, G.: The patent dispute over the breakthrough mRNA technology. Front. Bioeng. Biotechnol. **10**, 1049873 (2022)
82. Maier-Hauff, K., Rothe, R., Scholz, R., Gneveckow, U., Wust, P., Thiesen, B., et al.: Intracranial thermotherapy using magnetic nanoparticles combined with external beam radiotherapy: results of a feasibility study on patients with glioblastoma multiforme. J. Neurooncol. **81**, 53–60 (2007)
83. Tinkle, S., McNeil, S.E., Mühlebach, S., Bawa, R., Borchard, G., Barenholz, Y., et al.: Nanomedicines: addressing the scientific and regulatory gap. Ann. N. Y. Acad. Sci. **1313**(1), 35–56 (2014)
84. Hare, J.I., Lammers, T., Ashford, M.B., Puri, S., Storm, G., Barry, S.T.: Challenges and strategies in anti-cancer nanomedicine development: an industry perspective. Adv. Drug Deliv. Rev. **108**, 25–38 (2017)
85. Wang, A.Z., Langer, R., Farokhzad, O.C.: Nanoparticle delivery of cancer drugs. Annu. Rev. Med. **63**, 185–198 (2012)
86. Sitia, L., Sevieri, M., Signati, L., Bonizzi, A., Chesi, A., Mainini, F., et al.: HER-2-targeted nanoparticles for breast cancer diagnosis and treatment. Cancers **14**(10), 2424 (2022)
87. Nuhn, L.: Artificial intelligence assists nanoparticles to enter solid tumours. Nat. Nanotechnol. **18**, 550–551 (2023). https://doi.org/10.1038/s41565-023-01382-7
88. Torchilin, V.P. (ed.): Handbook of Nanobiomedical Research: Fundamentals, Applications, and Recent Developments. World Scientific Publishing Co (2020)
89. Kumar, C.S.S.R. (ed.): Nanomaterials for Cancer Therapy. Wiley-VCH Verlag GmbH & Co (2010)
90. Ozpolat, B., Sood, A.K., Lopez-Berestein, G.: Liposomal siRNA nanocarriers for cancer therapy. Adv. Drug Deliv. Rev. **66**, 110–116 (2014)
91. Yang, M., Li, J., Gu, P., Fan, X.: The application of nanoparticles in cancer immunotherapy: targeting tumor microenvironment. Bioact. Mater. **6**(7), 1973–1987 (2021)
92. Paulis, L.E., Mandal, S., Kreutz, M., Figdor, C.G.: Dendritic cell-based nanovaccines for cancer immunotherapy. Curr. Opin. Immunol. **25**(3), 389–395 (2013)
93. Beik, J., Khateri, M., Khosravi, Z., Kamrava, S.K., Kooranifar, S., Ghaznavi, H., Shakeri-Zadeh, A.: Gold nanoparticles in combinatorial cancer therapy strategies. Coord. Chem. Rev. **387**, 299–324 (2019)
94. Broadbent, M., Chadwick, S.J., Brust, M., Volk, M.: Gold Nanoparticles for Photothermal and Photodynamic Therapy. ACS Omega (2024)
95. Halder, J., Pradhan, D., Kar, B., Ghosh, G., Rath, G.: Nanotherapeutics approaches to overcome P-glycoprotein-mediated multi-drug resistance in cancer. Nanomedicine: Nanotechnol., Biol. Med. **40**, 102494 (2022)
96. Bray, F., Ferlay, J., Soerjomataram, I., Siegel, R.L., Torre, L.A., Jemal, A.: Global cancer statistics 2018: GLOBOCAN estimates of incidence and mortality worldwide for 36 cancers in 185 countries. CA: Cancer J. Clin. **68**(6), 394–424 (2018)

97. Rais, A., Varshney, I., Kumar, S., Kumar, V., Prasad, T.: Green nanotechnology: applications in Medicine. In: Innovations in Green Nanoscience and Nanotechnology, pp. 167–207. CRC Press (2022)
98. Jain, K.K.: Advances in the field of nanooncology. BMC Med. **8**, 1–11 (2010)
99. Rashidi, N., Davidson, M., Apostolopoulos, V., Nurgali, K.: Nanoparticles in cancer diagnosis and treatment: progress, challenges, and opportunities. J. Drug Deliv. Sci. Technol. 105599 (2024)

Economic and Ethical Considerations in Nanoparticle Therapies

Ankita Doshi, Jigar Y. Soni, and Brijesh Shah

Abstract Pharmaceutical industries are transforming rapidly due to continuous research and innovation. Nanoparticle based therapies provide growth engines to this sector. It offers targeted, efficient and customizable treatment options. This chapter provides detailed overview on various types of nanoparticles and discusses the underlying mechanisms of action. It also explains the economic aspects of nanoparticles therapies, it's role in economics, it also explains challenges related to manufacturing scalability and its market accessibility. Furthermore, ethical considerations are also examined, including environmental impacts, patient consent, and equitable access to treatment. The chapter concludes by discussing emerging trends in nanoparticle research and proposing strategies to address current economic and ethical gaps, offering a forward-looking perspective on the responsible advancement of nanomedicine.

Keywords Nanoparticles · Drug delivery · Drug administration · Targeted drug delivery · Administration route · Therapeutics

1 Introduction

Twenty-first century is considered as the century of material sciences. As materials play a very pivotal role in our day-to-day life to uplift human lifestyles. In material science materials having particle size between 1 and 100 nm are known as nanoparticles. This nanosized material possesses extra ordinary properties than its

A. Doshi
Department of Biomedical and Life Sciences, School of Science, Navrachana University, Vadodara, India

J. Y. Soni
Department of Chemistry, Madhav University, Sirohi, Rajasthan, India

B. Shah (✉)
Department of Chemistry, School of Sciences, Navrachana University, Vadodara, GJ, India
e-mail: brijeshshah27@gmail.com

N. J. Joshi and G. Patel (eds.), *Nanoparticle Delivery in Cancer Care*,
Series in BioEngineering, https://doi.org/10.1007/978-981-95-0457-2_13

bulk counterparts due to higher surface to volume ratio. These properties of nanoparticles (NPs) make them suitable for wide range of applications from agriculture to advanced medical technologies. To make materials having one-billionth of a meter size, we required a multidisciplinary approach, which includes engineering principles along with advanced manufacturing nanotechnology.

At nanoscale the materials possess unique physical and chemical characteristics which make them highly adaptable and valuable for various biomedical applications [1]. In the area of drug delivery systems, diagnostics, imaging and regenerative medicines nanoparticles play an important role. As they offer a promising solution to patients with effective treatment and lesser side [2].

Nanoparticles possess high surface area relative to volume which makes them suitable for targeted drug delivery system. It also allows them to react efficiently with encapsulated therapeutic agents and improve the precision [3] for target tissues or cells. Nano size materials make them suitable to react with different biological molecules and cellular structures, which make them stand out differently from traditional materials and open new possibilities for material scientists for the research and innovation [4] for the industrialist.

Surface area of nanoparticles can be modified or functionalized with targeting ligand or by antibodies, which make them more efficient for targeted therapy with minimal off-target effects. This capability holds immense promise for personalized medicine, as nanoparticles can be designed to deliver therapeutics to individual patients based on their unique molecular [5] profiles. This targeted approach is especially promising for advancing personalized medicine, as nanoparticles can be customized to deliver therapeutic agents according to an individual's distinct molecular [5] characteristics.

In addition to their surface properties, nanoparticles display distinctive optical and magnetic features that make them ideal for applications in fluorescence imaging and magnetic resonance imaging (MRI). Their imaging capabilities surpass those of conventional agents. These cutting-edge techniques offer clinicians enhanced resolution for visualizing biological structures and pathological conditions, enabling early detection, accurate diagnosis, and the development of personalized treatment [6] plans.

1.1 Different Types of Nanoparticles Used in Therapies

The ongoing discoveries led to the inventing remarkable diversity of nanoparticles, categorizing based on their size, shape, characteristics and compositions. Few examples are as below,

Metallic Nanoparticles: These are made up of various metals like gold, silver, iron, or platinum, demarcating with different optical, electrical, and catalytic properties. Such nanoparticles are widely used as catalyzers, sensors, imagine tool and as potent delivery vehicles [7].

Polymeric Nanoparticles: Poly (lactic-coglycolic acid) (PLGA), polyethylene glycol (PEG), or polystyrene are also finding its utility in drug delivery systems due to their biodegradable and customizable properties, compatibility with biological systems [8].

Lipid-Based Nanoparticles: Lipids like phospholipids, cholesterol, or triglycerides are utilized in the production of such nanoparticles that liposomes, solid lipid nanoparticles (SLNs), nanostructured lipid carriers (NLCs) which are notably potent drug delivery, gene therapy, and vaccine delivery vehicles [9].

Carbon-Based Nanoparticles: These nanoparticles are used in electronics, sensors and biomedical imaging techniques due to their exceptional mechanical, electrical, and optical properties. Examples include fullerenes, carbon nanotubes (CNTs), graphene, and carbon dots [10].

Ceramic Nanoparticles: These are made up of inorganic materials like silica, alumina, titania, or zirconia. Its high thermal stability, mechanical strength and chemical inertness makes them valuable resources to be used as catalyzers, biomedical imagine tool and in tissue engineering.

Quantum Dots: Quantum dots are very well-known semiconductor nanoparticles that are widely used in biological imaging, biosensing and photovoltaics due to its unique optical properties and high photostability.

Magnetic Nanoparticles: Here, mostly oxides of iron is used to prepare magnetic nanoparticles and are used in magnetic resonance imaging (MRI), targeted drug delivery, and hyperthermia therapy [11].

Composite Nanoparticles: Two or more components such as metal-polymer, lipid-polymer, carbon-metal are usually hybridized to make such composite nanoparticles to enhance the overall properties of the final products that can be utilized in various applications as discussed above. The use of different nanoparticles has been explored in Fig. 1.

Nanotechnology being the continual and demanding field, the researchers thrived to excel with the new discoveries like novel of materials and synthesis methods to peculiar characteristics exhibiting nanoparticles for specified functions such as biomedicine, electronics, energy, and environmental cleanup etc. For instance, the development of the nanodrugs that can cross the blood–brain barrier, opening the preview of potent treatment regimen for neurological disorder like Alzheimer's disease and brain tumors [12].

Additionally, for the efficient use of the nanoparticle for controlled and targeted drug delivery, the controlled drug release mechanism has been developed which ultimately reduces the systemic drug toxicity. These will be beneficial to optimize the ADME properties, pharmacokinetics, drug efficacy and eventually patient adherence and trust in the nanoparticle-based treatment regimens [13].

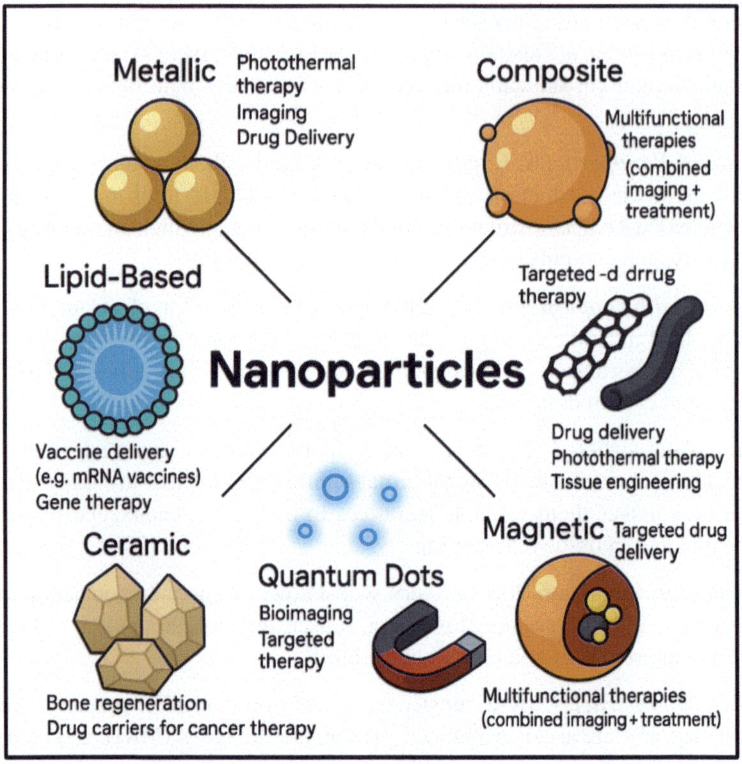

Fig. 1 Different types of nanoparticles and their medical uses

1.2 General Therapeutic Mechanism of Nanoparticles

The mechanisms of action of nanoparticles vary depending on circumstances like their composition, size, shape, and surface characteristics, as well as the intended application. Nonetheless, certain overarching mechanisms can be outlined in Fig. 2.

Drug Delivery: In pharmaceutical transport system nanoparticles serve as delivery agents, in which therapeutic drug is attached to nanoparticles either by encapsulation or by functional attachment. This combination travels through the bloodstream and collects at the targeted tissues or cell either via passive or active targeting methods. Finally, the drug molecules release via diffusion or breakdown or by simulation, offering therapeutics effects and reducing toxicity [14].

Imaging: Nanoparticles exhibit unique optical and magnetic properties which make them suitable candidates for the diagnostic medical imaging equipment industries. It acts as contrast agents for MRI, CT scan, PET and fluorescence imaging. Any change in intensity of electromagnetic radiation or magnetic fields resulted into the improved resolution emission. It will enhance contrast resulting into clear image as

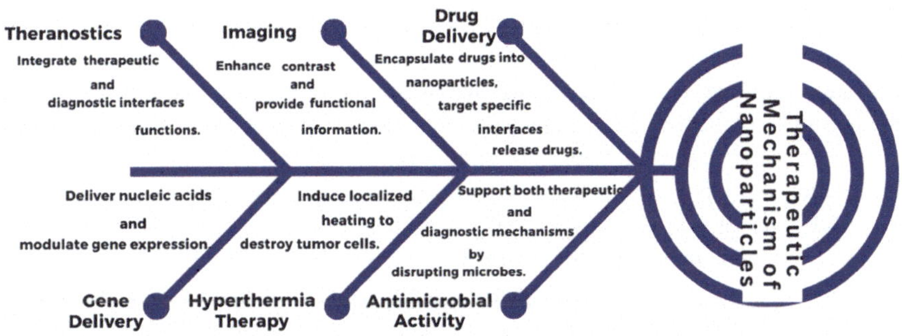

Fig. 2 Schematic representation of nanotherapeutic mechanism

well as it also gives functional information about the biological structures processes [15].

Theranostics: Nanoparticles can combine with therapeutic drug agents as well as with diagnostic functions use for imaging probes which results into effectiveness in treatment and in administration [16].

Gene Delivery: Different nucleic acids such as DNA, RNA and siRNA are the important parts of the human body. Different nanoparticles are used for transportation of these nucleic acids. Further they also use it to prevent degradation and increase the absorption of these acids in cells. When nanoparticle reacted nucleic acid entered it simply discharge the acid towards the targeted cells and influence the regulation of gene expression. These methods are nowadays used for the potential treatment of genetic disorders, cancers, and infectious diseases [17].

Hyperthermia Therapy: Nanoparticles have different magnetic properties than bulky molecules. When magnetic field is applied to them, they absorb energy from the given magnetic field and convert them into heat energy as result temperature of tissues increase which cause cancer cells to die. Magnetic nanoparticles are wieldy involves into hyperthermia therapy which have a targeted and minimally invasive cancer therapy [18, 19].

Antimicrobial Activity [20]: Nanoparticles exhibit smaller size and extensive surface area which allow them to react more effectively with microbial cells. As results they can disrupt microbial cell membranes, create reactive oxygen species and suppress enzyme functions. Further, they can also use wound dressing, medical equipment coatings and as disinfectants for water treatments.

Further, fundamental research in nanotechnology can enhance their applications in healthcare and other medical sectors. As we know, modern medicinal science evolved with the discovery of new versatile nanoparticles which can be used in drug delivery systems, improved imaging, and combined diagnostic and therapeutic treatments.

Nanotherapeutics can transform the healthcare sector more effectively only when these therapies practices are performed on human clinical trials. This trial helps to understand the behaviors of nanoparticles inside the body. In clinical trials safety, efficacy and practicality of nanoparticles has been explored in real-world patient groups. Research can optimize the formulation of nanoparticles under systematic clinical environments. They can consider the adjust dosage regimen or optimal therapeutics ranges or pinpoint adverse effects or drug interactions [21] for the formulations. To validate nanoparticles performance research is highly dependent on clinical trials. This try provides clarity for the scaling up of the synthesis, replicability of results and adherence to regulatory requirements [22]. Still a breakthrough is imminent in the field of nanotherapeutics due to the advent of nanoparticles, potentially revolutionizing the field by providing novel treatment alternatives for a variety of illnesses. To maximize the advantages of nanotherapeutics cooperation between various disciplines is required which also includes meticulously designed clinical trials are crucial. To maximize the benefits of nanoparticle-based treatments, further investigation, cooperation among specialists from various disciplines, and rigorous assessment. Clinical studies of nanoparticles can be boosted by improving patient care and precision medicine treatment [23].

Nanotherapeutics have the potential to reduce hospital stays by increasing the efficiency and pinpoint targeted drug delivery characteristics. It will also decrease the treatment duration and lower the healthcare expenditure. As result it helps to sustain healthcare systems and provide cost-effective medical care. The progress in nanoparticles therapeutics provides economic growth, high-tech employment opportunities and investment options for pharmaceutical and biotechnology sectors. However, from an ethical perspective, the rapid evolution of such technologies raises important concerns. Nanotherapeutics is cutting-edge medical treatment which will wealthier populations or nations by providing affordable and low-cost effective treatment. In corporation of nanoparticle therapeutics into healthcare required robust ethical guidelines and comprehensive policies. This chapter explores the complex relationships between research and innovation with economical and ethical aspects of nanoparticle therapies. The aim is to provide a comprehensive understanding of how innovation in nanoparticle therapies can be scientifically robust and ethical by addressing critical issues such as manufacturing challenges, regulatory frameworks, and equitable distribution.

2 Economic Considerations for Nanoparticles Therapies

2.1 Cost of Research and Development

Designing a nanomaterial for drug delivery system involves a interdisciplinary approach between nanotechnology, advanced materials and clinicians. Over the past few decades this topic has gained more attention or forefront for material science.

The development in these areas created more interest due to their unlocking potential industrial valuation. Which could create substantial economic effects [24]. Further, any surface modification of nanoparticles resulted in new materials with enhancing stability, reduction in non-specific interaction and improved in drug delivery precision make them superior then conventional drug system. Ultimately, this can reduce the costs associated with pharmaceutical development and delivery [25]. Continuous research and development in the field of nanotherapeutics resulted into advancement into nanotechnology and cutting-edge material synthesis which became a driving force for economic development and also fueling innovation in this growing sector. This development and innovation help to lower down the costs, boost operational efficiency, extend the durability of drug and optimize the consumption dose which also reduced environmental impact, the emergence of new market segments, increased employment opportunities, and the rise of innovative industries.

The development in nanotechnology and advanced materials has created new market opportunities, which are currently causing disruption in a established industries. As nanomedicines provide more targeted drug delivery lesser side effects companies try to take advantage of this distinct characteristics of nanomaterials which will transform the whole healthcare sector.

To more development in nanotherapeutics we must focus on "3I", Investigations, innovation, and industrialization of these fields which required more skilled and proficient workforce which includes materials science, clinicians' doctors and industrialist, related to these areas. Physical infrastructure such as laboratories, manufacturing facilities, and research institutions have been set up to meet the increasing need for skilled professionals in these fields. Research, investment, and collaboration have become concentrated in nanotechnology parks and innovation hubs to drive economic growth. The expansion of industries such as nanomedicine and nanomanufacturing has resulted in the formation of value chains that create employment opportunities at different skill levels, encompassing research and development, production, and marketing. Nanotechnology and advanced materials have a significant and broad economic impact. These areas have shown that they can lead to cost reductions, enhance product performance, lessen environmental impact, open new market opportunities, produce jobs, and encourage the growth of new industries.

2.2 Manufacturing and Scalability Challenges

The growing nano pharmaceutical market highlights the significant potential of nanotechnology to advance medical treatment across a wide range of diseases [26]. Several nanomedicines have already been approved by the U.S. Food and Drug Administration (FDA) and the European Medicines Agency (EMA) for various therapeutic indications. Over the past few decades, numerous nanoparticle-based drug delivery systems have gained clinical approval in both the USA and the EU, demonstrating the transition of nanomedicine from experimental research to commercial application. These delivery systems are generally categorized into lipid-based, protein-based, nano emulsion, polymer-based, and metal-based formulations. Among them, lipid-based systems are particularly prominent. Figure 3 illustrates examples of commercially available lipid-based nanoparticle therapeutics used to treat various diseases. While nanotechnology and advanced materials show promise in the laboratory, transitioning these innovations to large-scale production presents scalability challenges. Processes that work well in research settings may face obstacles when attempting to meet the demands of mass production. Achieving consistent quality, quantity, and reproducibility on an industrial scale is a formidable task that necessitates innovation in manufacturing techniques. Developing scalable synthesis methods and manufacturing processes for nanomaterials requires a deep understanding of materials science and engineering.

Collaboration between researchers and engineers is essential to bridge the gap between fundamental discoveries and practical applications, ensuring that new materials can be efficiently and cost-effectively incorporated into industries.

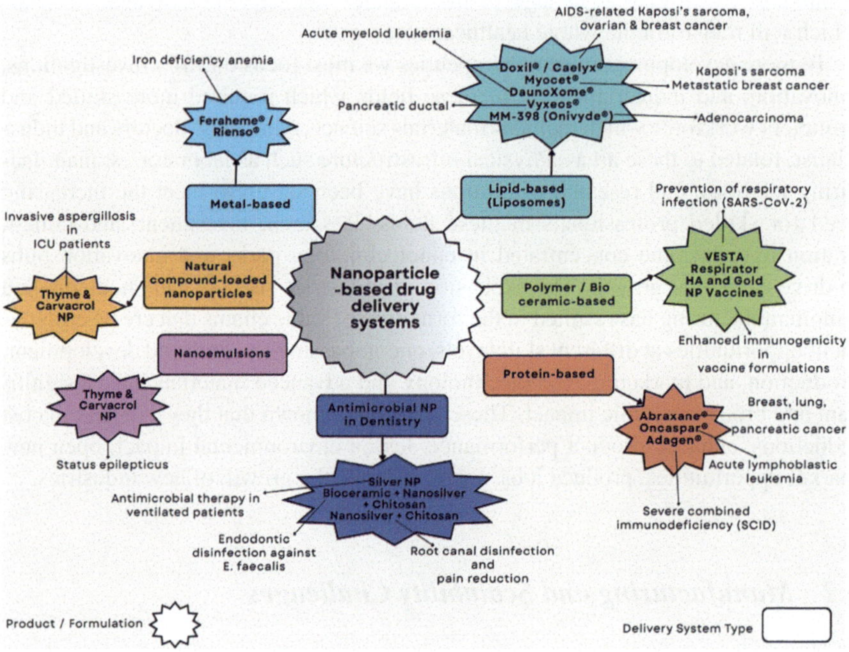

Fig. 3 Schematic representation of commercially available nano therapeutics with their delivery system types

2.3 *Market Accessibility and Affordability*

India is known as word pharmacy as it is third largest producer of medicines in terms of volume and on 14th rank in worldwide in terms of valuation. As per last financial year 2023–2024 the valuation of India's pharmaceutical market was USD 50 billion. Which was the sum of 23.5 billion from domestic sales and 26.5 billion from exports. India manufactured generic, bulk medicines over counter products, vaccines, biologics, and biosimilars. In recent years nanotherapeutics drugs are known for their targeted delivery, improved bioavailability and lesser side effects which makes them more feet for effective treatment of some diseases like cancer, tuberculosis, diabetes, and infectious diseases. As results of these Nano pharmaceuticals is the emerging and transformative market for the India healthcare and pharmaceutical industries. Some Indian leading research institutions such as IITs, IIsc and AIIMs are actively involved in the formulations of nanoscale drugs for more targeted delivery, improved bioactivity with reduction in side effect through their academic research. At the same time several biotech companies are exploring commercial applications and synthesis of these nanotherapeutics drugs. The Indian government also play a crucial role for the nanotherapeutics via different initiatives such as nano mission and make in India to encourage the researcher and industrialists for development and manufacturing of nanomedicine.

Various advantageous properties of nano pharmaceuticals drugs such as low dose, less side effect and less frequent administration, make them more suitable for addressing the rule healthcare gaps of India. Further, regulatory frameworks for nanomedicine in India make him a positioning itself as a key player in the global nanopharmaceutical landscape.

Revolutionize drug delivery properties of nano pharmaceutics drugs and its improved bioavailability make them a potential candidate for treatment of chronic and complex diseases like cancer, diabetes, and infections. Several leading Indian pharmaceutical companies such as Glenmark, Aurobindo Pharma, Biocon, Zydus Lifesciences, and Intas Pharmaceuticals are actively exploring or incorporating nanotechnology into their drug development pipelines. These companies are focusing on nano-formulations to enhance the performance of both generic and specialty drugs. With growing government support, research collaborations, and demand for advanced therapeutics, India is poised to become a significant player in the global nanopharmaceutical landscape.

3 Ethical Considerations in Nanoparticle Therapies

Nanotechnology and nanoparticle based medical therapeutics have proved their immense potential in the improvement in the area across the healthcare sector with quick advancements from bench to bed side. However, it has also posed a variety of ethical challenges that need careful consideration. The key concern includes, patient safety, regulatory compliance, environmental impacts, economical and social implications, data privacy, and dual-use potential (Fig. 4). Nanoparticle accumulation in the tissues, its toxic effects followed by unpredictable immune response poses concern towards the patient's safety (Table 1). For instance, some studies have shown toxic effects of silver nanoparticles and long-term health hazards in case of quantum dots. Many environmental problems can be addressed by the advancing nanotechnology such as resource depletion, pollution and many more by developing effective catalysts, water and air filters and solar cells, which come along with its pros and cons. Disposal of the nanoparticles and its release into environment may cause accumulation in the water, air and soil which may affect the biodiversity and overall ecosystem by entering into food chain (Table 2). Acquiring eco-friendly synthesis methods with proper disposal strategies, these issues can be mitigated. Another pressing concern about nanotechnology is its dual-use potential (Table 3). Advancements in the field are effective in bringing profound positive changes in all fields like energy sector, medical and pharmaceuticals and material science. The same holds equal potential to be misused. Scientists and policymakers must be vigilant enough to keep a check on such activities and encourage the constructive approach to mitigate such threats. Furthermore, data privacy with the increasing usage of AI is a growing concern (Table 4). Efficient functioning of AI, huge amounts of data collection of sensitive data and storage are required which also poses risk of data breach and its misuse. Strengthening the ethical guidelines for sharing the data and enhancing the efficacy of the protection tools can safeguard the data. Ensuring regulatory compliances is also a pivotal factor in the development nanomedicine. Current regulatory frameworks like FDA and EMA offers some sort of guidance but with respect to use of nanotechnology, comprehensive policies and guidelines that harmonize with international regulations are required to bridge the gap to promote safe applications. Furthermore, animal testing poses an ethical concern as usually the nano-formulations are usually tested on animal models to assess their potency and biocompatibility which raise concern regarding animal welfare. Development of In Vitro models and computational skills can reduce animal use, aligning research with ethical standards. Additionally, informed consent is also equally important because the advanced nature of technologies like nanoparticle-based gene therapy may create barriers as it becomes difficult for patients to understand the mechanism and associated risks. Coming up with simplified communication tailored to non-expert audience can improve decision-making and trust of the patient which ultimately lead to the success of nanotechnology. Disparity in access to nanodrug designed for rare diseases remain quite expensive for low-resource settings. Subsidizing the prices for such nano-formulations strengthens and fosters research collaborations globally which in turn will ensure equitable access. Such practices may enhance the cultural and social acceptance in adoption of nanotechnology and nanomedicine. Addressing all these concerns facilitates the development and deployment of nano-formulations responsibly and equitable to ensure patient well-being.

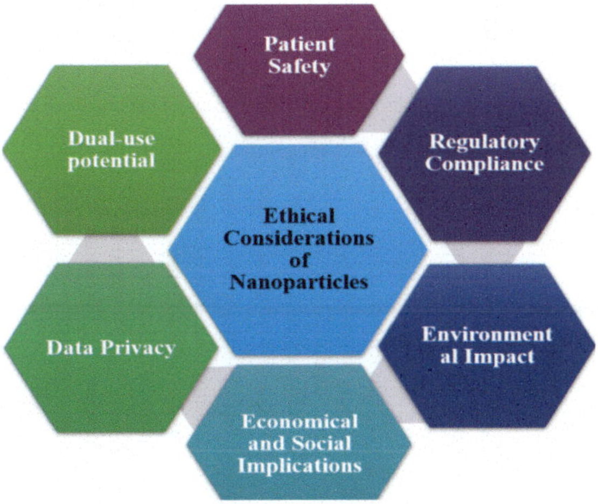

Fig. 4 Ethical considerations of nanoparticles

3.1 Environmental Impact

Nanoparticles (NPs) from laboratories and hospitals, such as silver nanoparticles (AgNPs) used in wound dressings, are frequently discharged into wastewater, posing risks of bioaccumulation and toxicity to aquatic organisms [28]. Gold nanoparticles have also found to be accumulated in the aquatic species, posing ecological concern [29]. Implementing advanced membrane filtration techniques and establishing exposure guidelines with developing biocompatible coating can help mitigate the risks. Additionally, the stable and persistent NPs like titanium dioxide in soil and water ecosystems found to be disruptive for microbial communities [30]. Furthermore, besides various uses of carbon nanotubes have been reported to hamper algal growth and photosynthesis. Soil microbiota is known to be affected by silver nanoparticles by disrupting nitrogen fixation, negatively impacting soil fertility and agricultural productivity [35]. To reduce the risk, the development of biodegradable and eco-friendly NP products should be promoted for agricultural aspects. Moreover, safety protocol, filtration systems and closed loop manufacturing processes can be utilized to reduce the risk of airborne NP emissions. For instance, while manufacturing medical applications and polymer-based drug carriers, NPs are released into the atmosphere which may pose health risks to workers [33]. Furthermore, NP production is highly energy-intensive, especially for carbon-based materials, leading to a significant carbon footprint and resource depletion [31] and on the other hand degradation of NP can generate harmful byproducts like zinc oxide decomposes into zinc ions, a potent toxicant [32]. Concerns like these can be addressed by refining the microbes and plant mediated NP synthesis to elevate reproducibility and large-scale applications.

Table 1 Ethical considerations in nanoparticle-based patient safety

Aspect	Description	Examples	Observed challenges	Proposed solution	Refs.
Toxicity	Toxic effects due to nanoparticle composition, size, shape, and surface properties	– Silver nanoparticles inducing cytotoxicity in human cells – Carbon nanotubes causing pulmonary toxicity	– Cellular damage – Organ accumulation – Long-term systemic effects	– Optimize nanoparticle size, shape, and surface properties to minimize cytotoxicity – Use surface coatings (e.g., polyethylene glycol) to reduce toxicity and improve biocompatibility – Conduct extensive in vitro and in vivo toxicological studies	[27, 28]
Immune response	Activation of the immune system leading to hypersensitivity or inflammation	– PEGylated nanoparticles causing complement activation – Liposomal formulations triggering cytokine release	– Hypersensitivity reactions – Anaphylaxis risk	– Employ immune-modulatory coatings (e.g., PEGylation or zwitterionic polymers) to reduce immune activation – Carefully select lipid and polymer materials for liposomal formulations to minimize cytokine release	[29]
Genotoxicity	Damage to genetic material caused by nanoparticles	– Titanium dioxide nanoparticles leading to DNA damage in fibroblast cells	– Potential mutagenesis – Carcinogenicity concerns	– Screen nanoparticles for genotoxic effects using advanced genomic and transcriptomic assays – Modify surface chemistry to reduce interaction with genetic material – Establish safety thresholds for exposure	[30]

(continued)

Table 1 (continued)

Aspect	Description	Examples	Observed challenges	Proposed solution	Refs.
NPs accumulation	Biopersistence and accumulation in organs such as the liver, spleen, and lungs	– Gold nanoparticles accumulate in the liver during chronic exposure	– Reduced organ function – Toxic build-up in excretory systems	– Develop biodegradable nanoparticles to prevent long-term accumulation – Use targeted delivery systems to ensure organ-specific clearance – Monitor chronic exposure outcomes through regular assessments	[31]
Biodegradability issues	Limited or slow biodegradability of synthetic nanoparticles, leading to prolonged presence in the body	– Non-biodegradable quantum dots used for imaging applications	– Long-term tissue irritation – Persistence in sensitive tissues	– Shift toward naturally degradable materials such as biopolymers (e.g., chitosan or alginate) – Improve synthetic strategies for designing environmentally degradable nanoparticles	[32]
Off-target effects	Unintended effects of nanoparticles on non-target tissues	– Doxorubicin-loaded nanoparticles affect healthy cardiac tissue	– Increased adverse effects – Reduced therapeutic index	– Use advanced targeting strategies such as ligand-conjugated nanoparticles – Incorporate stimuli-responsive nanoparticles for site-specific drug release – Validate efficacy and minimize side effects through controlled clinical studies	[33]

(continued)

Table 1 (continued)

Aspect	Description	Examples	Observed challenges	Proposed solution	Refs.
Size-dependent toxicity	Toxic effects vary with particle size due to differential cellular uptake or surface area interactions	– Zinc oxide nanoparticles <50 nm causing more oxidative stress compared to larger particles	– Difficulty in optimizing size for safety and efficacy	– Identify size ranges that balance safety and efficacy – Utilize hierarchical screening approaches for different nanoparticle sizes – Develop computational models for predicting size-dependent toxicity	[34]
Surface functionalization risks	Functional groups added to nanoparticles for targeting may interact adversely with biological systems	– Amine-functionalized nanoparticles causing hemolysis	– Adverse interactions with blood or cellular proteins	– Conduct extensive surface chemistry characterization to predict adverse interactions – Use biocompatible functional groups for nanoparticle stabilization – Perform hemocompatibility studies before clinical application	[35]
Metabolite interactions	Unpredictable interactions of nanoparticles with endogenous metabolites in vivo	– Nanoparticles altering metabolic pathways in hepatocytes	– Changes in drug metabolism – Potential toxicity	– Apply metabolomic studies to predict nanoparticle-metabolite interactions – Optimize nanoparticle formulations to minimize disruption of metabolic pathways	[36]

Table 2 Ethical considerations in nanoparticle-based on regulatory compliance

Regulatory compliance

Aspect	Description	Examples	Observed challenges	Proposed solution	References
Approval pathways	Regulatory pathways for nanoparticle-based drugs and therapies	– FDA approval of Abraxane (paclitaxel albumin-bound nanoparticles) for cancer treatment	– Lack of specific guidelines for nanomedicines	– Develop specific regulatory frameworks tailored for nanoparticle-based therapies – Collaborate with regulatory agencies to expedite approvals through adaptive pathways	[27]
Toxicity evaluation standards	Requirements for safety and toxicity testing of nanoparticles	– EU REACH regulations for evaluating nanomaterials' toxicity	– Lack of standardized testing protocols for nanoparticles	– Establish standardized testing protocols for nanoparticle toxicity – Implement multi-parameter toxicity assays for comprehensive safety assessments	[28]
Clinical trial regulations	Guidelines for conducting clinical trials involving nanomedicine	– Phase III clinical trials for liposomal doxorubicin formulations	– Complex protocols due to uncertainty in nanoparticle behavior in vivo	– Simplify and streamline clinical trial protocols for nanomedicine – Incorporate adaptive trial designs to address uncertainties in nanoparticle behavior	[35]
Good Manufacturing Practices (GMP)	Standards for nanoparticle production to ensure product quality and consistency	– GMP-compliant production of lipid nanoparticles for COVID-19 mRNA vaccines	– Complex manufacturing requirements for reproducibility	– Establish GMP guidelines specific to nanoparticle production – Automate and monitor production processes to ensure consistency and scalability	[31]

(continued)

Table 2 (continued)

Regulatory compliance					
Aspect	Description	Examples	Observed challenges	Proposed solution	References
Post-market surveillance	Monitoring nanoparticle-based products for safety and efficacy after market release	– Post-market studies for liposomal formulations in oncology treatments	– Lack of long-term surveillance frameworks for nanomedicines	– Develop robust frameworks for long-term monitoring of nanoparticle-based products – Use real-world evidence and pharmacovigilance systems for safety assessments	[32]
Labeling and consumer information	Accurate labeling requirements for nanoparticle-containing products	– Labeling of sunscreens containing zinc oxide nanoparticles	– Challenges in communicating nanoparticle risks to consumers	– Mandate clear labeling of products containing nanoparticles – Educate consumers on the benefits and potential risks of nanoparticles through public awareness campaigns	[33]

(continued)

Table 2 (continued)

Regulatory compliance

Aspect	Description	Examples	Observed challenges	Proposed solution	References
International regulatory harmonization	Efforts to align nanoparticle regulations across regions	– Coordination between FDA and EMA for nanoparticle drug approvals	– Regulatory differences leading to approval delays	– Establish international task forces to align nanoparticle regulations – Promote knowledge sharing between regulatory agencies such as FDA, EMA, and WHO	[30]
Ethical considerations in compliance	Balancing ethical obligations with regulatory requirements for nanoparticles	– Informed consent for clinical trials involving nanoparticle-based drug delivery	– Ensuring transparency while complying with complex regulations	– Ensure transparency in clinical trials involving nanoparticles – Obtain informed consent with clear communication of nanoparticle risks and benefits	[29]

Table 3 Ethical considerations in nanoparticle-based on dual-use potential

Aspect	Description	Examples	Observed challenges	Proposed solution	References
Nano-bioweapons development	The potential use of nanoparticles to create more effective and targeted biological weapons	Encapsulation of pathogens in nanoparticles for aerosol dispersion	Difficulty in tracking illicit production and distribution; international security threats	Strengthening international regulatory frameworks and early threat detection systems	[40]
Illicit drug enhancement	Use of nanotechnology to develop performance-enhancing drugs or recreational substances	Nanocarriers that improve drug bioavailability or mask detection in the bloodstream	Ethical concerns in competitive sports and difficulty in forensic detection	Enhanced doping control measures; development of nanoparticle detection technologies	[41]
Gene editing for malicious intent	Use of nanoparticle-based delivery systems for unauthorized or harmful gene editing	CRISPR-Cas9 delivery via lipid nanoparticles for cover genetic modifications	Potential misuse in bioterrorism or unethical human experiments	Ethical guidelines for genetic research; stringent monitoring of nanoparticle-based gene technologies	[42]
Stealthy surveillance devices	Nanoparticles used in advanced surveillance technologies for tracking individuals or collecting sensitive information	Nano-sensors are embedded in clothing or devices to monitor personal activities	Breach of privacy; unauthorized data collection	Development of privacy-protective regulations for nanotechnology applications	[43]
Biomedical espionage	Misuse of nanoparticle-enabled diagnostic devices to obtain sensitive biomedical information from individuals	Implantable nano-sensors that transmit health data without user consent	Ethical and legal dilemmas regarding data ownership and consent	Strengthening data protection frameworks and ethical guidelines for nanodevices	[44]
NPs as toxic agents	Development of toxic nanoparticles that target specific biological systems	Functionalized nanoparticles that interfere with cellular functions	Difficulty in detecting and neutralizing such agents	Establishing detection protocols and research on neutralization strategies	[39]

(continued)

Table 3 (continued)

Aspect	Description	Examples	Observed challenges	Proposed solution	References
Weaponized drug delivery	Repurposing drug delivery nanoparticles for harmful applications	Use of nanoparticles to deliver lethal doses of toxins or harmful chemicals	Dual-use dilemma in research and drug manufacturing	International collaboration on oversight of nanoparticle production and distribution	[45]
Cyber-nanotech integration	Potential misuse of nanotechnology in conjunction with cyber technologies for malicious purposes	Nanoparticle-embedded microchips for unauthorized data transmission or hacking	Security vulnerabilities and lack of comprehensive regulations	Implementation of robust cybersecurity frameworks for nanotech-based systems	[46]
Smart weapon systems	Incorporation of nanoparticles into advanced weapon systems for enhanced targeting and efficiency	Nano-coatings for armor penetration or radar evasion	Militarization of nanotechnology; arms race concerns	Diplomatic efforts and treaties regulating the military use of nanotechnology	[47]
Biological data exploitation	Unauthorized use of nanoparticle-based diagnostics for collecting population health data	Health monitoring devices capturing biometric data for unauthorized research	Ethical issues related to data sovereignty and unauthorized research practices	Transparent consent protocols and international data governance standards	[48]

Table 4 Ethical considerations in nanoparticle-based on data privacy

Data privacy					
Aspect	Description	Examples	Observed challenges	Proposed solution	References
Data collection in diagnostics	Nanoparticle-enabled devices collect and transmit large amounts of health data	Nanosensors for continuous glucose monitoring	Risk of unauthorized data access and data breaches	Implementation of end-to-end encryption and secure storage systems	[55]
Data ownership and consent	Defining who owns the data generated by nanotechnology-based devices	Wearable health trackers embedded with nanoparticles	Ambiguities in legal frameworks concerning data ownership	Clear legal guidelines defining data ownership and informed consent protocols	[56]
Cross-border data transfers	Nanoparticle-based systems often transmit health data across different jurisdictions	Remote health monitoring systems relying on international cloud servers	Legal conflicts due to varying data protection regulations	Harmonization of international data protection laws	[57]
Data security in implantable devices	Nanoparticle-embedded devices implanted in the human body for medical purposes	Smart pacemakers or drug delivery implants	Vulnerability to cyberattacks compromising patient health and safety	Development of robust cybersecurity protocols for medical devices	[58]
Ethical dilemmas in data usage	Ethical concerns related to the use of patient data for research without consent	Use of health data from nanoparticle-based diagnostics in AI-driven research	Potential misuse or unauthorized use of sensitive data	Strengthening ethical guidelines and transparent data usage agreements	[59]
Data privacy in personalized medicine	Nanoparticle-based treatments generate individualized patient data	Nanocarriers are tailored to deliver specific medications	Risk of exposure of sensitive genetic and health information	Implementation of strict privacy policies and anonymization protocols	[33]
Wearable devices and IoT integration	Nanoparticle-based wearables are integrated into the Internet of Things (IoT) for health monitoring	Smart fitness bands and biosensor patches	Vulnerability to hacking and data interception	Adoption of secure communication protocols and regular firmware updates	[60]

(continued)

Table 4 (continued)

Data privacy					
Aspect	Description	Examples	Observed challenges	Proposed solution	References
Anonymization challenges	Difficulty in anonymizing nanoparticle-based health data due to unique biomarkers	Identification through nanoparticle-enabled genomic data	Risk of re-identification despite anonymization efforts	Advanced data masking and encryption techniques	[57]
Regulatory gaps in data privacy	Lack of comprehensive data protection frameworks for nanotechnology in healthcare	Limited legal guidance on nanoparticle-based diagnostic data management	Inconsistent and outdated data privacy regulations	Development of nanoparticle-specific data privacy regulations and international harmonization efforts	[61]

3.2 Informed Consent

Key issue with the nanoparticle-based treatments is the difficulties for the patients to fully understand its underlying mechanism, potential risk factors and other treatment details is the major challenge for getting ethical considerations regarding informed consent. On contrary to generalized approaches, personalized and tailored nanomedicinal approaches are quite complicated to explain due to their patient-specific biomarkers specificity [35, 36]. Furthermore, as per WHO ethics report dealing with assessing consent from vulnerable patients from the group of elderly and children becomes even more crucial as involvement of the legal guardians may also be required.

This can be solved by providing patients with simplified, clear educational material, providing awareness, possible explanation and addressing insightful discussions [37, 38]. Above that, the linguistic barrier can be an additional challenge that may hinder the informed consent, highlighting need to upgrade the sensitive and robust communication strategies [38]. Patients should also be provided with the provision to revise and update their consent on getting the ongoing updates about clinical progress [39]. The usage of wearable nano-sensor devices raises concern for data privacy whereas off-label applications need transparency in informing patients and clarifying all the regulatory status [31, 32]. By addressing such issues, process of getting informed consent with transparency, will improve patients' autonomy and ease the acceptance of nanomedicines.

3.3 Equality in Access

Making nanotechnology-based drugs accessible to the larger population faces multiple challenges such as, the high cost of liposomal therapies like Doxil makes it affordable to only high-income regions [49]. On the other hand, developing countries required national as well as international funding to fulfill high equipment cost and to get trained personnel in order to upgrade their research findings [33, 50].

For quick availability of nanotechnology-based products, regional manufacturing hubs and technology transfer can be helpful to mitigate challenges related to global supply chain [51]. Other major concern is the intellectual property restrictions on the products like nanoparticle-based mRNA vaccine which can be mitigated either by open licensing or patent pooling approach [51]. Approval of the nanodrugs from regulatory firms also results in limiting their access whereby the proactive steps from the global policy makers become mandatory [52].

Additionally, awareness campaigns, community engagements and the overcoming of the educational gaps in adopting nanomedicines can be addressed to avoid skepticism towards nanotechnology and its acceptance [53, 54]. Addressing all these challenges might help in mitigating the challenges and wider accessibility and acceptability of the nanomedicines.

4 Emerging Trends and Innovation in Nanoparticle Therapeutics

Medical researchers and materials scientists play an important role in demonstrating the important role of nanoparticles in cancer treatments. Effective collaboration between two different fields results in the transformation of laboratory innovation into practical solutions. Which ultimately helps society to provide better and improved cancer treatments. Nanoparticle therapeutics play a pivotal role in various cancer treatments. Some different nanotherapeutics treatment involve liposomal based nanoparticles which revolutionized the drug delivery system. Chemotherapy drugs encapsulated with this type of nanoparticles improve the targeted delivery to tumor cells and reduce the damage of healthy tissues which results into minimizing the side effect as well as improve the therapeutic efficacy. Further, it also reduces the high doses of chemotherapy drugs, which help to reduce economic burden on to patients and experience improved quality of life due to less side effects, lesser hospitalization timing and faster recovery. Different types of nanoparticles were utilized and demonstrated their applications in more advantages way summarized in Table 5.

Based on clinical trials and materials modification may result into advancement of nanotherapeutics more safety, efficient and adaptable. Furthermore, integration of artificial intelligence and Immunotherapeutics will provide opportunities to design next-generation nanoparticle platforms.

5 Addressing Economic and Ethical Gaps

Designing a nanoparticles having biodegradable and sustainable synthesis methods are crucial to minimizing environmental impacts. Further, Nanotherapeutics involving NPs possess several limitations, such as stability, scalability, biodegradability, or their transport across biological membranes affect their effectiveness in treatments and also affect the reproducibility and consistency in outcomes. This limitation can be overcome by doing surface modification or by polymeric coating which can provide stability and targeted delivery. Further, large scale synthesis NPs required higher costs which limits its availability to a common patient. This can be overcome by taking interdisciplinary approach with robust regulatory frameworks including optimization of NPs synthesis parameters and their design, surface modification to enhance its efficiency which ultimately result in improved clinical feasibility. Further these modifications support sustainability as well as advancement in healthcare systems.

Patient safety and ethical standards must be maintained while developing NP-based therapies. Development of clear guidelines and operational protocols helps to mitigate risks and side effects consequences related to NP-based therapeutics. Presently FDA and European Medicines Agency are involved with the safety and efficacy of these therapeutics. These organizations address unique challenges related

Table 5 Overview of different types of nanoparticles, their structure/composition, applications and advantages in different treatments

Nanoparticle type	Structure/composition	Applications	Advantages	References
Liposomes	Outer lipid bilayer with hydrophilic/hydrophobic drug core	– Delivery of chemotherapeutics (e.g., doxorubicin, paclitaxel) – Cancer therapy (breast, prostate) – Combination therapy	– Mimic biophysical features of cells – Reduce cardiotoxicity (e.g., liposomal doxorubicin) – Enhance drug bioavailability – Reverse drug resistance	[23, 62–64]
Polymeric nanoparticles (PLGA)	Polylactic-co-glycolic acid copolymer	– Controlled drug delivery – Cancer therapy	– Biocompatible – Biodegradable – Exploits EPR effect	[65–67]
Dendrimers	Branched, 3D macromolecules with surface functional groups	– Drug and gene delivery – Targeted nanomedicine	– High drug loading – Biocompatible – Controlled release	[68, 69]
Polymeric micelles	Amphiphilic block copolymers self-assembled into core–shell structures	– Solubilization and delivery of hydrophobic anticancer drugs	– Improve circulation time – Reduce RES uptake – Enhance stability	[70, 71]
Gold nanoparticles (AuNPs)	Inert gold core, functionalized surface	Drug delivery (e.g., doxorubicin); cancer therapy; gene, photothermal, and immunotherapy	Non-toxic; enhance tumor accumulation; overcome drug resistance	[52, 71–73]
Carbon Nanotubes (CNTs)	Cylindrical carbon structures	Thermal ablation via NIR radiation; anticancer drug delivery (e.g., methotrexate)	Unique physical properties; thermal response	[71]
Silica Nanoparticles (SNPs)	Mesoporous structure with internal pore volume	Controlled release of anticancer drugs; immunotherapy	High loading capacity; improved pharmacokinetics; immunoadjuvant potential	[74–76]
Magnetic nanoparticles (MNPs)	Metal/metal oxide core, coated with polymers or fatty acids	Chemotherapy; gene therapy; magnetic hyperthermia	Targeted delivery; improved stability; thermal ablation	[77, 78]

(continued)

Table 5 (continued)

Nanoparticle type	Structure/composition	Applications	Advantages	References
Hybrid nanoparticles	Organic–inorganic combinations (e.g., lipid-polymer, liposome-silica)	Pancreatic, breast, prostate, and lung cancer therapy; multidrug delivery	Combine biocompatibility and stability; overcome resistance; prolonged circulation	[23, 75, 79–81]
Cell membrane-coated NPs	Natural cell membranes (e.g., leukocyte, RBC, cancer cell) coated on NPs	Cancer drug delivery; immune evasion; prolonged circulation	Immune evasion; tumor targeting; increased stability	[82]

to NP-based therapeutics. Which involves NP characterization and biological interactions, to ensure patient safety. Further they also update guidelines to reflect how the NP behaves in various biological systems.

Nanotechnological innovations bring new social dimensions along with scientific and clinical challenges. Public awareness campaigns and media releases help to set public perceptions of nanomedicine. Their acceptance and trust in this technology also increases which, influences the broader acceptance and trust in nanotechnological innovations. Transparent communication strategies are crucial for educating the public about both the potential benefits and risks associated with nanotechnology. By fostering informed dialogue and promoting trust in regulatory frameworks, stakeholders can help build a socially responsible pathway for the adoption of nanomedicine. Ultimately, transparency, accessibility of information, and ongoing engagement with the public are essential for securing societal support and ensuring ethical implementation.

6 Conclusion

These book chapters offer in depth exploration of the role of nanotherapeutics in medicinal science. The introduction part shows how physicochemical properties of nanoparticles make them different from conventional bulk molecules. The modification of surface chemistry of nanomaterials resulted in materials which can be used in cutting-edge diagnostics, targeted therapeutics, and personalized medicine. Further it also explains how the different nanoparticle systems—ranging from metallic, polymeric, and lipid-based formulations to advanced carbonaceous and composite materials—has enabled their integration into diverse biomedical applications, including drug delivery, imaging, gene therapy, and theragnostic. Second and third part explain the investment in interdisciplinary research and innovation can transform technologies from the laboratory to clinical and commercial realities involve significant economic, regulatory, and ethical considerations. It also talk about role and support from national initiatives and a strong pharmaceutical infrastructure for research and innovation. Further it also explains how the integration of nanoparticle therapeutics into global healthcare systems must be guided by equitable access, cost-effectiveness, and stringent ethical frameworks to avoid widening existing disparities.

The fourth and fifth parts explain the emerging trend and innovation taking place in nanotherapeutics. Further it also explains the ethical gaps in the implementation of nanotherapeutics in public health sectors. The future of healthcare is increasingly intertwined with nanotechnology. Sustained collaboration among scientists, clinicians, industry leaders, and policymakers is essential to harness the full potential of nanoparticles, ensuring that innovation is not only scientifically rigorous but also economically viable, ethically grounded, and globally accessible.

Nanoparticles based therapeutics involve integration between material scientists, clinicians and doctors. To transform these technologies from laboratory to public domain, it required rigorous ethical scrutiny from all the domains. Like environmental sustainability aspect from material scientists, data privacy, dual-use risks, informed consent and patient safety from a doctor and clinicians' aspect. All these constraints can be resolved by robust regulatory, transparent communication and formation of global policies. Formation and advancement of such eco-systems can be achieved by global collaboration which can mitigate risks and improve accessibility. Ethical principles at every stage of research and development and deployment is essential to ensure societal benefits of nanotechnology can reach to a common man of society with safety and sustainability.

References

1. Pei, J., Palanisamy, C.P., Alugoju, P., Anthikapalli, N.V., Natarajan, P.M., Umapathy, V.R., Swamikannu, B., Jayaraman, S., Rajagopal, P., Poompradub, S.: A comprehensive review on bio-based materials for chronic diabetic wounds. Molecules **28**(2) (2023). https://doi.org/10.3390/molecules28020604
2. Palanisamy, C.P., Cui, B., Zhang, H., Gunasekaran, V.P., Ariyo, A.L., Jayaraman, S., Rajagopal, P., Long, Q.: A critical review on starch-based electrospun nanofibrous scaffolds for wound healing application. Int. J. Biol. Macromol. **222**, 1852–1860 (2022). https://doi.org/10.1016/j.ijbiomac.2022.09.274
3. Lavanya, M., Krishnamoorthy, R., Alshuniaber, M.A., Manoharadas, S., Perumal Palanisamy, C., Priya Veeraraghavan, V., Jayaraman, S., Rajagopal, P., Padmini, R.: Formulation, characterization and evaluation of gelatin-syringic acid/zinc oxide nanocomposite for its effective anticancer, antioxidant and anti-inflammatory activities. J. King Saud Univ. Sci. **35**(8), 102909 (2023). https://doi.org/10.1016/j.jksus.2023.102909
4. Manimaran, D., Elangovan, N., Mani, P., Subramanian, K., Ali, D., Alarifi, S., Palanisamy, C.P., Zhang, H., Rangasamy, K., Palanisamy, V., Mani, R., Govarthanan, K., Aruni, W., Shanmugam, R., Srinivasan, G.P., Kalirajan, A.: Isolongifolene-loaded chitosan nanoparticles synthesis and characterization for cancer treatment. Sci. Rep. **12**(1), 19250 (2022). https://doi.org/10.1038/s41598-022-23386-4
5. Pei, J., Yan, Y., Palanisamy, C.P., Jayaraman, S., Natarajan, P.M., Umapathy, V.R., Gopathy, S., Roy, J.R., Sadagopan, J.C., Thalamati, D., Mironescu, M.: Materials-based drug delivery approaches: recent advances and future perspectives. **13**(1) (2024). https://doi.org/10.1515/gps-2023-0094
6. Rezaei, B., Yari, P., Sanders, S.M., Wang, H., Chugh, V.K., Liang, S., Mostufa, S., Xu, K., Wang, J.-P., Gómez-Pastora, J., Wu, K.: Magnetic nanoparticles: a review on synthesis, characterization, functionalization, and biomedical applications. Small **20**(5), 2304848 (2024). https://doi.org/10.1002/smll.202304848
7. Sharma, R.K., Yadav, S., Dutta, S., Kale, H.B., Warkad, I.R., Zbořil, R., Varma, R.S., Gawande, M.B.: Silver nanomaterials: synthesis and (electro/photo) catalytic applications. Chem. Soc. Rev. **50**(20), 11293–11380 (2021). https://doi.org/10.1039/D0CS00912A
8. Pulingam, T., Foroozandeh, P., Chuah, J.-A., Sudesh, K.: Exploring various techniques for the chemical and biological synthesis of polymeric nanoparticles. Nanomaterials **12**(3) (2022). https://doi.org/10.3390/nano12030576
9. Sheoran, S., Arora, S., Samsonraj, R., Govindaiah, P., Vuree, S.: Lipid-based nanoparticles for treatment of cancer. Heliyon **8**(5) (2022). https://doi.org/10.1016/j.heliyon.2022.e09403
10. Sharma, S., Shekhar, S., Gautam, S., Sharma, B., Kumar, A., Jain, P.: Chapter 13—Carbon-based nanomaterials as novel nanosensors. In: Pal, K., Gomes, F. (eds.) Nanofabrication for Smart Nanosensor Applications. Micro and Nano Technologies, pp. 323–347. Elsevier (2020). https://doi.org/10.1016/B978-0-12-820702-4.00014-3

11. Montiel Schneider, M.G., Martín, M.J., Otarola, J., Vakarelska, E., Simeonov, V., Lassalle, V., Nedyalkova, M.: Biomedical applications of iron oxide nanoparticles: current insights progress and perspectives. Pharmaceutics **14**(1) (2022). https://doi.org/10.3390/pharmaceutics14010204
12. Huang, Q., Chen, Y., Zhang, W., Xia, X., Li, H., Qin, M., Gao, H.: Nanotechnology for enhanced nose-to-brain drug delivery in treating neurological diseases. J. Control. Rel. **366**, 519–534 (2024). https://doi.org/10.1016/j.jconrel.2023.12.054
13. Moradi Kashkooli, F., Soltani, M., Souri, M.: Controlled anti-cancer drug release through advanced nano-drug delivery systems: static and dynamic targeting strategies. J. Control. Rel. **327**, 316–349 (2020). https://doi.org/10.1016/j.jconrel.2020.08.012
14. Raj, S., Khurana, S., Choudhari, R., Kesari, K.K., Kamal, M.A., Garg, N., Ruokolainen, J., Das, B.C., Kumar, D.: Specific targeting cancer cells with nanoparticles and drug delivery in cancer therapy. Semin. Cancer Biol. **69**, 166–177 (2021). https://doi.org/10.1016/j.semcancer.2019.11.002
15. Kalra, J., Krishna, V., Reddy, B.S.V., Dhar, A., Venuganti, V.V.K., Bhat, A.: 10—Nanoparticles in medical imaging. In: Gopinath, S.C.B., Gang, F. (eds.) Nanoparticles in Analytical and Medical Devices, pp. 175–210. Elsevier (2021). https://doi.org/10.1016/B978-0-12-821163-2.00010-8
16. Llop, J., Lammers, T.: Nanoparticles for cancer diagnosis, radionuclide therapy and theranostics. ACS Nano **15**(11), 16974–16981 (2021). https://doi.org/10.1021/acsnano.1c09139
17. Yadav, K., Sahu, K.K., Sucheta, Gnanakani, S.P.E., Sure, P., Vijayalakshmi, R., Sundar, V.D., Sharma, V., Antil, R., Jha, M., Minz, S., Bagchi, A., Pradhan, M.: Biomedical applications of nanomaterials in the advancement of nucleic acid therapy: mechanistic challenges, delivery strategies, and therapeutic applications. Int. J. Biol. Macromol. **241**, 124582 (2023). https://doi.org/10.1016/j.ijbiomac.2023.124582
18. Jose, J., Kumar, R., Harilal, S., Mathew, G.E., Parambi, D.G.T., Prabhu, A., Uddin, M.S., Aleya, L., Kim, H., Mathew, B.: Magnetic nanoparticles for hyperthermia in cancer treatment: an emerging tool. Environ. Sci. Pollut. Res. **27**(16), 19214–19225 (2020). https://doi.org/10.1007/s11356-019-07231-2
19. Peñate Medina, T., Gerle, M., Humbert, J., Chu, H., Köpnick, A.-L., Barkmann, R., Garamus, V.M., Sanz, B., Purcz, N., Will, O., Appold, L., Damm, T., Suojanen, J., Arnold, P., Lucius, R., Willumeit-Römer, R., Açil, Y., Wiltfang, J., Goya, G.F., et al.: Lipid-iron nanoparticle with a cell stress release mechanism combined with a local alternating magnetic field enables site-activated drug release. Cancers **12**(12) (2020). https://doi.org/10.3390/cancers12123767
20. Sayed, F.A.-Z., Eissa, N.G., Shen, Y., Hunstad, D.A., Wooley, K.L., Elsabahy, M.: Morphologic design of nanostructures for enhanced antimicrobial activity. J. Nanobiotechnol. **20**(1), 536 (2022). https://doi.org/10.1186/s12951-022-01733-x
21. Halwani, A.A.: Development of pharmaceutical nanomedicines: from the bench to the market. Pharmaceutics **14**(1) (2022). https://doi.org/10.3390/pharmaceutics14010106
22. Đorđević, S., Gonzalez, M.M., Conejos-Sánchez, I., Carreira, B., Pozzi, S., Acúrcio, R.C., Satchi-Fainaro, R., Florindo, H.F., Vicent, M.J.: Current hurdles to the translation of nanomedicines from bench to the clinic. Drug Deliv. Transl. Res. **12**(3), 500–525 (2022). https://doi.org/10.1007/s13346-021-01024-2
23. Hu, X., Zhang, Y., Ding, T., Liu, J., Zhao, H.: Multifunctional gold nanoparticles : a novel nanomaterial for various medical applications and biological activities. **8**(August), 1–17 (2020). https://doi.org/10.3389/fbioe.2020.00990
24. Chaikittisilp, W., Yamauchi, Y., Ariga, K.: Material evolution with nanotechnology, nanoarchitectonics, and materials informatics: what will be the next paradigm shift in nanoporous materials? Adv. Mater. **34**(7), 2107212 (2022). https://doi.org/10.1002/adma.202107212
25. Manoharan, M.: Research on the frontiers of materials science: the impact of nanotechnology on new material development. Technol. Soc. **30**(3), 401–404 (2008). https://doi.org/10.1016/j.techsoc.2008.04.016
26. Petros, R.A., DeSimone, J.M.: Strategies in the design of nanoparticles for therapeutic applications. Nat. Rev. Drug Discov. **9**(8), 615–627 (2010). https://doi.org/10.1038/nrd2591

27. Shrader-Frechette, K.: Nanotoxicology and ethical conditions for informed consent. NanoEthics **1**(1), 47–56 (2007). https://doi.org/10.1007/s11569-007-0003-x
28. Cheng, B., Chen, P., Zhang, X., Fang, K., Qin, X., Liu, W.: Personalized privacy protection-preserving collaborative filtering algorithm for recommendation systems. Appl. Sci. **13**(7) (2023). https://doi.org/10.3390/app13074600
29. Barreto, A., Luis, L.G., Pinto, E., Almeida, A., Paíga, P., Santos, L.H.M.L.M., Delerue-Matos, C., Trindade, T., Soares, A.M.V.M., Hylland, K., Loureiro, S., Oliveira, M.: Effects and bioaccumulation of gold nanoparticles in the gilthead seabream (Sparus aurata)—single and combined exposures with gemfibrozil. Chemosphere **215**, 248–260 (2019). https://doi.org/10.1016/j.chemosphere.2018.09.175
30. Marchant, G.E., Sylvester, D.J., Abbott, K.W., Danforth, T.L.: International harmonization of regulation of nanomedicine. **3**(3) (2010). https://doi.org/10.2202/1941-6008.1120
31. Sivakumar, C.L.V., Mone, V., Abdumukhtor, R.: Addressing privacy concerns with wearable health monitoring technology. WIREs Data Min. Knowl. Discov. **14**(3), e1535 (2024). https://doi.org/10.1002/widm.1535
32. Dabas, R., Musyuni, P., Aggarwal, G.: An overview of regulatory perspectives of nanoparticles and liposomal products in India and USA. J. Gener. Med. **19**(4), 166–178 (2023). https://doi.org/10.1177/17411343231206469
33. Kumarasamy, R.V., Natarajan, P.M., Umapathy, V.R., Roy, J.R., Mironescu, M., Palanisamy, C.P.: Clinical applications and therapeutic potentials of advanced nanoparticles: a comprehensive review on completed human clinical trials. (October), 1–25 (2024). https://doi.org/10.3389/fnano.2024.1479993
34. Khan, F., Shariq, M., Asif, M., Siddiqui, M.A., Malan, P., Ahmad, F.: Green nanotechnology: plant-mediated nanoparticle synthesis and application. Nanomaterials **12**(4) (2022). https://doi.org/10.3390/nano12040673
35. Foulkes, R., Man, E., Thind, J., Yeung, S., Joy, A., Hoskins, C.: The regulation of nanomaterials and nanomedicines for clinical application: current and future perspectives. Biomater. Sci. **8**(17), 4653–4664 (2020). https://doi.org/10.1039/D0BM00558D
36. Latte-Naor, S.: Managing patient expectations: integrative, not alternative. Cancer J. **25**(5) (2019). [Online]. https://journals.lww.com/journalppo/fulltext/2019/09000/managing_patient_expectations__integrative,_not.2.aspx
37. Haik, Y., Bantekas, I.: Nano-drug clinical trials: informed consent and risk management through blockchain. Pittsburgh J. Technol. Law Policy **21**(1) SE-Articles (2021). https://doi.org/10.5195/tlp.2021.243
38. Sharma, S., Parveen, R., Chatterji, B.P.: Toxicology of nanoparticles in drug delivery. Curr. Pathobiol. Rep. **9**(4), 133–144 (2021). https://doi.org/10.1007/s40139-021-00227-z
39. Holdstock, D.: Quarterly of healthcare ethics. Cambridge Q. Healthc. Ethics **15**(4), 356–365 (2006). [Online]. https://doi.org/10.1017/S0963180106060476
40. Nasu, H., Faunce, T.: Nanotechnology and the international law of weaponry: towards international regulation of nano-weapons. J. Law, Inf. Sci. **20**(1), 23–60 (2009). [Online]. https://search.informit.org/doi/10.3316/informit.617926207449707
41. Evans, R., Guy, O.: Ethics, nanobiosensors and elite sport: the need for a new governance framework. Sci. Eng. Ethics **23**(6), 1487–1505 (2017). https://doi.org/10.1007/s11948-016-9855-1
42. Kropf, M.: The ethically significant difference between dual use and slippery slope arguments, in relation to CRISPR-Cas9: philosophical considerations and ethical challenges. Res. Ethics **21**(2), 346–361 (2024). https://doi.org/10.1177/17470161241240587
43. Cilliers, L.: Wearable devices in healthcare: privacy and information security issues. Heal. Inf. Manag. J. **49**(2–3), 150–156 (2019). https://doi.org/10.1177/1833358319851684
44. Thwala, L.N., Ndlovu, S.C., Mpofu, K.T., Lugongolo, M.Y., Mthunzi-Kufa, P.: Nanotechnology-based diagnostics for diseases prevalent in developing countries: current advances in point-of-care tests. Nanomaterials **13**(7) (2023). https://doi.org/10.3390/nano13071247

45. Rickli, J.M., Vllasi, G.: Connect. Q. J. **24**(1), 91–112 (2025). [Online]. https://doi.org/10.11610/Connections.24.1.07
46. Assessing the factors influencing cybersecurity effectiveness: a PLS-SEM approach. Int. Trans. Educ. Technol. **2**(1), 49–58 (2023)
47. Kosal, M.E.: The security implications of nanotechnology. Bull. At. Sci. **66**(4), 58–69 (2010). https://doi.org/10.2968/066004006
48. Shajar, F., Saleem, S., Mushtaq, N.U., Shah, W.H., Rasool, A., Padder, S.A., Tahir, I., Rehman, R.U.: Regulatory and Ethical Issues Raised by the Utilization of Nanomaterials BT—Interaction of Nanomaterials with Living Cells, pp. 899–924. In: Sheikh, F.A., Majeed, S., Beigh, M.A. (eds.) Springer Nature Singapore, Singapore (2023). https://doi.org/10.1007/978-981-99-2119-5_31
49. Bosetti, R., Jones, S.L.: Cost-effectiveness of nanomedicine: estimating the real size of nano-costs. Nanomedicine **14**(11), 1367–1370 (2019). https://doi.org/10.2217/nnm-2019-0130
50. Udegbe, F.C., Nwankwo, E.I., Igwama, G.T.: Advancing point-of-care diagnostics through nanotechnology: a focus on low-cost solutions for rural healthcare. **5**(10), 724–746 (2023). https://doi.org/10.51594/ijarss.v5i10.1484
51. Square, P., Algorithm, O., Fang, Y., Shen, X.: Nanoinformatics: opportunities and challenges in the development and delivery of healthcare products in developing countries. https://doi.org/10.1088/1755-1315/655/1/012018
52. Pratish, D., Bipasa, C., Arshia, D., Vaishnavi, T., Srijita, V.: Advancing Autonomous Nanomedicine: Bridging the Gap from Concept to Potential Clinical Studies, pp. 2607–2635 (2024).
53. Abbas, J.J., Smith, B., Poluta, M., Velazquez-Berumen, A.: Improving health-care delivery in low-resource settings with nanotechnology: challenges in multiple dimensions. Nanobiomedicine **4**, 1849543517701158 (2017). https://doi.org/10.1177/1849543517701158
54. Pollitzer, E.: Why gender is relevant to materials science and engineering. MRS Commun. **11**(5), 656–661 (2021). https://doi.org/10.1557/s43579-021-00093-1
55. Bao, R., Chen, Z., Obaidat, M.S.: Challenges and techniques in Big data security and privacy: a review. Secur. Priv. **1**(4), e13 (2018). https://doi.org/10.1002/spy2.13
56. Knight, T.: Data ownership: legalities concerning wearable technologies. (2021). [Online]. https://doi.org/10.4018/978-1-7998-3487-8.ch003
57. Olatunji, I.E., Rauch, J., Katzensteiner, M., Khosla, M.: A review of anonymization for healthcare data. Big Data **12**(6), 538–555 (2022). https://doi.org/10.1089/big.2021.0169
58. Sabra, M.N.A.: Cyberthreats on implantable medical devices. J. Inf. Secur. Cybercrimes Res. **4**(1 SE), 36–42 (2021). https://doi.org/10.26735/XVJR7905
59. Wasti, S., Lee, I.H., Kim, S., Lee, J., Kim, H.: Ethical and Legal Challenges in Nanomedical Innovations: A Scoping Review, no. May, pp. 1–11. (2023). https://doi.org/10.3389/fgene.2023.1163392
60. Singh, K.R.B., Nayak, V., Singh, J., Singh, R.P.: Nano-enabled wearable sensors for the Internet of Things (IoT). Mater. Lett. **304**, 130614 (2021). https://doi.org/10.1016/j.matlet.2021.130614
61. Gleason, K., Suen, J.J.: Going beyond affordability for digital equity: closing the 'Digital Divide' through outreach and training programs for older adults. J. Am. Geriatr. Soc. **70**(1), 75–77 (2022). https://doi.org/10.1111/jgs.17511
62. Zylberberg, C., Matosevic, S.: Pharmaceutical liposomal drug delivery: a review of new delivery systems and a look at the regulatory landscape. Drug Deliv. **23**(9), 3319–3329 (2016). https://doi.org/10.1080/10717544.2016.1177136
63. Hua, S., Wu, S.Y.: The use of lipid-based nanocarriers for targeted pain therapies. Front. Pharmacol. **4** (2013). https://doi.org/10.3389/fphar.2013.00143
64. Satsangi, J.: The Montreal classification of inflammatory bowel disease: controversies, consensus, and implications. Gut **55**(6), 749–753 (2006). https://doi.org/10.1136/gut.2005.082909
65. Acharya, S., Sahoo, S.K.: PLGA nanoparticles containing various anticancer agents and tumour delivery by EPR effect. Adv. Drug Deliv. Rev. **63**(3), 170–183 (2011). https://doi.org/10.1016/j.addr.2010.10.008

66. Saneja, A., Kumar, R., Mintoo, M.J., Dubey, R.D., Sangwan, P.L., Mondhe, D.M., Panda, A.K., Gupta, P.N.: Gemcitabine and betulinic acid co-encapsulated PLGA–PEG polymer nanoparticles for improved efficacy of cancer chemotherapy. Mater. Sci. Eng. C **98**, 764–771 (2019). https://doi.org/10.1016/j.msec.2019.01.026
67. Amreddy, N., Babu, A., Muralidharan, R., Panneerselvam, J., Srivastava, A., Ahmed, R., Mehta, M., Munshi, A., Ramesh, R.: Recent Advances in Nanoparticle-Based Cancer Drug and Gene Delivery, pp. 115–170. (2018). https://doi.org/10.1016/bs.acr.2017.11.003
68. Nanjwade, B.K., Bechra, H.M., Derkar, G.K., Manvi, F.V., Nanjwade, V.K.: Dendrimers: emerging polymers for drug-delivery systems. Eur. J. Pharm. Sci. **38**(3), 185–196 (2009). https://doi.org/10.1016/j.ejps.2009.07.008
69. Sherje, A.P., Jadhav, M., Dravyakar, B.R., Kadam, D.: Dendrimers: a versatile nanocarrier for drug delivery and targeting. Int. J. Pharm. **548**(1), 707–720 (2018). https://doi.org/10.1016/j.ijpharm.2018.07.030
70. Cagel, M., Tesan, F.C., Bernabeu, E., Salgueiro, M.J., Zubillaga, M.B., Moretton, M.A., Chiappetta, D.A.: Polymeric mixed micelles as nanomedicines: achievements and perspectives. Eur. J. Pharm. Biopharm. **113**, 211–228 (2017). https://doi.org/10.1016/j.ejpb.2016.12.019
71. Yao, Y., Zhou, Y., Liu, L., Xu, Y., Chen, Q., Wang, Y., Wu, S., Deng, Y., Zhang, J., Shao, A.: Nanoparticle-based drug delivery in cancer therapy and its role in overcoming drug resistance. Front. Mol. Biosci. **7** (2020), Frontiers Media S.A. https://doi.org/10.3389/fmolb.2020.00193
72. Han, G., Ghosh, P., Rotello, V.M.: Functionalized gold nanoparticles for drug delivery. Nanomedicine **2**(1), 113–123 (2007). https://doi.org/10.2217/17435889.2.1.113
73. Jiang, C., Zhu, J., Li, Z., Luo, J., Wang, J., Sun, Y.: Chitosan–gold nanoparticles as peroxidase mimic and their application in glucose detection in serum. RSC Adv. **7**(70), 44463–44469 (2017). https://doi.org/10.1039/C7RA08967H
74. Wang, L., Ning, C., Pan, T., Cai, K.: Role of silica nanoparticles in abiotic and biotic stress tolerance in plants: a review. Int. J. Mol. Sci. **23**(4), 1947 (2022). https://doi.org/10.3390/ijms23041947
75. Zhang, T., Prasad, P., Cai, P., He, C., Shan, D., Rauth, A.M., Wu, X.Y.: Dual-targeted hybrid nanoparticles of synergistic drugs for treating lung metastases of triple negative breast cancer in mice. Acta Pharmacol. Sin. **38**(6), 835–847 (2017). https://doi.org/10.1038/aps.2016.166
76. Fontana, F., Shahbazi, M., Liu, D., Zhang, H., Mäkilä, E., Salonen, J., Hirvonen, J.T., Santos, H.A.: Multistaged nanovaccines based on porous silicon@acetalated dextran@cancer cell membrane for cancer immunotherapy. Adv. Mater. **29**(7) (2017). https://doi.org/10.1002/adma.201603239
77. Basoglu, H., Bilgin, M.D., Demir, M.M.: Protoporphyrin IX-loaded magnetoliposomes as a potential drug delivery system for photodynamic therapy: fabrication, characterization and in vitro study. Photodiagn. Photodyn. Ther. **13**, 81–90 (2016). https://doi.org/10.1016/j.pdpdt.2015.12.010
78. Hoopes, P.J., Mazur, C.M., Osterberg, B., Song, A., Gladstone, D.J., Steinmetz, N.F., Veliz, F.A., Bursey, A.A., Wagner, R.J., Fiering, S.N.: Effect of intra-tumoral magnetic nanoparticle hyperthermia and viral nanoparticle immunogenicity on primary and metastatic cancer, p. 100660G. In: Ryan, T.P. (ed.) (2017). https://doi.org/10.1117/12.2256062
79. Zhao, X., Li, F., Li, Y., Wang, H., Ren, H., Chen, J., Nie, G., Hao, J.: Co-delivery of HIF1α siRNA and gemcitabine via biocompatible lipid-polymer hybrid nanoparticles for effective treatment of pancreatic cancer. Biomaterials **46**, 13–25 (2015). https://doi.org/10.1016/j.biomaterials.2014.12.028
80. Cheow, W.S., Hadinoto, K.: Factors affecting drug encapsulation and stability of lipid–polymer hybrid nanoparticles. Colloids Surf. B Biointerfaces **85**(2), 214–220 (2011). https://doi.org/10.1016/j.colsurfb.2011.02.033

81. Cirillo, G., Cappello, A.R., Curcio, M., Fiorillo, M., Frattaruolo, L., Avena, P., Scorzafave, L., Dolce, V., Nicoletta, F.P., Iemma, F.: Novel CD44-targeted albumin nanoparticles: an innovative approach to improve breast cancer treatment. Int. J. Mol. Sci. **25**(19), 10560 (2024). https://doi.org/10.3390/ijms251910560
82. Fang, R.H., Kroll, A.V., Gao, W., Zhang, L.: Cell membrane coating nanotechnology. Adv. Mater. **30**(23) (2018). https://doi.org/10.1002/adma.201706759

The manufacturer's authorised representative in the EU is Springer Nature Customer Service Centre GmbH, Europaplatz 3, 69115 Heidelberg, Germany. If you have any concerns regarding our products, please contact ProductSafety@springernature.com

Printed and bound by CPI Group (UK) Ltd, Croydon, CR0 4YY
01/10/2026
02225915-0012